MODERN SYNTHETIC METHODS

Volume 3 · 1983

MODERN SYNTHETIC METHODS

Editor: R. Scheffold

Volume 1 · 1976

M. Mąkosza, Naked Anions—Phase Transfer

A. Patchornik, Polymer Supported Reagents

D. Seebach, S- and Se-Reagents

Volume 2 · 1980

L. Eberson, Electro-Organic Synthesis

Chiral Building Blocks in Enantiomer Synthesis:

D. Seebach – ex Tartaric Acid

A. Vasella – ex Sugars

A. Fischli – Using Enzymatic Transformations

MODERN SYNTHETIC METHODS

Volume 3 · 1983

Transition Metals in Organic Synthesis

Edited by R. Scheffold
Institut für organische Chemie der Universität Bern
Freiestrasse 3, CH-3012 Bern

J. K. Stille · Principles and Rules
L. S. Hegedus · Group VIII Transition Metals
J. F. Normant · Copper and Manganese
D. Seebach · Titanium, Zirconium, Vanadium
R. Scheffold · Macrocyclic Cobalt Complexes

Conference Papers
of the International Seminar
on Modern Synthetic Methods 1983
Interlaken, May 5th/6th, 1983

Sponsored and organized by
The Association of Swiss Chemists

Otto Salle Verlag
Frankfurt am Main · Berlin · München

Verlag Sauerländer
Aarau · Frankfurt am Main · Salzburg

John Wiley & Sons
Chichester · New York · Brisbane · Toronto · Singapore

Permissions:

Selected experimental procedures have been included with the permission of the authors and the copyright holders. We gratefully acknowledge the permissions given by:

Prof. J.P. Collman, Stanford University; Prof. E.J. Corey, Harvard University; P.D. Dr. H. Eckert, Techn. Universität München; Prof. H. Horino, Tohoku University; Prof. M.P. Mertes, University of Kansas; Prof. E.-I. Negishi, Purdue University; Prof. R. Noyori, Nagoya University; Prof. A.J. Pearson, University of Cambridge, GB; Prof. M. Rosenblum, Brandeis University; Prof. B.M. Trost, University of Wisconsin; Prof. K.P.Ch. Vollhardt, University of California, Berkeley.

Rolf Scheffold (Editor)
Modern Synthetic Methods
Volume 3 · 1983

Cover by Stephan Bundi
Order No. 5527
ISBN 3-7935-5527-5 (Salle)
ISBN 3-7941-2377-8 (Sauerländer)

Coedition by John Wiley & Sons Ltd., Chichester
Wiley Ltd. ISBN 0-471-90190-3

Printed in Switzerland by Sauerländer AG, Aarau

British Library Cataloguing in Publication Data:

International Seminar on Modern Synthetic Methods:
(1983: Interlaken)
Modern synthetic methods.
vol. 3: 1983: Transition metals in organic synthesis
1. Chemistry, Organic—Synthesis—Congresses
I. Title II. Stille, J. K.
547'.2 QD 262

ISBN 0 471 90190 3

Preface

Continous up-dating and understanding of current prospects in organic synthesis, as well as the early recognition of important trends, are difficult and laborious. It is the aim of the triennial Interlaken seminars on MODERN SYNTHETIC METHODS to provide participants with an easy access to important and rapidly developing areas of synthetic organic chemistry.

The 1983 conference is dedicated entirely to transition metals in organic synthesis. This takes into account the recent impactuous developments in this field related to synthetic applications. Contributions compiled in this volume include an introductory chapter on the basic concepts as well as chapters addressing selected important transition metal compounds and catalysts. These chapters also include as an integral part representative experimental procedures, which enable immediate application of these principles in the laboratory.

The purpose of this volume is twofold. It should aid participants in following the lectures. The main reason, however, is essentially to serve as a guide for chemists interested in transition metal chemistry applied to organic synthesis.

Rolf Scheffold

TRANSITION METALS IN ORGANIC SYNTHESIS

CONTENTS:

TRANSITION METALS IN ORGANIC SYNTHESIS

PRINCIPLES OF TRANSITION METAL CHEMISTRY

Professor John K. Stille
Department of Chemistry
Colorado State University
Fort Collins, Colorado 80523, U.S.A.

INTRODUCTION

The organometallic chemistry of transition metals has grown explosively in the past decade. Homogeneous catalysis by soluble transition metal complexes has become a major synthetic tool in industrial processes and in the laboratory. Today over 10 million tons of organic chemicals are produced in the United States by homogeneous catalytic processes. These catalytic reactions span a wide range of compounds and reaction types, from the synthesis of L-DOPA by an asymmetric hydrogenation reaction to the synthesis of bulk chemicals and products such as acetic acid from methanol, acetaldehyde from ethylene, oxoalcohols by hydroformylation and polyolefins from ethylene and butadiene. The acceptance of homogeneous catalysis is a result of the fact that most homogeneous transition metal reactions are low energy processes--the reaction temperatures and pressures are low--and they are selective, resulting in pure products in high yields.[1]

Our understanding of the mechanisms of the organometallic reactions of transition metals today is at about the same state as our understanding of reaction mechanisms in organic chemistry 30 years ago. As usual, the development of the catalytic processes preceeded our mechanistic understanding of the individual steps in the catalytic cycle. Nevertheless, the level of understanding today is sufficient to allow the intelligent development of new homogeneous catalytic processes, and to develop new and useful transition metal assisted organic syntheses.

The difficulty in understanding organotransition metal reaction mechanisms is derived from several factors. In most organic reactions catalyzed by homogeneous transition metal complexes, a carbon-metal σ-bond is formed and broken during the course of the catalytic cycle. In order for a catalytic reaction to be efficient, the carbon-metal σ-bond cannot be particularly stable, since it is required to undergo some facile subsequent reaction, leading to product with the regeneration of the original catalytic species. Consequently, in many catalytic reactions, the intermediate organometallic complexes cannot be isolated so that their reaction to the next intermediate can be studied.

Unlike the main group metals, transition metals, because of their many oxidation states and coordination numbers, can assemble organic units as ligands and bring about their reaction. Transition metal complexes with almost every organic functional group, and upon complexation, the chemistry of the functional group can be dramatically changed, reacting by mechanisms available to the transition metal. Futhermore, the chemistry of the transition metal can be changed by changing the "spectator" ligands on the metal. Thus, a reaction may be fine-tuned by making subtle adjustments in the ligands.

The emphasis in discussing the principles of organic transition metal chemistry is placed on reaction mechanisms, particularly oxidative addition, reductive elimination, insertion, nucleophilic attack on coordinated ligands and transmetallation. Before discussing these key reaction mechanisms, however, an understanding of the structure and bonding in transition metal complexes is necessary.

STRUCTURE AND BONDING OF ORGANOTRANSITION METAL COMPLEXES[2]

In order to understand the reactions of organometallic complexes and the reasons they react in a certain manner, it is important to know such things the geometry of a complex, the oxidation state of the metal, whether it is coordinatively saturated or unsaturated, and what effect the ligands will have on the reaction. The vast array of transition metal complexes is not as random and disorganized as might be seen by the casual observer. There are, in fact, rational bases for the structure and reactivity of organotransition metal complexes.

Transition metals are those triads from titanium to copper, i.e. those that contain partially filled d-shells in some or all of their oxidation states.[3] In the transition series, the (n+1) s level is higher in energy than the nd level. Up to the first triad of the Group VIII metals, the group number is the same as the number of d-electrons in the zero valent metal. Thus, copper(II) has 9 d-electrons, and the copper triad is the last of the transition metal series.

The "oxidation state" of the metal is defined as the charge left on the metal when all the ligands are removed in their normal, closed-shell configuration, or the charge left on the metal when each shared electron pair is given to the more electronegative atom. In this sense, everything is more electronegative than the metal, including carbon and hydrogen, even though many metal hydrides are in fact, strong Bronsted acids. Once the formal oxidation state of a metal has been assigned, then the d-configuration can be determined. This is done simply by substracting the appropriate number of electrons from the d^n configuration in the zero oxidation level of the metal as is necessary to obtain the oxidation state. Thus, the formal oxidation state of a transition metal gives its d^n configuration.

The d^n configuration can be used to determine the coordination number (CN, the number bonds to the metal) and the expected geometry of a transition metal complex. For monometallic, diamagnetic compounds, there is a maximum coordination number allowed for each d^n. The "eighteen electron rule", gives the relationship between the d^n configuration and the coordination number, allowing an assessment of the maximum coordination. This coordination number

$$n + 2(CN)_{max} = 18$$

is equal to the number of empty s-, p-, and d-orbitals of a transition metal. Filling these orbitals brings the metal to a rare gas configuration. There are few exceptions to the rule; some organometallics appear to be 20 electron complexes. Table 1 gives some examples of common ligands that are found in organotransition-metal complexes.

Iron pentacarbonyl, is a zero-valent complex in which iron has the d^8 configuration. Its maximum coordination number, therefore is five, and the complex is coordinatively saturated, achieving the eighteen electron config-

$$CN = \frac{18 - n}{2}$$

OC, CO, Fe—CO, OC, CO

Ph_3P, Cl, Pt, Ph_3P, Cl

uration. Bis(triphenylphosphine)dichloropalladium(II), however, is coordinatively unsaturated, since its oxidation number is +2, yielding a d^8 configuration and a maximum coordination number of five. A molecular orbital description of transition metal complexes is omitted here because of the limitation of scope of this chapter. However, the basis of the 18-electron rule can be described in terms of M.O. theory.

The geometry of organometallic complexes is determined largely by steric factors, i.e. the ligands are placed at the maximum distance from one another. Some examples are shown in Table II. The square planar geometry is found among coordinatively, unsaturated, diamagnetic d^8 complexes of the nickel triad (Ni^{2+}, Pd^{2+}, Pt^{2+}) and the cobalt triad (Co^{1+}, Rh^{1+}, Ir^{1+}) instead of the sterically more favorable tetrahedral geometry. In these examples the more favorable electronic situation (diamagnetic) dictates the geometry. Also, the square pyramidal geometry is known for a number of transition metal complexes, but the activation barrier from trigonal bipyramid to square pyramidal is low, particularly for d^8 configurations.

Because olefins, carbon monoxide and phosphines are important ligands in organometallic complexes, and because olefin and carbonyl bonded ligands often undergo subsequent reactions on the transition metal in the synthesis of organic products, a further description of their bonding is warranted. The electronic effects of the ligand are transmitted to the metal and to the ligand trans to it.

Table I

Classification of Common Ligands In Organotransition Metal Chemistry

Type of Ligand	Ligand Name	Representation of Complex	Ligand Representation for Electron Counting Purposes
2-Electron Ligands			
η^1-anionic	Hydrido	M-H	$H^{\ominus}$
	Halo	M-X	$:\ddot{X}:^{\ominus}$
	Cyano	M-CN	$:N\equiv C:^{\ominus}$
	Alkyl(aryl,vinyl)	M-C≡	$R_3C:^{\ominus}$
	η^1-Acyl	M-C(=O)-R	$R-C(=O):^{\ominus}$
	η^1-Cyclopentadienyl	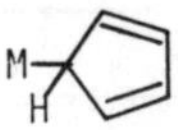 M-C_5H_5 (M and H on the same carbon)	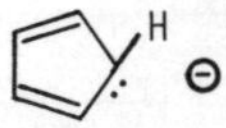 $C_5H_5^{\ominus}$ (H on the anionic carbon)
	η^1-Allyl	$M-CH_2-CH=CH_2$	$CH_2=CH-CH_2^{\ominus}$
η^1-neutral	Phosphine	$M-PR_3$	$R_3P:$
	Carbonyl	$M-C\equiv O$	$O\equiv C:$
	Nitrile	M-NCR	$RC\equiv N:$
	Carbene	$M=CH_2$	$H_2C:$
η^2-neutral	Alkene	$R_2C=CR_2$ (arrow from C=C to M)	$R_2C=CR_2$
η^1-cationic	Nitrosyl	$M-NO^{\oplus}$	$:N\equiv O^{\oplus}$
η^3-cationic	Cyclopropenyl	 $\oplus$ ring above M	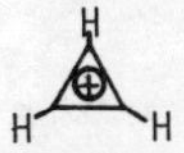 $C_3H_3^{\oplus}$ (ring with three H)
4-Electron Ligands			
η^2-anionic	Acyl	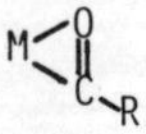 M bonded to both O and C of C(=O)-R	$R-C(=\ddot{O}:):^{\ominus}$
	Acetoacyl	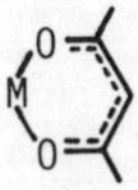 M chelated by both O atoms of $CH_3C(O)CHC(O)CH_3$	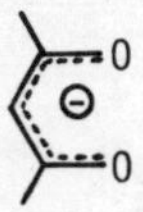 $CH_3C(O)CHC(O)CH_3^{\ominus}$ (delocalized)
η^3-anionic	Allyl	 M bonded to allyl	 allyl$^{\ominus}$ (delocalized)

TABLE I (cont'd.)

Type of Ligand	Ligand Name	Representation of Complex	Ligand Representation for Electron Counting Purposes
η^4-neutral	1,3-Diene		
6-Electron Ligands			
η^4-anionic	Trimethylenemethyl		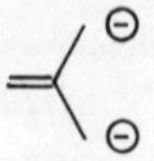
η^5-anionic	Cyclopentadienyl		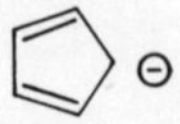
η^6-neutral	Arene		
η^6-cationic	Cycloheptatrienyl	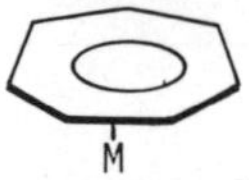	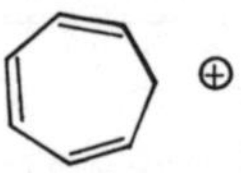

A number of ligands, including carbon monoxide and alkenes, have low-lying vacant orbitals having the appropriate symmetry to overlap with filled d-orbitals on the metal. Thus, in addition to having a filled orbital that will donate an electron pair to an empty d-orbital in a metal, some electron return is experienced by the "backbonding" of these π-acids. In the orbital picture of the bonding of an olefin to a metal (Figure 1) the arrows indicate the donation of electrons from a filled orbital to an empty orbital. It is the empty orbital on the olefin through which back-bonding can occur,

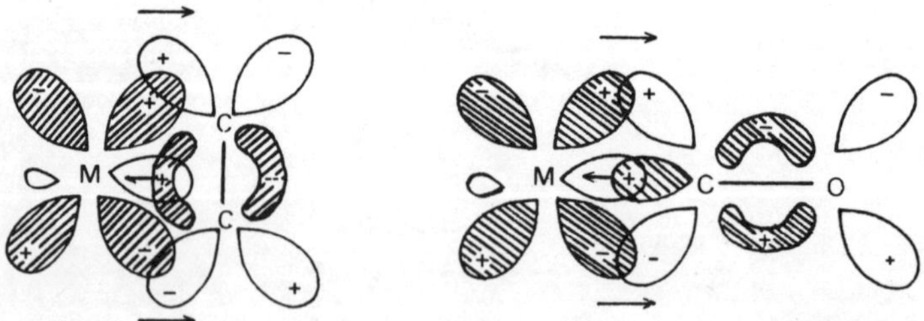

Figure 1

and the geometry through which this takes place requires a side-bound olefin ligand. Carbon monoxide provides an example through which end-bonding can take place (Figure 1).

TABLE II
GEOMETRIES OF SOME TRANSITION
METAL COMPLEXES

Coordination Number	Geometry	Example
2	Linear	Ph_3P-Au-Cl
3	Trigonal	$Pt(PCy_3)_3$ (PCy₃ above Pt; Cy_3P, PCy_3 below)
4	Tetrahedral	$Ni(CO)_4$ (OC, CO, OC, CO)
4	Square Planar	$Pd(PPh_3)_4$ (PPh_3 / Ph_3P-Pd-PPh_3 / PPh_3)
5	Trigonal Bipyramidal	$Fe(CO)_5$ (CO / OC—Fe, CO, CO / CO)
5	Square Pyramidal	NO / Cl, Ir, PPh_3 / Ph_3P, CO
6	Octahedral	$Mo(CO)_6$ (CO / OC, Mo, CO / OC, CO / CO)

The degree of electron transfer to a metal is variable, and is dependent on the back-bonding. Carbon monoxide, with a high amount of back-bonding, is a strong π-acid ligand, lowering the electron density in the filled d-orbitals and thereby stabilizing a metal toward oxidative addition, a reaction requiring a high electron density on the metal. Phosphine ligands, on the other hand are primarily donors and increase the electron density on the metal. Phosphines readily transmit electron effects to other ligands through the metal, particularly to those ligands trans to it. In Wilkinson's complex, $(Ph_3P)_3RhCl$, the phosphine ligands mutually trans to each other are labilized, and the phosphine metal bonds are 0.1Å longer than that trans to the chloro ligand.

Steric effects are perhaps more important with phosphine ligands than the electronic effects. The bulk of the phosphine has been measured by the "cone angle", (Figure 2), the magnitude of which determines the steric interaction with a complex of a certain geometry. As a result, in the tetrahedral $Ni[P(OMe)_3]_4$, the cone angle of the $P(OMe)_3$ ligand is 107°, smaller than the tetrahedral bond angle. Dissociation of a phosphine ligand in this nickel complex does not take place as readily as it does in the square planar 16-electron (already coordinatively unsaturated) Wilkinson's complex, $(PPh_3)_3RhCl$, in which the phosphine cone angle is 145°.

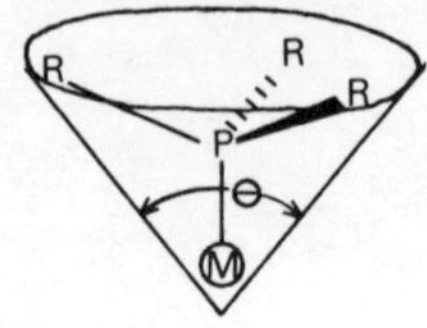

Figure 2

There are several other useful generalizations that can be made concerning organo-transition metal complexes. Stronger metal-carbon and metal-hydrogen bonds are formed with 5d and 4d metals than with 3d metals. The early transition metals are more electropositive and form strong bonds with hard donor atoms such as oxygen. Oxidation catalysts usually are high valent ions containing hard ligands such as water or carboxylate ions. The early transition metals do not bind strongly to π-acid ligands such as carbon monoxide and olefins, so that migratory insertion reactions to yield acyl or alkyl ligands, respectively, are facile and the equilibrium lies on the acyl or alkyl side. By contrast, the late transition metals, particularly those of Group VIII, form strong bonds with soft, polarizable donor atoms such as phosphorus and π-acids. Thus, the reverse of the insertion reactions, decarbonylation and β-hydride elimination are more prevalent. The organometallic complexes in Group VIII with d^8 and d^{10} configurations tend to under oxidative-addition and reductive elimination reactions readily.

OXIDATIVE ADDITION[4-6]

Oxidative addition is a term given to a general class of reactions in which the metal is oxidized by a substrate XY. The increase in the oxidation state is usually accompanied by an increase in the coordination number. Low spin transition metal complexes, particularly Group VIII d^8 and d^{10} complexes, undergo two electron oxidative addition reactions, although this type of oxidative addition is known for most all even d^n configurations. For example, the five coordinate 18-electron complexes of the d^8 configuration undergo oxidative addition by prior dissociation of a ligand. The dissociation produces a coordinatively unsaturated complex which, after oxidative addition achieves a d^6 configuration.

d^8 (18e) $\xrightarrow{-L}$ d^8 (16e) $\xrightarrow{XY}$ d^6 (18e) (trans or cis)

The overall two electron change may take place by a concerted two electron transformation or sequential one electron changes. For this reaction, there is a family of mechanisms that lie close enough to one another in energy such that subtle changes in the substrate (XY), the ligands and different metals result in mechanistic changes. There are a smaller number of oxidative additions that take place by one electron changes.

As illustrated above, either a cis or a trans octahedral complex can be obtained from the oxidative addition to the d^8, 16-electron complex. In most reactions studied, it is not known whether the isomer observed is the kinetic product. The 18-electron d^{10} square planar complex undergoes ligand dissociation of at least one ligand before oxidative addition takes place. The product of oxidative addition is a 16-electron complex. Phosphine

$$L_4M \underset{+L}{\overset{-L}{\rightleftharpoons}} ML_3 \underset{+L}{\overset{-L}{\rightleftharpoons}} ML_2$$

$ML_3 \xrightarrow{XY} [XML_3]^{\oplus}Y^{\ominus} \xrightarrow{-L}$ trans-$MXYL_2$ or cis-$MXYL_2$; $ML_2 \xrightarrow{XY}$ $MXYL_2$

ligands (L) provide a high electron density on the metal, making it a good nucleophile, thereby favoring oxidative addition. High electron density on the metal also favors dissociation of the ligands to a coordinatively unsaturated complex.

If a d^8 complex is coordinatively saturated, strong oxidizing agents usually are required, and the reaction proceeds through an intermediate octahedral cationic d^6 complex.

$$ML_5 + XY \longrightarrow [MXL_5]^{\oplus} Y^{\ominus} \xrightarrow{-L} MXYL_4$$

Substrates (XY) that undergo oxidative addition can be classified as polar or non-polar, polar substrates being susceptible to nucleophilic attack by the metal. (This is analagous to an S_N2 displacement). Polar substituents include organic halides, Bronsted acids and halogens. Examples of non-polar oxidants are dihydrogen and carbon hydrogen bonds. The overall addition to the metal in the latter case is almost always stereospecifically cis as could be expected from a concerted, three centered addition. A special class of non-polar substrates are those that retain at least one bond in the reaction product to give a metallocyclic product. Strained ring hydrocarbons fall into this category.

Organic Halides

Some representative oxidative addition reactions of organic halides, are illustrated particularly with Group VIII transition metals since most of the synthetic organic chemistry involving an oxidative addition step has been accomplished with this group. The mechanisms of the reactions are quite diverse, and a wide variety of techniques and mechanistic tools have been involved in establishing the reaction mechanisms, the details of which will not be discussed here.

The reaction of d^{10} organocuprates with vinyl and alkyl halides probably takes place by a sequence of oxidative addition, reductive elimination reactions,[7] although the oxidative addition product has not been isolated. The cross coupling reactions proceed with predominate (84-92%) inversion of configuration at the allyl halide carbon and retention of geometry in the vinyl halide.[8,9] This second order reaction requires that oxidative addition gives a d^8 square planar trans Cu(III) complex followed by an exclusive reductive elimination of cis partners. Retention of geometry in the vinyl halide suggests two retention reactions (oxidative addition and reductive elimination), while in the alkyl halide inversion of configuration in the oxidative addition reaction and retention in the reductive elimination must take place.

$$R_2Cu^{I}Li + R'(X)C{=}CH_2 \xrightarrow{\text{rate}} \left[R\text{—}Cu(X)(CH{=}CHR')\text{—}R \right]^{\ominus} Li^{\oplus} \xrightarrow{\text{fast}} R'CH{=}CHR + RCu^{I}$$

$$R_2Cu^{I}Li + \text{Me}\cdots C(R')(H)-Br \xrightarrow{\text{rate}} \left[\begin{array}{c} R'(Me)(H)C \\ | \\ R-Cu-R \\ | \\ X \end{array} \right]^{\ominus} Li^{\oplus} \xrightarrow{\text{fast}} R-C(R')(Me)(H) + RCu^{I}$$

Although the Cu(III) intermediate product of oxidative addition is not isolated, it can be with certain gold complexes.[10] A trans complex is obtained, but isomerization can take place, particularly in the presence of phosphine ligands.

$$RX + Me_2Au^{I}(L)^{\ominus}Li^{\oplus} \longrightarrow R-\overset{\displaystyle Me}{\underset{\displaystyle Me}{Au}}-L + LiX \qquad L = PPh_3$$

Oxidative addition to d^{10} complexes of the nickel triad takes place with a broad spectrum of organic halides. Vinyl halides react with all three members of the triad with retention of geometry at the double bond. The analogous platinum complexes are less reactive than those of nickel or palladium. For example, tetrakis(triphenylphosphine)platinum(0) will not undergo oxidative addition with vinyl chloride whereas the nickel and palladium analogs will. Substitution of a second halogen on the olefin usually produces a reaction in the case of platinum.

$$L_4M + (R)(H)C{=}C(X)(H) \longrightarrow \left[L_2M\langle C(H)(R)-C(H)(X) \rangle \right] \longrightarrow (H)(R)C{=}C(H)-\overset{\displaystyle L}{\underset{\displaystyle L}{M}}-X$$

The mechanism proposed for these reactions involves prior formation of a metallacyclopropane. Indeed, with platinum, such intermediate complexes have been isolated and they rearrange to the d^8 vinyl platinum product.

Aryl halides also oxidatively add to complexes of the nickel triad. Chlorobenzene adds slowly to tetrakis(triphenylphosphine)nickel(0) but instantaneously to the more nucleophilic tetrakis(triethylphosphine) analog. Chlorobenzene does not add to the triphenylphosphine complexes of palladium or platinum. The reaction probably proceeds analogously to a nucleophilic aromatic substitution reaction, at least with palladium, while a one electron transfer mechanism appears to be an alternate path in the case of nickel and platinum.

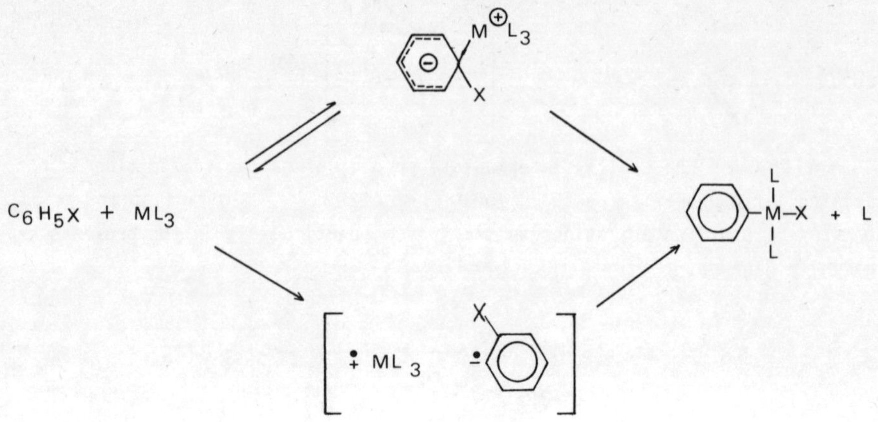

$C_6H_5X + ML_3$

Allyl halides react rapidly with complexes of the nickel triad, even with weakly nucleophilic complexes. Nickel tetracarbonyl, for example, yields the π-allyl halogen bridged dimer (a useful reagent for effecting cross coupling reactions).[11] The nmr spectra of such π-allyl nickel com-

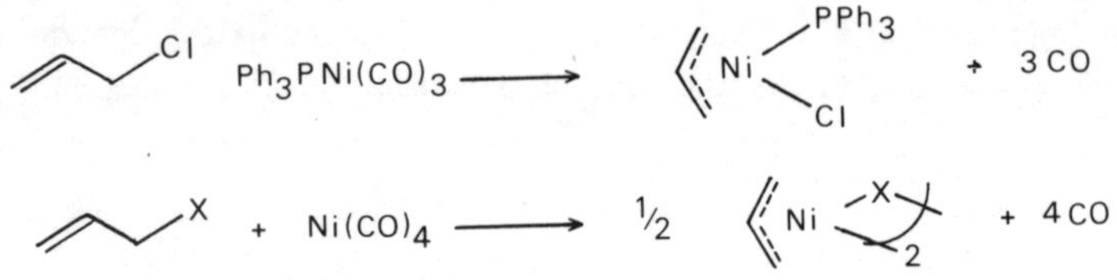

$Ph_3PNi(CO)_3 \longrightarrow$ + 3 CO

+ $Ni(CO)_4 \longrightarrow \frac{1}{2}$ + 4 CO

plexes show a rapid, dynamic equilibrium between σ- and π-allyl complexes in solution. X-ray structures show that the allyl group lies in a plane orthogonal to the plane of the nickel and the other two ligands.[12]

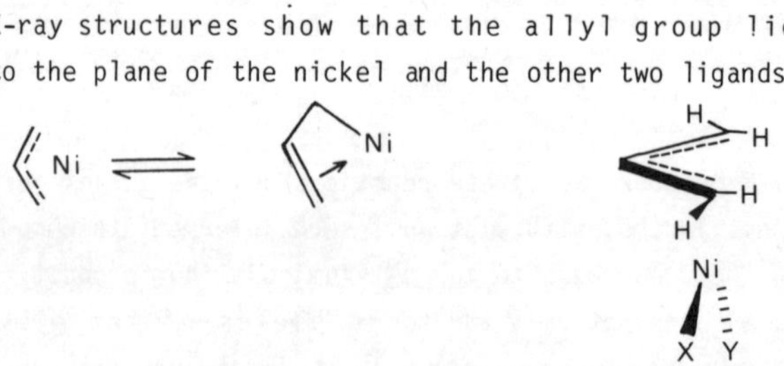

The reaction of allyl chlorides with tetrakis(triphenylphosphine)palladium(0), has been reported to give a rapidly equilibrating σ-allyl complex. However, when the π-allyl chloro bridged dimer is allowed to react with excess phosphine, either a cationic π-allyl complex or a sigma complex may be formed, depending on the reaction solvent.[13]

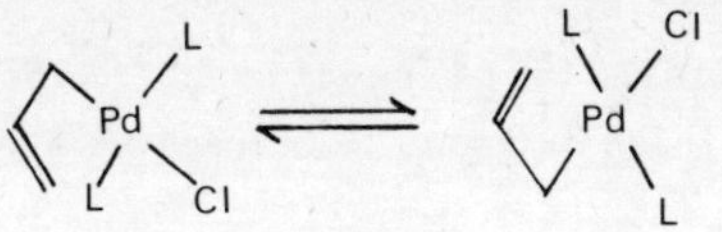

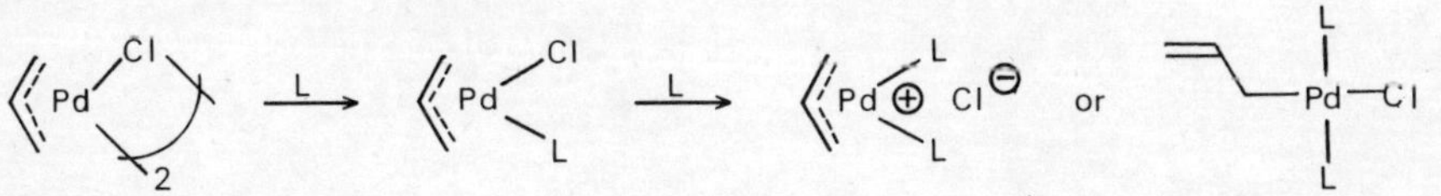

A useful and important oxidative addition reaction with allyl acetates takes place with palladium(0) species. The oxidative addition products react with a variety of nucleophiles, yielding a nucleophilic substitution product of overall retention of configuration at the allylic carbon as a result of inversion both during the oxidative addition and the nucelophilic displacement.[14]

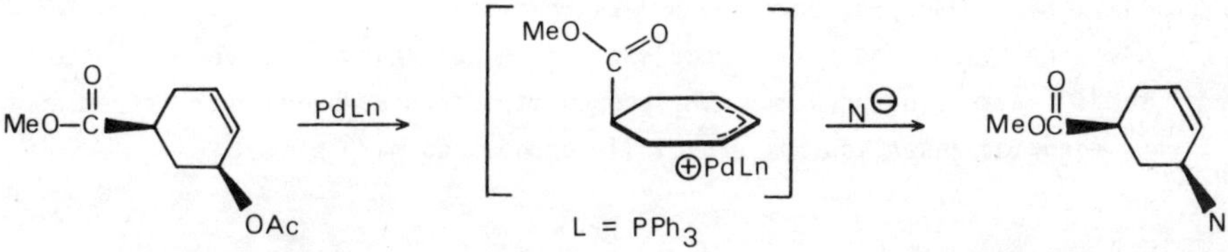

L = PPh_3

Alkyl halides oxidatively add to complexes of the nickel triad, but if the alkyl group contains an sp^3 hydrogen β- to the metal, β-elimination of metal hydride is usually rapid and the metal alkyl product of oxidative addition usually is not isolated. The oxidative addition can be easily studied and the oxidative addition product isolated in reactions with substrates that do not contain β-hydrogens, such as benzyl, methyl, and neopentyl halides.

Several mechanisms have been proposed (Figure 3), the three most cited being an S_N2-type (a), a radical pair (c) and a radical chain (d). Nickel(0) appears to react both by S_N2 and radical pair mechanisms. Palladium, on the other hand, can undergo oxidative addition by an S_N2-type mechanism since chiral benzyl halides react with tetrakis(triphenylphosphine)-, tris(triethylphosphine)- and carbonyltris(triphenylphosphine)palladium(0) complexes to yield products resulting from inversion of configuration at carbon.[16] A key step in determining the stereochemistry of the oxidative addition

Figure 3

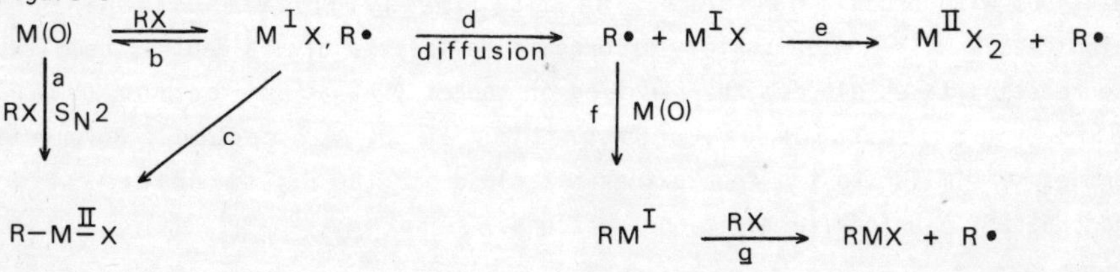

$$M(0) \underset{b}{\overset{RX}{\rightleftharpoons}} M^{I}X, R\bullet \xrightarrow[\text{diffusion}]{d} R\bullet + M^{I}X \xrightarrow{e} M^{II}X_2 + R\bullet$$

$$M(0) \xrightarrow[S_N2]{a,\ RX} R{-}M^{II}{-}X \qquad M^{I}X, R\bullet \xrightarrow{c} R{-}M^{II}{-}X$$

$$R\bullet + M(0) \xrightarrow{f} RM^{I} \xrightarrow[g]{RX} RMX + R\bullet$$

reaction is an "insertion" reaction of carbon monoxide (to give an acyl palladium complex) which is known to proceed with retention of configuration

$$\text{Ph(H)(R)C–X} + \text{LnPd} \longrightarrow \text{Ph(R)(H)C–Pd(L)}_2\text{–X} \xrightarrow{\text{CO}}$$

$$\text{Ph(R)(H)C–C(=O)–Pd(L)}_2\text{–X} \xrightarrow{\text{EtOH}} \text{Ph(R)(H)C–CO}_2\text{Et}$$

at carbon (vide infra). Existing evidence supports a radical process for the reaction of platinum(0) complexes with most alkyl halides (Figure 3, paths b-g). Alkylplatinum halides containing β-hydrogens can be isolated since the β-elimination reaction occurs much less readily.

Acid chlorides react with complexes of the nickel triad to yield acyl oxidative addition product. The propensity for decarbonylation (reverse of carbon monoxide insertion) is especially pronounced with nickel.

$$\text{RC(=O)–Cl} + \text{ML}_n \longrightarrow \text{R–C(=O)–M(L)}_2\text{–X} \underset{+\text{CO}}{\overset{-\text{CO}}{\rightleftharpoons}} \text{R–M(L)}_2\text{–X}$$

Further oxidative addition of organic halides can take place to d^8 complexes of the nickel triad, particularly with platinum and palladium. The oxidative addition products to platinum(II) complexes, the d^6 platinum(IV) complexes, are stable enough to be isolated in many cases. Palladium(IV) oxidative addition products are implicated in a number of catalytic

$$\text{R–Pt}^{\text{II}}\text{(L)}_2\text{–X} + \text{R}'\text{X}' \longrightarrow \text{R}'\text{(R)(L)(L)(X)(X}'\text{)Pt}^{\text{IV}}$$

d^6 (18e)

The cobalt and iron triads undergo some important oxidative addition reactions with organic halides. The bis(dimethylglyoximato)cobalt(I) anion reacts with a wide variety of organic halides to give a neutral complex. The reaction takes place with retention of geometry at an sp^2 carbon (vinyl halide) and inversion of configuration at an sp^3 carbon. Porphyrin complexes, similar to the cobaloxines but closer to the B_{12} structure, also undergo facile oxidative addition reactions.

$$\equiv \ Co(DMGH)_2L^{\ominus} + RX \longrightarrow RCo(DMGH)_2L + X^{\ominus}$$

The oxidative addition reactions of chlorotris(triphenylphosphine)rhodium(I), Wilkinson's complex, with organic halides[17] and the decarbonylation reactions of acid chlorides and aldehydes[18] (Figure 13) have become important synthetically. A key step in the acetic acid synthesis from methanol and carbon monoxide is an oxidative addition reaction of methyl iodide to a d^8 rhodium(I) species (Figure 8). Halocarbonylbis(phosphine)rhodium complexes yield, on oxidative addition the octahedral d^6 18-electron product. The reaction takes place with alkyl, allyl and acyl halides.

$$X{-}Rh(L)_2{-}CO + RX' \longrightarrow Rh(R)(X)(X')(L)_2(CO)$$

Oxidative addition to Wilkinson's complex takes place only with the most reactive organic halides. The oxidative addition of acid chlorides is the first step in catalytic and stoichiometric decarbonylation reactions of acid chlorides.[17,18] Both square pyramidal and trigonal bipyramidal complexes have been isolated, depending on the acyl group.

$$L_3RhCl + RCOCl \longrightarrow Rh(COR)Cl_2L_2 \underset{k_{-1}}{\overset{k_1}{\rightleftharpoons}} Rh(R)Cl_2(CO)L_2$$

Rearrangement of the acyl complex to the rhodium carbonyl takes place reversibly when R is benzyl (k_1/k_{-1} 7 x 10^{-2}), but when R is aryl, the carbonyl complex is formed irreversibly.

The oxidative addition reaction to iridium(I) complexes is initiated by radicals, apparently including trace amounts of oxygen. Loss of stereochemistry occurs in the oxidative addition of alkyl halides; a radical chain mechanism is most consistent with the data. Generally, iridium is not a good catalyst in reactions that require an oxidative addition reaction since the product of oxidative addition is much more stable than the analagous rhodium compounds.

$$\mathrm{Cl{-}IrL_2{-}CO} + \mathrm{RX} \longrightarrow \mathrm{Ir(R)(X)(OC)(L)_2Cl}$$

Initiation:

$$\mathrm{i} + \mathrm{Ir}^{I} \longrightarrow \mathrm{iIr}^{II}$$

$$\mathrm{iIr}^{II} + \mathrm{RX} \longrightarrow \mathrm{iIrX} + \mathrm{R}\bullet$$

Propagation:

$$\mathrm{R}\bullet + \mathrm{Ir}^{I} \longrightarrow \mathrm{RIr}^{I}\bullet$$

$$\mathrm{RIr}^{I}\bullet + \mathrm{RX} \longrightarrow \mathrm{RIrX} + \mathrm{R}\bullet$$

The disodium tetracarbonyl ferrate "supernucleophile", which is $\sim 10^8$ times more nucleophilic than the thiophenoxide anion, reacts rapidly with organic halides to give the organoiron product.[20] Alkyl halides and tosylates can be coupled to give ketones and converted to aldehydes or carboxylic acid dertivatives. Ketones are synthesized by the successive additions of alkyl halides. The question of whether alkyl migration (CO in-

$$\underset{d^{10}}{\mathrm{Fe(CO)_4^{2-}}} \longrightarrow \underset{d^{8}}{[\mathrm{RFe(CO)_4}]^{\ominus}} \xrightarrow{R'X} \underset{d^{6}}{[\mathrm{Fe(R)(R')(OC)(CO)_3}]}$$

$$\underset{d^{8}}{[\mathrm{R{-}C(=O){-}Fe(CO)_3}]^{\ominus}} \xrightarrow{R'X} \underset{d^{6}}{[\mathrm{Fe(C(=O)R)(R')(CO)_3}]} \longrightarrow \mathrm{R{-}C(=O){-}R'}$$

sertion) takes place before a second oxidative addition occurs is unsettled. All intermediates except the 16-electron d^8 acyl iron(0) anion are coordinatively saturated. The ^{5}-cyclopentadienyldicarbonyl iron anion (Fp^-) undergoes oxidative addition reactions with a wide variety of organic halides and is a useful reaction in a number of syntheses.[21] The mechanisms of these reactions involving anionic iron complexes are all S_N2-like.

Carbon-Carbon Bonds

Transition metals catalyze a number of pericyclic reactions, including valence isomerization.[22] An early, key step in the sequence of reactions for an isomerization is the oxidative addition of a hydrocarbon to a transition metal by breaking a carbon-carbon bond to form a metallocycle.[23] Generally, the reaction occurs with small ring, strained hydrocarbons, but there are a few examples in which the reaction occurs with strain-free hydrocarbons.

Although many transition metal complexes catalyze these rearrangements, the number of isolable compounds containing a carbon-metal-carbon bond formed from a carbon-carbon bond cleavage by a transition metal is surprisingly narrow. The subsequent reactions, particularly the reductive elimination step, in most of the catalytic rearrangements evidently have lower activation energies, such that the oxidative addition intermediates are not isolated. Most of the isolated products are those of platinum(II and IV), rhodium(III) and iron(II). Platinacyclobutanes are formed from the reaction of cyclopropanes with platinum(0) and palladium(0) complexes.[24] Platinum(II) complexes

L_nM + [cyclopropane] ⟶ L_2M[cyclobutane ring]

undergo this reaction, in which the mechanism can be compared to ring opening reaction of cyclopropanes with acids.

Rhodium(I) complexes, particularly chlororhodium diene and chlororhodium carbonyl dimers catalyze the isomerization reactions of nortricyclene to norbornadiene and cubane to the syn-cyclobutadiene dimer. In the reaction of the chlororhodium carbonyl dimer with cubane, the 6-membered rhodacycle can be isolated.[25] The same type of reaction takes place between diironnonacarbonyl and a variety of strained ring hydrocarbons.

[cubane] + $[Rh(CO)_2Cl]_2$ ⟶ Cl–Rh(CO)(C=O) rhodacycle

Carbon-Hydrogen Bonds

Although the intramolecular oxidative addition reaction of carbon-hydrogen bonds on ligands attached to a transition metal is observed quite frequently, the intermolecular oxidative addition reaction of hydrocarbons is more the exception. The transition metal hydrides resulting from such a reaction usually are not isolated, nor can these unstable intermediates be observed, even in the intramolecular reaction. There are a few such examples in which the hydride products have been isolated.

Cyclometallation[27] is a widely occurring reaction type in which a carbon-hydrogen bond of a ligand is cleaved to give a chelate ring containing a carbon metal bond. Many of these reactions that occur on an aromatic ring are electrophilic substitution reactions in which the metal is not oxidized at any stage of the reaction. Nucleophilic attack by the metal however, results at least in an intermediate oxidative addition product which in most cases is followed by reductive elimination.

The ligand Z is a group V or VI donor and Y usually is alkyl or halogen. Most cyclometallations occur with Group VI-VIII transition metals and with the d^8 zero valent complexes of the iron triad as well as with rhodium(I) and iridium(I) complexes. The cyclometallated oxidative addition products often are isolated. For orthometallations of aromatic rings that are part of the ligand, electron donors enhance the reaction rates when the reaction is an electrophilic substitution reaction (not oxidative addition). The positively charged metal simply acts as an electrophile. True oxidative addition involving an aromatic carbon-hydrogen bond is favored by $+\sigma$ substituents since the reaction proceeds by nucleophilic attack by the metal on the ring.

Decarbonylation of aldehydes to hydrocarbons by Wilkinson's complex takes place under mild conditions to give a high yield of the corresponding alkane and in some cases minor amounts of alkene.[17,18] The mechanism requires a rate-determining oxidative addition of aldehyde followed by alkyl migration (reverse of CO insertion) and reductive elimination. The reaction is stereospecific, taking place with retention of configuration (sp^3) or geometry (sp^2) at carbon. Such an oxidative addition intermediate has been isolated in an intramolecular insertion reaction of 8-quinoline aldehyde.[30]

8-quinolinecarboxaldehyde (CHO) + $RhClL_3$ $\xrightarrow[95\%]{CH_2Cl_2}$ acyl(hydrido)rhodium complex; L = PPh_3

$\xrightarrow{-CO,\ \Delta}$ quinoline + $RhClL_3$

Probably the most important recent development in CH activation is the discovery that the trimethylphosphine-η^5-pentamethylcyclopentadienyliridium-(I) complex reacts with benzene, cyclohexene, and neopentane by oxidative addition of a CH bond to iridium.[31] The reactive 16-electron iridium(I) complex is generated by the photolysis of the corresponding iridium(III)

$$Cp^*(Me_3P)IrH_2 \ (d^6\ (18e)) \xrightarrow{h\nu} [Cp^*(Me_3P)Ir] \ (d^8\ (16e)) \xrightarrow{RH} Cp^*(Me_3P)Ir(H)(R)$$

hydride. The iridium complex $[IrH_2S_2L_2]^+BF_4^-$ (S=acetone or water; L=PPh_3) reacts with cyclopentane in the presence of an olefin to give η^5-cyclopentadienyl iridium hydride.[32] The function of the olefin is to remove hydrogen from iridium giving the iridium(I) complex which oxidatively adds cyclopentane, and, in a series of steps, carries it to cyclopentadiene which gives the observed product of oxidative addition.

$$IrH_2S_2L_2^{\oplus}\,BF_4^{\ominus} + C_5H_{10} + 3\,Bu^tCH{=}CH_2 \longrightarrow L_2Ir(Cp)^{\oplus}(H) + 3\,Bu^tCH_2CH_3$$

Dihydrogen

The oxidative addition of hydrogen to a transition metal complex takes place in catalytic hydrogenations (Figures 11,12), hydroformylations (Figure 14), and the Fischer-Tropsch reaction. There are a number of examples in which the product of oxidative addition of hydrogen can be isolated, most of which are additions to d^8 complexes. Dihydrogen readily adds to coordinatively unsaturated transition metal complexes. Alternatively, the heterolytic cleavage of hydrogen by two metal centers may occur.

$$M + H_2 \rightleftharpoons \overset{H}{\overset{|}{M}}-H \qquad M-L + H_2 \longrightarrow \overset{H}{\overset{|}{M}}-H + L$$

$$2M + H_2 \rightleftharpoons 2M-H$$

For example, Vaska's complex adds dihydrogen to give the cis-complex. In a number of reactions involving oxidative addition of dihydrogen, the

$$Cl-Ir(L)_2-CO + H_2 \longrightarrow Ir(CO)(L)_2(Cl)(H)_2 \qquad L = PMe_2Ph$$

dihydride is not isolated, but reductive elimination of a molecule such as a hydrocarbon or a hydrogen acid ultimately yields the monohydride.

$$R-Rh(PPh_3)_3 + H_2 \longrightarrow [R-RhH_2(PPh_3)_3] \longrightarrow HRh(PPh_3)_3 + RH$$

One Electron Oxidative Additions[33]

One electron oxidative additions take place usually with odd-electron complexes. The reaction with organic halides takes place in two steps, the

$$2\ L_nM + XY \longrightarrow L_nMX + L_nMY$$

first of which is rate determining. This type of reaction is well

$$L_nM + RX \xrightarrow{\text{rate}} L_nM-X + R\bullet$$

$$L_nM + R\bullet \xrightarrow{\text{fast}} L_nM-R$$

documented with Co(II) (17e) and Cr(II) (17e) complexes.

REDUCTIVE ELIMINATION

Although a carbon-transition metal bond is not necessarily weak, its lability is derived from the variable oxidation states and coordination numbers of the transition metals that are not available to the main group analogs. The breaking of a transition metal-carbon σ-bond usually involves other coordination sites in addition to the σ-bond under consideration. The stability of transition metal-carbon σ-bonds and the reactions by which they undergo scission has been the subject of a number of reviews.[34] The dissociation energy of such a bond is an important property of organometallic compounds, but the factors that influence the bond strength and the overall stability of transition metal complexes containing carbon-metal σ-bonds are not well understood, in part because there is scant reliable data on bond energies.[35] Some recent data has been obtained, but the dissociation energies obtained by different methods, ICR[36] and kinetic determinations,[37] differ considerably. The cobalt carbon bond dissociation energies, for example, range from 61/Kcal/mole to 20/Kcal/mole, respectively.

It is recognized that one of the low energy pathways for alkyl-transition metal complex decomposition is β-elimination. If this pathway can be blocked, by construction of an alkyl portion such that β-hydrogens are absent, by preparing a complex in which β-hydrogen elimination is geometrically unfavorable, or by preventing the generation of an open coordination site (a requirement for β-elimination), then the transition metal σ-bonded complex may be quite stable. The coupling reactions of organic compounds catalyzed by transition metals is an important method of generating carbon-carbon bonds, the final step of which requires the elimination of the organic partners from the transition metal. The elimination can take one or more paths, categorized according to the mechanism (and products), including heterolytic as well as homolytic or concerted α-elimination, β-elimination, 1,1-reductive elimination and dinuclear elimination (Figure 4).

It has been stated[34b] that these reactions are usually concerted, proceeding by electron pair processes, and that no high energy intermediates such as free radicals are involved, particularly in reductive elimination.

Figure 4

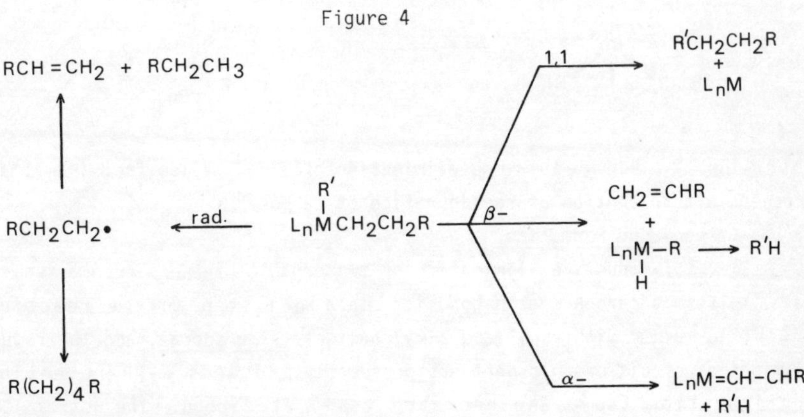

This appears to be the case with most even electron transition metal complexes, and although these reactions appear to be concerted, low energy transformations, there is evidence for free radical pathways in a number of systems. In odd electron complexes, decomposition by homolytic metal-carbon scission generally appears to be the mode of decomposition.[38]

$$\mathrm{L(DMGH)Co^{III}{-}R} \longrightarrow [\mathrm{L(DMGH)_2Co\bullet\ \bullet R}] \xrightarrow{\text{fast}} \mathrm{CoH} + \mathrm{PhCH{=}CH_2}$$

$$\mathrm{CoH} \longrightarrow \mathrm{Co^{II}} + 1/2\,\mathrm{H_2} \qquad \mathrm{R} = -\mathrm{CH(CH_3)}-\mathrm{Ph}$$

In the 1,1-reductive elimination of an organo group and its partner (halogen, hydrogen or another organo group) the formal oxidation state and the coordination number of the metals are reduced by two; bond breaking is accompanied by bond making. The reductive elimination reaction frequently follows an oxidative addition reaction, and this combination is required in stoichiometric and catalytic coupling reactions, catalytic hydrogenation reactions of alkenes, decarbonylation reactions of acid chlorides and aldehydes, etc.

Carbon-Halogen Bond Formation.

An important type of 1,1-reductive elimination occurs between an organic group and halogen. This step which takes place in the stoichiometric and catalytic decarbonylation of acid chlorides by Wilkinson's catalyst (Figure 13),[17,18] is first order with an activation energy of 23 kcal/mole ($R=CH_3$). A concerted reaction is favored; retention of configuration at carbon is

$$\mathrm{Rh(R)(CO)(Cl)_2(PPh_3)_2} \longrightarrow \mathrm{RCl} + \mathrm{Cl{-}Rh(PPh_3)_2{-}CO}$$

observed.[39] The reductive elimination of alkyl halide from iron(II) also proceeds with retention of configuration at carbon.[40]

Carbon-Carbon Bond Formation.

The 1,1-reductive elimination of two organic partners from a transition metal to form a carbon-carbon bond is the final step in the reaction of organocuprates with vinyl and alkyl halides (vide supra), and the reductive elimination of cis-organic partners presumably proceeds with retention of configuration (sp^3) and geometry (sp^2) at carbon. The intramolecular reductive elimination of ethane from trimethyl(triphenylphosphine)gold(III) requires prior dissociation of phosphine and only cis-methyl groups undergo coupling.[41]

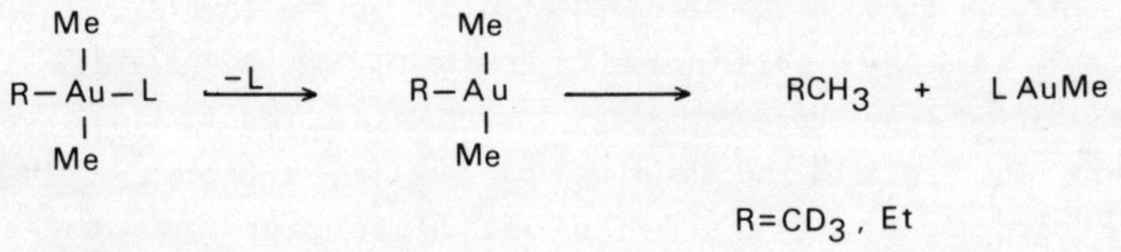

$$R{-}Au(Me)_2{-}L \xrightarrow{-L} R{-}Au(Me)_2 \longrightarrow RCH_3 + LAuMe$$

$R = CD_3$, Et

Reductive elimination takes place from a 3-coordinate intermediate, calculated to be a "Y"- or "T"-shaped 14-electron complex, but not through a trigonal planar complex, which is a high energy species. Isomerization from cis- to trans-"T"-shaped minima takes place through a "Y"-shaped saddle points, which are channels for reductive elimination. When R is large, the trans-complex is favored, thereby assuring cross-coupling.

The concerted coupling of alkyl groups bound to transition metals of the nickel triad is an allowed process for adjacent partners,[42] but may proceed by a variety of mechanistic pathways, depending on the metal, the ligands and the organic groups. A summary of the various pathways (Figure 5) shows that

Figure 5

Elimination Pathways for the Nickel Triad

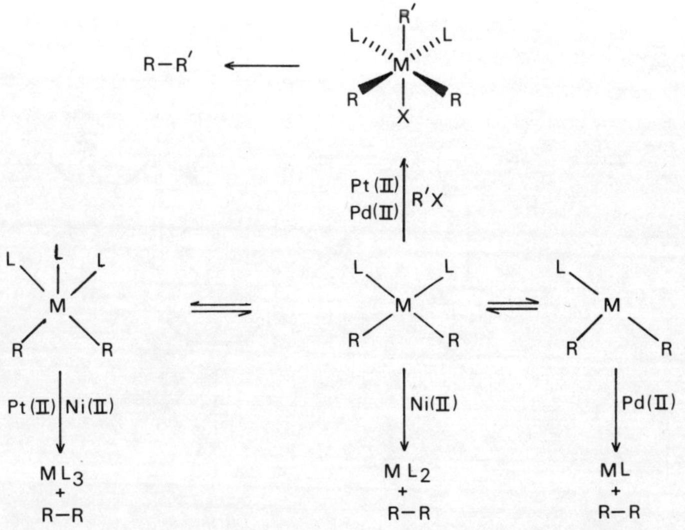

both associative pathways and dissociative pathways (18-, 16- and 14-electron complexes) are available for nickel. A dissociative path is preferred for dialkylpalladium(II) complexes, while both platinum and palladium undergo reductive elimination through the tetravalent, 18-electron complex. Although stable di- and trialkyl platinum(IV) complexes can be isolated, and the reductive elimination from these complexes can be studied, palladium(IV) complexes are not isolated and the coupling reaction of organic partners is only by implication, particularly in catalytic coupling reactions. Dialkyl-platinum(II) complexes do not undergo 1,1-reductive elimination reactions readily.

The situation with nickel(II) is best illustrated with nickelacyclopentane complexes.[43] Although a β-elimination reaction is supressed because of the difficulty in achieving a coplanar transition state, this mode of elimination is favored for the 14-electron complex (Figure 6). The 1,1-reductive elimination takes place predominately from the 16-electron complex, while the five coordinated nickel serves as an intermediate to a new 4-coordinate geometry that yields ethylene. A square planar geometry is favorable for the generation of cyclobutane.[44] The reverse cyclometallation reaction to give ethylene from a planar complex is forbidden, but is allowed from a geometry in which the ligands are orthogonal to the plane of the nickelacyclopentene. Direct interconversion of these two 4-coordinate geometries are symmetry forbidden, and must proceed through the 18-electron 5-coordinate complex.

Figure 6

Elimination Reactions of Nickelacyclopentanes

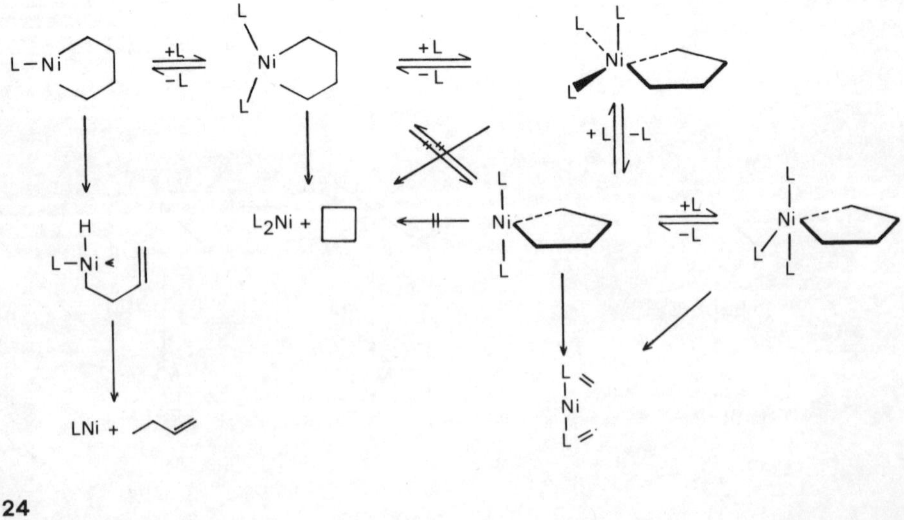

The reductive elimination of ethane from bisphosphinedimethyl complexes, takes place only with the cis-complex, trans to cis isomerization being required first.[45] The elimination is first order, and the palladium complex generated is the bisphosphinepalladium(0) complex, trapped as its tolan

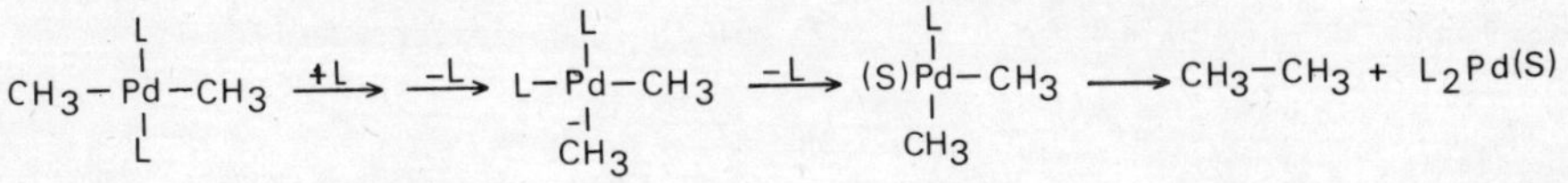

$$CH_3-\underset{L}{\overset{L}{Pd}}-CH_3 \xrightarrow{+L} \xrightarrow{-L} L-\underset{CH_3}{\overset{L}{Pd}}-CH_3 \xrightarrow{-L} (S)\underset{CH_3}{\overset{L}{Pd}}-CH_3 \longrightarrow CH_3-CH_3 + L_2Pd(S)$$

adduct. The TRANSPHOS dimethylpalladium complex fails to undergo reductive elimination, and is stable in solution above 80°C. The reaction is intra-

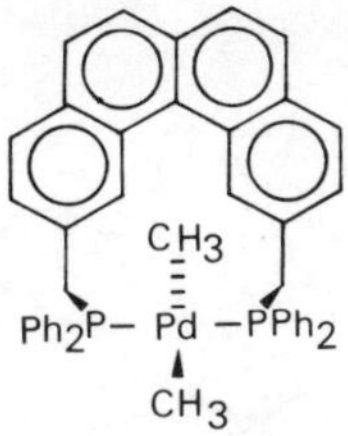

molecular and first order. The energies and entropies of activation for the reductive elimination in solvents of varying polarity are consistent with an elimination reaction from a 3-coordinate complex that produces a coordinatively unsaturated palladium(0) complex, and a late transition state having some of the characteristics of the product.[46]

Calculations[47] of the energetics of reductive elimination of ethane from bis(phosphine)dimethylpalladium support this mechanism. Several important conclusions have emerged from these calculations. First, strong donor ligands trans to the leaving groups increase the barrier to elimination. Second, the reductive elimination from the four coordinate cis-bis(phosphine)palladium is symmetry allowed, but is controlled by the energy of an antisymmetric b_2 orbital. The energy barrier for elimination from this complex is substantial and is much higher than that of the corresponding nickel complex containing a lower energy b_2 orbital. Third, reductive elimination from a three-coordinate phosphinedimethylpalladium complex containing methyls in adjacent positions has a substantially lower activation energy than that of the four-coordinate complex. Fourth, the "Y" and "T"-shaped geometries of phosphinedimethylpalladium are more stable than the trigonal geometry (Figure 7). Finally, the barrier to isomerization from one "T"-shaped geometry to another is substantial, much higher than the energy barrier for reductive elimination from either a "T" or a "Y"-shaped complex. Thus, there is a substantial energy barrier around the Jahn-Teller wheel. A "T"-shaped trans-$PdLR_2$, which might be produced by liberating L from the trans-four coordinate PdL_2R_2, will encounter a substantial energy barrier to rearrangement to cis-$PdLR_2$, which can undergo reductive elimination. Thus trans-PdL_2R_2 must first isomerize to the cis isomer (by an associative mechanism). Dissociation of phosphine from the cis isomer generates the "T" geometry directly (4 or 8 o'clock on the Jahn-Teller wheel). It is not

Figure 7

Interconversion of "T" and "Y" $LPdR_2$ Complexes

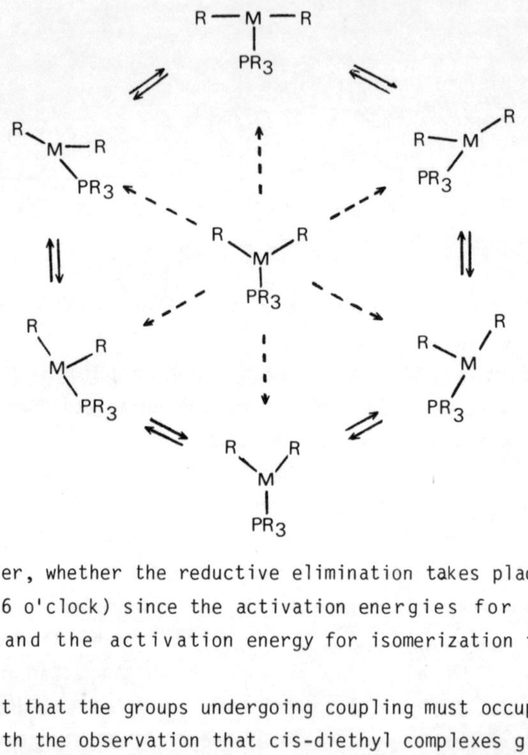

clear, however, whether the reductive elimination takes place from the "T" or "Y" complex (6 o'clock) since the activation energies for each are nearly identical, and the activation energy for isomerization from "T" to "Y" is low.

The fact that the groups undergoing coupling must occupy cis-position is consistent with the observation that cis-diethyl complexes give n-butane, but the trans isomer yields ethene and ethane, as a result of β-elimination followed by reductive elimination.[48] Also, this required geometry is consistent with the observation that carbonylation of a cis-diethylpalladium complex yields ethylene and propanal, while the trans isomer produces diethyl ketone.[49] The sequence of reactions requires first that carbon monoxide occupies the position of one of the dissociated phosphines, and a second, that alkyl migration to carbonyl takes place. There is some precedent for this stereochemistry and these requirements. The reductive elimination of propenyl from (E)- and Z-styrylmethylbis(diphenylmethylmethylphosphine)palladium(II) give E- and Z-propenylbenzene respectively.

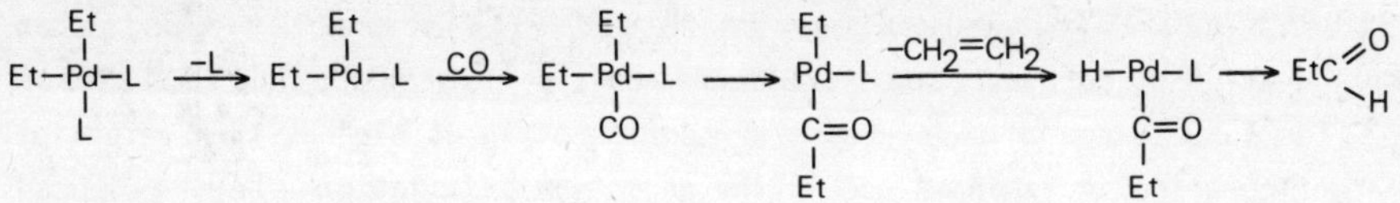

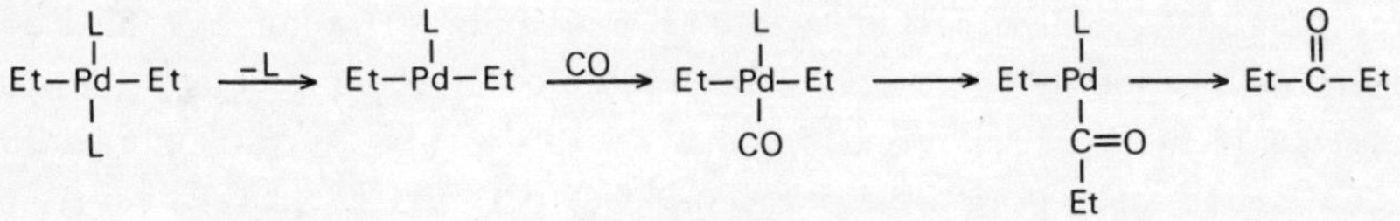

The same requirements are necessary for reductive elimination from these complexes as are necessary for the dimethylpalladium complexes.

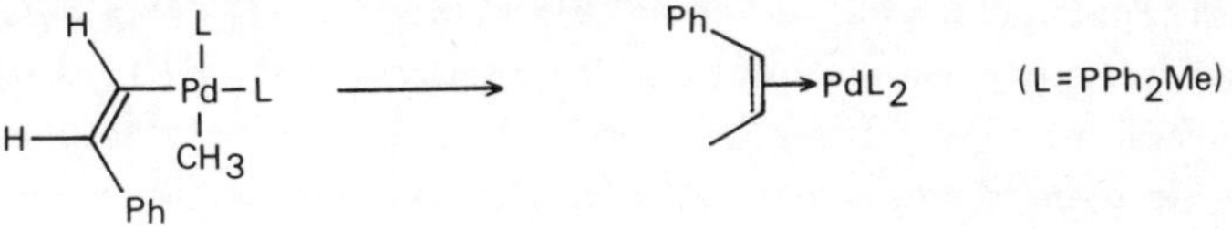

$(L = PPh_2Me)$

Palladium catalyzes the coupling of organic halides with organometallics, and the reaction gives only cross-coupled products (Figures 9,10). A catalytic cycle involving Pd(0) ⇌ Pd(II) transformations usually is written for this type of coupling (Figure 9). In a number of couplings, however, the rates of reductive elimination observed in the stoichiometric reactions are much slower than can be accomodated by a catalytic process. Thus, palladium(II) ⇌ palladium(IV) transformation have been implicated, since the rate of oxidative addition of methyl iodide to dimethylpalladium-(II) complexes is consistent with the catalytic rate.[46]

Although the TRANSPHOS dimethylpalladium complex is stable toward the reductive elimination of ethane, its reaction with methyl iodide-d_3 produces ethane-d_3 immediately, in direct support of a palladium(IV) intermediate.[45]

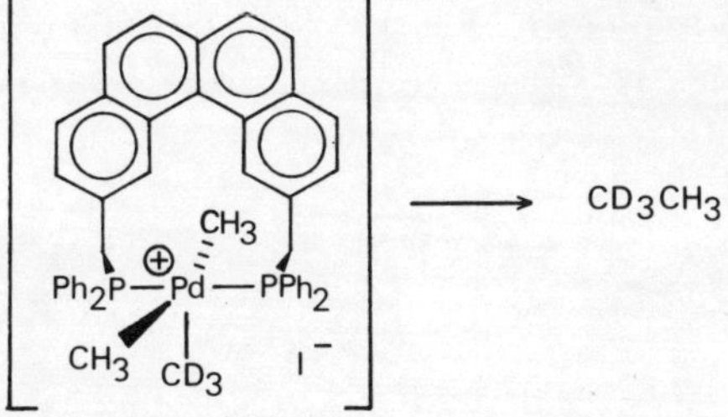

CD_3CH_3

In addition, when _cis_-dimethylbis(triphenylphosphine)palladium(II) reacts with benzyl bromide, ethyl benzene is the only product.[51] When the reaction is carried out with optically active α-deuteriobenzylbromide, optically active α-deuterioethylbenzene is formed with overall inversion of configuration at carbon. Since oxidative addition to low valent palladium species takes place with inversion of configuration at carbon, then retention of configuration is taking place in the reductive elimination step. Chiral chloro-(α-deuteriobenzyl)bis(triphenylphosphine)palladium(II) also yields, upon reaction with tetramethyltin, (in the presence of a benzyl halide) chiral α-deuterioethylbenzene in which retention of configuration at carbon has resulted from the reductive elimination.[51,52] It is not clear why selected cross coupling always takes place, but this could be a result of complex geometry, relative metal-carbon bond strengths, and the ligands trans to the metal carbon-bond.

There are a wide variety of dialkylplatinum(II) complexes that undergo reductive elimination to yield coupled products, but β-elimination is the preferred reaction pathway, even when the dihedral angle between platinum and the β-hydrogen is not favorable.[53] Cyclobutane can be the major product from the 1,1-reductive elimination of a platinacyclopentane bearing phosphines

with bulky alkyl groups.[54] The reductive elimination of biaryls from a cis-platinum complex is a concerted first order process.[55]

Di- or triorganoplatinum(IV) complexes undergo 1,1-reductive elimination much more readily, however. For example, cyclobutane is obtained from platinacyclopentanes.[56] Halobis(phosphine)platinum(IV) complexes containing mutually cis-alkyl groups show a preference for reductive elimination partners, depending on the ligands trans to the alkyl groups and the electron withdrawing or donating ability of the alkyl group[57-59] ($CH_3 > C_6H_5 > CF_3$).

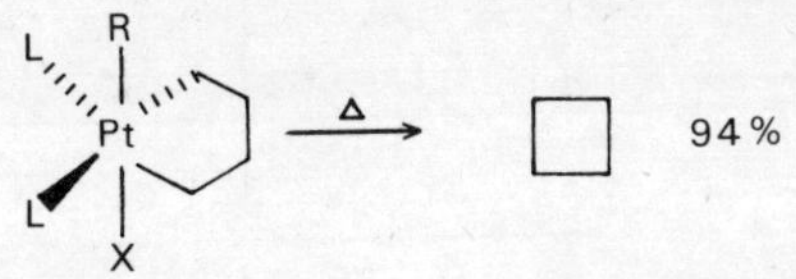

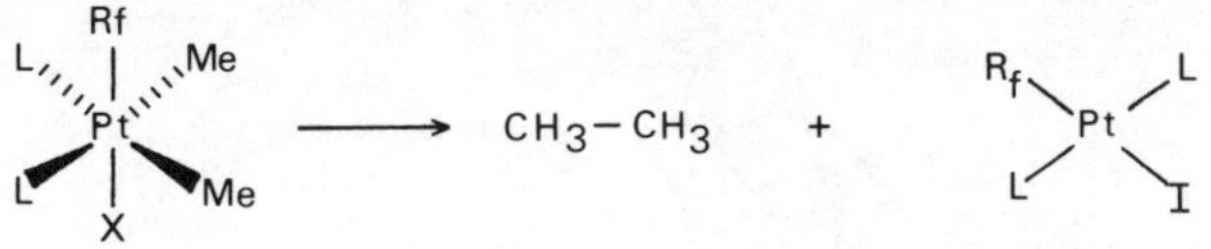

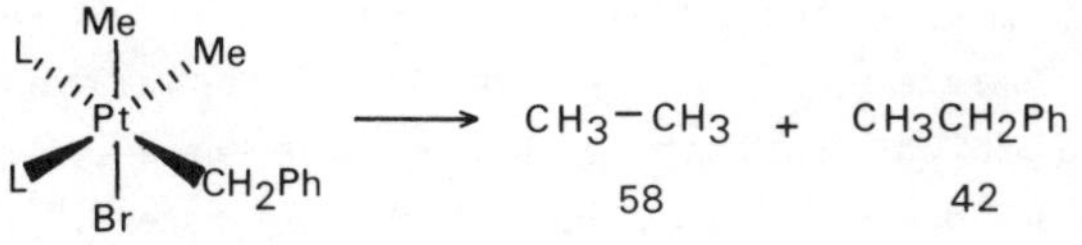

Reductive elimination of methyl and vinyl groups from rhodium(III) takes place with retention of geometry at the sp^2 carbon, indicative of a concerted reaction.[60]

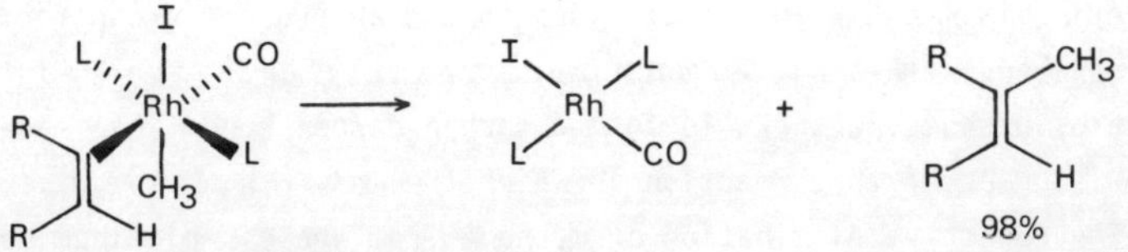

The 1,1-reductive elimination of alkyl groups from d^6 iron(II) complexes does not give high yields of coupled products. Thermolysis of (DIPHOS) dimethyliron yields minor amounts of ethane. The majority of the product is methane, which suggest an α-elimination mechanism.[61]

The 1,1-reductive elimination of organic partners from early transition metals is rare, homolysis being observed in most elimination reactions. Most of the early transition metals bearing methyl groups give methane, usually below ambient temperature. Where β-elimination in tetraalkyl derivatives of

$$Fe(P\text{-}P)(CH_3)_2 \longrightarrow [Fe(P\text{-}P)(H)(CH_3)(=CH_2)] \longrightarrow Fe(P\text{-}P)(=CH_2) + CH_4$$

minor: $CH_3{-}CH_3$

P–P = $Ph_2PCH_2CHPPh_2$

the early transition metal is prevented, for example in certain 1-norbornyl derivatives, coupling products are observed.

$M(\text{1-norbornyl})_4$ M=Hf, Zr, Ti, V, Cr

Titanacyclopentane and titanacyclohexane yield only minor amounts of cyclobutane and cyclopentane, respectively.[62] The introduction of carbon monoxide into solutions containing titanacyclopentane, however, yields cyclopentanone _via_ an insertion, reductive elimination process.[62,63]

$$Cp_2Ti(CH_2)_4 \xrightarrow{CO} Cp_2Ti(C(O)(CH_2)_4) \longrightarrow \text{cyclopentanone}$$

Carbon-Hydrogen Bond Formation

The reductive elimination of hydrogen and an organic group forming a carbon-hydrogen bond, takes place much more readily than the reductive elimination of organic partners to form a carbon-carbon bond. Consequently, there are few studies of this reaction in which the metal hydride has been isolated. The reductive elimination of methane from the _cis_-platinum complex takes place at rapidly at -25°C, so the study of this reaction was carried out by generating the complex _in situ_ at -50°C.[64] The elimination is intramolecular and first order.

$$(Ph_3P)_2Pt(H)(CH_3) \longrightarrow CH_4 + [(Ph_3P)_2Pt]$$

Alkyl hydrides that are trans are relatively stable, but elimination will take place if trans to cis rearrangement can take place. The planar nicke(II) complex, trans $NiH(CH_3)(PCy_3)_2$ is inert kinetically, since the large phosphine ligands must remain trans to each other.

The elimination of acetone from acetonylchlorohydridotris(triethylphosphine)rhodium(III) is an intramolecular, first order reaction that requires prior dissociation of a phosphine.[65]

$$L_3RhClH(CH_2COCH_3) \underset{+L}{\overset{-L}{\rightleftharpoons}} [L_2RhClH(CH_2COCH_3)] \xrightarrow{+L} CH_3COCH_3 + L_3RhCl$$

There are several well documented examples of intermolecular reductive elimination, the most well studied example being that of osmium(II) carbonyls.[66] Two requirements appear to be necessary for this type of reac-

$$Os(CO)_4(CH_3)_2 + Os(CO)_4H_2 \longrightarrow CH_4 + HOs(CO)_4Os(CH_3)(CO)_4$$

tion to take place: 1) One metal complex must be a hydride, 2) the other metal may contain a hydride, alkyl or acyl, but must be able to produce a vacant coordination site. The vacant coordination site may be produced by dissociation of a ligand or a CO "insertion" reaction. The mechanism for the elimination involves a hydride bridged species.

$$\underset{18e}{Os(CO)_4Me_2} \longrightarrow \underset{16e}{Os(CO)_3(COMe)Me} \xrightarrow{H_2Os(CO)_4} [(OC)_3(MeCO)(Me)Os(\mu\text{-}H)Os(CO)_4H] \longrightarrow CH_4 + CH_3CHO$$

Perhaps the most important example of intermolecular elimination is that which occurs in the oxo process (Figure 15).

INSERTION REACTIONS

An insertion reaction is one in which one ligand (R) bound to a metal migrates to form a bond to a second ligand (X) which also is bound to the metal. The group X is at least diatomic and contains multiple bonds between at least two of the atoms in the ligand. Insertion reactions in which X is carbon monoxide, sulfur dioxide, isonitriles, alkenes and alkynes are some of the most commonly observed; however, carbon monoxide and alkene insertions are the most important insertion reactions. Consequently, the discussion will be limited to these two.

Carbon Monoxide[67]

The carbon monoxide insertion reaction generally is reversible and the reverse, decarbonylation reaction also provides information about the mechanism of the reaction. Both carbonylations and decarbonylations will be

$$\underset{\displaystyle M-CO}{\overset{R}{|}} \rightleftharpoons M-\overset{O}{\overset{\|}{C}}-R \underset{-L}{\overset{+L}{\rightleftharpoons}} \overset{L}{\underset{M}{|}}-\overset{O}{\overset{\|}{C}}-R$$

discussed. The reaction is a key step in decarbonylations of aldehydes and acid chlorides (Figure 13)) and in the reverse reactions, hydroformylation reactions (Figure 14) and carbonylations of organohalides (Figure 8).

In the insertion reactions of carbon monoxide, the question of whether the R group migrates to the bound carbonyl or the carbonyl actually inserts into the carbon-metal bond has been answered at least in the insertion reactions of methylmanganesepenta carbonyl. When the carbonyl, labeled with one ^{13}C carbonyl cis to the methyl group is allowed to react with a ligand (phosphine or ^{12}C carbon monoxide) the two ^{13}CO stretching frequencies observed in the infrared spectrum were consistent only with methyl migration.[68] This observation has been confirmed more recently utilizing ^{13}C NMR at low temperatures.[69] The analysis presumes that CO or methyl can migrate only to a cis position, and cannot migrate to a trans position.

In the pseudo tetrahedral iron system, the reaction of one enantiomer will proceed with inversion of configuration if the alkyl group migrates and retention if carbonyl inserts. The steric course of the reaction depends on the solvent used. Without solvent intervention, inversion of configuration takes place, consistent with alkyl migration.

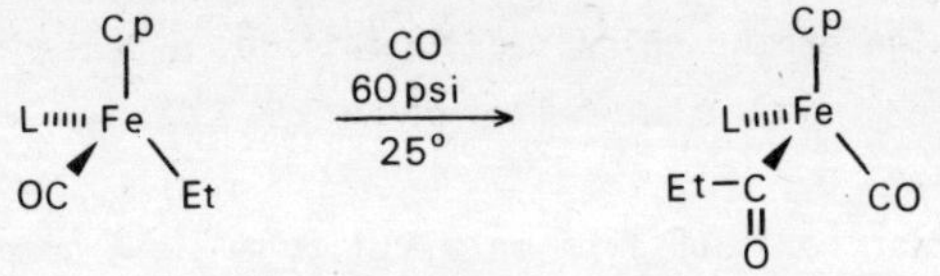

The question of the stereochemistry at carbon has been determined with many more transition metals. In every case examined, retention of configuration at carbon has been observed. Each of the two palladium(II) complexes containing a chelating olefin ligand undergoes insertion of carbon monoxide with retention of configuration at carbon.[71] The stereochemistry of alkyl-acyl migratory insertion also has been determined to proceed with retention of configuration at carbon by examination the reactions of erythro and threo cyclopentadienyliron carbonyl complexes.[72]

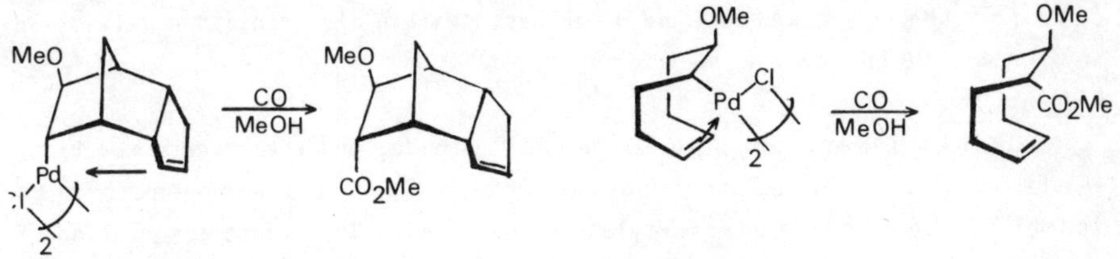

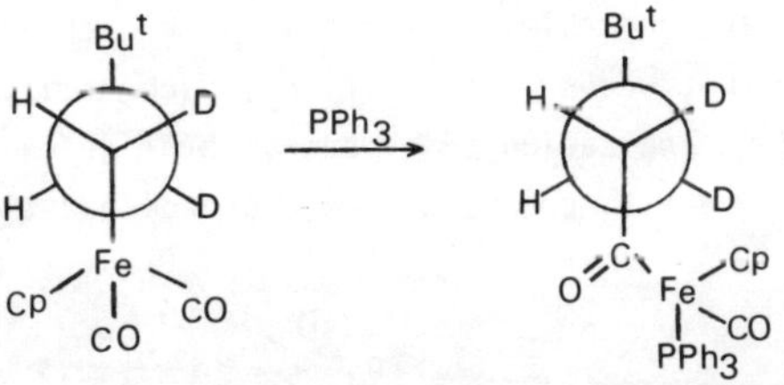

Decarbonylation reactions also are stereospecific, occuring with retention of configuration at an sp^3 center. In the decarbonylation reactions of aldehydes, retention or racemization can be observed depending on the structure of the aldehyde.[73]

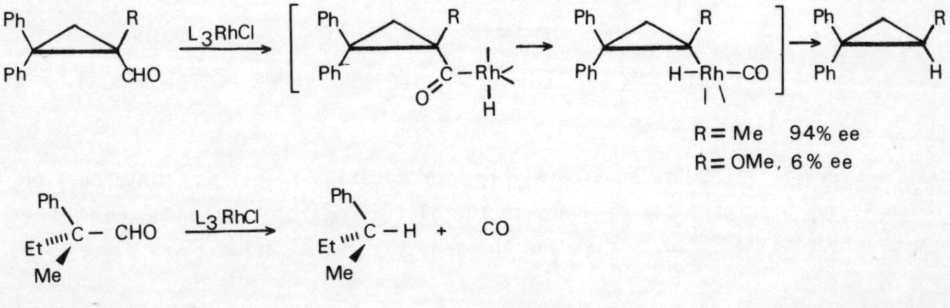

R = Me 94% ee
R = OMe, 6% ee

The decarbonylation of the chiral acid chloride, α-deuteriophenylacetyl chloride gave benzyl-α-d-chloride with only 20-27% net retention of configuration at carbon.[74]

In the rearrangement reaction from an alkyl to an acyl complex, there is evidence in certain systems that solvent participates during the alkyl migration and assists the migration, even though the alkyl complex is coordinatively saturated.

From the available data, several general conclusions can be made concerning migratory aptitudes. For the group undergoing migration on the carbonyl, CH_3CH_2->CH_3>$PhCH_2$->CH_2=CH->Ph- $ROCH_2$->$HOCH_2$->>H, CH_3CO-, CF_3-. For the last three members of this series, migratory insertions are known only for the lanthanides and actinides. Within the transition metals, 3d akyls > 4d alkyls > 5d alkyls.

Alkenes

Olefin insertions into a metal hydride, and the reverse reaction, β-elimination, represent important steps in hydrogenation, olefin isomerization, and hydroformylation reactions. The direct observation of this reaction, i.e. observation of a metal hydride containing a coordinated olefin proceeding to a metal alkyl has been observed only in a few examples. The reaction is extremely rapid and so most metal hydrides bearing a coordinated olefin are unstable. Stability can be achieved, however, if the hydride and alkene do not occupy cis positions, such as is the case with the platinum cation.[75] The cationic molybdenum hydride containing an ethylene also is stable, but insertion can be achieved by the addition of phosphine.[76]

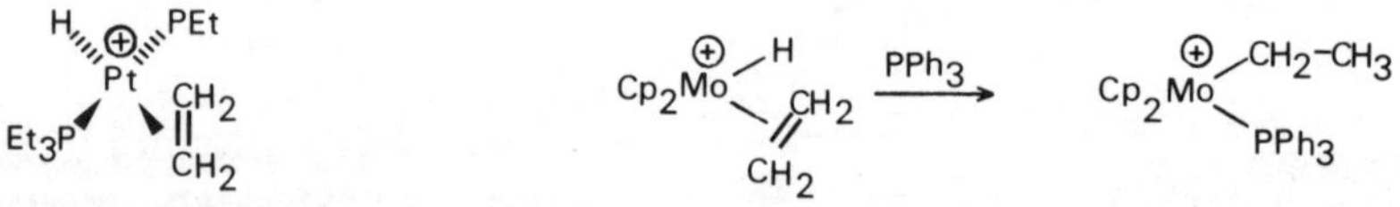

An equilibrium between the rhodium hydride and rhodium alkyl can be observed by NMR.[77]

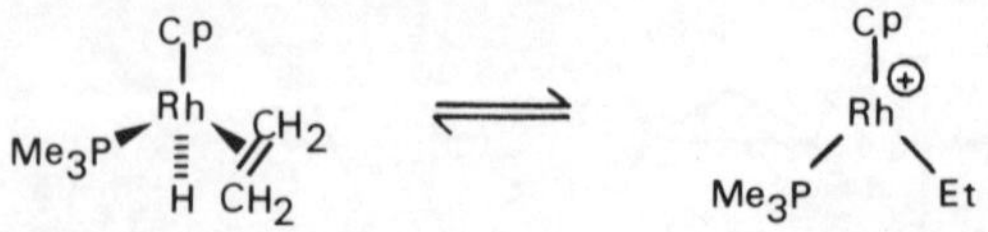

Overall, in homogeneous catalytic hydrogenation reactions, dihydrogen or dideuterium undergoes cis delivery to the olefin, which suggests that the metal hydride addition is cis, and the reductive elimination takes place with retention of configuration at carbon. In the insertion reaction, the hydride and olefin occupy cis positions, and in hydrogenations with platinum, a concerted reaction of a four-coordinate complex is favored.[78]

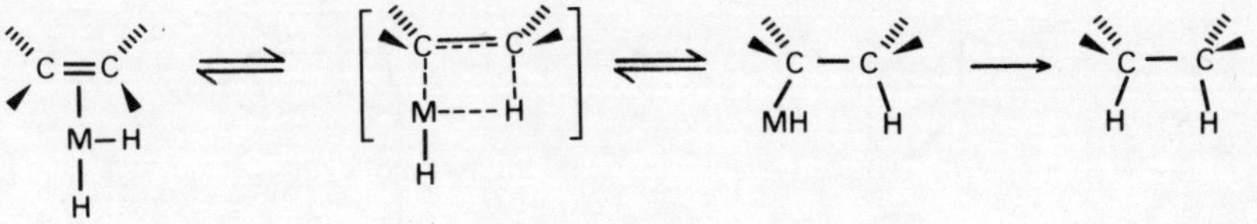

The zirconium hydride, Cp_2ZrClH adds to the labeled t-butyl-ethylene to yield zirconium alkyl products of syn addition,[79] supporting the stereochemistry in the first step of hydrogenation reactions. This stereochemistry is consistent with the mechanism of asymmetric olefin hydrogenations with rhodium catalyst that has been so elegantly elucidated by Halpern (Figures 11, 12).[80]

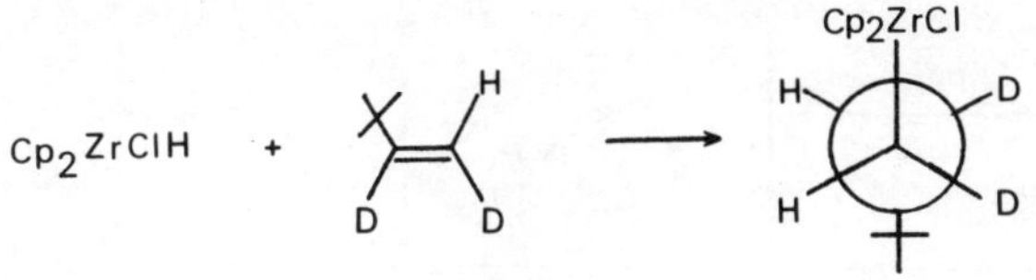

The reverse reaction, β-elimination is readily observed, although the stereochemistry is difficult to assess since readdition (insertion) takes place, randomizing the stereochemical label. In one case, the stereochemistry has been observed since facile readdition with the reverse regiochemistry does not occur. Oxidative addition of erythro or threo-2,3-diphenyl butanoyl chloride to Wilkinson's catalyst yields the corresponding acyl derivative which decarbonylates, on warming, to the rhodium alkyl. Assuming retention of configuration in decarbonylation, then β-elimination proceeds with the syn geometry.[39d]

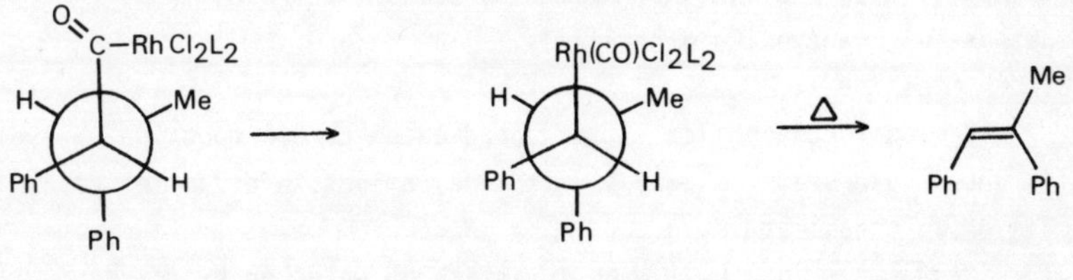

Insertion of an alkene into a carbon-metal bond occurs in olefin dimerization (oligomerization) and is the propagation step in the Cossee mechanism for Ziegler-Natta polymerization of 1-alkenes. There are few reports of the direct observation of this reaction. The palladium complex containing a metal-carbon σ-bond and a chelating olefin undergoes syn addition with retention of configuration at the migratory center,[81] but there is little other choice in this reaction.

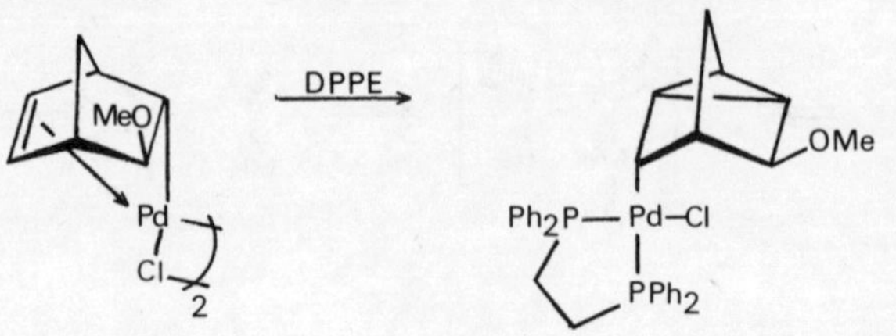

The reaction of the diperdeuteriomethyl cobalt complex with ethene yields 3,3,3-trideuteriopropene and trideuteriomethane, as a result of an insertion, β-elimination and 1,1-reductive elimination sequence.[82]

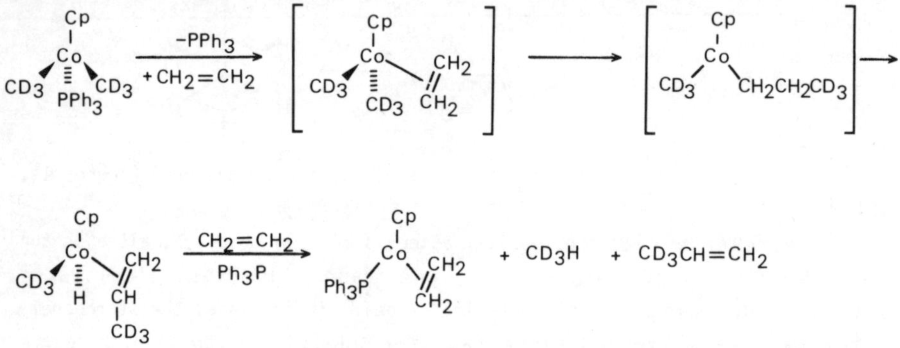

NUCLEOPHILIC ADDITION TO COORDINATED LIGANDS[83]

In the reaction of a nucleophile with an organotransition metal complex, attack can occur at the metal center or at one of the ligands. The second type of reaction can occur when the ligand is an isonitrile, carbon monoxide an olefin or an acyl, but the present discussion is limited to reactions at carbon monoxide and olefins, primarily because there have proven to be the most useful reactions synthetically.

Carbon Monoxide

Various nucleophiles attack coordinated carbon monoxide including oxygen (in water, hydroxide, alcohols, alkoxide) amines, hydride and carbanions from main group organometals.

Metal carbonyls undergo attack by water or hydroxide to yield metal carboxylate complexes that are usually not isolated but decarboxylate, giving carbon dioxide and metal hydride.[84,85] This is the important step in the water-gas shift reaction. The proclivity of the metal carbonyl to undergo attack by nucleophile is inversely proportional to the extent of back bonding.

$$\mathrm{M{-}CO} + \begin{matrix} OH^{\ominus} \\ \text{or} \\ H_2O \end{matrix} \longrightarrow \left[\mathrm{M{-}\overset{O}{\overset{\|}{C}}{-}OH} \right] \longrightarrow \mathrm{M{-}H} + CO_2$$

Also, cationic complexes generally will react with water while neutral carbonyls, such as iron pentacarbonyl require base.

$$Fe(CO)_5 + OH^{\ominus} \longrightarrow \left[HFe(CO)_4\right]^{\ominus} + CO_2$$

Methanol, in the presence of base, will attack a carbonyl ligand that is bound to a Group VIII transition metal, usually platinum or palladium. Thus, platinum(II) carbomethoxy complexes can be formed from the reaction of the bisphosphine dichloride with carbon monoxide in methanol in the presence of a tertiary amine base.[86] This mode of generation of carbomethoxy groups bound

$$L_2MX_2 \xrightarrow{CO} \left[(L)(L)M(CO)(X) \right]^{\oplus} X^{\ominus} \xrightarrow[Et_3N]{MeOH} (L)(X)M(CO_2Me)(L) + Et_3N{\cdot}HCl$$

to palladium in crucial in the 1,2-dicarboxylation of alkenes.[87] Similarly, the reaction of amines with coordinated carbonyl yields amide complexes.[88] The reaction takes place readily with cationic carbonyl complexes of the Group VII B and VIII transition metals, but fails with the neutral complexes.

$$\mathrm{M{-}CO} + R_2NH \longrightarrow \mathrm{M{-}\overset{O}{\overset{\|}{C}}{-}NR_2}$$

Organolithium reagents react with metal carbonyls to yield acyl derivatives. Nickel carbonyl reacts to give acyl derivatives that have not been well characterized, but whose chemistry is consistent with the acyl structure. Iron pentacarbonyl reacts similarly, as do Group VI metal carbonyls.

$$RLi + Ni(CO)_4 \longrightarrow \left[R\overset{O}{\overset{\|}{C}}{-}Ni(CO)_3 \right]^{\ominus} ; \quad RLi + Fe(CO)_5 \longrightarrow \left[R\overset{O}{\overset{\|}{C}}{-}Fe(CO)_4 \right]^{\ominus}$$

Hydrides, particularly alkoxyborohydrides generate transition metal formyls from carbonyls by hydride attack at the carbonyl carbon.

$$HB(OR)_3^{\ominus} + L_nMCO \longrightarrow L_nM{-}\overset{O}{\overset{\|}{C}}{-}H \quad (M = Fe, Cr, W)$$

Alkenes and π-Allyl Complexes

The reaction of a nucleophile at an sp^2 carbon of an alkene coordinated to a transition metal takes place from outside the coordination sphere. The stereochemistry of the reaction in which a nucleophile to carbon and a metal to carbon bond is generated is a trans addition across the olefin. This is the opposite stereochemistry of the olefin "insertion" reaction in which a ligand bound to the metal and the metal add to the carbon-carbon double bond. The susceptibility of the alkene to nucleophilic attack depends, in part, on the extent of bonding to the metal and the charge on the metal. This nucleophilic addition reaction has been observed primarily with palladium(II), platinum(II), and iron(II) complexes. Mono- or disubstituted olefins react rapidly while more highly substituted olefins are too weakly coordinated. Oxygen nucleophiles (such as hydroxide, alkoxy and acetoxy), amines, chloride and certain carbanions undergo this trans addition. Under certain conditions, acetoxy and chloride become coordinated to the metal center first, in which case a cis-addition is observed. With simple alkenes, nucleophilic attack occurs at the most substituted sp^2 carbon. Calculations show that the olefin is activated toward nucleophilic attack by slipping from a position in which the metal is symmetrically η^2 bonded (midway between the two carbons of the double bond) to a position closer to one of the carbons.[89]

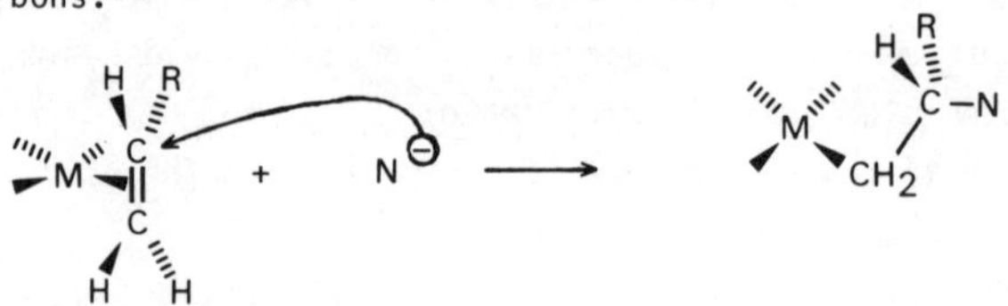

Palladium(II) and platinum(II) complexes of chelating diolefins add methanol, amines, and soft carbanions (acetylacetonate, acetoacetic ester, malonate) from the face of the olefin opposite to that which is coordinated to the metal.[90] In the reaction of dicyclopentadiene, with methanol, for

M = Pd, Pt

example, the more reactive norbornene double bond undergoes addition, while with dipentene, the terminal methylene undergoes a reaction with the nucleophile becoming attached to the most highly substituted position (Markovnikov addition). Water also adds trans to the cyclooctadiene complex yielding a stable β-hydroxypalladium species[91] which is similar to the proposed intermediate in the palladium catalyzed oxidation of ethylene to acetaldehyde (vide infra).

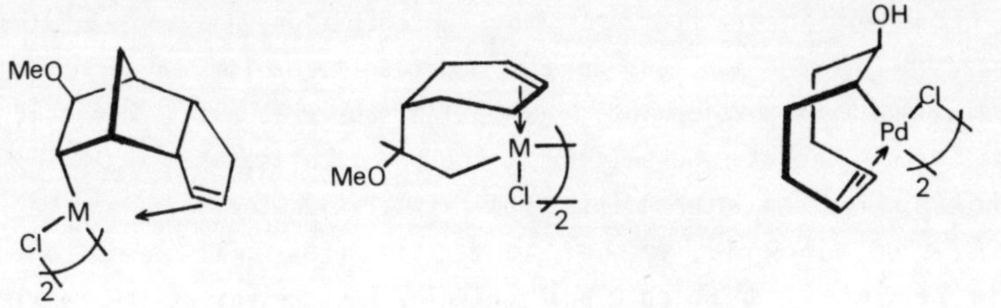

Monoolefins coordinated to platinum or palladium also undergo nucleophilic attack, and the stereochemistry of the reaction in this case is not subject to steric effects that may be imposed by chelating diolefins. Although the products from nucleophilic attack on the chelating diolefins are stable, the β-substituted complexes from the monoolefins are unstable such that the stereochemistry of the reaction must be obtained by means other than a direct inspection of an isolated complex.

When the methoxypalladation of cis- or trans-2-butene is carried out in the presence of carbon monoxide, the intermediate carbon-palladium σ-bonded complex is trapped by CO "insertion" which takes place with retention of configuration at carbon (vide supra). Stereospecific trans methoxypalladation is observed, yielding exclusively the threo- and erythro-β-methoxy esters.[92] In one example, the methoxypalladation product could be isolated, and was determined by ^{1}H NMR to be the product resulting from trans-addition.[93]

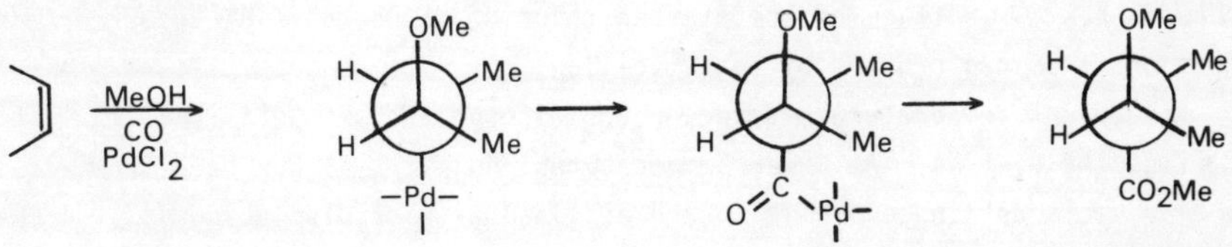

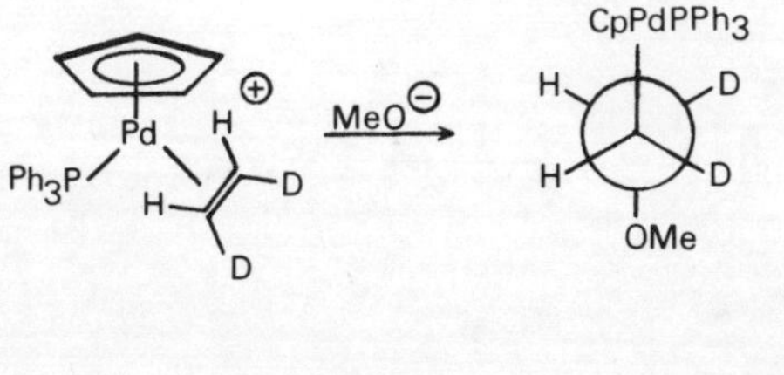

The addition of water or hydroxide to palladium-coordinated ethylene is the first step in the Wacker oxidation. On the basis of kinetic arguments, a syn attack of coordinated hydroxide was suggested as mechanism for this reaction.[94] The reaction is first order each in ethylene and $PdCl_4^{2-}$, inverse first order in hydrogen ion concentration, and at least inverse second order in chloride ion, but an anti addition mechanism in which water attacks ethylene from outside the coordination sphere is also consistent with the kinetic data. The stereochemistry of the reaction is, in fact, anti.[95] The hydroxypalladium intermediate from cis-1,2-dideuterioethylene has been trapped by carbon monoxide, either in a stoichiometric reaction or a catalytic reaction (in which cupric chloride was present as the reoxidant) to give exclusively trans-2,3-dideuteriopropiolactone. Notable is the

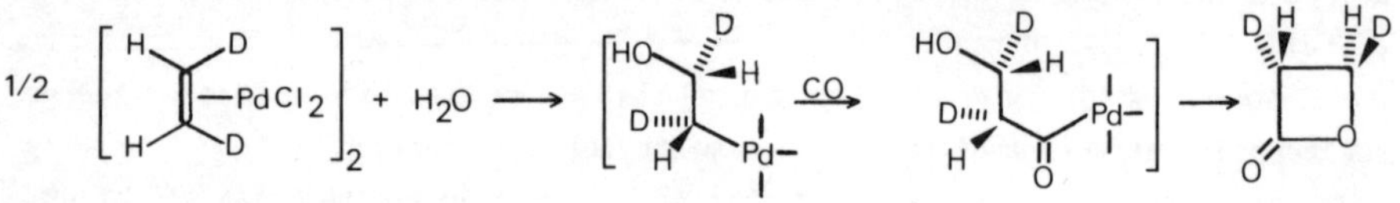

observation that the reaction of propene under the same conditions resulted in the formation of the lactone and acetone, demonstrating that the Wacker oxidation was competitive in this case. Also consistent with these results is the observation that the oxidation of trans-1,2-dideuteroethylene gave cis-chlorohydrin in the presence of high concentrations of chloride ion and copper(II)--conditions know to generate chlorohydrin as a by product in the Wacker oxidation. The chlorohydrin was analyzed as the oxirane[96]. This analysis depends upon inversion of configuration by chloride at the carbon bearing

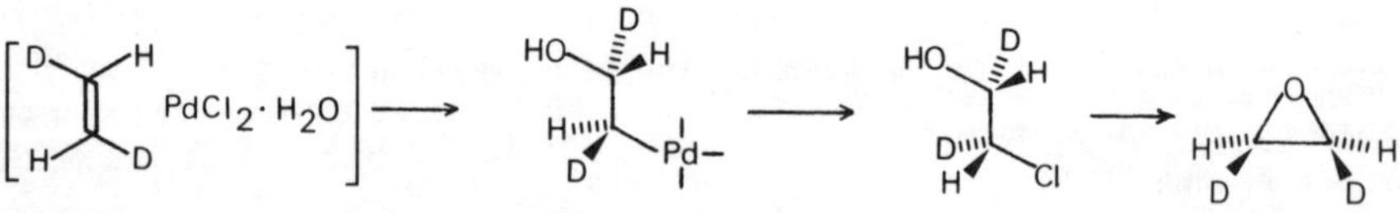

palladium. The attack of acetate and chloride at monoolefins under certain conditions also can proceed by anti addition.[97]

At ambient temperatures, amines react with palladium olefin complexes to displace the olefins. At lower temperatures, however, it has been shown that the amine acts as a nucelophile toward ethylene without displacing it.[98] The stereochemistry of the addition to a trans-2-butene palladium complex shows that an anti addition takes place.[99]

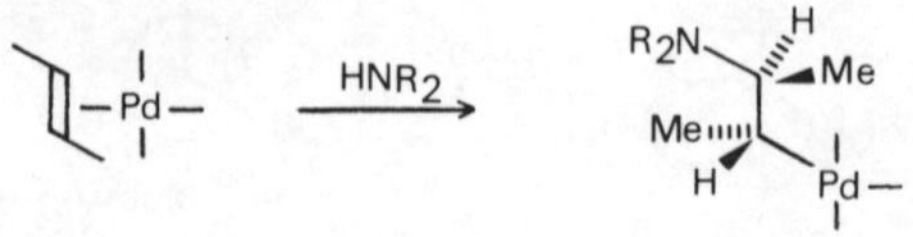

The reaction requires three equivalents of amine for good yields, and no olefin amination takes place with the addition of just one equivalent.[100] The β-aminoalkyl palladium complex can be characterized by NMR at temperatures below -20°C.

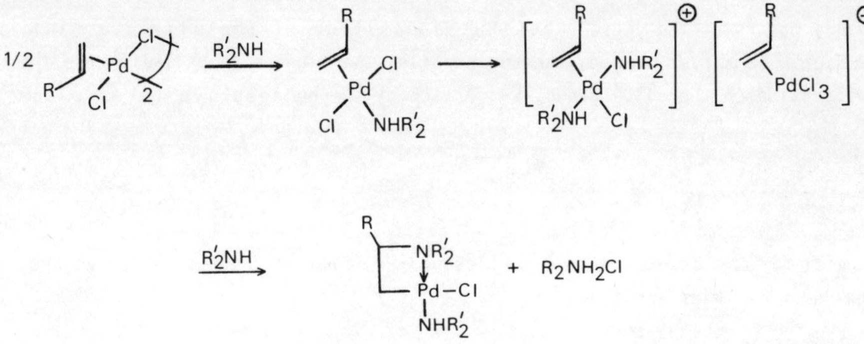

Platinum olefin complexes also add amines anti. The platinum complex containing a prochiral olefin and a chiral amine, resolved into its diastereomers, reacts with a second mole of amine (diethylamine) to give the addition product, which was decomposed by protonolysis.[101] The (S)-N,N-diethyl-sec-butylamine obtained from the enantiomer of the 1-butene complex shown, requires attack by the amine from the face of the olefin anti to the platinum.

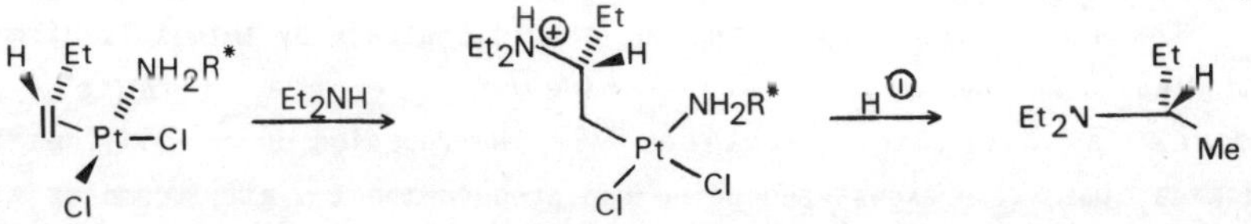

Nucleophilic attack of alkoxide, amines, mercaptans and phosphines on $CpFe(CO)_2(olefin)^+$ gives stable products, in which nucleophilic attack has occurred of the most substituted carbon.[102] Soft carbanions (acetoacetate, malonate, cyanoacetate, nitromethane) and enamines also react, but the regiospecificity depends on the coordinated olefin and the anion.

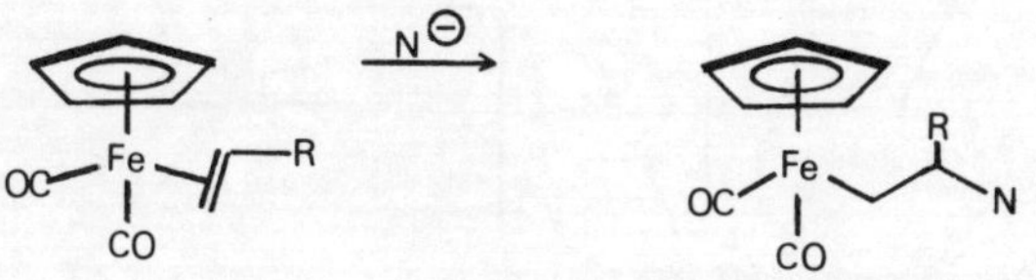

A wide variety of π-allyl palladium complexes are attacked by nucleophiles at the π-allyl ligand. The π-allyl palladium halide complexes are relatively inert to nucleophilic attack. However, in the presence of strongly coordinating nucleophiles such as phosphine, nucleophilic attack by stabilizied carbanions, amines or ketone enamines takes place. Alkylation usually occurs at the least substituted position of the allyl termini, although this depends on the nucleophile, the ligands on palladium and the substitution on the allyl group.[103] Attack by the nucleophile takes place

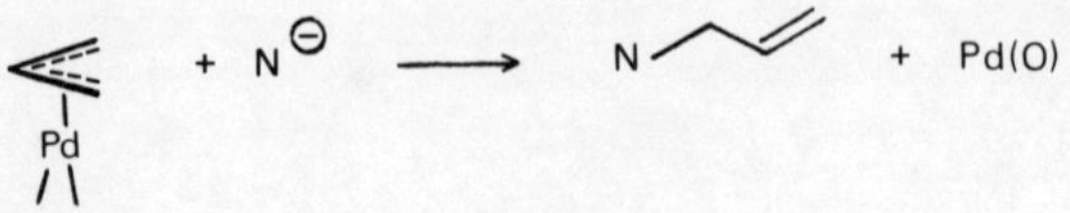

from the face opposite the palladium.[104] Similar results also have been obtained with amine addition.

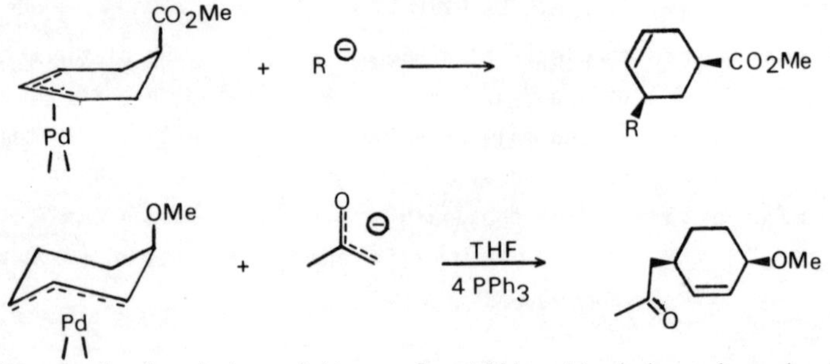

The regiochemistry of the nucleophilic attack by amines is different, depending on whether a σ-allyl phosphine complex or a cationic π-allyl complex is undergoing reaction.[105] The reaction of a π-allyl palladium chloride dimer with excess phosphine can produce the σ-allyl complex.

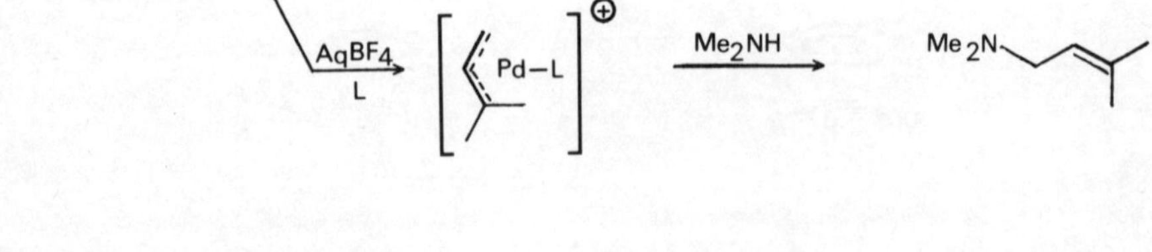

π-Allyl iron complexes react with stabilized carbanions while a cationic molybdenum complex reacts with hard nucleophiles to give the product of allylic substitution.[83]

TRANSMETALLATION

Transmetallation reactions of a main group organometallic such as a Grignard reagent or an organolithium reagent with transition metal complexes serve to form a transition metal-carbon bond. The reaction takes place with nearly all transition metals, but has enjoyed utility in synthetic organic chemistry, particularly with reactions catalyzed by Group VIII transition metals. The mechanism of transfer of the organic portion from the main group metal to the transition metal is not known, but in a number of cases, the reaction has been shown to be stereospecific, proceeding with retention of configuration or geometry at the carbon bound to the metal.

$$-\overset{|}{\underset{|}{M}}-X \quad + \quad m-R \longrightarrow -\overset{|}{\underset{|}{M}}-R \quad + \quad mx$$

M= transition metal ; m = main group metal

Lithium and Grignard reagents, which are configurationally stable at -78°C, react with copper reagents to yield alkyl copper or dialkyl copper lithium reagents in which the transfer of the organic group to copper has taken place with retention of configuration. The configuration of the copper reagent was determined by subsequent protonolysis, coupling or deuteride reduction reactions which presumably take place with retention. For example, the cyclopropyl lithium reagent undergoes transfer to copper and subsequent protonation of the copper reagent to yield the cyclopropane product resulting from overall retention of configuration at carbon.[106] The

$$\text{Ph}_2\text{C(cyclopropyl)C(Li)(CN)} \longrightarrow [R_2Cu]\ Li \xrightarrow{H^+} \text{Ph}_2\text{C(cyclopropyl)C(H)(CN)}$$

reaction with a similar cyclopropyllithium reagent with 3-iodo-2-cyclohexenane gave a coupling product with retention of configuration at the carbon-metal bond.[107]

Organomercury halides are some of the more widely utilized main group organometallic in transmetallation reactions.[108] The mechanism of the transmetallation is regarded as an electrophilic cleavage of the mercury carbon bond by the transition metal. In the transfer of an alkyl group to palladium(II), retention of configuration at carbon has been observed,[109] threo- or erythro-configurations being retained in the final product.

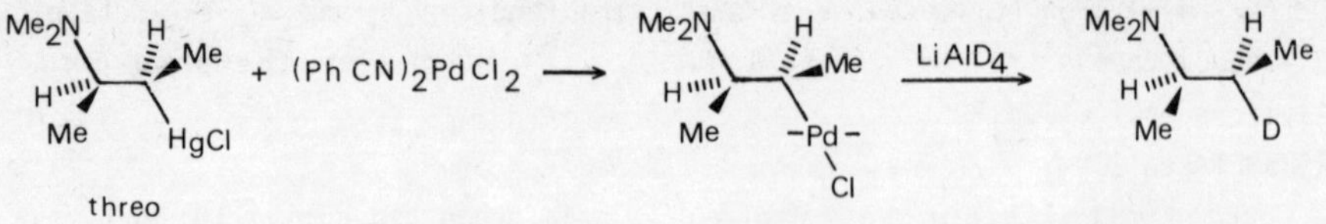

threo

Similarly, a transmetallation-carbonylation sequence also shows overall retention.[110]

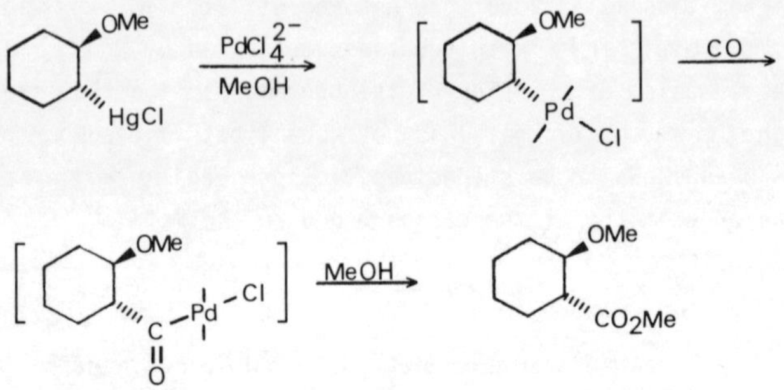

Vinyl allanes, obtained by the hydroalumination of acetylenes, and vinyl halides can be coupled catalytically by nickel or palladium with retention of geometry at the double bond.[111] This reaction requires, sequentially, retention of geometry during the oxidative addition of the vinyl halide, retention of geometry at the sp^2 carbon during transmetallation, and retention of geometry in both partners in reductive elimination.

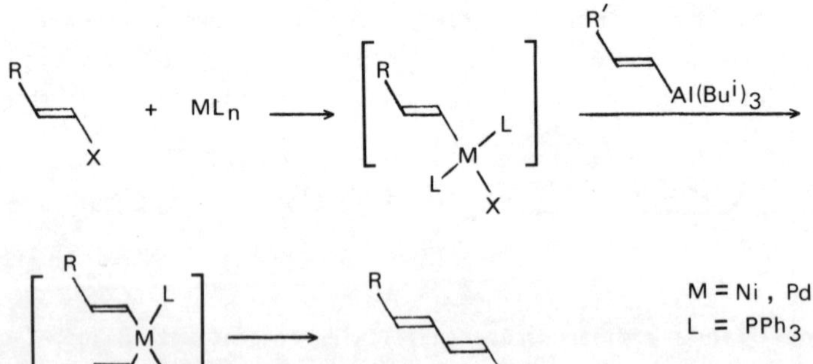

M = Ni, Pd
L = PPh_3

Retention of geometry during transmetallation--as observed in the coupling or carbonylation products--occurs with organoboron, organozirconium, and organozinc reagents. Organotin reagents and palladium(II) complexes undergo transmetallation reactions with retention and are especially useful in the synthesis of a variety of organic compounds in palladium catalyzed coupling reactions with organic halides.[52,53,112]

HOMOGENEOUS CATALYTIC REACTIONS

In Figures 8-17 are the catalytic cycles for some homogeneous catalytic reactions. Represented in these cycles are the fundamental types of organometallic reactions, oxidative addition, reductive elimination, olefin and CO insertion, nucleophilic addition to olefins and transmetallation. Each step in the cycle is marked as such, except when a step is not labelled, it is a ligand association or dissociation step. Under each complex is its valence and, in parentheses, its electron number.

In the Monsanto process for the conversion of methanol to acetic acid (Figure 8),[113] the methyl iodide is generated by the reaction of methanol with hydrogen iodide. Thus, the reaction must be started with catalytic quantities of hydrogen iodide or methyl iodide. The reductive elimination of acetyl iodide is shown, although another as plausible sequence, attack of water at the acetyl group bound to rhodium is possible.

Two catalytic cycles (Figures 9 and 10) are shown for the coupling of organic halides with main group organometals. Which catalytic process occurs depends on the Group VIII metal, the reactants and the reaction conditions. It is not unreasonable that both these catalytic processes may be taking place simultaneously.

Two different catalytic cycles are shown for the hydrogenation of olefins with rhodium catalysts. Hydrogenation with Wilkinson's catalysts containing monophosphines (Figure 11) proceeds by oxidative addition of hydrogen prior to coordination of the olefin. When the phosphine is a chelating phosphine, such as is the case in Monsanto's L-DOPA synthesis,[114] (Figure 12) hydrogen will not add to the rhodium(I) complex until olefin coordination takes place.[80] Oxidative addition of hydrogen is the rate determining step. When the alkene is N-acylaminocinammic acid, at -40°C, k_2 is slower than k_1 so that the alkylrhodium hydride can be observed by NMR. It is stable at -78°C and does not undergo reductive elimination at that temperature. The complex of the amino acid precursor with rhodium has been isolated and its structure determined by X-ray.

When the phosphines are chiral, such that an asymmetric hydrogenation is achieved, the enantioselection has been shown[80] not to be due to the olefin complexation step but to the rates at which these complexes are hydrogenated. When the substrate is prochiral, there are a number of diastereomeric olefin complexes obtained, depending on which enantioface the olefin presents to the metal, and which coordination site it occupies. These complexes are present in unequal amounts, depending on their relative stabilities. It is not the most abundant complex that produces the chiral amino acid in high enantiomeric excess, but rather the diastereomeric complex which undergoes hydrogenation the fastest; i.e. the most unstable complex which may be present in such small amounts that it cannot be detected.

The decarbonylation of acid chlorides by Wilkinson's complex can be carried out stoichiometrically or catalytically (Figure 13). In the stoichiometric reaction benzyl and certain aryl halides undergo the reductive elimination step readily, while the elimination with aryl halides is much more complicated and difficult, temperatures greater than 150°C being required. In order for the catalytic reaction to proceed, higher temperatures (~200°C) are necessary, the difficult steps being oxidative addition to the bis(triphenylphosphine)carbonylrhodium(I) chloride and the loss of carbon monoxide.

In the catalytic hydroformylation reaction (Figure 14), two pathways to a rhodium alkyl are shown, one a dissocciative path and the other an associative path.[115] The associative path is that which has been suggested to be responsible for the production of n-aldehyde. The dissociative path is that which yields branched aldehyde, as the result of less sterically crowded rhodium being able to accommodate a branched alkyl group. Overall, the stereochemistry of this hydroformylation is <u>cis</u>, resulting from the syn addi-

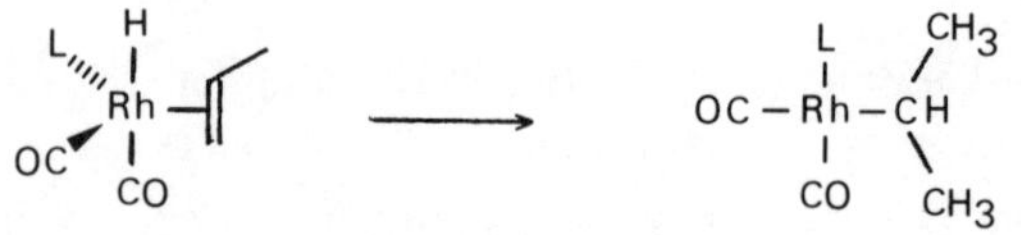

H
L
Rh
OC
CO
L
CH_3
OC — Rh — CH
CO
CH_3

tion of rhodium hydride to the alkene and CO insertion with retention of configuration at carbon.

There are similarities in the oxo reaction (Figure 15), the most distinctive departure being that the source of hydride for the reductive elimination of the aldehyde and regeneration of dicobaltoctacarbonyl is a result of a dinuclear elimination.[66]

The nickel catalyzed hydrocyanation of butadiene to yield adiponitrile (DuPont) first yields a 2:1 mixture of pentenenitriles, but tetrakis(phosphite)nickel(0) catalyst isomerizes the branched to the linear product.[1] The nickel complex then catalyzes the addition of a second mole of hydrogen cyanide to yield the adiponitrile. the catalytic cycle for the hydrocyanation of a simple olefin is shown (Figure 16).

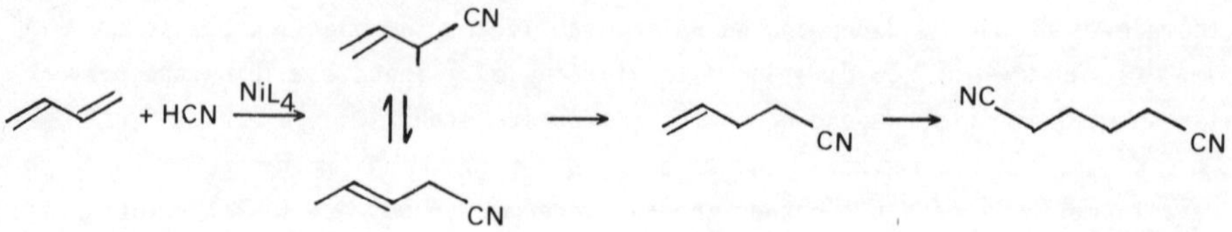

+ HCN $\xrightarrow{NiL_4}$
CN
CN
CN
NC
CN

Finally, in the conversion of ethylene to acetaldehyde catalyzed by palladium, a cycle in which the nucleophilic attack by water from the face opposite palladium is shown (Figure 17), rather than that usually written.[94] Ignoring all the different ligand associations and dissociations, the key steps are the following:

$$H_2O + CH_2{=}CH_2 + Pd^{2+} \longrightarrow CH_3CHO + Pd(0) + 2H^+$$

$$Pd(0) + 2Cu^{2+} \longrightarrow Pd^{2+} + 2Cu^+$$

$$2Cu^+ + 1/2\ O_2 + 2H^+ \longrightarrow 2Cu^{2+} + H_2O$$

$$CH_2{=}CH_2 + 1/2\ O_2 \longrightarrow CH_3CHO$$

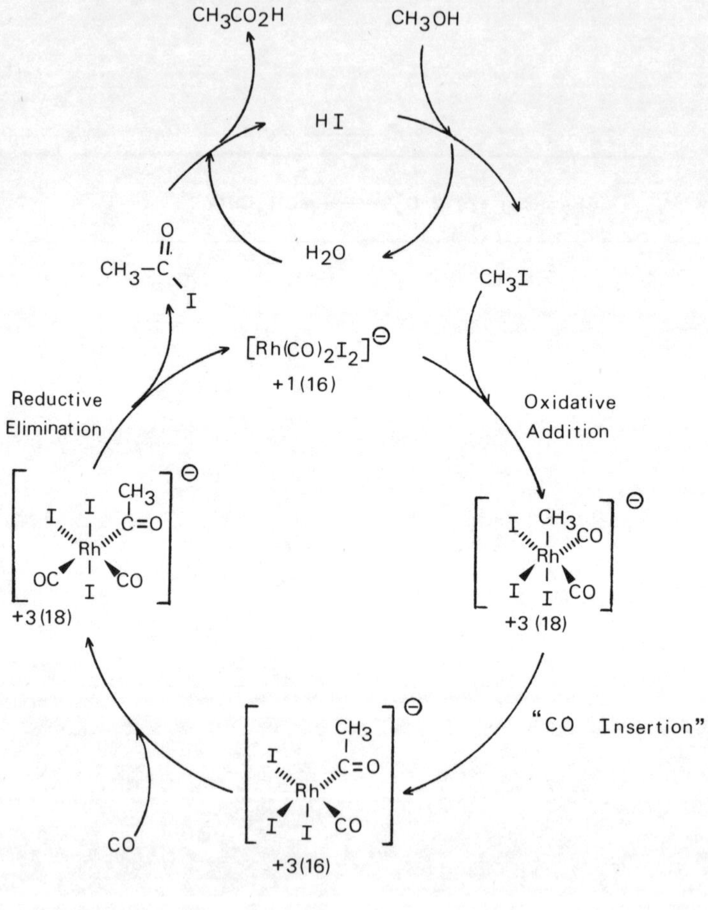

$$CH_3OH + CO \longrightarrow CH_3CO_2H$$

Figure 8: Methanol Carbonylation

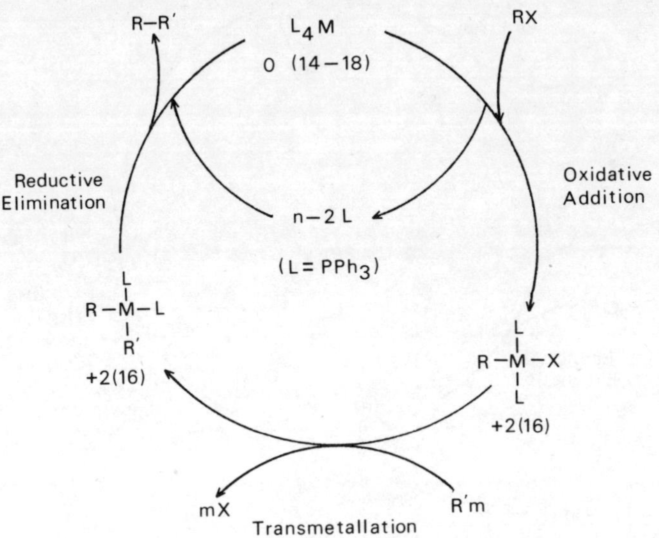

Figure 9: Transition Metal Catalyzed Coupling Reaction, M(II) ⇌ M(0)

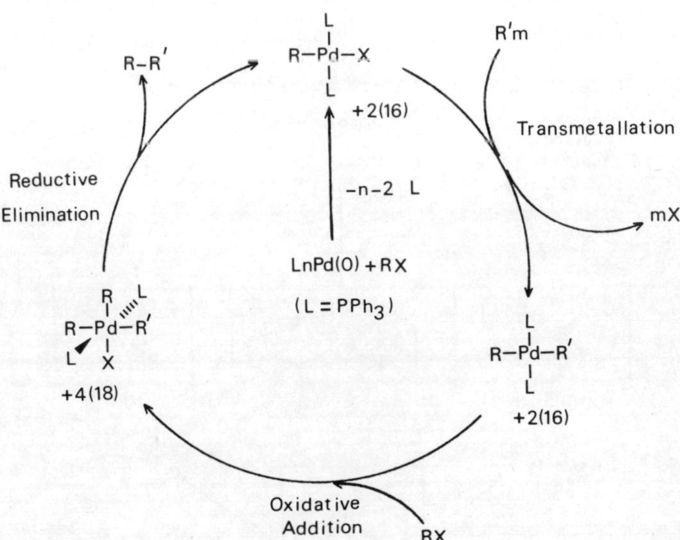

Figure 10: Palladium Catalyzed Coupling Reaction Pd(II) ⇌ Pd(IV)

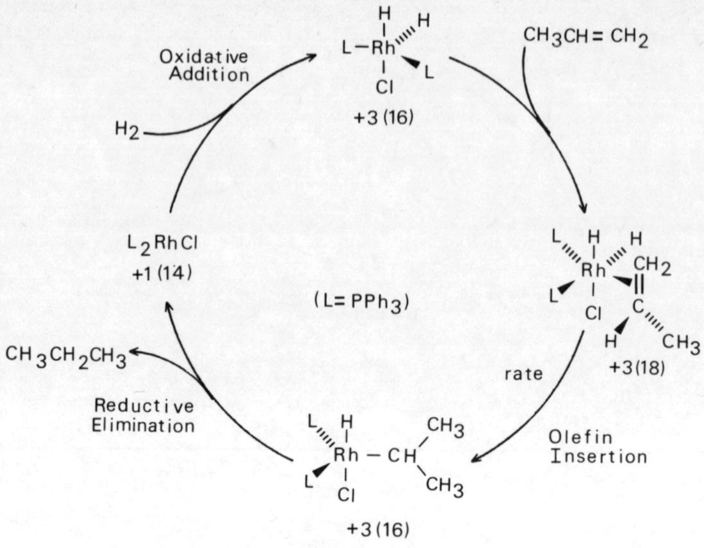

Figure 11: Rhodium Catalyzed Hydrogenation (Monophosphine)

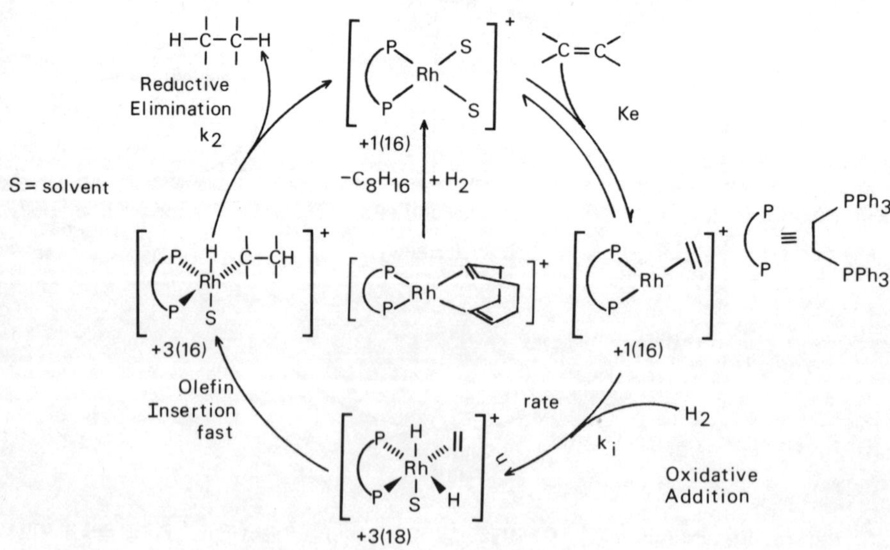

Figure 12: Rhodium Catalyzed Hydrogenation (Chelating Phosphine)

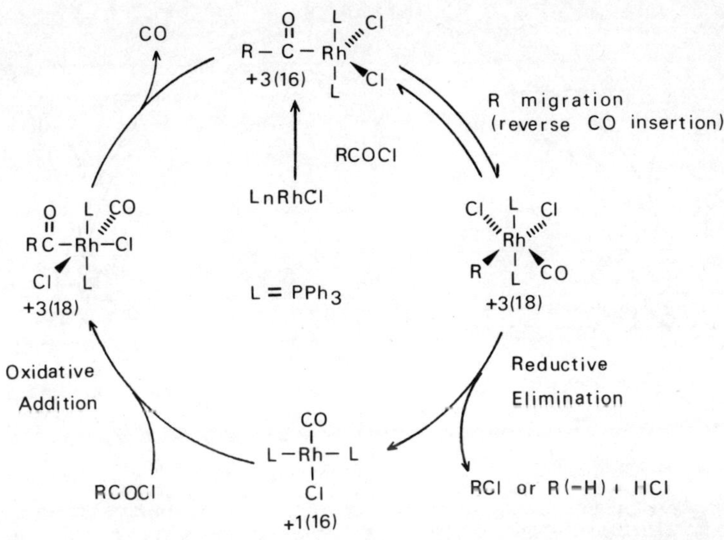

Figure 13: Rhodium Catalyzed Decarbonylation of Acid Chlorides

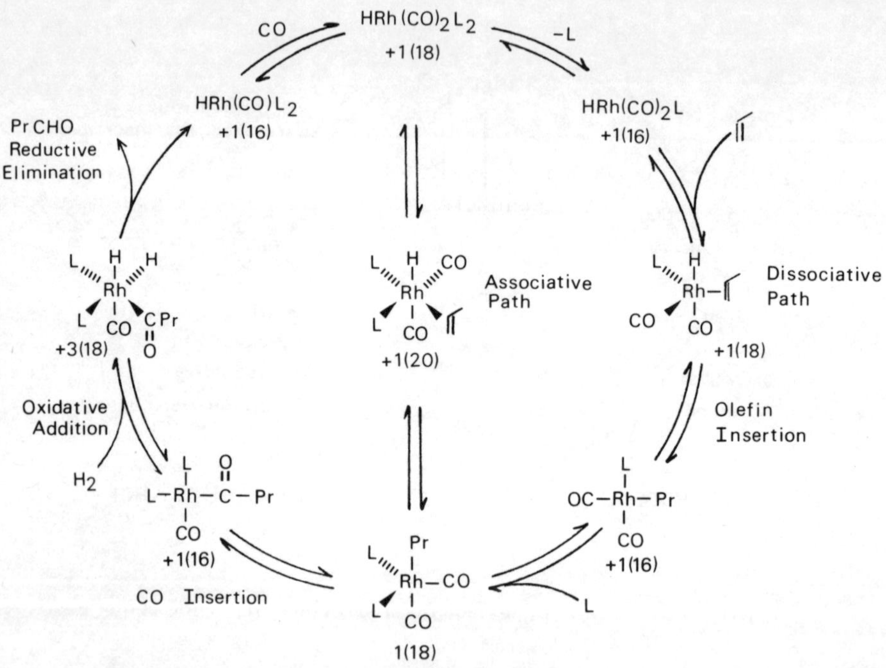

Figure 14: Rhodium Catalyzed Hydroformylation

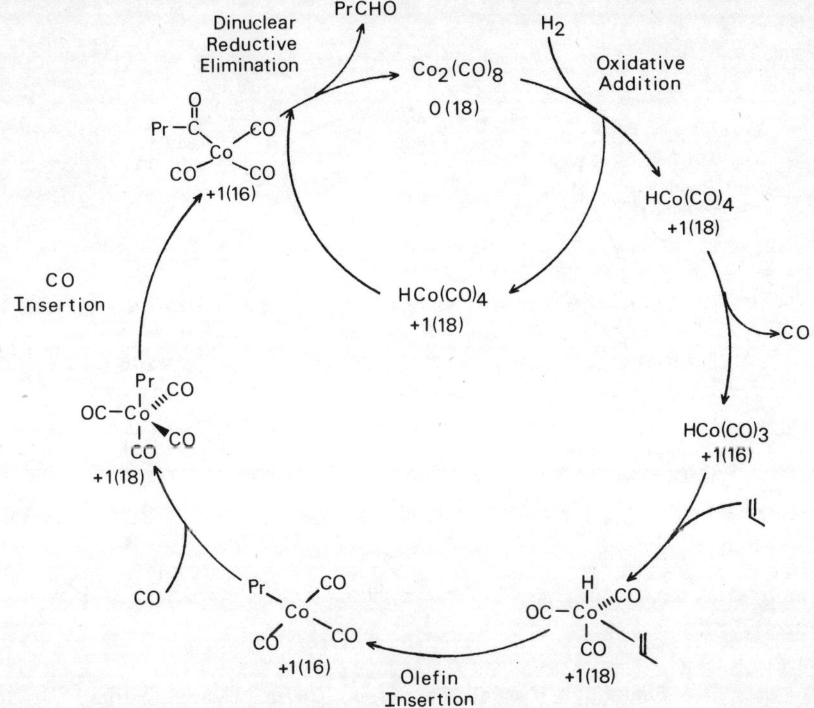

Figure 15: The Oxo Reaction

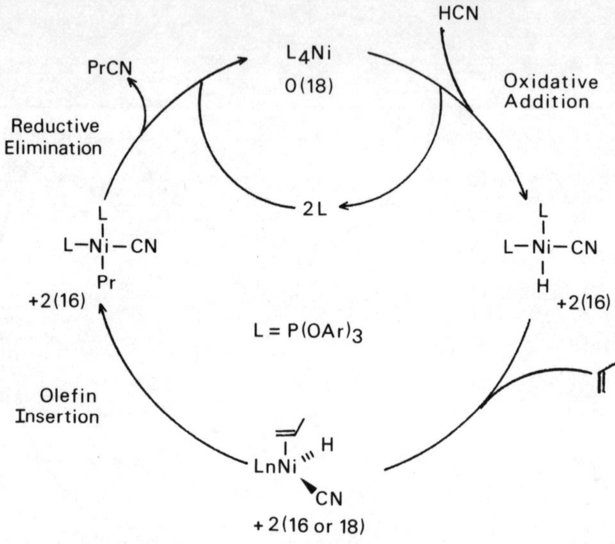

Figure 16: Nickel Catalyzed Olefin Hydrocyanation

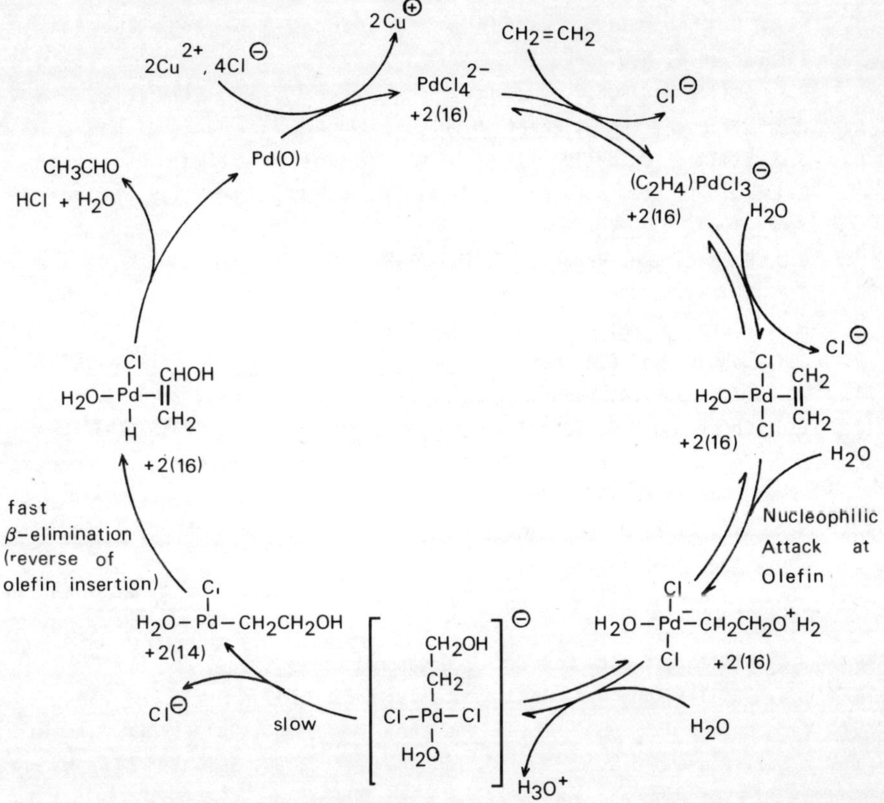

Figure 17: Palladium Catalyzed Ethylene Oxidation

References

1. An excellent survey of industrial homogeneous catalytic processes can be found in G.W. Parshall "Homogeneous Catalysis", John Wiley and Sons, New York, 1980.
2. For a comprehensive survey, see J.P. Collman and L.S. Hegedus, "Principles and Applications of Organotransition Metal Chemistry", University Science Books, Mill Valley, CA, 1980.
3. F.A. Cotton and G. Wilkinson, "Advanced Inorganic Chemistry", Interscience, New York, 1980.
4. J.K. Stille and K.S.Y. Lau, Acc. Chem. Res., 10, 434, (1977).
5. J.L. Davidson, Inorg. React. Mech., 5, 398 (1977).
6. J.K. Stille, "Oxidative Addition and Reductive Elimination", in S. Patai, The Chemistry of the Metal Carbon Bond, John Wiley and Sons, New York, 1983.
7. G.H. Posner, Org. React., 22, 253 (1974).
8. G.M. Whitesides, W.F. Fischer, Jr., J. San Filippo, Jr., R.W. Bashe and H.O. House, J. Am. Chem. Soc., 91, 4871 (1969).
9. C.R. Johnson and G.A. Dutra, J. Am. Chem. Soc., 95, 7777, 7783 (1973).
10. A. Tamaki and J.K. Kochi, J. Chem. Soc., Dalton Trans., 2620 (1973).
11. E.J. Corey and M.F. Semmelhack, J. Am. Chem. Soc., 89, 2755 (1967).
12. P.W. Jolly and G. Wilke, "The Organic Chemistry of Nickel", VI, Academic Press, New York, 1974.
13. B. Akermark, G. Akermark, L.S. Hegedus and K. Zetterberg, J. Am. Chem. Soc., 103, 3037 (1981).
14. B.M. Trost, Tetrahedron, 33, 2615 (1977); Acc. Chem. Res., 13, 385 (1980).
15. J.K. Stille and A.B. Cowell, J. Organomet. Chem., 124, 253 (1977).
16. J.K. Stille, Ann. N.Y. Acad. Sci., 295, 52 (1977); K.S.Y. Lau, R.W. Fries and J.K. Stille, J. Am. Chem. Soc., 96, 4983 (1974); P.K. Wong, K.S.Y. Lau and J.K. Stille, J. Am. Chem. Soc., 96, 5956 (1974); K.S.Y. Lau, P.K. Wong and J.K. Stille, J. Am. Chem. Soc., 98, 5832 (1976); J.K. Stille and K.S.Y. Lau, J. Am. Chem. Soc., 98, 5841 (1976); Y. Becker and J.K. Stille, J. Am. Chem. Soc., 100, 838, 845 (1978).
17. F.H. Jardine, Progr. Inorg. Chem., 28, 63-202 (1981).
18. M.C. Baird in "The Chemistry of Acid Derivatives", S. Patai, Ed., Wiley, New York, 1979, Supplement B, Part 2, p. 825.
19. J.A. Labinger and J.A. Osborn, Inorg. Chem., 19, 3230 (1980); J.A. Labinger, J.A. Osborn and N.J. Coville, Inorg. Chem., 19, 3236 (1980).

20. J.P. Collman, Acc. Chem. Res., 8, 342 (1975).
21. See T.C.T. Chang and M. Rosenblum, J. Org. Chem., 46, 4626 (1981), and references cited therein.
22. F.D. Mango, Coord. Chem. Rev., 15, 109 (1975).
23. K.C. Bishop, III, Chem. Rev., 76, 461 (1976).
24. R.J. Puddephatt, Coord. Chem. Rev., 33, 149 (1980).
25. L. Cassar, P.E. Eaton and J. Halpern, J. Am. Chem. Soc., 92, 3515 (1970).
26. G.W. Parshall, Catalysis, 1, 335 (1977).
27. A.E. Shilov and A.A. Shteinman, Coord. Chem. Rev., 24, 97 (1977); A.E. Shilov, Pure Appl. Chem., 50, 725 (1978).
28. D.E. Webster, Adv. Organomet. Chem., 18, 147 (1977).
29. A.J. Deeming and I.P. Rothwell, Pure Appl. Chem., 52, 649 (1980).
30. J.W. Suggs, J. Am. Chem. Soc., 100, 640 (1978).
31. A.H. Janowicz and R.G. Bergman, J. Am. Chem. Soc., 104, 352 (1982).
32. R.H. Crabtree, M.F. Mellea, J.M. Mihelcic and J.M. Quirk, J. Am. Chem. Soc., 104, 107 (1982).
33. J. Halpern, Acct. Chem. Res., 3, 386 (1970).
34. For recent reviews, see a) R.R. Schrock and G.W. Parshall, Chem. Rev., 76, 243 (1976); b) P.J. Davidson, M.F. Lappert and R. Pearce, Chem. Rev., 76, 219 (1976); c) P.S. Braterman, Topics in Current Chemistry, (Fort. Chem. Forsch.), 92, 149 (1980); d) J. Halpern, Acc. Chem. Res., 15, 332 (1982).
35. J.A. Connor, Topic Current Chem., 71, 71 (1977).
36. See L.F. Halle, P.B. Armentrout and J.L. Beauchamp, Organometallics, 1, 963 (1982) and references therein.
37. T.-T. Tsou, M. Loots and J. Halpern, J. Am. Chem. Soc., 104, 623 (1982).
38. F.T.T. Ng, G.L. Rempel and J. Halpern, J. Am. Chem. Soc., 104, 621 (1982).
39. a) J.K. Stille and M.T. Regan, ibid., 96, 1508 (1974); b) J.K. Stille and R.W. Fries, ibid., 96, 1514 (1974); c) J.K. Stille, F. Huang and M.T. Regan, ibid, 96, 1518 (1974).
40. D.A. Slack and M.C. Baird, J. Am. Chem. Soc., 98, 5539 (1976).
41. A. Tamaki, S.A. Magennis and J.K. Kochi, J. Am. Chem. Soc., 96, 6140 (1974); S. Komiya, T.A. Albright, R. Hoffmann and J.K. Kochi, ibid, 98, 7255 (1976).
42. B. Akermark and A. Ljungqvist, J. Organomet. Chem., 182, 59 (1979).
43. R.H. Grubbs and A. Miyashita, J. Am. Chem. Soc., 100, 1300, 7416, 7418 (1978); R.H. Grubbs, A. Miyashita, M. Liu and P. Burk, ibid., 2418 (1978).

44. R.J. McKinney, D.L. Thorn, R. Hoffmann and A. Stockis, J. Am. Chem. Soc., **103**, 2595 (1981).
45. A. Gillie and J.K. Stille, J. Am. Chem. Soc., **102**, 4933 (1980).
46. A. Moravskiy and J.K. Stille, J. Am. Chem. Soc., **103**, 4182 (1981).
47. K. Tatsumi, R. Hoffmann, A. Yamamoto and J.K. Stille, Bull. Chem. Soc. Jpn., **54**, 1857 (1981).
48. F. Ozawa, T. Ito, Y. Nakamura and A. Yamamoto, Bull. Chem. Soc. Jpn., **54**, 1868 (1981).
49. F. Ozawa and Y. Yamamoto, Chem. Letters, 289 (1981).
50. M.K. Loar and J.K. Stille, J. Am. Chem. Soc., **103**, 4174 (1981).
51. D. Milstein and J.K. Stille, J. Am. Chem. Soc., **101**, 4981 (1979).
52. D. Milstein and J.K. Stille, J. Am. Chem. Soc., **101**, 4992 (1979).
53. J.X. McDermott, J.F. White and G.M. Whitesides, J. Am. Chem. Soc., **95**, 4451 (1973).
54. J.X. McDermott, J.F. White and G.M. Whitesides, J. Am. Chem. Soc., **98**, 6521 (1976).
55. P.S. Braterman, R.J. Cross and G.B. Young, J. Chem. Soc., Dalton Trans., 1892 (1977) and earlier references.
56. G.B. Young and G.M. Whitesides, J. Am. Chem. Soc., **100**, 5808 (1978).
57. T.G. Appleton, H.C. Clark and L.E. Manzer, J. Organomet. Chem., **65**, 275 (1974).
58. H.C. Clark and J.D. Ruddick, Inorg. Chem., **9**, 2556 (1970).
59. M.P. Brown, R.J. Puddephatt, C.E.E. Upton and S.W. Lavington, J. Chem. Soc., Dalton Trans., 1613 (1974).
60. J. Schwartz, D.W. Hart and J.L. Holden, J. Am. Chem. Soc., **94**, 9269 (1972).
61. T. Ikariya and A. Yamamoto, J. Organomet. Chem., **118**, 65 (1976).
62. J.X. McDermott, M.E. Wilson and G.M. Whitesides, J. Am. Chem. Soc., **98**, 6529 (1976); J.X. McDermott and G.M. Whitesides, ibid, **96**, 947 (1974).
63. R.H. Grubbs and A. Miyashita, J. Chem. Soc., Chem. Comm., 864 (1977).
64. L. Abisg, A. Sen and J. Halpern, J. Am. Chem. Soc., **100**, 2915 (1978).
65. D. Milstein, J. Am. Chem. Soc., **104**, 5227 (1982).
66. J.R. Norton, Acct. Chem. Res., **12**, 139 (1979).
67. E.J. Kuhlmann and J.J. Alexander, Coord. Chem. Rev., **33**, 195 (1980).
68. K. Noack and F. Calderazzo, J. Organomet. Chem., **10**, 101 (1967).
69. T.C. Flood, J.E. Jensen and J.A. Statler, J. Am. Chem. Soc., **103**, 4410 (1981).
70. T.C. Flood, K.D. Campbell and H.H. Downs, unpublished results; T.C. Flood, F.J. DiSanti and D.L. Miles, Inorg. Chem., **15**, 1910 (1976).

71. J.K. Stille and L.F. Hines, J. Am. Chem. Soc., 92, 1798 (1970); L.F. Hines and J.K. Stille, ibid, 94, 485 (1972).
72. P.I. Bock, D.J. Boschetto, J.R. Rasmussen, J.P. Demers and G.M. Whitesides, J. Am. Chem. Soc., 96, 2814 (1974).
73. H.M. Walborsky and L.E. Allen, J. Am. Chem. Soc., 93, 5465 (1971).
74. K.S.Y. Lau, Y. Becker, F. Huang, N. Baenziger and J.K. Stille, J. Am. Chem. Soc., 99, 5664 (1977).
75. A. J. Deeming, B.F.G. Johnson and J. Lewis, J. Chem. Soc. (D), 1848 (1973).
76. F.W.S. Benfield and M.L.H. Green, J. Chem. Soc. (D), 1324 (1974).
77. H. Werner and R. Feser, Angew. Chem., Int. Ed. Engl., 18, 157 (1979).
78. D.L. Thorn and R. Hoffmann, J. Am. Chem. Soc., 100, 2079 (1978).
79. J.A. Labinger, D.W. Hart, W.E. Seibert, III and J. Schwartz, J. Am. Chem. Soc., 97, 3851 (1975).
80. A.S.C. Chan and J. Halpern, J. Am. Chem. Soc., 102, 838 (1980); A.S.C. Chan, J.J. Pluth and J. Halpern, ibid., 102, 5952 (1980); P.S. Chua, N.K. Roberts, B. Bosnich, S.J. Okrasinki and J. Halpern, J. Chem. Soc., Chem. Comm., 1278 (1981).
81. D.R. Coulson, J. Am. Chem. Soc., 91, 200 (1969).
82. E.R. Evitt and R.G. Bergman, J. Am. Chem. Soc., 102, 7003 (1980).
83. L.S. Hegedus, "Nucleophilic Attack on Transition Metal Organometallic Compounds", in S. Patai, The Chemistry of the Metal-Carbon Bond, John Wiley and Sons, New York, 1983.
84. M. Catellani and J. Halpern, Inorg. Chem., 19, 566 (1980).
85. N. Grice, S.C. Kao and R. Pettit, J. Am. Chem. Soc., 101, 1627 (1979).
86. H.C. Clark and K. von Werner, Syn. React. Inorg. Metal-Org. Chem., 4, 355 (1974).
87. D.E. James, L.F. Hines and J.K. Stille, J. Am. Chem. Soc., 98, 1806 (1976); J.K. Stille, D.E. James and L.F. Hines, J. Am. Chem. Soc., 95, 5062 (1973); D.E. James and J.K. Stille, J. Am. Chem. Soc., 98, 1810 (1976).
88. R.J. Angelici, Acct. Chem. Res., 5, 335 (1972).
89. O. Eisenstein and R. Hoffmann, J. Am. Chem. Soc., 103, 4308 (1981).
90. J.K. Stille, R.A. Morgan, D.D. Whitehurst and J.R. Doyle, J. Am. Chem. Soc., 87, 3282 (1965); J.K. Stille and R.A. Morgan, J. Am. Chem. Soc., 88, 5135 (1966); J.K. Stille and D.B. Fox, Inorg. Nucl. Chem. Letters, 5, 157 (1969); J.K. Stille and D.B. Fox, J. Am. Chem. Soc., 92, 1274 (1970).

91. J.K. Stille and D.E. James, J. Am. Chem. Soc., 97, 674 (1975); J.K. Stille and D.E. James, J. Organomet. Chem., 108, 401 (1976).

92. J.K. Stille, D.E. James and L.F. Hines, J. Am. Chem. Soc., 95, 5062 (1973); D.E. James, L.F. Hines and J.K. Stille, J. Am. Chem. Soc., 98, 1806 (1976).

93. T. Majima and H. Kurosawa, J. Chem. Soc., Chem. Comm., 610 (1977).

94. For an early discussion of this mechanism see P.M. Henry, Adv. Organometal. Chem., 13, 363 (1975), and references therein.

95. J.K. Stille and R. Divakaruni, J. Am. Chem. Soc., 100, 1303 (1978); J.K. Stille and R. Divakaruni, J. Organometal. Chem., 169, 239 (1979).

96. J.-E. Backvall, B. Akermark and S.O. Ljunggren, J. Chem. Soc., Chem. Comm., 264 (1977); J. Am. Chem. Soc., 101, 2411 (1979).

97. P.M. Henry, J. Org. Chem., 37, 2443 (1972); P.M. Henry, J. Am. Chem. Soc., 94, 7305 (1972).

98. B. Akermark, J.-E. Backvall, L.S. Hegedus and K. Zetterberg, J. Organomet. Chem., 72, 127 (1974); B. Akermark, J.-E. Backvall and K. Siirala-Hansen, Tet. Lett., 1363 (1974).

99. J.-E. Backvall, J. Chem. Soc., Chem. Comm., 413 (1977); Tet. Lett., 163 (1978).

100. L.S. Hegedus, B. Akermark and K. Zetterberg, manuscript in preparation.

101. A. Panunzi, A. DeRenzi and G. Paiaro, J. Am. Chem. Soc., 92, 3488 (1970).

102. P. Lennon, M. Madhavarao, A. Rosan and M. Rosenblum, J. Organometal. Chem., 108, 93 (1976); P. Lennon, A. Rosan and M. Rosenblum, J. Am. Chem. Soc., 99, 8426 (1977).

103. B.M. Trost, Acc. Chem. Res., 13, 385 (1980) and references therein.

104. B. Akermark and A. Jutland, J. Organometal. Chem., 217, C41 (1981).

105. B. Akermark, G. Akermark, L.S. Hegedus and K. Zetterberg, J. Am. Chem. Soc., 103, 3037 (1981).

106. M.P. Periasamy and H.M. Walborsky, J. Am. Chem. Soc., 97, 5930 (1975).

107. E. Piers and E.H. Ruediger, J. Chem. Soc., Chem. Comm., 166 (1979).

108. R.C. Larock, Angew Chem. Int. Ed. Engl., 17, 27 (1978).

109. J.-E. Backvall and B. Akermark, J. Chem. Soc., Chem. Comm., 82 (1975).

110. J.K. Stille and P.K. Wong, J. Org. Chem., 40, 335 (1975).

111. S. Baba and E. Negishi, J. Am. Chem. Soc., 98, 6729 (1976).

112. D. Milstein and J.K. Stille, J. Org. Chem., 44, 1613 (1979).

113. D. Forster, Adv. Organomet. Chem., 17, 255 (1979).

114. B.D. Vineyard, W.S. Knowles, M.J. Sabacky, G.L. Bachman and D.J. Weinkauff, J. Am. Chem. Soc., 99, 5946 (1977).

115. R.L. Pruett, Ann. N.Y. Acad. Sci., 295, 239 (1977).

GROUP VIII TRANSITION METALS IN ORGANIC SYNTHESIS

Professor Louis S. Hegedus
Department of Chemistry
Colorado State University
Fort Collins, Collorado 80523, U.S.A.

Louis S. Hegedus

Department of Chemistry, Colorado State University, Fort Collins, Colorado, USA, 80523

I. INTRODUCTION

The group VIII transition metals lie to the right hand side of the transition series, and consist of the iron, cobalt, and nickel triads (Figure 1). They typically form electron-rich complexes having from six to ten d electrons and stable formal oxidation states from -2 to +3. Most reactions of use in organic synthesis involve the first and second row members of this group (Fe, Co, Ni, Rh, Pd), while complexes of third row members, (because of their stability) are often used to answer mechanistic questions. Since it is the chemistry of the complexed organic ligand that is of most interest to synthetic chemists, reactions will be considered by ligand type rather than by metal.

Figure I. The Transition Series

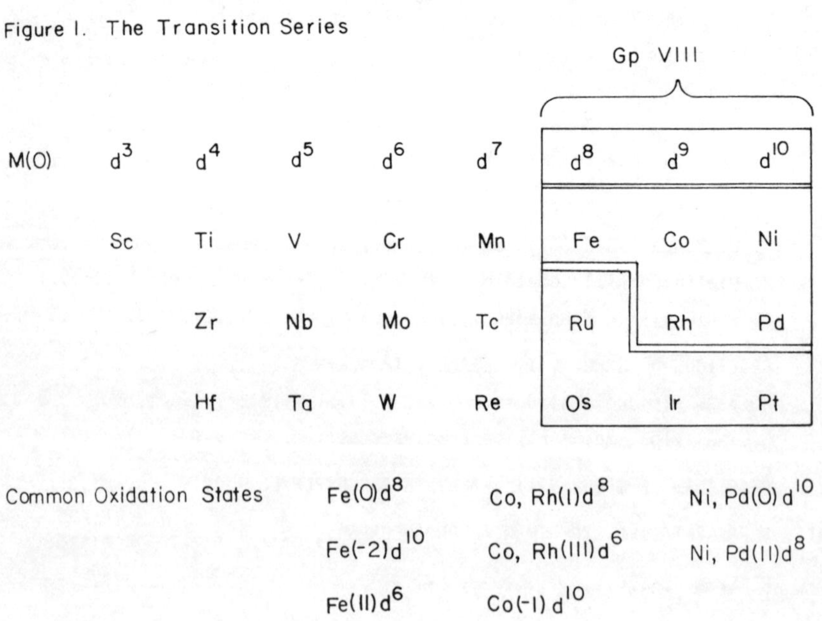

II. REACTIONS OF GROUP VIII METAL CARBONYLS

As a ligand, carbon monoxide is a strong π-acceptor which stabilizes low oxidation states of transition metals. Thus many Group VIII metal carbonyls can be reduced to give anionic species, $[M(CO)n]^{-Y}$, which are often strong nucleophiles and reducing agents. Metal carbonyls participate in two classes of reactions of interest to synthetic chemists: coupling reactions and carbonylation (acylation) reactions.[1]

A. Nickel Carbonyl Coupling of Allylic Halides

Nickel carbonyl is a volatile (BP 43°), toxic liquid which rapidly reacts with allylic halides in polar solvents such as dimethylformamide (DMF) or hexamethylphosphoramide [HMPA, $(Me_2N)_3PO$] to form biallyls (equation 1).[2] This is a very general reaction and tolerates a wide variety of functional groups on the allyl moiety (alkyl, aryl, olefinic, ester, ether, amido, cyano). With unsymmetrical allyl halides, coupling occurs predominately at the <u>least</u> substituted terminus of the allyl system, although small amounts of the other possible regioisomers are often obtained. Since the reaction proceeds through a π-allylnickel intermediate, the stereochemistry of the double bond may be lost. Although the exact mechanism of this coupling reaction is not known, it is thought to involve the reaction of a π-allylnickel halide species with the allylic halide, in a radical-chain process (see section VI A). This reaction <u>cannot</u> be used efficiently to cross-couple two different allylic halides since all possible self- and cross-coupled products are obtained.

(1)

$Ni(CO)_4 \rightleftharpoons CO + Ni(CO)_3$ ⇌ (allyl bromide) [allyl–Ni(CO)₂Br] ⇌ π-allyl Ni(Br)(CO) + CO ⇌ [π-allyl Ni(μ-Br)]₂ + 2 CO; with allyl bromide → $NiBr_2$ + 1,5-hexadiene

Intramolecular versions of this coupling reaction have been used to synthesize a number of macrocyclic dienes. In initial studies simple 12, 14, and 18-carbon straight-chain α,ω-*bis*-allylic halides cyclized under high dilution conditions to produce the corresponding cyclic 1,5-diene in excellent yields.[3] Subsequently, this coupling reaction has been used to synthesize macrocyclic natural products including humulene,[4] cembrene,[5] macrocyclic lactones,[6] and casbene (equation 2).[7] In these intramolecular coupling reactions, coupling again invariably occurs at the *less* substituted allylic terminus.

Br Br $\xrightarrow{Ni(CO)_4}$ (2)

B. Iron Carbonyl Coupling of α,α'-Dihaloketones with Dienes and Enamines

The above reactions are formally reductive couplings of allyl halides, in that the low valent nickel(0) species is oxidized to nickel(II) in the process. Iron carbonyl promotes a related reductive coupling of dienes with α,α'-dihaloketones to provide facile entry into a number of normally difficult-to-synthesize seven membered ring systems.[8] This chemistry relies upon the reaction of "$Fe(CO)_4$" (from $Fe_2(CO)_9$, an orange air stable solid) with the dihaloketone to give an iron stabilized oxallylcation (equation 3), which is never isolated, but rather generated and used *in situ*. The cationic nature of this species has been confirmed by a number of experimental observations,[9] including its behavior in cycloaddition reactions. Allyl cations are three carbon two electron systems which should undergo a 3 + 4 cyclocoupling reaction (formally a $\pi^2 + \pi^4$ cycloaddition) to give seven membered rings. Indeed, open chain dienes react with tertiary or secondary α,α'-dihaloketones and $Fe_2(CO)_9$ to produce 4-cycloheptenones in good yield (equation 4).[10] Cyclic dienes are even more efficient, giving bridged bicyclic systems in excellent yield (equation 5).

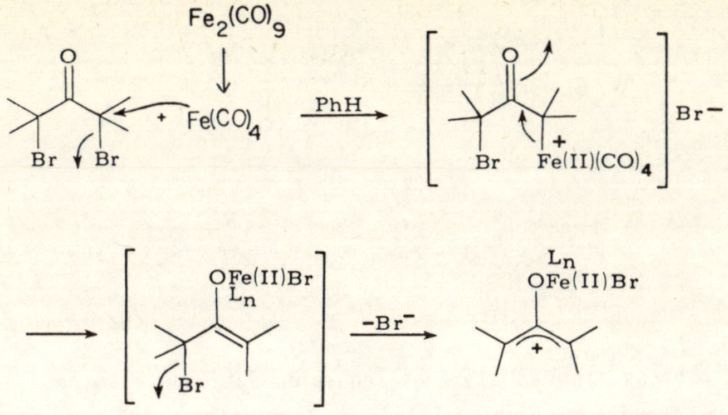

(3)

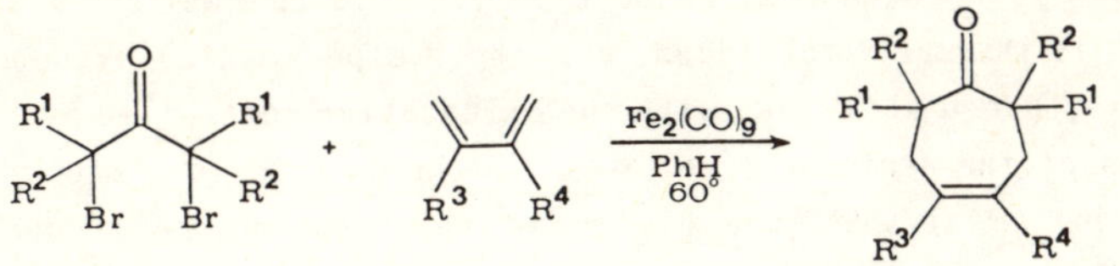

(4)

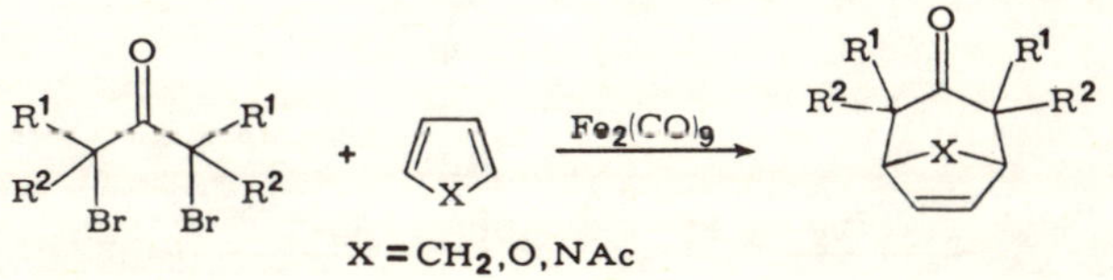

(5)

$X = CH_2, O, NAc$

A major limitation of this reaction was that dibromoacetone itself, as well as dibromides of other methyl ketones, failed to give cycloadducts in the above reactions, making unsubstituted cycloheptenones unavailable by this procedure. However, tetrabromoacetone was reactive, producing dibromo adducts which were readily dehalogenated by treatment with zinc/copper couple. This modification permits all substitution patterns (unsubstituted, mono, 1,1-di, 1,2-di, tri-, and tetrasubstituted) to react, and offers the most direct route to a number of natural products containing seven-membered rings. With furans as the diene component of this cycloaddition, nezukone, β-thujaplicin, α-thujaplicin,[11] and C-nucleosides[12] have been synthesized. With pyrroles as substrates, tropane alkaloids including tropine, scopine, and hyoscyamine were synthesized (equation 6).[13] These 3 + 4 cycloadditions are thought to procede by a concerted process in which the regioselectivity is controlled by the frontier molecular orbitals of the cycloadduct.[14]

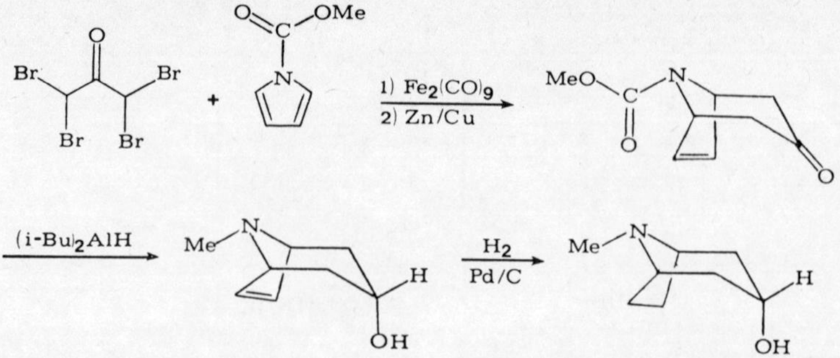

(6)

Electron-rich olefins such as styrenes (equation 7)[15] and enamines (equation 8)[16] react in a stepwise fashion with this oxallylcation species to produce cyclopentanones and cyclopentenones, respectively. This is formally a 3+2 cycloaddition, which is thermally forbidden, and ample experimental evidence indicates that it is indeed stepwise, with the orientation controlled by the relative stabilities of the ionic intermediates. Although this is a stepwise process, the stereochemistry of the olefin is retained in the product, apparently because free rotation in the initial adduct is prevented by strong ionic interactions. With cycloheptatrienones, addition to the carbonyl group occurs, also in a stepwise process (equation 9).[17]

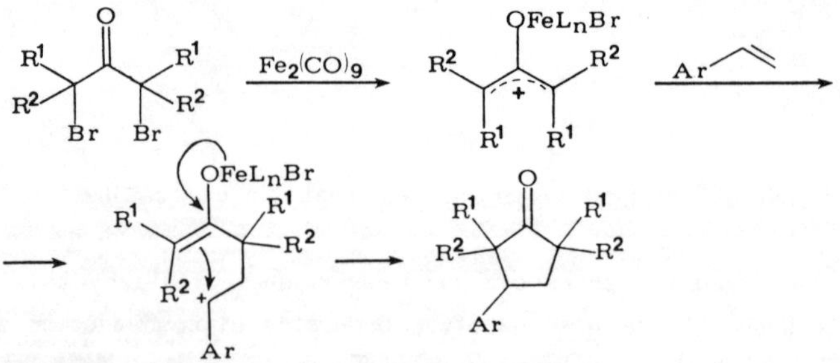

(7)

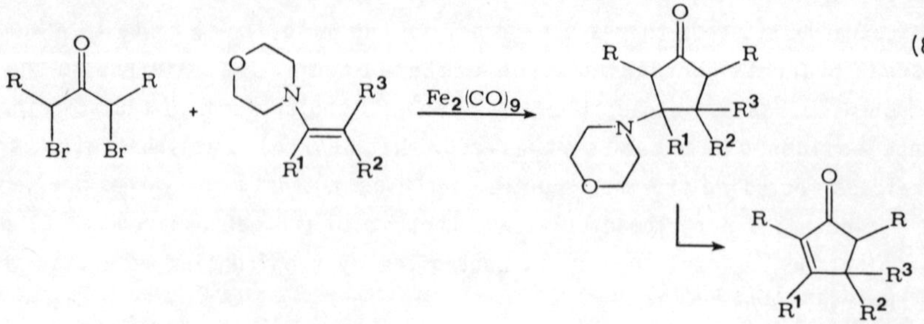

(8)

$$\text{2-R-tropone} + \text{(CH}_3)_2\text{CBr–CO–CBr(CH}_3)_2 + Fe_2(CO)_9 \longrightarrow \text{bicyclic adduct} \quad (\sim 50\%) \qquad (9)$$

C. Acylation Processes

Although the above reactions involved metal carbonyls as reactive species, the carbon monoxide ligand served simply to provide complexes of appropriate reactivity, but did not become incorporated into the product. Metal carbonyls also participate in a variety of carbonylation processes, in which one (or more) carbon monoxide becomes incorporated in to the organic product. One of the most generally useful acylation reagents is $Na_2Fe(CO)_4$, (a white, air-sensitive commercially available solid) prepared by the reduction of $Fe(CO)_5$.[18] This species, a d^{10}, iron(-2) complex is a strong base (pK_a = 14) and a strong nucleophile, and reacts with typical S_N2 kinetics, stereochemistry, and order of reactivity (Me > 1° > 2°; RI > RBr > ROTs > RCl; aryl, vinyl inert) with halides and tosylates to produce alkyliron(0) complexes. These readily insert carbon monoxide to form the much less nucleophilic acyliron complexes. These react with 1° iodides to give ketones, with acetic acid to give aldehydes, and with halogens to give carboxylic acid derivatives (Figure 2). This chemistry offers a convenient conversion of halides and tosylates to aldehydes, ketones, or carboxylic acid derivatives, and tolerates ester, ketone, nitrile, and olefin functionality elsewhere in the substrate. It will discriminate between bromide and chloride. As with other strongly basic nucleophiles, elimination processes compete with acylation with 2° substrates. The mechanisms of these reactions have been extensively studied, but are beyond the scope of this work.[19]

Other metal carbonyls are readily reduced to produce nucleophilic anionic complexes. Dicobalt octacarbonyl ($Co_2(CO)_8$) readily disproportionates when treated with base, or reduces when treated with sodium amalgam, to produce the Co(-1)d^{10} complex $NaCo(CO)_4$, an air sensitive colorless crystalline solid. This

Figure 2

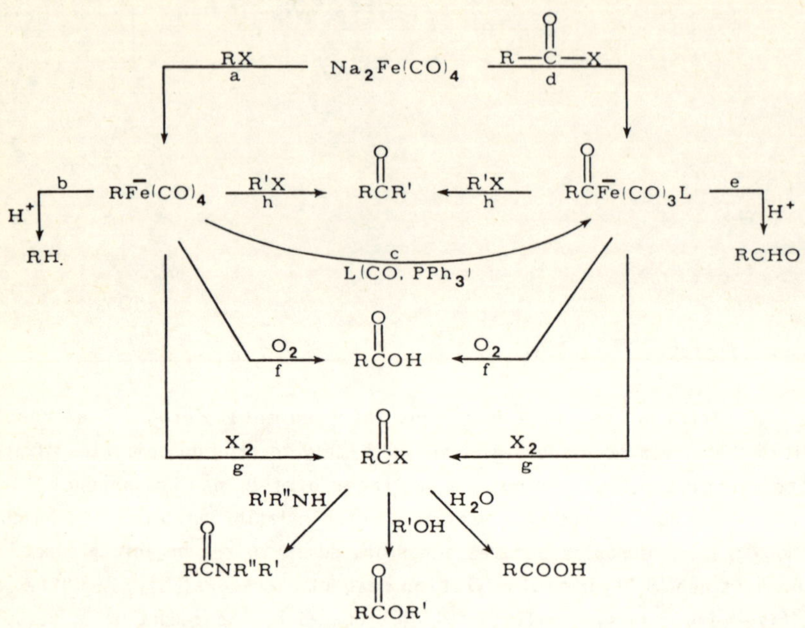

anionic metal carbonyl is a much weaker base (pKa = 1) than is $Na_2Fe(CO)_4$, but also reacts in an S_N2 process with reactive halides and tosylates (1°, 2°, allyl, benzyl, α-haloester) in the presence of alcohols to produce esters (equation 10).[20] This carbonylation reaction can be made catalytic in cobalt carbonyl by providing a base to regenerate $Co(CO)_4^-$.

$$Co(CO)_4^- + RX \longrightarrow RCo(CO)_4 \xrightarrow{CO} RC(=O)Co(CO)_4 \xrightarrow{R'OH} RC(=O)OR' + HCo(CO)_4 \qquad (10)$$

Neither the iron nor the cobalt carbonyl anions react with aryl or vinyl halides. However, treatment of nickel carbonyl with sodium methoxide in methanol produces in situ an uncharacterized acylnickel complex which efficiently carbonylates aryl iodides, olefinic halides, and allylic halides but is inert to alkyl halides (equation 11). With vinyl halides the stereochemistry of

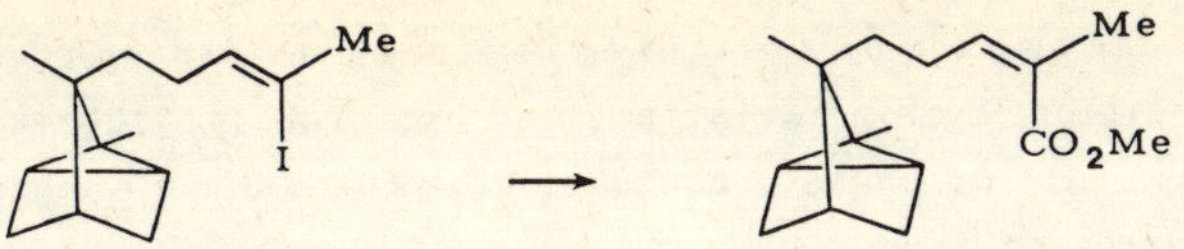

(11)

the double bond is maintained.[21] A related acylnickel complex, produced in situ by the reaction of organolithium reagents, with nickel carbonyl behaves somewhat differently. Aryl, acyl, and alkyl halides are inert to this reagent, whereas allylic halides undergo clean acylation without allylic transposition (equation 12). Olefinic halides also acylate, but the thus formed α,β-unsaturated carbonyl compound reacts further to produce a 1,4-dicarbonyl compound (equation 13). This Michael acylation of conjugated enones is general, and acylnickel species from methyl, n-butyl- and t-butyllithium (but not phenyllithium) react with benzalacetone, methyl cinnamate, 2-cyclohexenone, mesityl oxide, and methyl crotonate to give 1,4-dicarbonyl compounds in good yield (equation 14).[22]

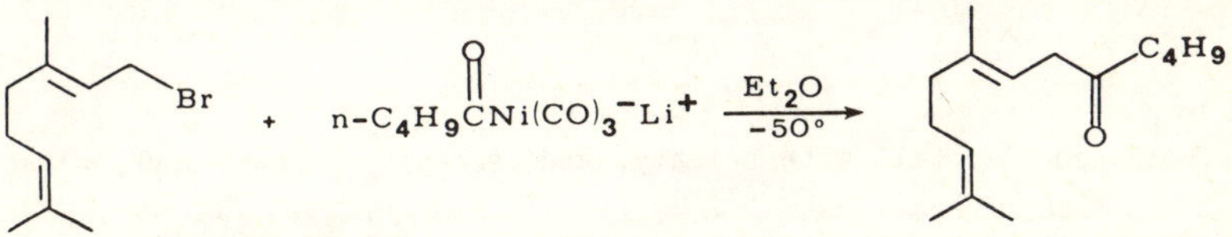

$$+\ n\text{-}C_4H_9\overset{O}{\overset{\|}{C}}Ni(CO)_3^-Li^+ \xrightarrow[-50^\circ]{Et_2O} \quad (12)$$

$$Me\overset{O}{\overset{\|}{C}}-Ni(CO)_3^-Li^+ + \ Ph\diagup\!\!=\!\!\diagdown Br \longrightarrow \quad (13)$$

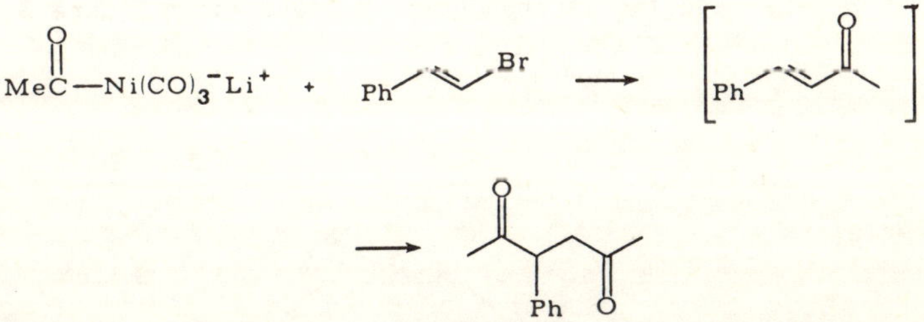

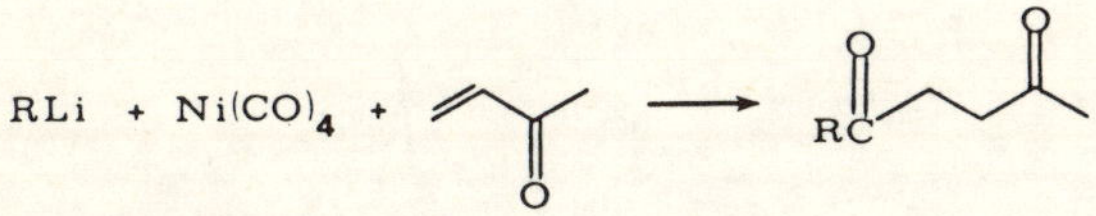

$$RLi + Ni(CO)_4 + \longrightarrow R\overset{O}{\overset{\|}{C}} \quad (14)$$

III. REACTIONS OF GROUP VIII METAL ALKYL COMPLEXES

Alkyl complexes of Group VIII metals are among the most useful organometallic reagents for organic synthesis, for they permit the formation of carbon-carbon bonds using both substrates and conditions not accessible to routine organic synthetic methodology. Virtually all of the Group VIII metal complexes of use in synthesis are either palladium or, to a lesser degree, nickel complexes. They can be made by a variety of methods, including transmetallation, oxidative addition, cyclometallation, and nucleophilic attack on metal-complexed olefins. All of these procedures have found use in organic synthesis, and will be discussed below.

A. Nickel- and Palladium-Catalyzed Grignard Reactions

Transition metal catalyzed Grignard reactions have long been studied, with the greatest emphasis being placed on copper-catalyzed reactions. Recently the selective cross coupling of Grignard reagents with aryl and vinyl halides catalyzed by nickel or palladium salts has been developed (equation 15).[23] The

$$RMgX + \begin{matrix} ArX' \\ \text{or} \\ CH_2{=}CHX \end{matrix} \xrightarrow{ML_2Cl_2} \begin{matrix} RAr \\ \text{or} \\ CH_2{=}CHR \end{matrix} + MgXX' \qquad (15)$$

reaction works equally well with primary, and secondary alkyl, aryl, alkenyl, and allylic Grignard reagents. A variety of simple, fused, substituted, and heteraromatic halides, and alkenyl halides are reactive, but simple alkyl halides are inert. The reaction is thought to proceed as in Figure 3 and

Figure 3

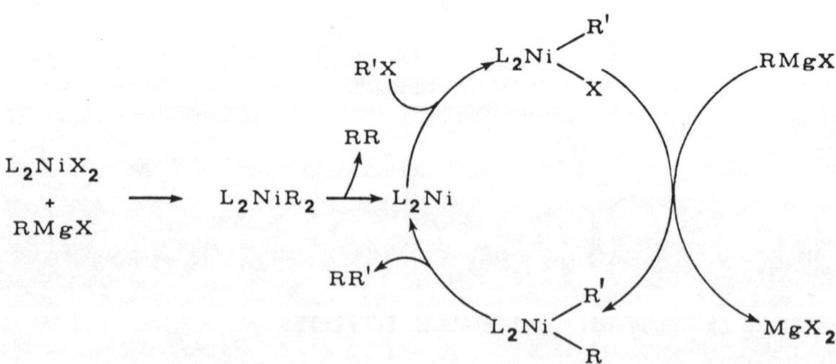

involve transmetallation from magnesium to nickel, and oxidative addition-reductive elimination chemistry. The synthetic value of this procedure was greatly enhanced when it was observed that the use of chiral ligands on the metal resulted in high optical induction in the coupling of secondary Grignard reagents with vinyl halides (equation 16).[24] The ligands most effective in this chiral coupling reaction were unusual (aminoalkyl)ferrocenyl phosphines and aminophosphines from chiral aminoacids (Figure 4).[25] It should be noted that this reaction is actually a kinetic resolution of the racemic Grignard reagent. Since the Grignard reagent undergoes inversion rapidly relative to coupling, high optical yields are obtained.

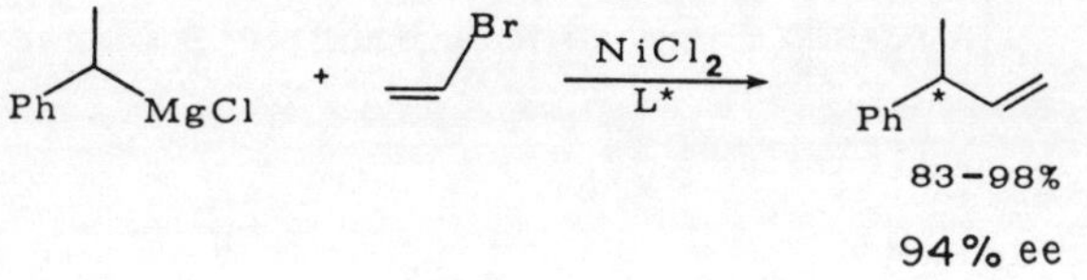

Ph MgCl + Br $\xrightarrow[L^*]{NiCl_2}$ Ph * (16)

83-98%

94% ee

Figure 4

$R\overset{*}{C}H\text{-}CH_2PPh_2$, NMe_2

Fe, PPh_2, C—NMe_2, Me, H (S),(R)

Me, C, H, Fe, NMe_2, PPh_2 (R),(S)

Fe, CH_2NMe_2, PPh_2 (S)

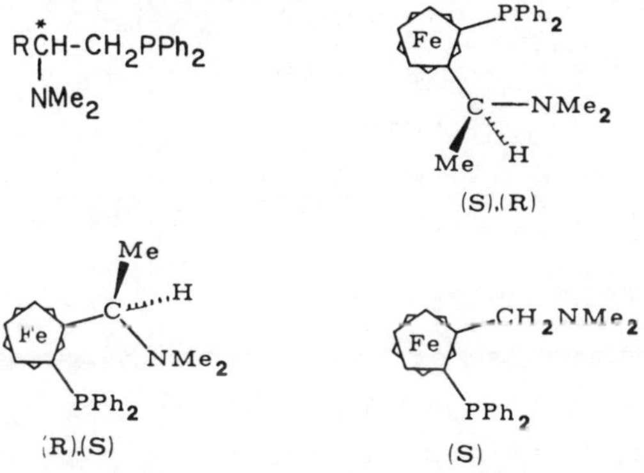

B. Coupling Reactions Involving Transmetallation to Palladium(II) Complexes

Grignard and organolithium reagents are of only limited use in transmetallation to Group VIII metals because they are potent reducing agents, and often lead to reduction of the transition metal rather than to formation of an alkylmetal species. Furthermore, they are intolerant of a variety of functional groups, limiting the complexity of both reagent and substrate. Fortunately other organometallics, including those of Hg, B, Al, Sn, Zn, and Zr, readily transfer alkyl groups to Group VIII metals, and a host of synthetically useful procedures involving transmetallation have been developed. Mercury is the most extensively used main group metal, primarily because of the ease of its direct introduction into unsaturated (particularly aryl) systems. Palladium(II) is the most popular Group VIII metal for this process because of the facility with which it undergoes insertion reactions with carbon monoxide and olefins (Figure 4).

Figure 4. Transmetallation-Insertion Chemistry of Palladium II

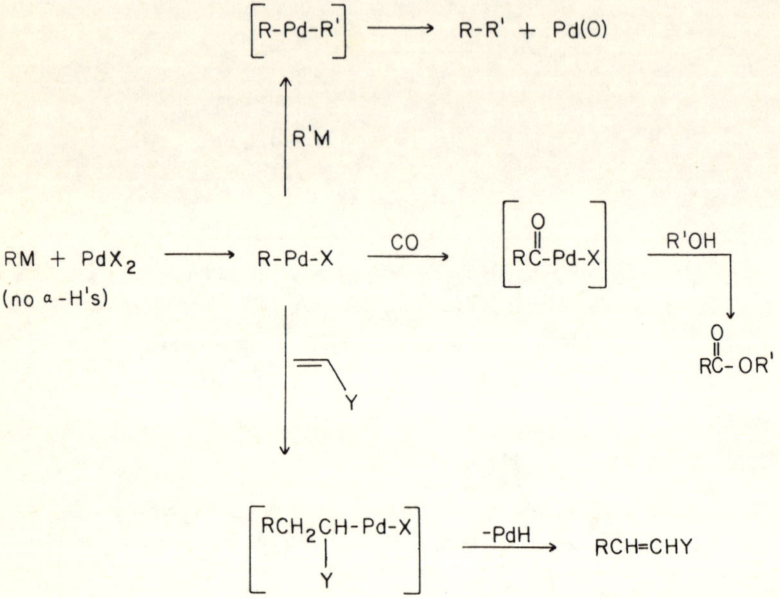

$$[\text{R-Pd-R}'] \longrightarrow \text{R-R}' + \text{Pd(0)}$$

$$\text{RM} + \text{PdX}_2 \ (\text{no } \alpha\text{-H's}) \longrightarrow \text{R-Pd-X}$$

$$\text{R-Pd-X} \xrightarrow{\text{R'M}} [\text{R-Pd-R}']$$

$$\text{R-Pd-X} \xrightarrow{\text{CO}} [\text{RC(=O)-Pd-X}] \xrightarrow{\text{R'OH}} \text{RC(=O)-OR}'$$

$$\text{R-Pd-X} \xrightarrow{\text{CH}_2\text{=CHY}} [\text{RCH}_2\text{CH(Y)-Pd-X}] \xrightarrow{\text{-PdH}} \text{RCH=CHY}$$

The early work in this area used simple mercurated arenes and olefins,[26] but recently much more highly functionalized systems have been studied. Any aromatic system which mercurates is a potential substrate for this chemistry. For example, isoquinolones were alkylated in the 4-position using this chemistry (equation 17).[27] This reaction has found extensive application in nucleoside

(17)

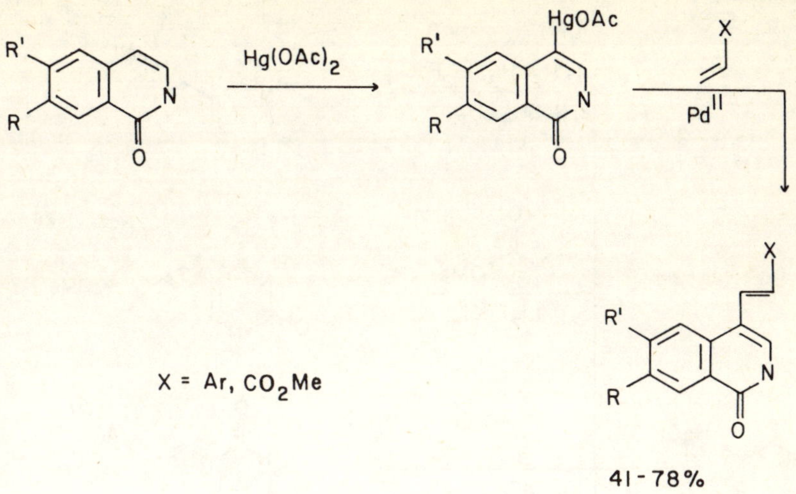

$Hg(OAc)_2$ → (HgOAc intermediate) → Pd^{II}

X = Ar, CO_2Me

41-78%

chemistry because the pyrimidine ring mercurated readily at the 5-position (Figure 5). Bicycloprostaglandin analogs were made using transmetallations from both mercury and copper (equation 18).[33] Propargyl alcohols were converted to butenolides using a chloromercuration-transmetallation-carbonylation sequence (equation 19).[34]

Nonhalogenated vinylmercurials are not directly accessible from olefins or alkynes. However, vinylboranes, (from hydroboration of alkynes) react with mercuric chloride to give vinylmercurials, which have then been used in organopalladium chemistry as above.[35] However, procedures for direct transmetallation from boron to palladium have recently been developed, making vinyl boranes (and hence alkynes) accessible to palladium insertion chemistry (equation 20). Transmetallation from tin to palladium is also efficient, and has been used to form carbocycles by insertion of thioenol ethers (equation 21).[37] (The mercuric chloride scavanged the thiomethyl group to prevent irreversible complexation to the palladium(II) catalyst). A requirement of all of these transmetallation reactions is that the organic group transferred to palladium must lack β-hydrogens since β-hydride elimination is competitive with the desired insertion reactions.

Figure 5. Pd(II) Assisted Alkylation of Mercurated Pyrimidines

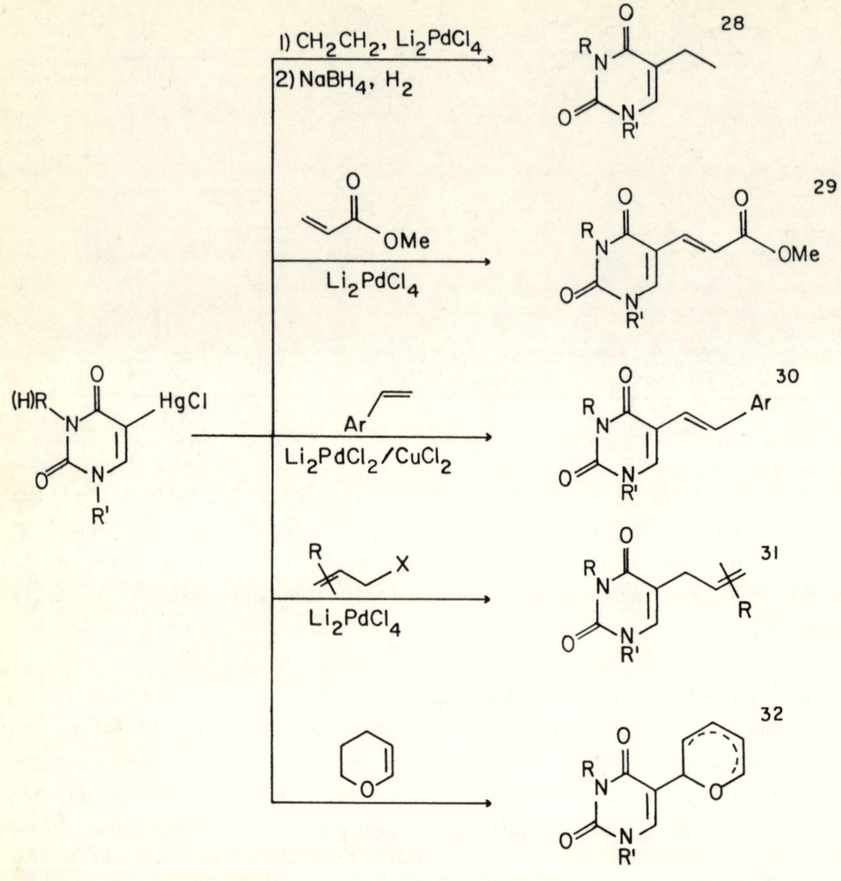

1) CH_2CH_2, Li_2PdCl_4
2) $NaBH_4$, H_2

28

OMe

Li_2PdCl_4

29

(H)R

HgCl

Ar

$Li_2PdCl_2/CuCl_2$

30

R X

Li_2PdCl_4

31

32

R' = Me, deoxyribose

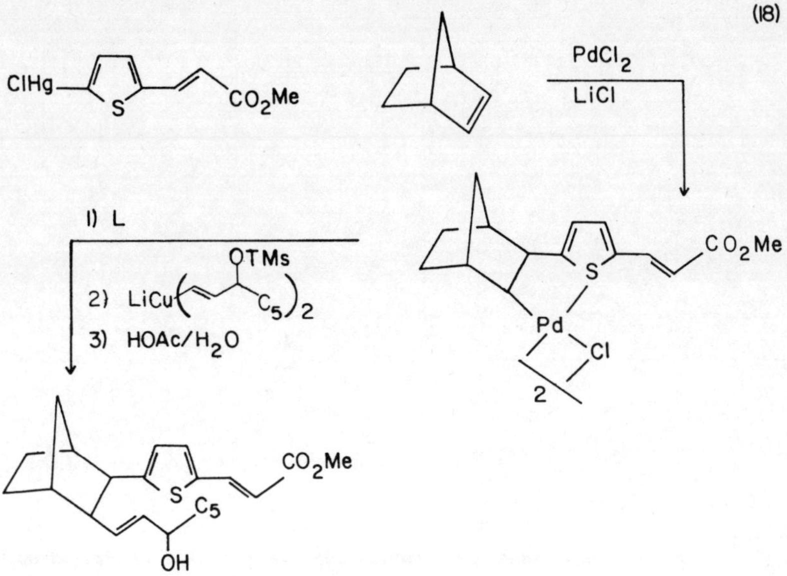

(18)

ClHg— S —CO_2Me

$PdCl_2$

LiCl

CO_2Me

S

Pd

Cl

2

1) L

2) LiCu(OTMs, C_5)$_2$

3) $HOAc/H_2O$

CO_2Me

S C_5

OH

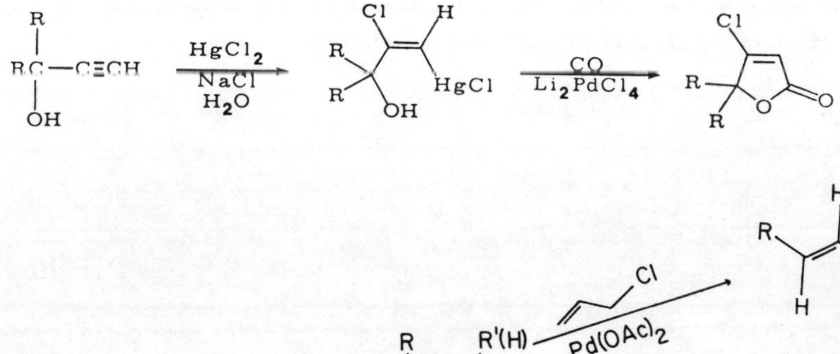

R

RC—C≡CH

OH

$HgCl_2$

NaCl

H_2O

Cl H

R

R HgCl

OH

CO

Li_2PdCl_4

Cl

R— O =O

R

(19)

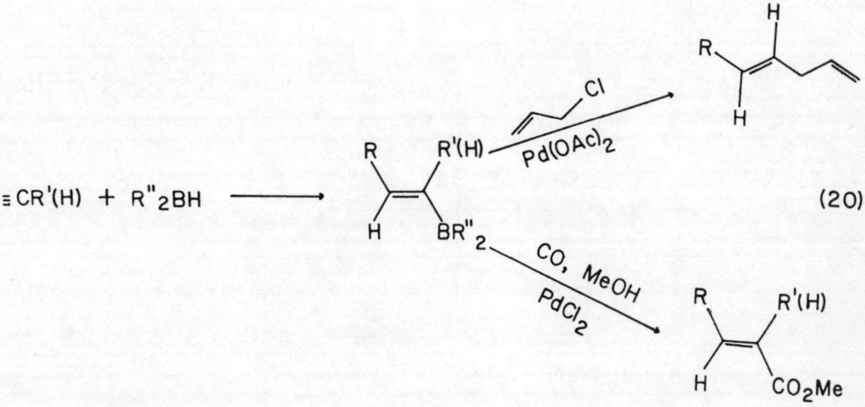

$RC{\equiv}CR'(H) + R''_2BH$ ⟶

R R'(H)

H BR''_2

Cl

$Pd(OAc)_2$

H

R

H

CO, MeOH

$PdCl_2$

R R'(H)

H CO_2Me

(20)

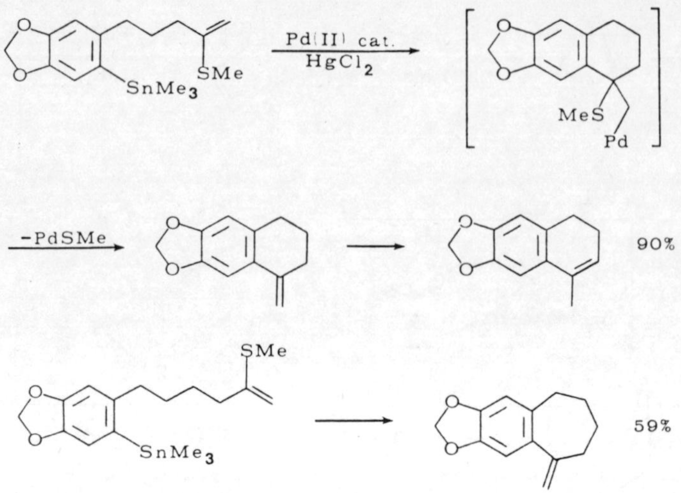

$SnMe_3$ SMe $\xrightarrow[HgCl_2]{Pd(II)\ cat.}$ [MeS Pd] (21)

$\xrightarrow{-PdSMe}$ $\longrightarrow$ 90%

SMe $SnMe_3$ $\longrightarrow$ 59%

C. Coupling Reactions Involving Oxidative Addition to Palladium(0) Complexes

Another approach to σ-aryl- or σ-vinylpalladium(II) complexes involves the oxidative addition of aryl or alkenyl halides to palladium(0) complexes. The resulting σ-aryl or σ-vinyl complex can then undergo a variety of insertion or transmetallation process of synthetic value (Figure 6).

Figure 6

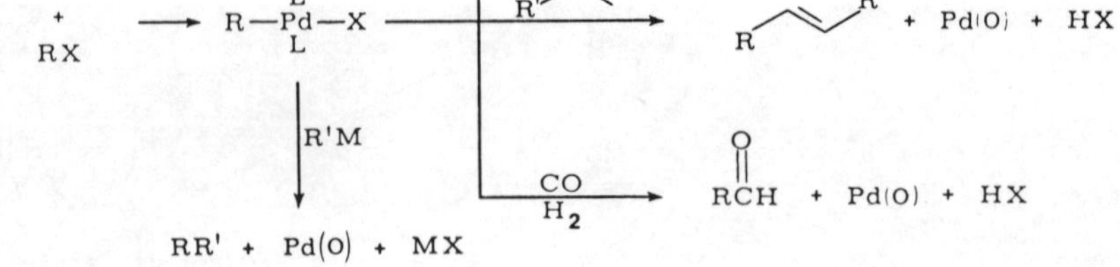

$L_nPd(0)$ + RX $\longrightarrow$ R—Pd(L)(L)—X

CO, R'OH: RCOR' (C=O) + Pd(0) + HX

CO, $R'NH_2$: RCNHR' (C=O) + Pd(0) + HX

R'CH=CH$_2$: RCH=CHR' + Pd(0) + HX

CO, H_2: RCH (C=O) + Pd(0) + HX

R'M: RR' + Pd(0) + MX

Oxidative addition-insertion chemistry has found extensive application in the synthesis of heterocycles. Carbon monoxide insertions have been used to make butenolides (equation 22),[37] benzolactams (equation 23)[38] including sendaverine (equation 24),[39] and even β-lactams (equation 25).[40] This same chemistry was used to synthesize benzodiazepines (equation 26)[41] and the yohimbane skeleton (equation 27).[42]

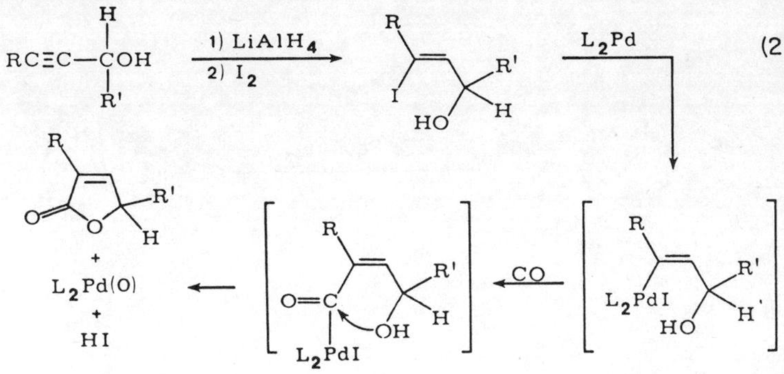

(22)

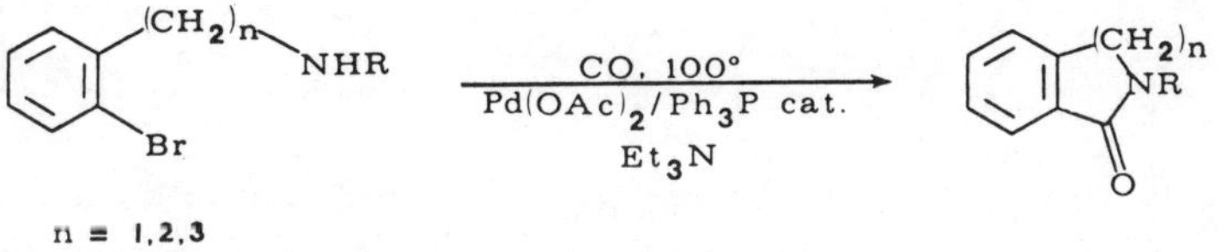

(23)

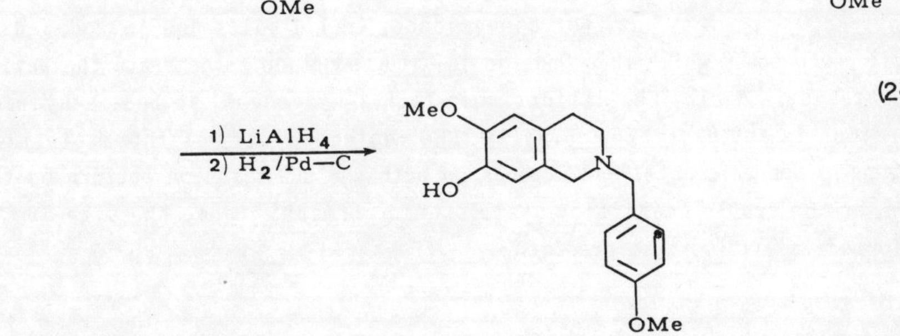

(24)

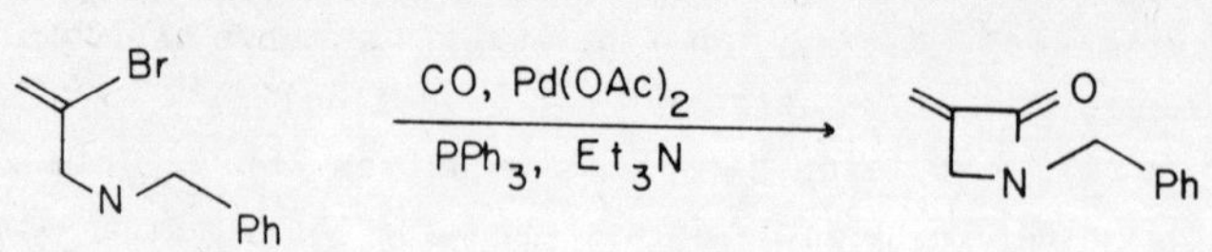

(25)

(26)

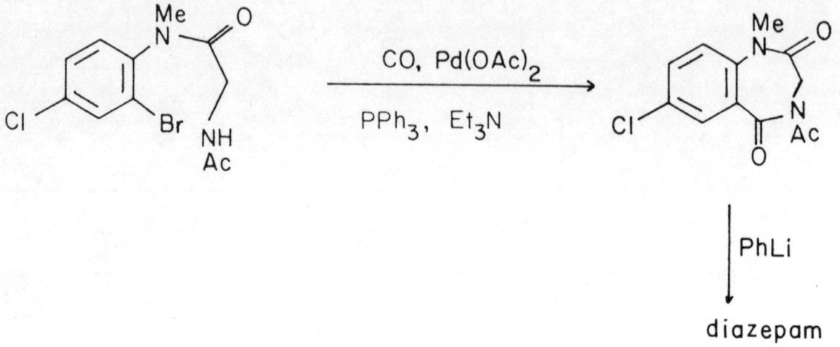

diazepam

(27)

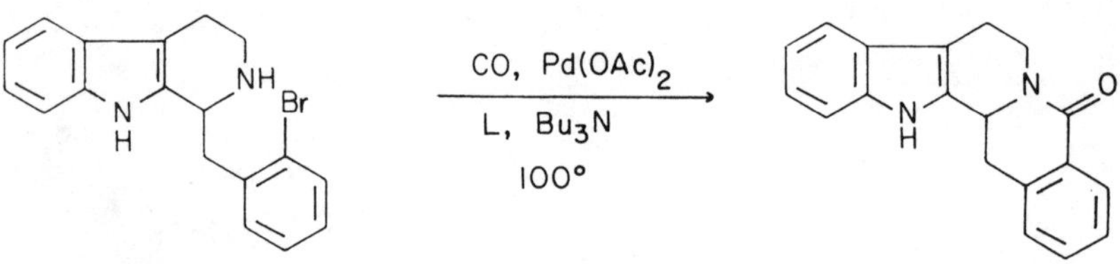

Olefin insertions have also been very useful in the synthesis of heterocycles, including indoles (equations 28[43] 29,[44] and 30[45]) quinolines (equation 31),[44] isoquinolines (equation 32),[43] and macrocyclic lactones (equation 33).[46] Several features of this reaction require comment. These reactions are catalyzed by Pd(0) species, but, for convenience, Pd(II) salts are introduced as catalyst precursors, which then undergo in situ reduction to generate the active Pd(0) catalyst. Again, the halides must lack β-hydrogens, so that β-hydride elimination is not possible. Finally, the regiochemistry of intramolecular insertion is unpredictable, and depends on both the substitution pattern on the olefin, and on subtle transition state steric effects. Hence the size of the ring formed is difficult to predict.

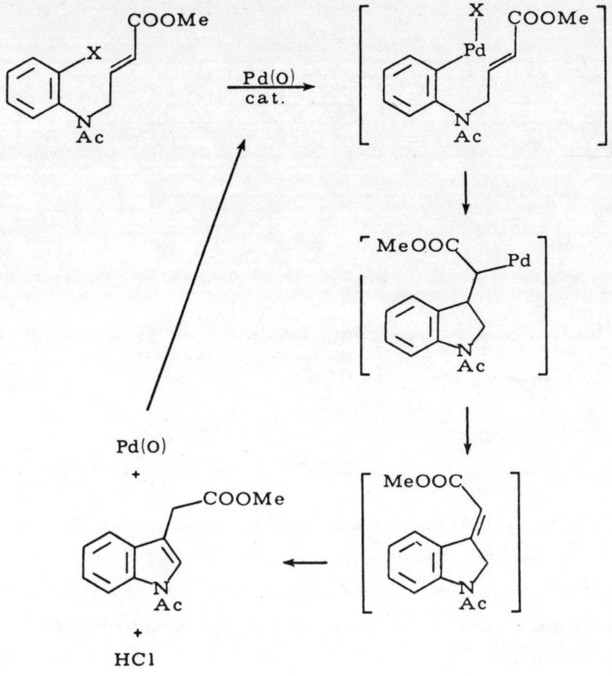

COOMe

X

N
Ac

Pd(O)
cat.

X
Pd
COOMe

N
Ac

MeOOC
Pd

N
Ac

(28)

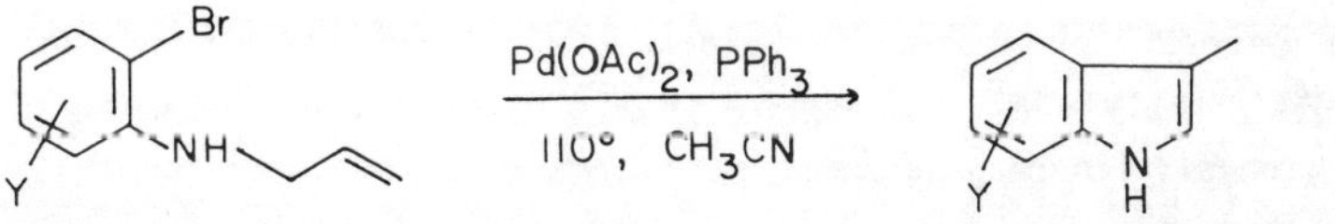

Pd(O)
+

COOMe

N
Ac
+
HCl

MeOOC

N
Ac

Br

Y
NH

$Pd(OAc)_2$, PPh_3

110°, CH_3CN

Y
N
H

(29)

(30)

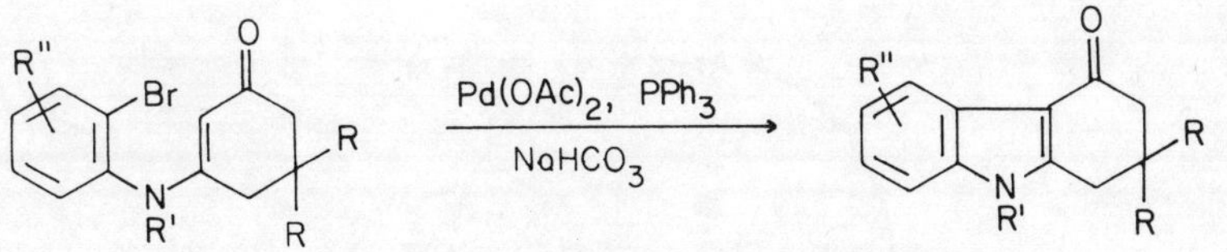

R''
Br
O
R
N
R'
R

$Pd(OAc)_2$, PPh_3

$NaHCO_3$

R''
O
R
N
R'
R

(31)

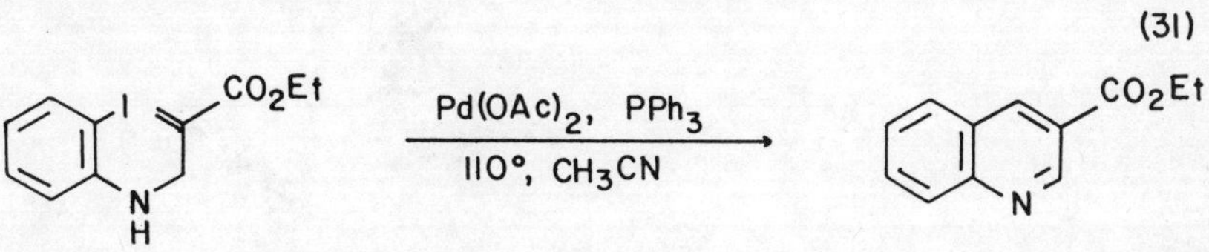

CO_2Et
I
N
H

$Pd(OAc)_2$, PPh_3

110°, CH_3CN

CO_2Et
N

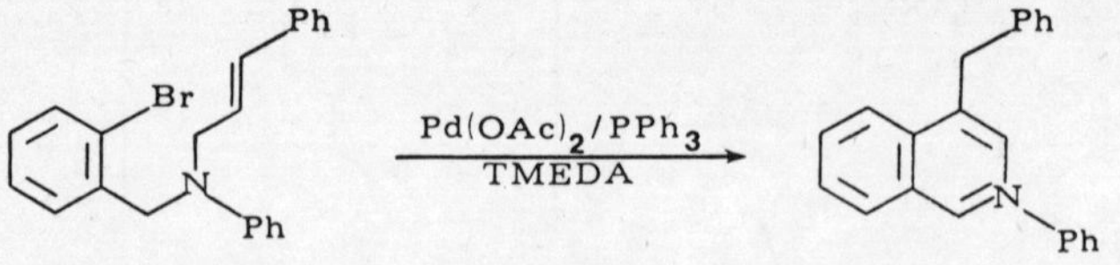

$$\xrightarrow[\text{TMEDA}]{Pd(OAc)_2/PPh_3} \qquad (32)$$

$$\xrightarrow[\text{HCOOH}]{\text{1 eq. } PdCl_2(MeCN)_2} \qquad (33)$$

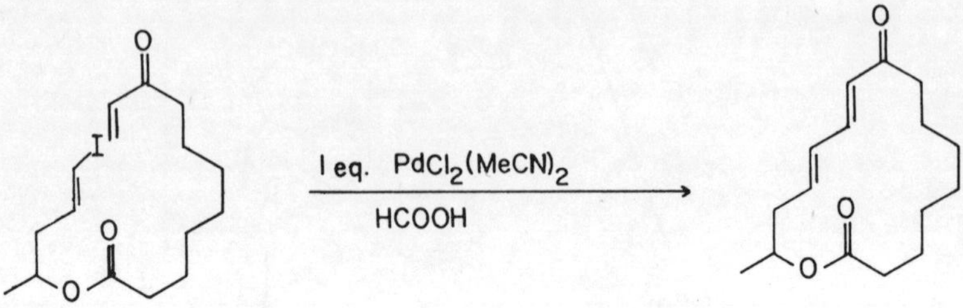

Combining transmetallation with oxidative addition provides a very general procedure for the coupling of aryl or alkenyl halides to virtually any other main group organometallic compound (equation 34). (Organolithium and magnesium halides tend to react poorly in these systems but conversion to the corresponding zinc chloride salt allays this problem.) This has been used in a number of ways. Organotin reagents cleanly alkylate acid halides to produce ketones in the presence of Pd(0) a catalyst (equation 35).[47] This is an extremely general process and tolerates a wide variety of functional groups. Similarly, biallyls are formed by the Pd(0) catalyzed coupling of allylic halides with allytin reagents.[48]

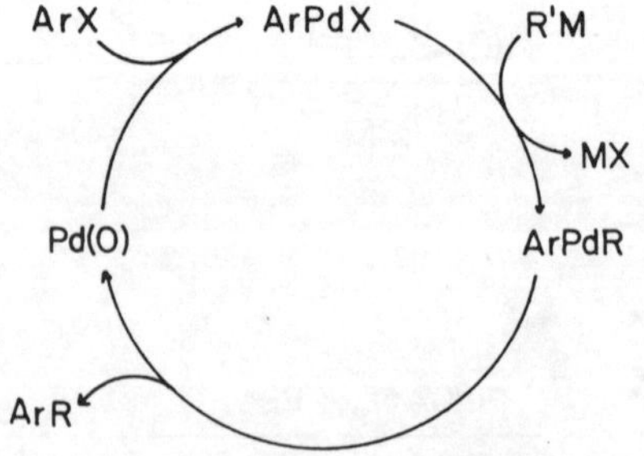

ArX → ArPdX; R'M → MX; ArPdR; ArR; Pd(O) (34)

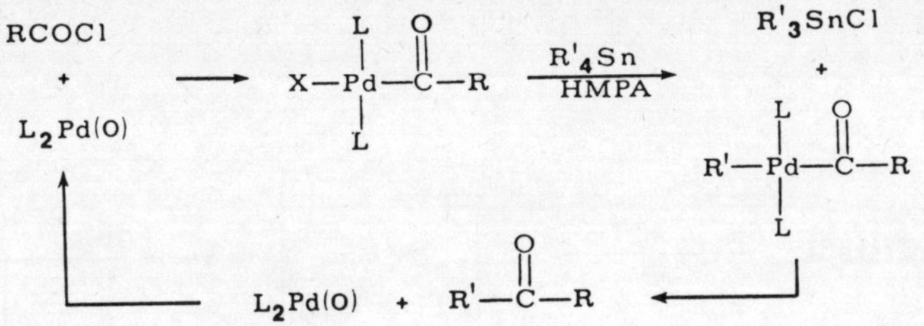

(35)

The real advantage of this procedure is the variety of sources for the main group metal component. For instance, it can come from the hydroboration (equation 36),[49] carboalumination (equation 37),[50] or hydrozirconation of alkynes (equation 38)[51] since boron, aluminum and zirconium cleanly transmetallate to palladium. Grignard reagents do not react cleanly, but conversion to the zinc halide salt permits efficient transmetallation (equation 39).[52] (Generally organozinc halides, whatever their source, efficiently transmetallate to alkylpalladium(II) halides, whatever their source).[53] Zinc halide salts of enol ether and allenic ether anions couple cleanly with aryl and vinyl halides, permitting the direct acylation of these substrates (equations 40 and 41).[54]

(36)

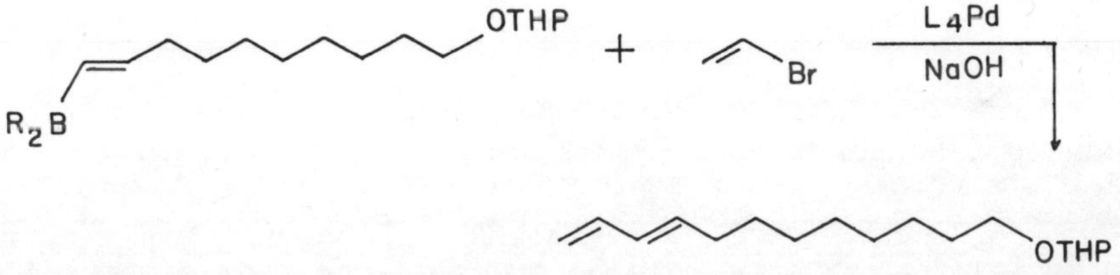

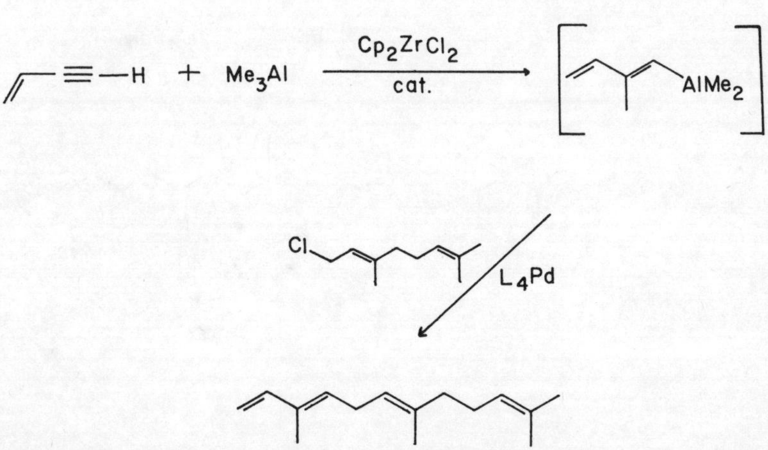

(37)

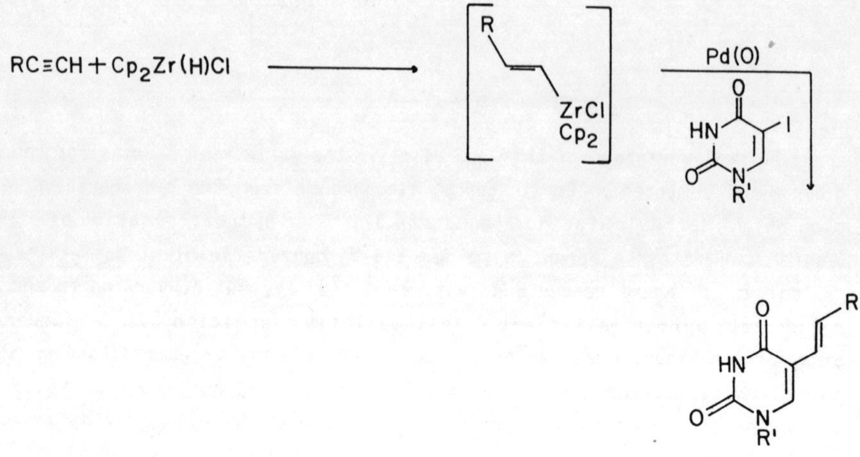

(38)

$RC{\equiv}CH + Cp_2Zr(H)Cl$ ⟶ [R–CH=CH–$ZrCl$ Cp_2] —Pd(0), HN, O, I, N, R'→ O, R, HN, O, N, R'

(39)

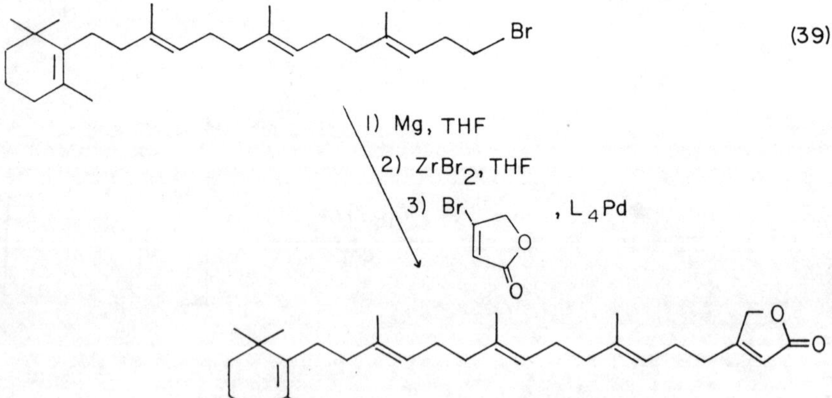

Br

1) Mg, THF

2) $ZrBr_2$, THF

3) Br, O, O, L_4Pd

O

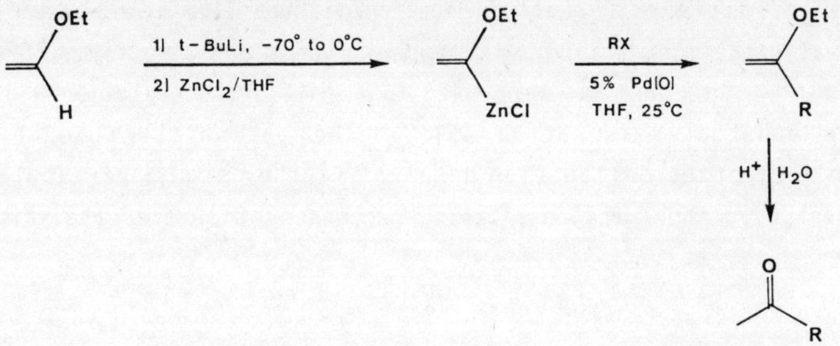

(40)

R = aryl, alkenyl

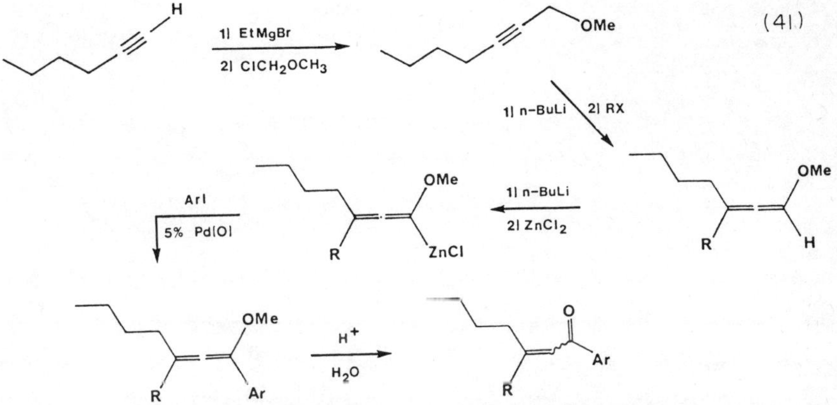

(41)

R = TMS, Ar = m-MePh 74%
R = Me, Ar = m-MePh 60%
R = n-Bu, Ar = m-MePh 50%

D. Coupling Reactions Involving Cyclopalladation Reactions

Appropriately substituted arenes having nitrogen ligands in a benzylic position undergo an electrophilic aromatic substitution with palladium(II) salts to generate stable arylpalladium(II) complexes (equation 42).[55] This is one of the few ways of making arylpalladium compounds directly from the hydrocarbon, and is of great potential use for the selective functionalization of arenes via insertion reactions. Early work studied carbonylation reactions of these "orthopalladation" products, and permitted ortho acylation of aromatics (equation 43).[56] More interesting is the insertion of conjugated enones into cyclopalladation complexes (equation 44).[57] Although five membered palladiacycles are generally the most stable, six member systems can also form. Orthopalladation of N-acyl anilines has been used to permit ortho acylation and alkylation of these substrates (equation 45).[58] The potentially useful chemistry accessible by ortho palladation deserves further study, particularly in the development of ortho functionalization processes which are catalytic in palladium.

(42)

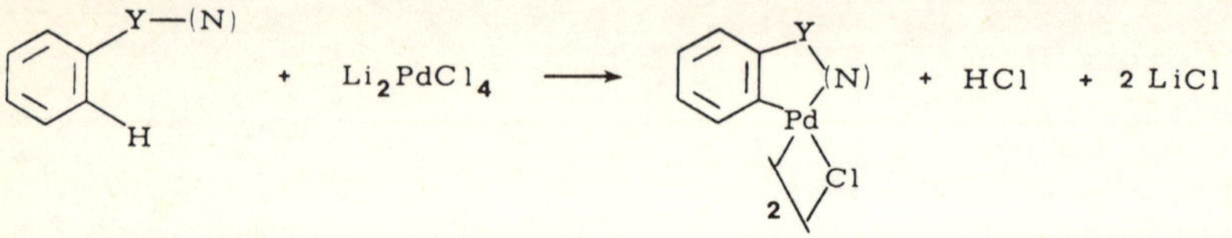

Y—(N) + Li_2PdCl_4 ⟶ (N) Pd Cl 2 + HCl + 2 LiCl

Y—(N) = $—CH_2NMe_2$, —CH=NOH, CH=NR, CH=N—NHR, 2-pyridyl, —N=NPh

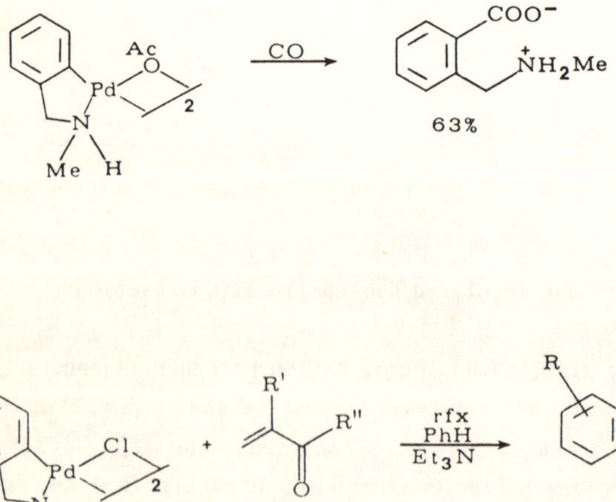

(43)

Pd Ac O 2 N Me H —CO→ COO^- $\overset{+}{N}H_2Me$ 63%

(44)

R Pd Cl 2 N + R' R'' O —rfx PhH Et_3N→ R O R'' R' N

(45)

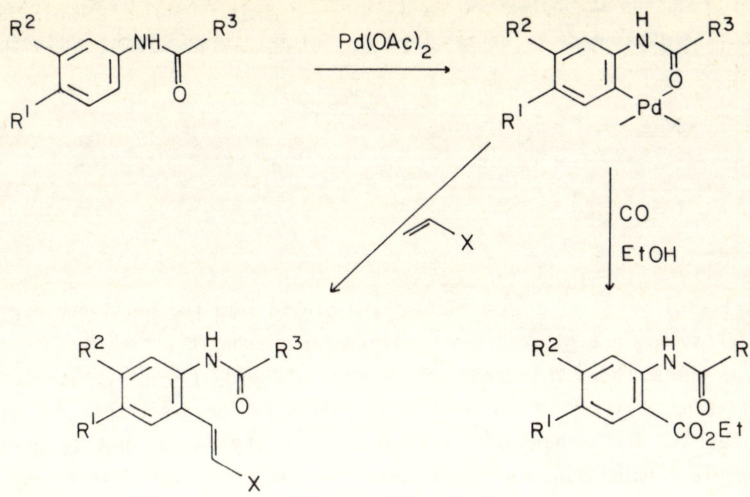

$Pd(OAc)_2$

CO

EtOH

IV. REACTIONS OF GROUP VIII METAL OLEFIN COMPLEXES

Virtually every transition metal forms complexes with olefins, but the chemical behavior of the complexed olefin is strongly dependent upon the particular metal to which it is complexed. Of the group VIII metals, palladium(II) and iron(II) have the most useful olefin chemistry because olefins complex rapidly to these metals, the complexed olefins are subject to nucleophilic attack (a reversal of their normal chemistry), and the thus-formed organic products are readily freed from the metal.

A. Olefinpalladium(II) Complexes

Palladium(II) salts have been extensively used to promote nucleophilic attack on olefins. Olefins complex very rapidly to palladium(II) salts (as nitrile solvates for solubility) in organic solvents, with the general order of olefin reactivity being: || > > > > >> . Enol esters and enol ethers complex strongly, while electron deficient olefins such as acrylonitrile or conjugated enones complex only weakly. Once complexed to palladium the olefin becomes very reactive towards nucleophiles, with nucleophiles ranging from Cl^- to Ph^- (a range of $\approx 10^{36}$ in basicity) capable of attacking the complexed olefin. Two modes of nucleophilic attack have been observed (equation 46). One involves external nucleophilic attack, without prior coordination of the nucleophile to the metal. This is a trans addition, with the nucleophile attacking the olefin from the face opposite the metal.

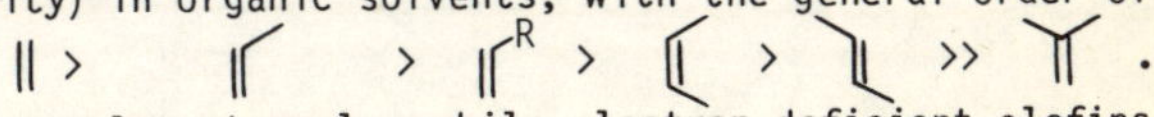

This mode of reaction strongly favors attack at the most substituted position of the olefin. The other mode of attack involves initial attack of the nucleophile

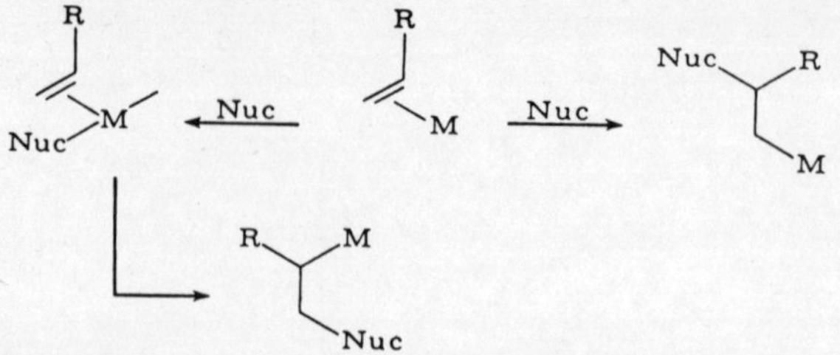

(46)

on the metal, followed by an insertion of the olefin into the metal-nucleophile bond. This results in a cis addition, with the nucleophile attacking from the same face as the metal. This mode of reaction strongly favors attack at the least substituted position of the olefin regardless of the electronic nature of the substituents. Both modes of attack produce σ-alkylpalladium(II) species which undergo β-elimination, hydrogenalysis, insertions, or oxidative cleavages, resulting in difunctionalization of the olefin (Figure 7). All of these processes ultimately produce palladium(0), and by developing in situ processes for reoxidation to palladium(II) all of these processes may be made catalytic in palladium.

Figure 7. Reactions of σ-Alkylpalladium(II) Complexes

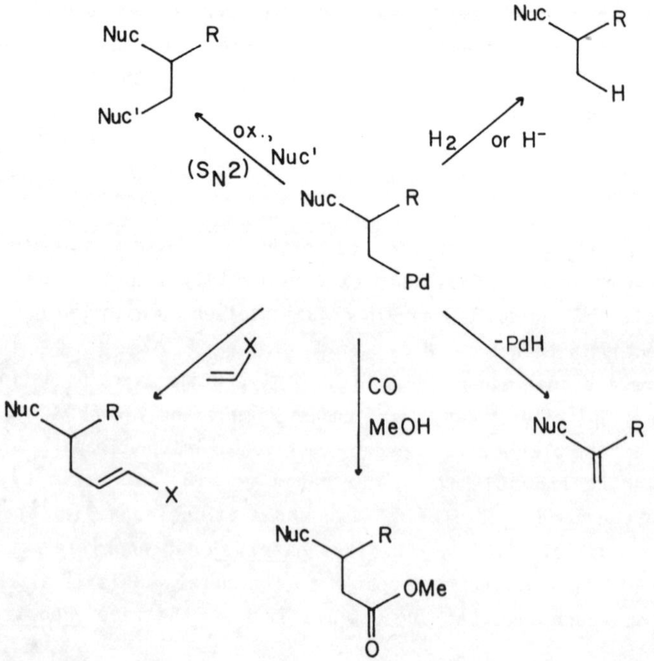

Early studies centered on the use of oxygen nucleophiles such as water, acetate, and alcohols, and led to the development of the commercially important "Wacker" oxidation processes for the production of acetaldehyde, vinyl acetate and enol ethers from ethylene.[59] This reaction is very sensitive to the nature of the nucleophile, and extension to nitrogen and carbon nucleophiles was slow. Under very mild conditions (-50°C, THF, 3 eq of amine) olefins undergo amination.[60] Terminal olefins react in highest yield. Secondary amines aminate olefins in almost quantitative yield, while primary amines react in somewhat lower yield (40-50%). The reaction proceeds through a σ-alkylpalladium complex which can be reduced to give the amine,[60] treated with CO and oxidatively cleaved to give β-aminoacid derivatives,[61] oxidatively cleaved to give oxamination or diamination products,[62] or, with adducts of primary amines, oxidatively cleaved to give aziridines[63] (Figure 8). The amination requires three equiva-

Figure 8. Amination of Olefins

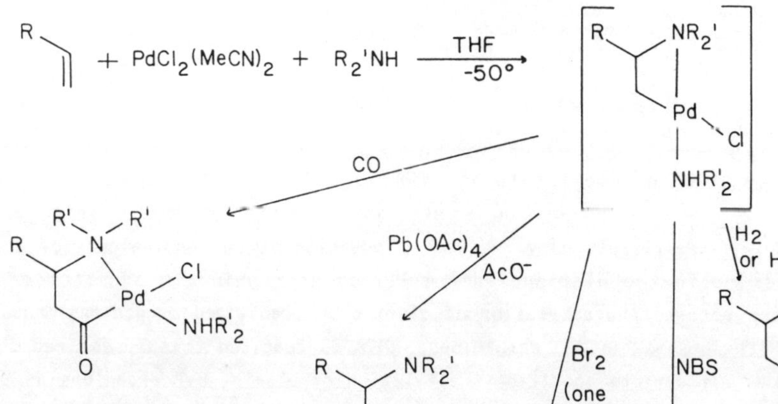

lents of amine, and the step involving carbon-nitrogen bond formation involves nucleophilic attack on a palladium-olefin species having at least one amine already in its coordination sphere. By using chiral tetriary amines as the initial ligand in this process, oxamination of olefins could be carried out with up to 60% ee. Optically active β-blockers were synthesized in this manner (equation 47).[64]

(47)

PhO–CH₂CH=CH₂ + $PdCl_2(MeCN)_2$ + 2 R^*NMe_2 ⟶ Me_2NH ⟶ ox., HOAc ⟶ PhOCH₂C*H(OAc)CH₂NMe₂

45% yield

11% ee

Alkylation of olefins is an even more difficult process, since most carbanions are strong reducing agents, and reduction of olefinpalladium(II) complexes is relatively easy. However, addition of two equivalents of a tertiary amine to the olefinpalladium(II) complex prior to addition of the carbanion permits the alkylation of olefins by stabilized carbanions (equation 48). With less stabilized carbanions, HMPA is required to suppress reduction, but under appropriate conditions, alkylation of olefins by carbanions as basic as phenyl- and benzyllithium is possible (equation 49).[65] The regiochemistry of alkylation is different for stabilized and nonstabilized carbanions, suggesting a change in mechanism. The process procedes through a σ-alkylpalladium complex, and, again CO insertion is possible, permitting the facile "carboacylation" of olefins.[66]

Intramolecular versions of these alkylation, amination, and alkoxylation reactions of olefins are particularly useful in organic synthesis. A variety of benzofused heterocycles are available from palladium(II) catalyzed cyclizations of unsaturated aromatic systems (equation 50). This is a very general process

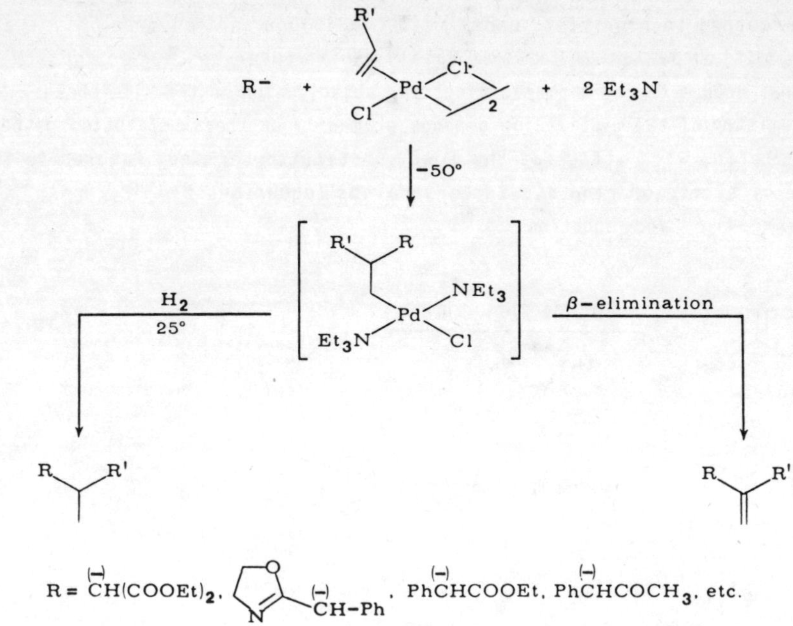

$$\text{R}^- + [(\text{R}'\text{CH=CH}_2)\text{PdCl}_2]_2 + 2\ \text{Et}_3\text{N} \xrightarrow{-50^\circ} [\text{R}'\text{CH(R)CH}_2\text{Pd(NEt}_3)_2\text{Cl}] \tag{48}$$

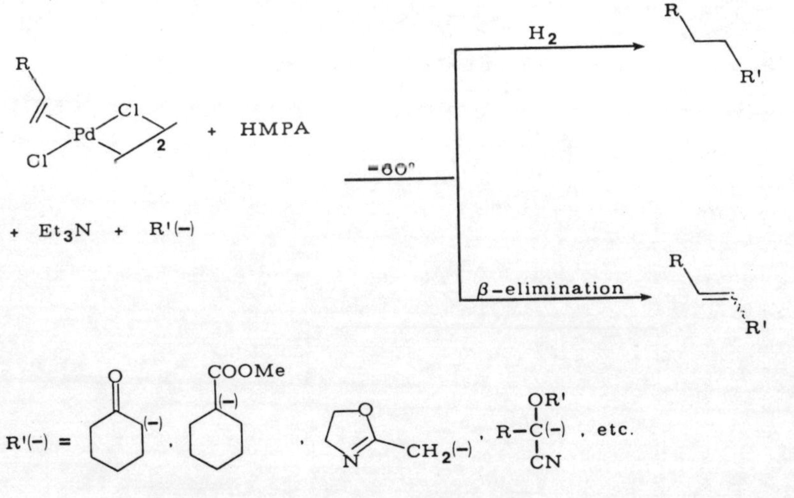

H_2, 25° → R–CH(CH$_3$)–R′

β-elimination → R–C(=CH$_2$)–R′

R = $\overset{(-)}{\text{C}}$H(COOEt)$_2$, 2-oxazolinyl-$\overset{(-)}{\text{C}}$H–Ph, Ph$\overset{(-)}{\text{C}}$HCOOEt, Ph$\overset{(-)}{\text{C}}$HCOCH$_3$, etc.

$$[(\text{RCH=CH}_2)\text{PdCl}_2]_2 + \text{HMPA} + \text{Et}_3\text{N} + \text{R}'^{(-)} \xrightarrow{-60^\circ} \tag{49}$$

H_2 → RCH$_2$CH$_2$R′

β-elimination → RCH=CHR′

R′(–) = 2-oxocyclohexyl(–), 1-(COOMe)cyclohexyl(–), 2-oxazolinyl-CH$_2$(–), R–C(–)(OR′)(CN), etc.

$$\text{X-C}_6\text{H}_3(\text{CH}_2\text{CH=CH}_2)(\text{YH}) \xrightarrow[\text{THF, benzoquinone}]{1\%\ \text{PdCl}_2(\text{MeCN})_2,\ \text{LiCl}} \text{2-methylbenzo-fused heterocycle (X, Y)} \tag{50}$$

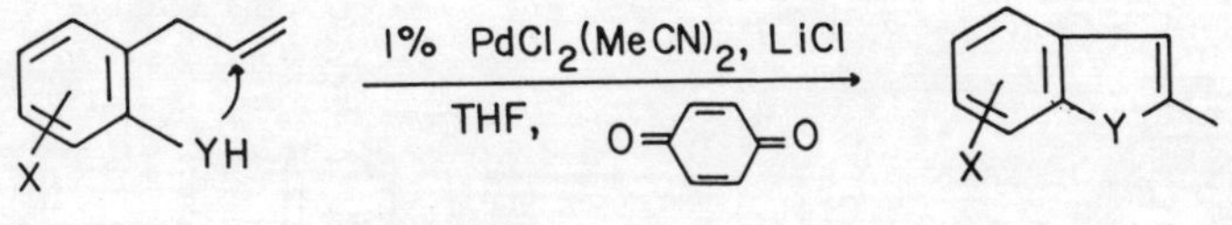

Y = O,[67] NH,[68] NAc,[68] CO$_2$,[69] CONH,[69] CONR[69]

which proceeds in high yield under mild conditions and requires only small amounts (1%) of palladium salts as catalyst. It tolerates a wide variety of functional groups in the aromatic ring, and substitution on the side chain. The regiochemistry of the cyclization depends primarily on the substitution pattern on the olefin, with attack at the most substituted terminus favored in the absence of steric or ring strain constraints (equations 51, 52, and 53),[68] (equation 54),[70] and (equation 55).[71]

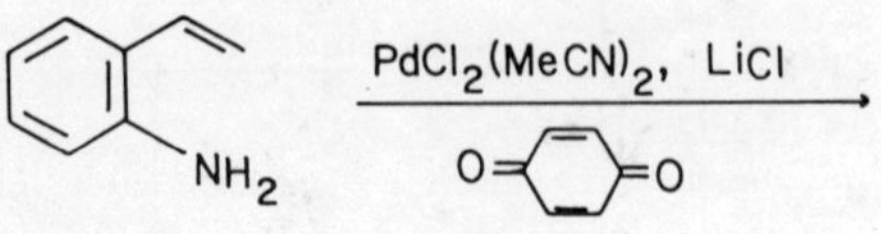

$PdCl_2(MeCN)_2$, LiCl (51)

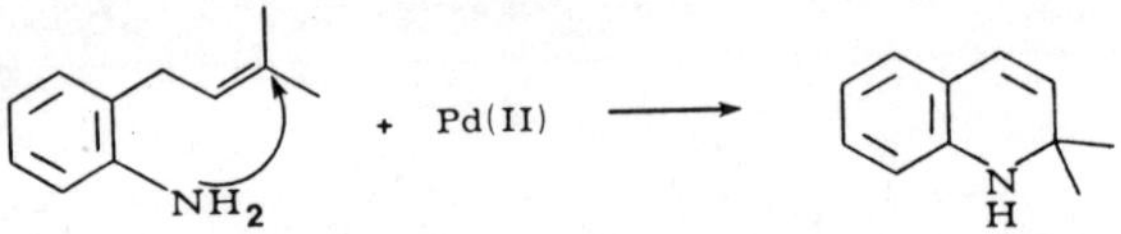

+ Pd(II) (52)

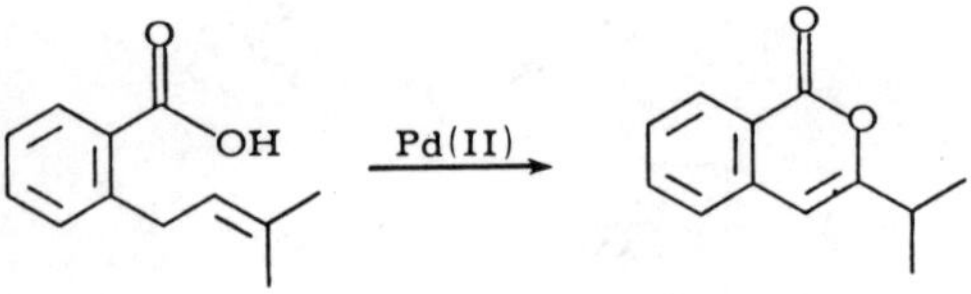

Pd(II) (53)

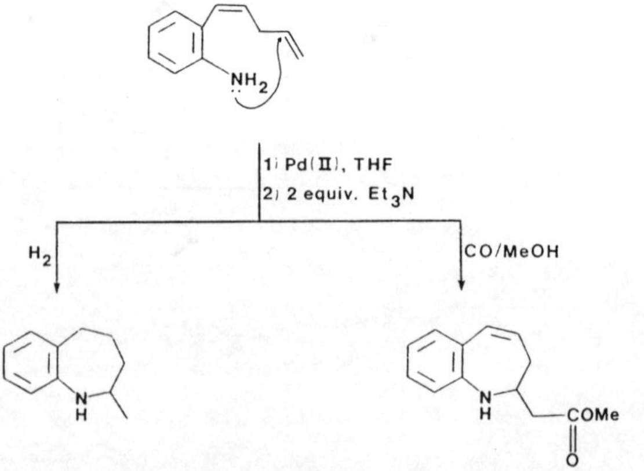

1) Pd(II), THF
2) 2 equiv. Et_3N (54)

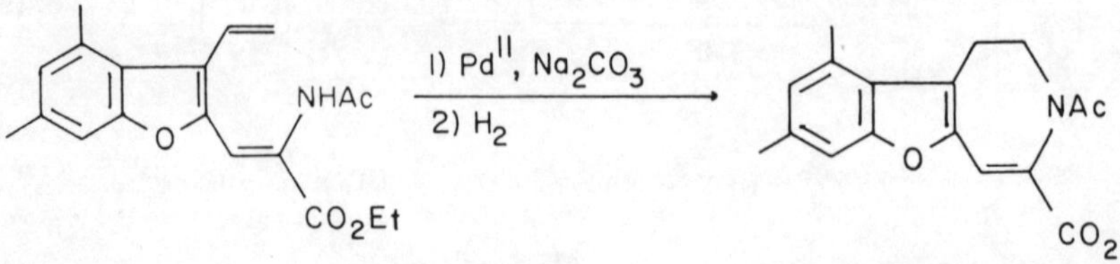

H_2 CO/MeOH

1) Pd^{II}, Na_2CO_3
2) H_2 (55)

This reaction is not restricted to benzofused systems. Provided that the nucleophile does not coordinate strongly to palladium(II), most unsaturated systems can be cyclized. Thus 2-vinyltetrahydrofurans form from β,γ-unsaturated alcohols,[72] lactones from 3- and 4-alkenoic acids,[73] pyrones from conjugated dienoic acids,[74] isoxazoles from α,β-unsaturated oximes,[75] 2-pyridones from dienic acid amides,[76] 3-pyrazolones from α,β-unsaturated acid hydrazides,[77] and uracil from unsaturated urea derivatives.[78] In contrast, simple aminoolefins do not cyclize; rather they coordinate too strongly to palladium(II) and essentially poison the catalyst. However, their corresponding tosamides cyclize readily to produce tosylated enamines which are easily reduced and detosylated (equation 56).[79] Again, all of these intramolecular aminations procede through σ-alkylpalladium complexes which can undergo insertion reactions (equations 57 and 58).[80]

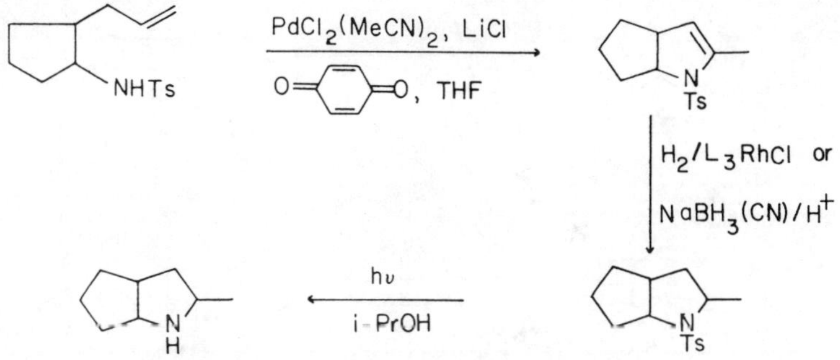

NHTs $\xrightarrow[\text{O=C}_6\text{H}_4\text{=O, THF}]{PdCl_2(MeCN)_2,\ LiCl}$ N-Ts (56)

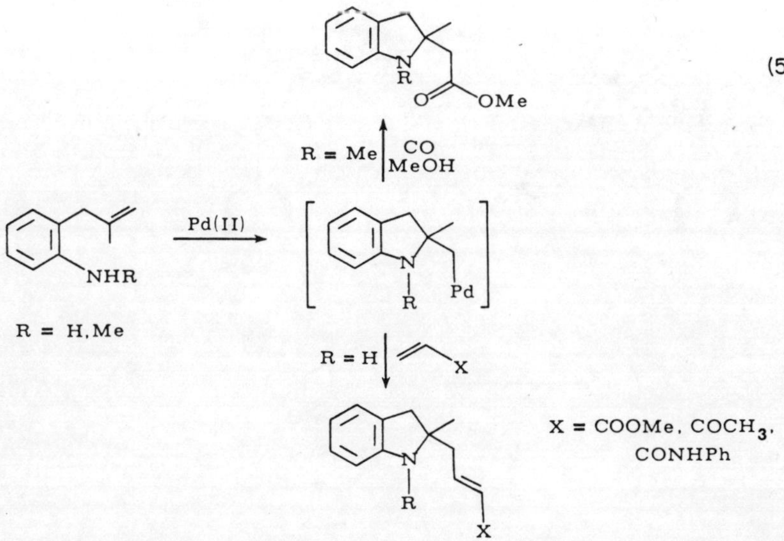

H_2/L_3RhCl or $NaBH_3(CN)/H^+$

N-H $\xleftarrow[\text{i-PrOH}]{h\nu}$ N-Ts

(57)

OMe

R = Me | CO, MeOH

NHR $\xrightarrow{Pd(II)}$ [N-R, Pd]

R = H, Me

R = H | X

X = COOMe, COCH$_3$, CONHPh

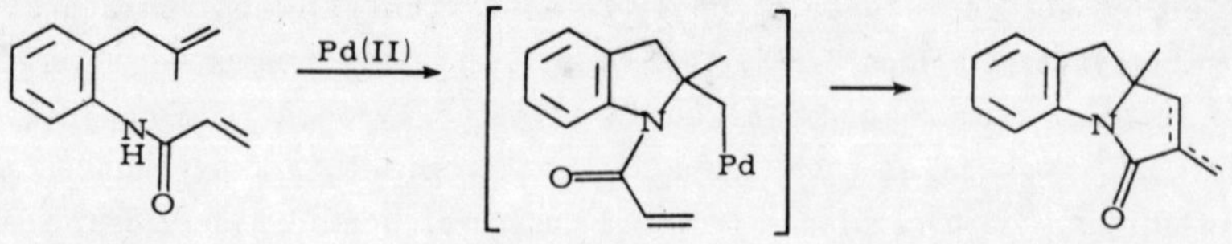
Pd(II)
N
H
O
N
O
Pd
N
O

(58)

Palladium-promoted intramolecular alkylations are considerably less common, but four recent examples (equations 59 and 60)[81] (equation 61 and 62)[82] illustrate the potential of this process for the synthesis of complex carbocycles.

(59)

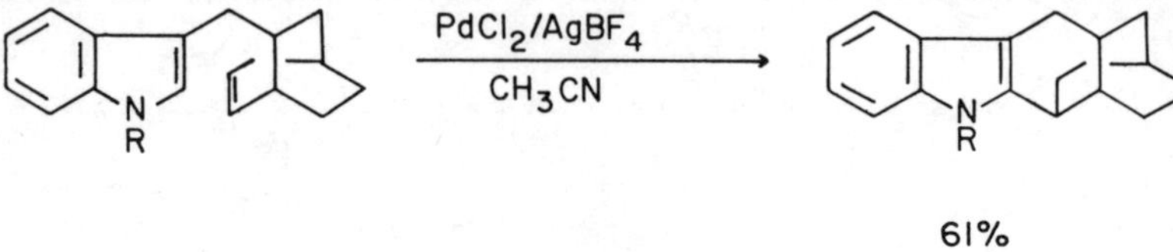
PdCl2/AgBF4
CH3CN
N
R
N
R
61%

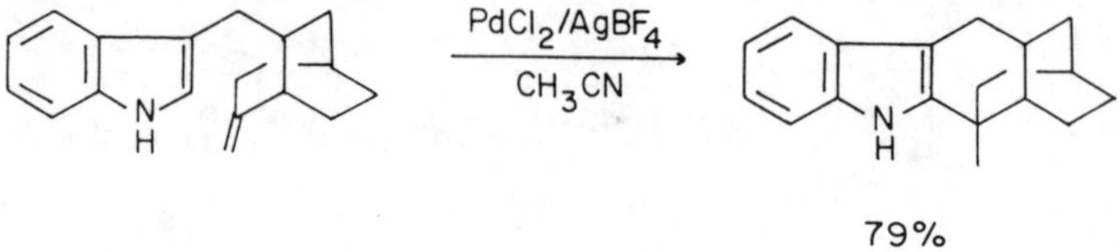
PdCl2/AgBF4
CH3CN
N
H
N
H
79%

(60)

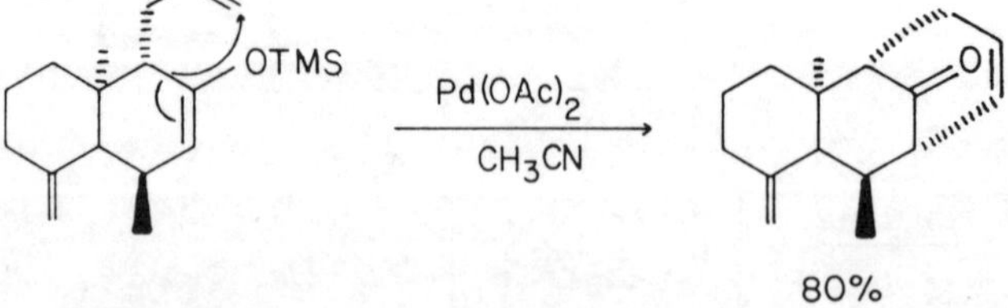
OTMS
Pd(OAc)2
CH3CN
O
80%

(61)

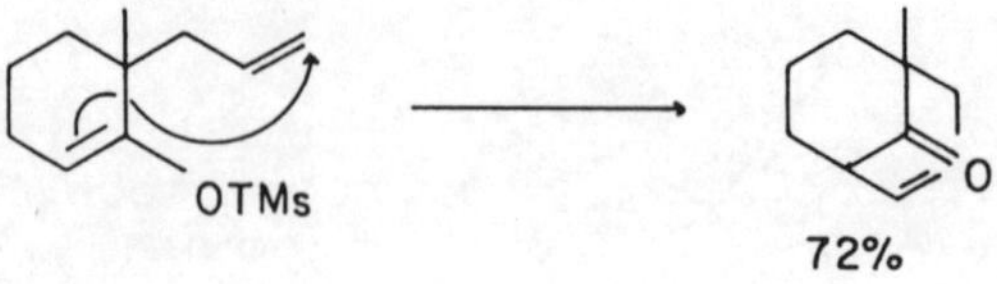
OTMs
O
72%

(62)

B. Olefiniron(II) Complexes

The most generally useful olefin complexes of iron are those containing the (η^5-cyclopentadienyl)Fe(CO)$_2^+$ group (abbreviated Fp$^+$). These are relatively easily synthesized, react with a wide variety of nucleophiles, and the resulting stable (in contrast to palladium) σ-alkyl complexes undergo further useful reactions. Alkoxides, mercaptans, phosphines, and phosphites all attack these cationic iron-olefin complexes to give relatively stable adducts,[83] as do a wide range of carbanions including phosphorous ylides and organocuprates (equation 63).[84]

[Scheme: $[\text{CpFe(CO)}_2(\text{CH}_2{=}\text{CHR})]^+$ + Nuc ⟶ CpFe(CO)$_2$CH$_2$CH(R)Nuc] (63)

R=H, Me, Ph, CHO, CH$_2$OMe

Nuc = MeO$^-$, RS$^-$, Ph$_3$P, (RO)$_3$P, R$_2$NH, R(-)

A problem with many of these is that the σ-alkyl complexes are too stable, and removal of the iron group is difficult. Oxidative cleavage is often used, but frequently unexpected reactions occur. This is the basis of a very clever β-lactam synthesis in which an oxidatively-driven CO insertion plays a key role (equations 64,[85] and 65[86]).

[Scheme: $[\text{CpFe(CO)}_2(\text{CH}_2{=}\text{CHR})]^+$ + PhCH$_2$NH$_2$ ⟶ CpFe(CO)$_2$CH$_2$CH(R)NHCH$_2$Ph; $\xrightarrow{\text{Ag}_2\text{O}}$ [CpFe$^+$(CO)(C(=O)CH$_2$CH(R)N(H)CH$_2$Ph)] ⟶ N-PhCH$_2$ β-lactam bearing R] (64)

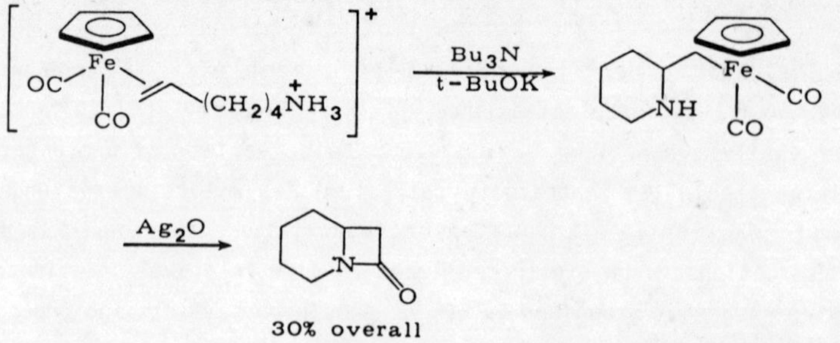

(65)

30% overall

Enol ethers complex well to the Fp^+ group and are also generally reactive toward nucleophiles. These complexes can be viewed as vinyl cation equivalents, and offer a procedure for the introduction of vinyl groups α to carbonyl groups (equation 66).[87] The enol ether of pyruvic ester was used in the synthesis of α-methylene lactones (equation 67).[88]

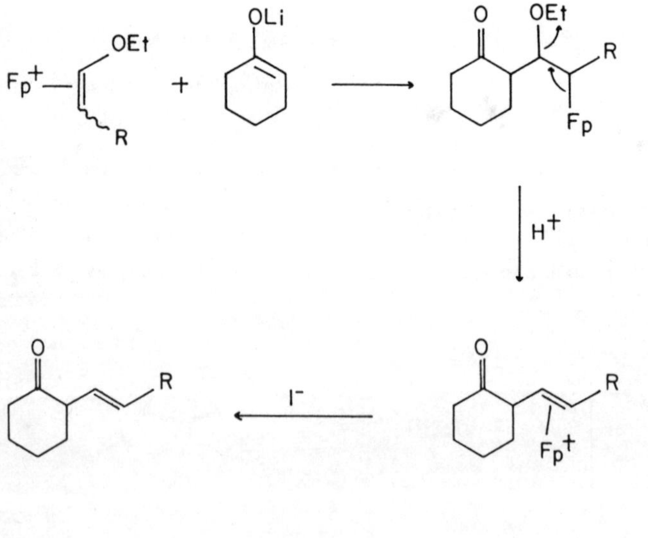

(66)

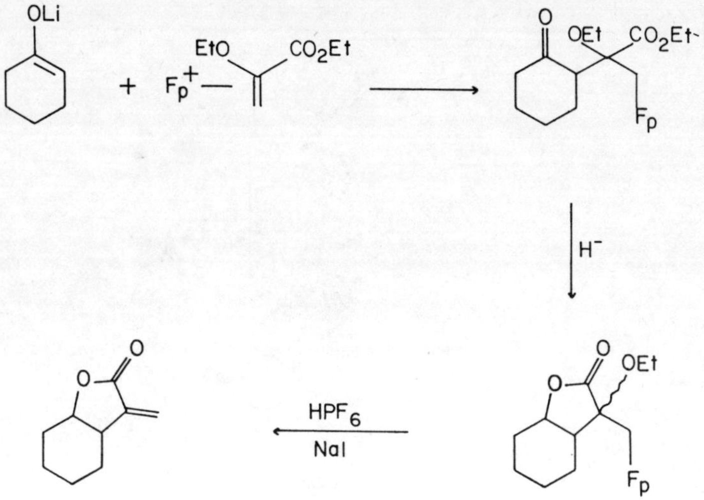

(67)

C. π-Dienyliron Complexes

Iron carbonyl reacts with conjugated dienes to form extremely stable complexes of the form (η^4-diene)$Fe(CO)_3$. These are air-stable yellow solids or liquids which are inert to catalytic hydrogenation, unreactive toward maleic anhydride in a Diels-Alder reaction, and undergo Friedel Crafts acylation without suffering decomplexation from the iron. Not remarkably, these complexes have found little use in organic synthesis. However, diene complexes of this type having allylic hydrogens react readily with hydride abstractors such as $Ph_3C(+)$ to produce cationic dienyliron complexes which are highly reactive toward a wide range of nucleophiles.[89] Cyclohexadienyl complexes have been most extensively studied and the nucleophiles which react range from organolithium, copper, cadmium, and -zinc reagents, through ketone enolates, malonates, and nitroalkyl anions, to nucleophilic aromatic compounds such as indoles and polyalkoxybenzenes (equation 68).[90] The resulting substituted diene can be released from the iron by oxidation with amine oxides. The nucleophile always attacks at a terminus of the dienyl system, and always from the face opposite the iron. Complexes of methoxycyclohexadienes (from Birch reduction of anisoles) are particularly useful since they undergo exclusive attack at the dienyl terminus remote from methoxy group, and thus serve as synthetic equivalents to various cyclohexenone cations (equations 69[89a] and 70).[91]

$$\text{1,3-cyclohexadiene} + Fe_2(CO)_9 \longrightarrow (C_6H_8)Fe(CO)_3 \xrightarrow{Ph_3C^+} (C_6H_7)^+Fe(CO)_3 \xrightarrow{Nuc} \text{Nuc-}(C_6H_7)Fe(CO)_3 \quad (68)$$

$$\text{(1-OMe-4-Me-}C_6H_5)^+Fe(CO)_3 + R(-) \longrightarrow \text{(1-OMe-4-Me-4-R-diene)}Fe(CO)_3 \xrightarrow[2)\ H_3O^+]{1)\ Me_3NO} \text{4-Me-4-R-cyclohex-2-enone} \quad (69)$$

R(-) = CN, $CH(CO_2R)_2$, etc.

$$\text{(2-OMe-}C_6H_6)^+Fe(CO)_3 + R(-) \longrightarrow (CO)_3Fe\text{(OMe-diene-R)} \xrightarrow[2)\ H_3O]{1)\ Me_3NO} \text{5-R-cyclohex-2-enone} \quad (70)$$

R(-) = t-Bu, $CH_2{=}C(OLi)CH_3$, $CH_2{=}C(OTMS)CH_3$

Very recently π-cyclohexadienyl complexes have found extensive use in the synthesis of natural products. Although diverse compounds have been made using this methodology, they all rely upon the nucleophilic attack of a methoxydienyl-iron complex as a key step (equations 71,[92] 72,[93] and 73[94]).

(71)

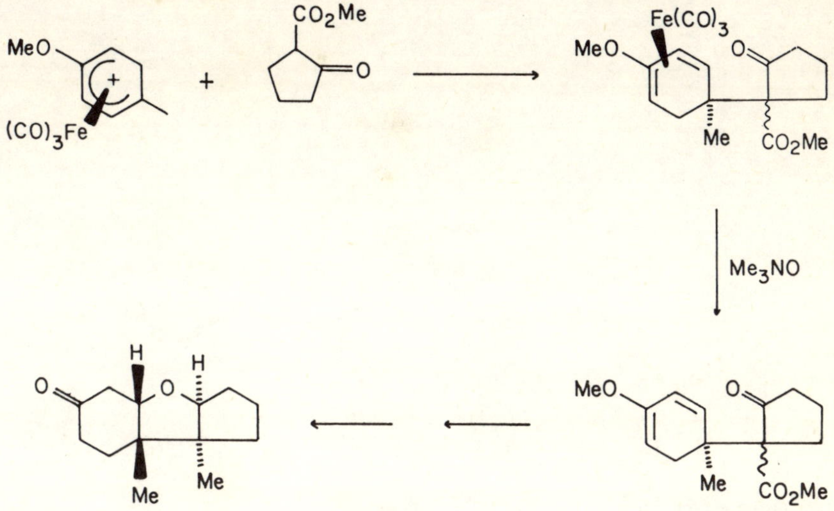

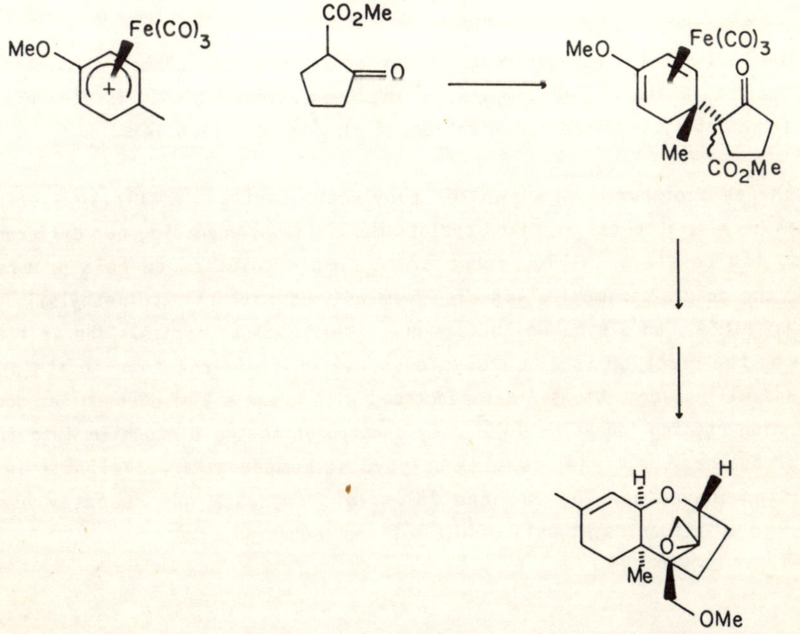

(72)

(73)

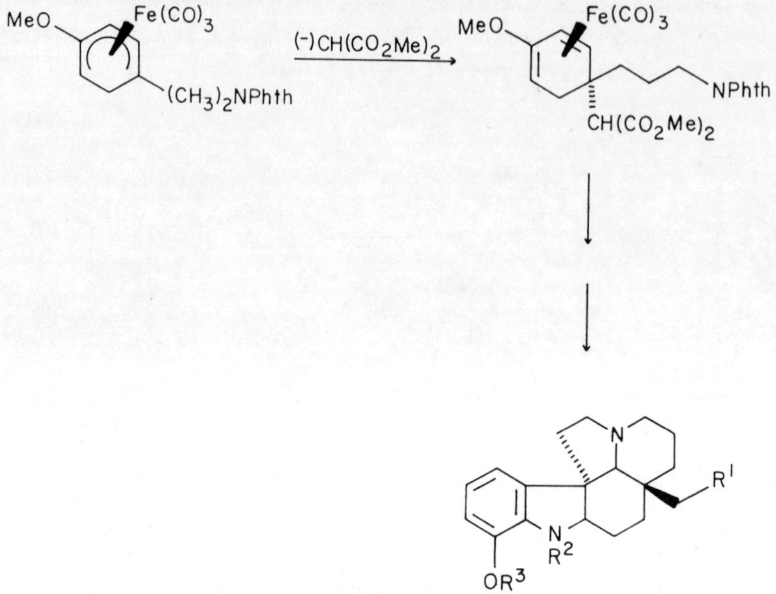

V. REACTIONS OF GROUP VIII METALS WITH ALKYNES

Group VIII metals are very reactive toward alkynes, and either oligomerize or cyclotrimerize them. The complex (η^5-cyclopentadienyl)$Co(CO)_2$ is very efficient in the catalytic cyclotrimerization of alkynes (equation 74).

This reaction has been known for many years, but was of little use for synthesis because attempts at mixed cyclotrimerizations, involving two different alkynes, led to all possible products. A simple solution to this problem involves the co-cyclotrimerization of 1,5-hexadiynes with bis-(trimethylsilyl)-acetylene (BTMSA), an alkyne which does not itself cyclotrimerize. The initial product of the reaction is a benzocyclobutane, which is reactive in its own right, and will undergo Diels-Alder reactions with a variety of dienophiles upon thermal ring opening (equation 75).[95] By incorporating the dienophile into the same molecule as the diyne, complex polycyclic compounds are available in a single step (equations 76, 77, and 78). This approach has recently been employed in an elegant synthesis of d,ℓ estrone (equation 79).[96]

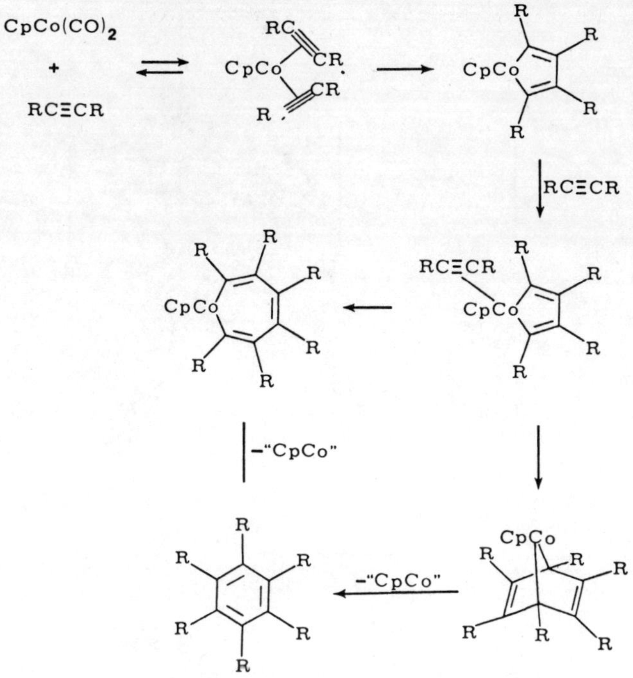

$CpCo(CO)_2$ + RC≡CR ⇌ CpCo(RC≡CR)(RC≡CR) → CpCo(cyclopentadiene, R₄) —RC≡CR→ (RC≡CR)CpCo(R₄ ring) → CpCo(R₆ ring) —–"CpCo"→ C_6R_6; (RC≡CR)CpCo(R₄ ring) → CpCo bicyclic (R₆) —–"CpCo"→ C_6R_6

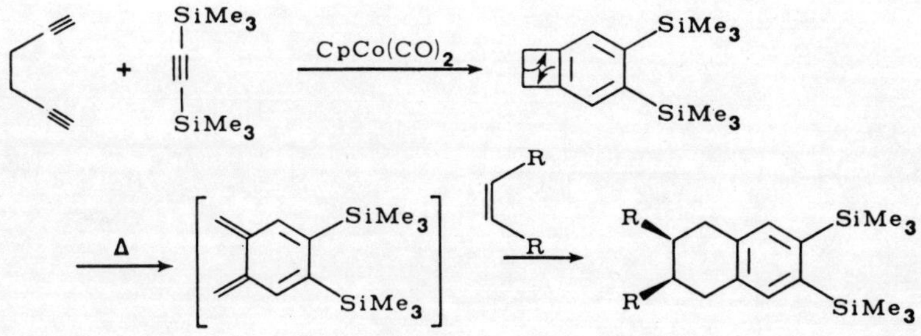

(74)

(hexa-1,5-diyne) + $Me_3SiC≡CSiMe_3$ —$CpCo(CO)_2$→ 4,5-bis(SiMe₃)benzocyclobutene —Δ→ [bis(SiMe₃)-o-xylylene] —RCH=CHR→ 2,3-R₂-6,7-bis(SiMe₃)tetralin

(75)

HC≡C– / HC≡C– + $Me_3SiC{\equiv}CSiMe_3$ —$CpCo(CO)_2$→

Me_3Si Me_3Si ←— Me_3Si Me_3Si (76)

Me_3Si H H Me_3Si 80%

HC≡C– HC≡C– —1) BTMSA $CpCo(CO)_2$ 2) Δ→ Me_3Si H H Me_3Si 60% (77)

HC≡C– HC≡C– N OMe —1) BTMSA $CpCo(CO)_2$ 2) Δ→ Me_3Si H H N OMe Me_3Si 45% (78)

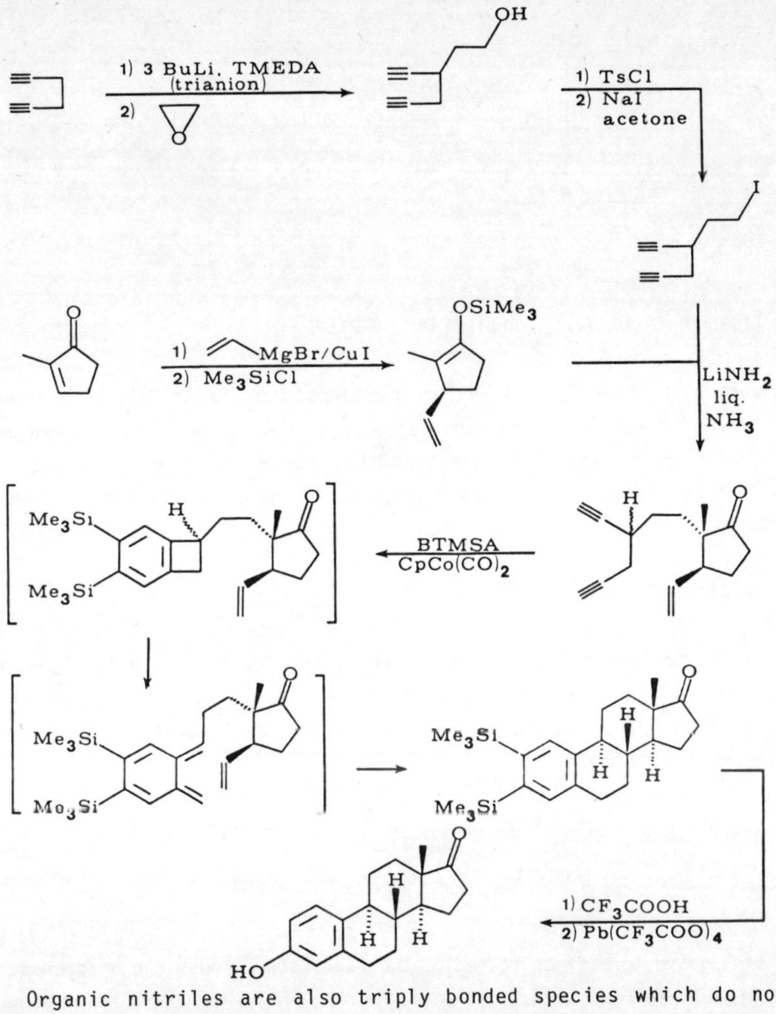

(79)

Organic nitriles are also triply bonded species which do not self-trimerize, but can be cotrimerized with alkynes to produce pyridines. With unsymmetrical alkynes, however, all possible isomers are obtained. On the other hand, when combined with diynes, the isoquinoline or quinoline nucleus is formed (equations 80 and 81).[95,97]

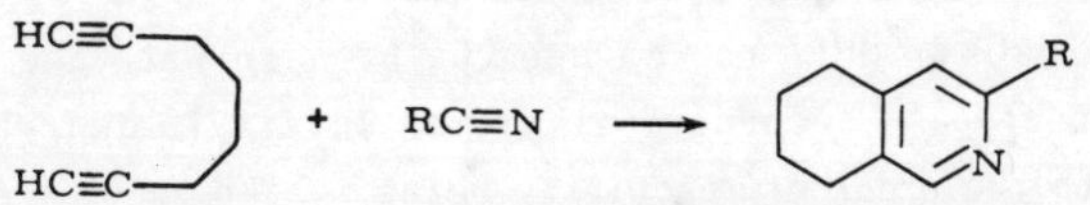

$$HC{\equiv}C{-}(CH_2)_4{-}C{\equiv}CH + RC{\equiv}N \longrightarrow \text{5,6,7,8-tetrahydro-3-R-isoquinoline} \qquad (80)$$

$$\text{(CH}_2\text{)}_n\text{(C≡CH)(C≡N)} + \text{RC≡CR} \xrightarrow[h\nu]{CpCo(CO)_2} \text{(CH}_2\text{)}_n\text{-fused 2,3-dialkylpyridine} \quad (81)$$

n = 3, 4, 5

VI. REACTIONS OF GROUP VIII π-ALLYLMETAL COMPLEXES

Although π-allylmetal complexes are known for most Group VIII metals, only the halo-bridged dimers of nickel and palladium, (π-allyl $MX)_2$, have found much use in organic synthesis. Although these halo-bridged dimers of nickel and palladium are isoelectronic and isostructural, they have very different chemical properties.

A. π-Allylnickel Halides[98]

π-Allylnickel halides are prepared by the reaction of allylic halides with nickel carbonyl or bis-(cyclooctadiene)nickel in nonpolar solvents such as benzene (equation 82). Functional groups including alkyl, aryl, vinyl (noncon-

$$\text{R-allyl-X} + \begin{matrix} Ni(CO)_4 \\ \text{or} \\ Ni(COD)_2 \end{matrix} \xrightarrow{PhH} (\pi\text{-R-allyl})Ni(\mu\text{-X})_2Ni(\pi\text{-R-allyl}) \quad (82)$$

jugated, conjugated, cross-conjugated), ether and ester groups can be present in the allyl group, and quite complex π-allylnickel halides, including those of steroids, have been made. The complexes are red-to-brown crystalline solids, quite air sensitive in solution, but stable for years in the solid state in the absence of air, and easily transferred and handled by using inexpensive glove bags.

π-Allylnickel halide complexes react with a wide variety of organic halides in polar solvents such as DMF or HMPA to replace the halogen with the allyl group (equation 83). Primary and secondary alkyl, aryl, vinyl, benzyl, and allylic halides react, with the order of reactivity being I > Br >> Cl. A very wide range of functional groups are tolerated in the substrate, including OH,

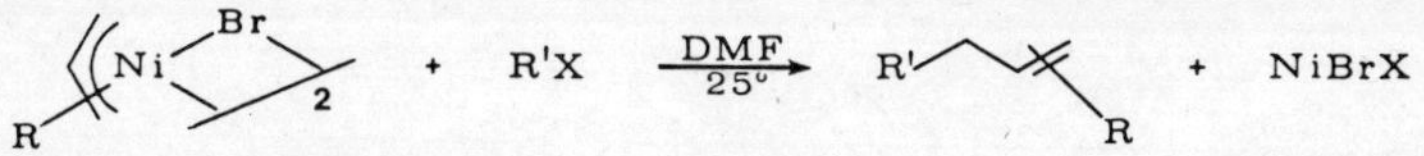

$$\text{(allyl-R)Ni(}\mu\text{-Br)}_2 + R'X \xrightarrow[25^\circ]{DMF} R'CH_2CH{=}CHR + NiBrX \qquad (83)$$

CO_2R, $CONH_2$, COCl, NH_2, and CN. Aldehydes and ketones react with π-allylnickel halides but higher temperatures (50°C) are required and the reagents can usually discriminate between halide and carbonyl groups. Nitro groups are <u>not</u> tolerated, since this coupling reaction appears to be a radical chain reaction which is inhibited by nitro compounds. With unsymmetrical allyl groups, coupling always occurs at the least substituted allyl terminus (for example, equation 84),[99] in contrast to most allylmetal (Mg, Li, Zn) systems. Some examples of its use in synthesis are shown in equations 85-91.

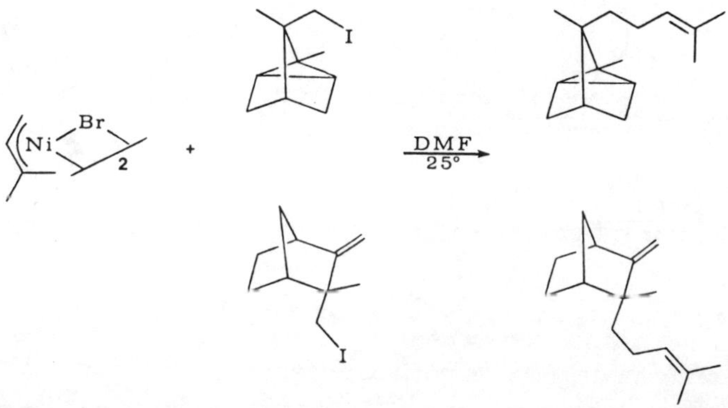

$$\xrightarrow[25^\circ]{DMF} \qquad (84)$$

I

I

$$MeO \text{–allyl–Ni(Br)}_2 + CH_2{=}C(CO_2Et)CH_2Br \longrightarrow CH_2{=}C(CO_2Et)CH_2CH_2C(OMe){=}CH_2 \qquad (85)[100]$$

$$\xrightarrow[H_2O]{H^+} CH_2{=}C(CO_2Et)CH_2CH_2COCH_3$$

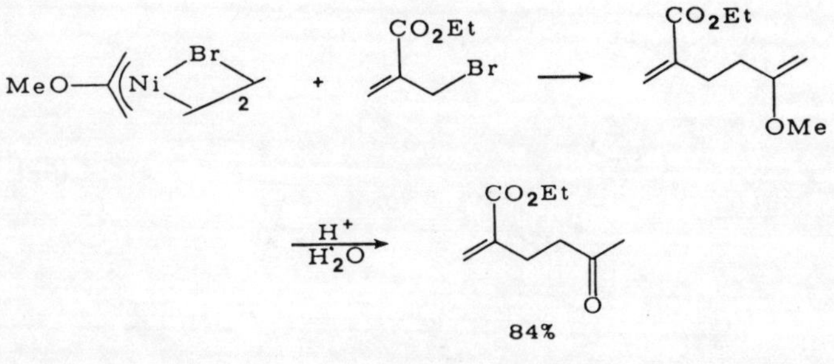

84%

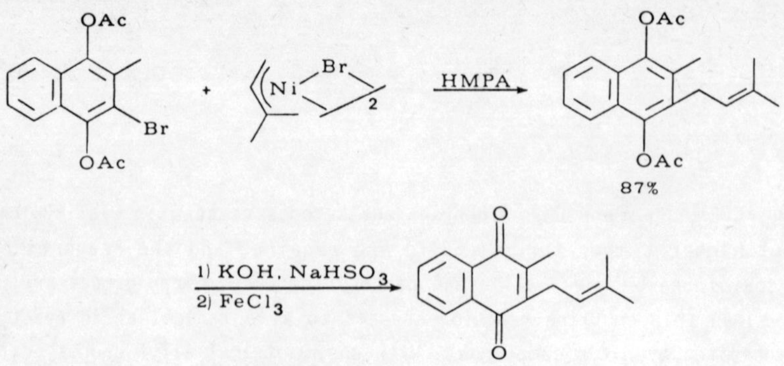

OAc

Br

OAc

+ Ni Br 2

HMPA

OAc

OAc

87%

(86)[101]

1) KOH, $NaHSO_3$

2) $FeCl_3$

O

O

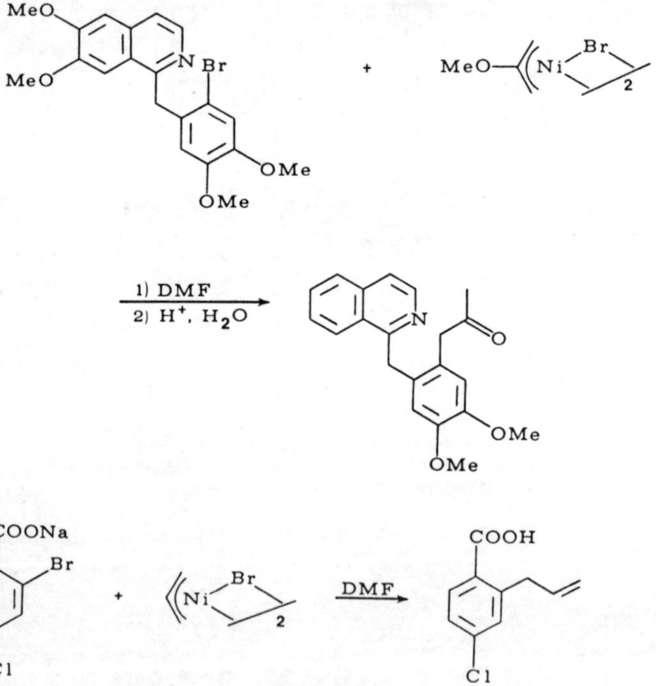

MeO

MeO

N Br

OMe

OMe

+ MeO Ni Br 2

(87)[100]

1) DMF

2) H^+, H_2O

N

O

OMe

OMe

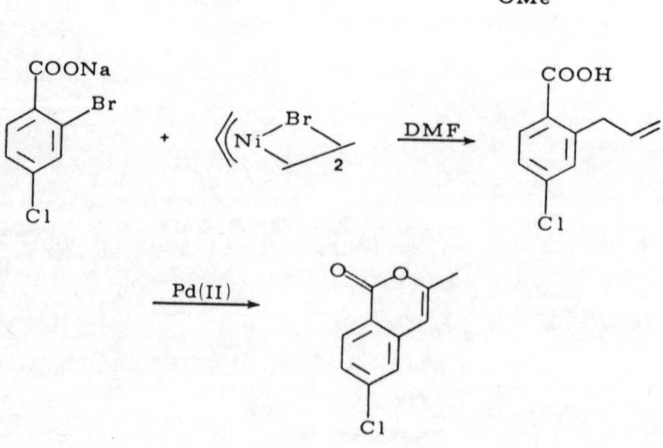

COONa

Br

Cl

+ Ni Br 2

DMF

COOH

Cl

(88)[69]

Pd(II)

O

O

Cl

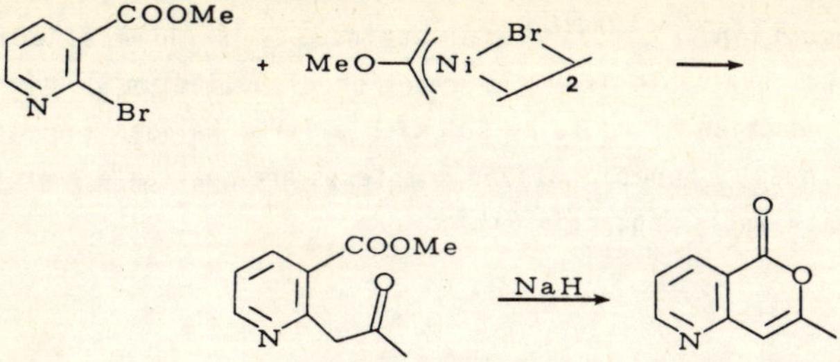

COOMe
N Br
+ MeO Ni Br 2

(89)[69]

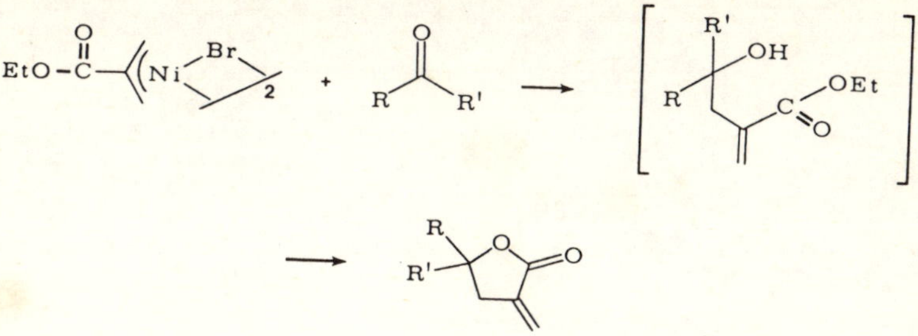

COOMe
N
NaH
O
O
N

O
EtO-C
Ni Br 2
+
O
R R'

R'
OH
R
OEt
C
O

(90)[102]

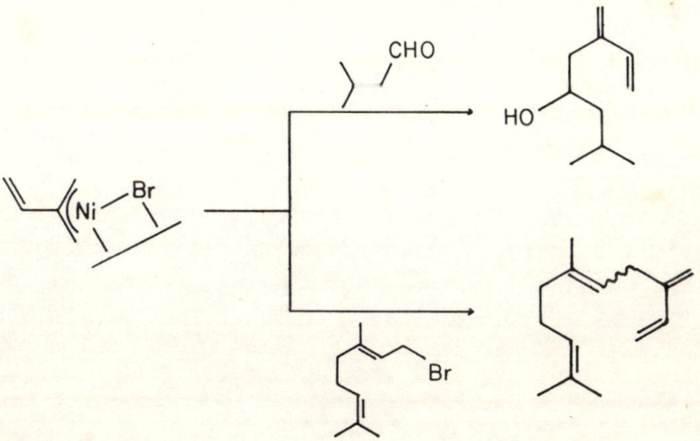

R
O
R'
O

CHO

HO

Ni Br

(91)[103]

Br

B. π-Allylpalladium Halide Complexes

π-Allylpalladium halide complexes are yellow, air stable, crystalline solids which are very useful in organic synthesis. They are prepared in several ways. In contrast to the corresponding nickel complexes, they can be prepared directly from olefins under a variety of conditions. With unsymmetrical olefins, mixtures of complexes, the composition of which depends strongly on the reaction conditions (equation 92),[104,105] are obtained. π-Allylpalladium halide complexes are also available from the reaction of palladium(0) species (generated by in situ reduction of $PdCl_2$ by CO) with allylic halides (equation 93)[106] or acetates. Finally, functionalized complexes are available from the reaction of butadiene with $PdCl_2$ (equation 94).[107]

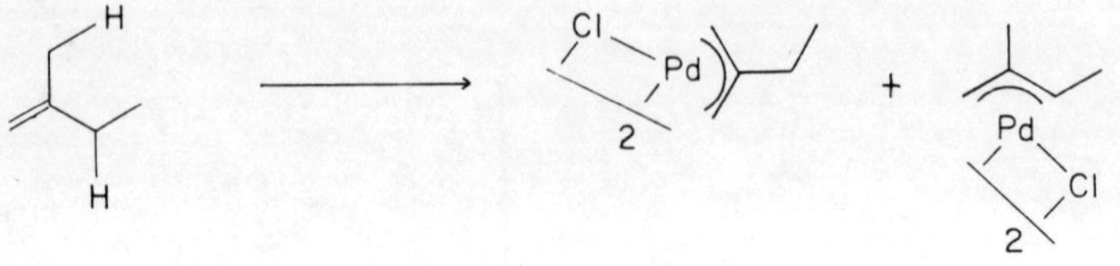

(92)

$PdCl_2$	29	71
$PdCl_2/CuCl_2$	74	26

But

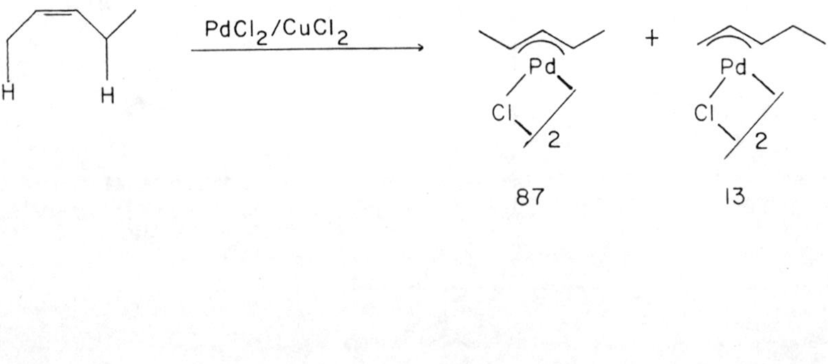

$PdCl_2/CuCl_2$

87 13

CO, $CHCl_3$ / Na_2CO_3

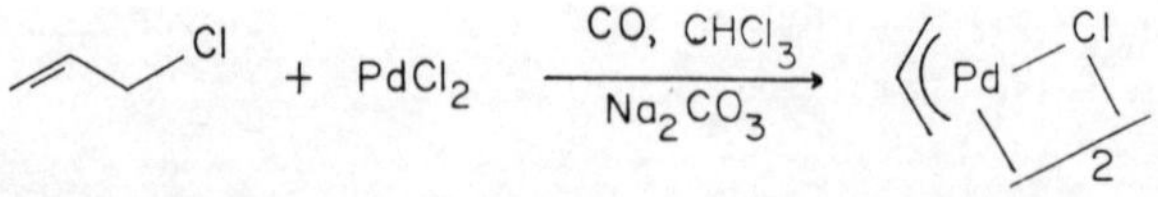

(93)

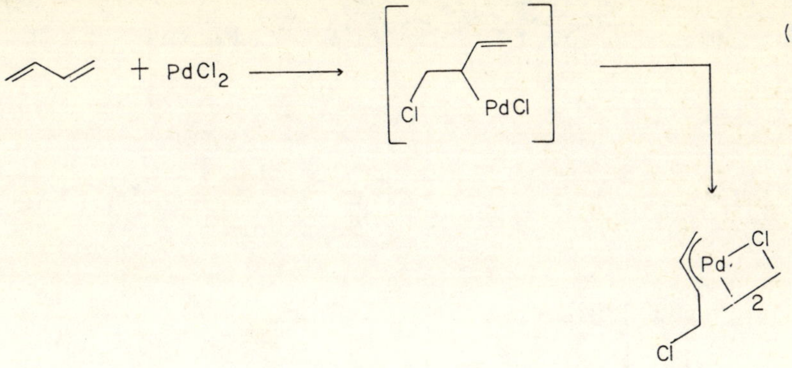

(94)

π-Allylpalladium halide complexes are important in synthesis because, in the presence of added ligands (usually phosphines), they react with nucleophiles ranging from acetate to amines to stabilized (pKa = 25) carbanions (equation 95). The regiochemistry depends on both the substrate and the nucleophile,

(95)

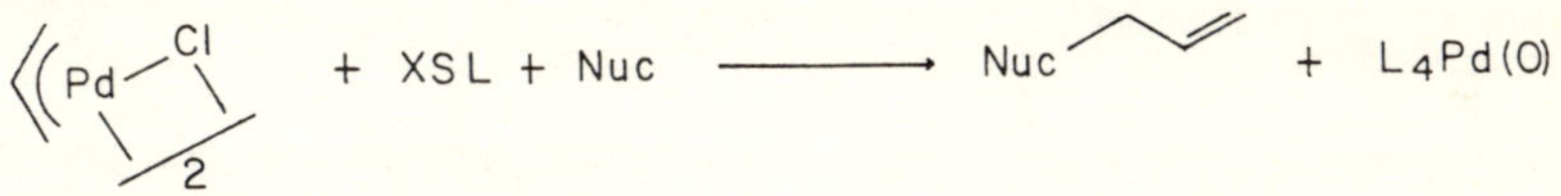

although attack at the less substituted terminus of the allyl group usually predominates. The <u>stoichiometric</u> reaction of nucleophiles with π-allylpalladium complexes has not been extensively used, because of the cost of palladium. However, it has been used for the alkylation of steroids at the 6-position (equation 96),[108] the synthesis of Vitamin A analogs (equation 97),[109] the introduction of the side chain at a steroid C-20 position (equation 98)[110] and the synthesis of long chain polyenes (equation 99).[111] Note that the last two examples involve transmetallation from zirconium to palladium, and probably involve initial attack at palladium, whereas the rest of the examples involve direct external attack at the π-allyl ligand.

(96)

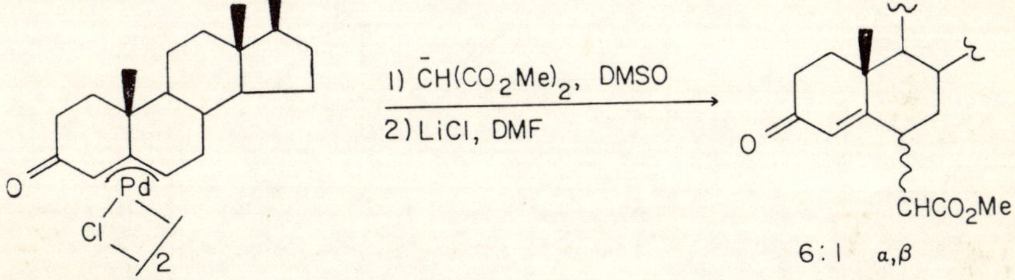

PPh$_3$ DMF (97)

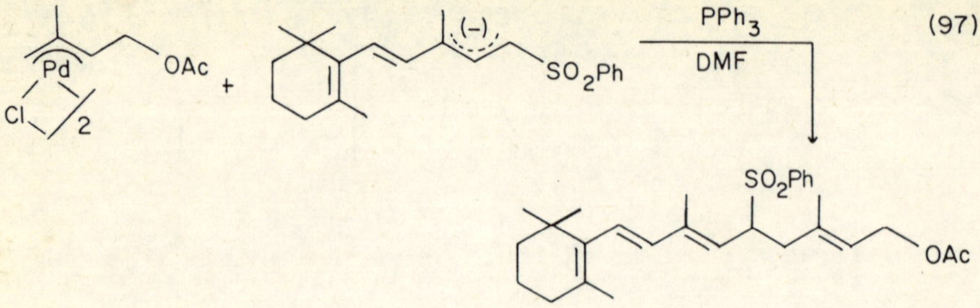

OAc SO$_2$Ph

Pd Cl 2

SO$_2$Ph OAc

(98)

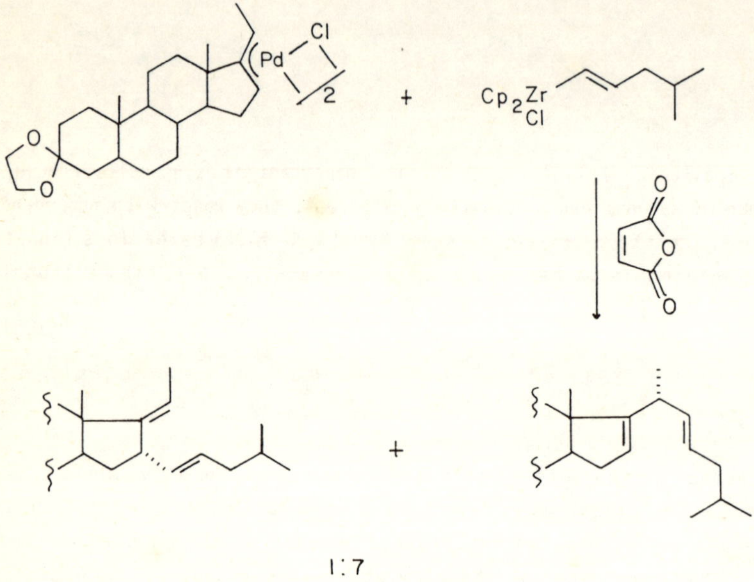

Pd Cl 2 Cp$_2$Zr Cl

O O

1:7

(99)

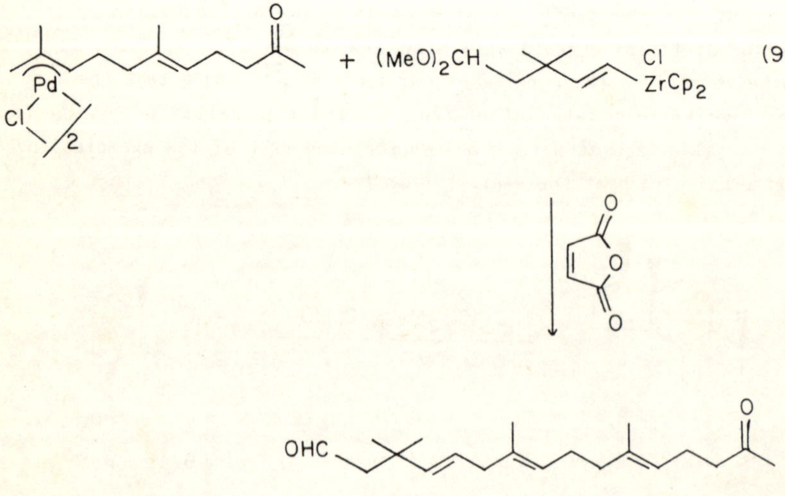

(MeO)$_2$CH Cl ZrCp$_2$

Pd Cl 2

OHC O

Of considerably greater use synthetically is the palladium(0) catalyzed nucleophilic displacement of a range of allylic leaving groups (equation 100).

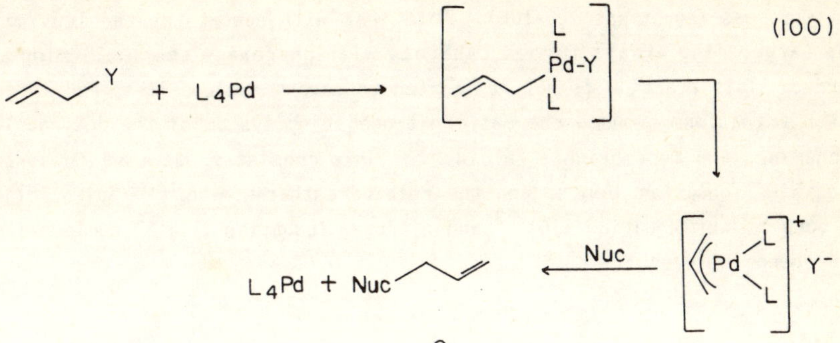

(100)

$Y = OAc, OPh, OH, NR_2, SO_2Ar, NO_2,$ [epoxide]

$Nuc = R(-), R_2NH$

The reaction is thought to procede by oxidative addition of the allylic substrate to Pd(0) to produce a π-allylpalladium species. Nucleophilic attack on this species gives the observed product, and the regenerates the catalytically active Pd(0) species. This is an important process because it permits nucleophilic displacement of allylic groups not normally reactive to S_N2 conditions. The observed stereochemistry is retention, the result of two inversions.

The literature is rife with examples of the use of this chemistry in organic synthesis,[112] but only a few illustrative examples will be presented here. Chrysanthemic acid derivatives were synthesized using two consecutive palladium catalyzed processes (equation 101).[113] Note that only three membered ring formation (attack at the less substituted terminus) was observed. Allylic

(101)

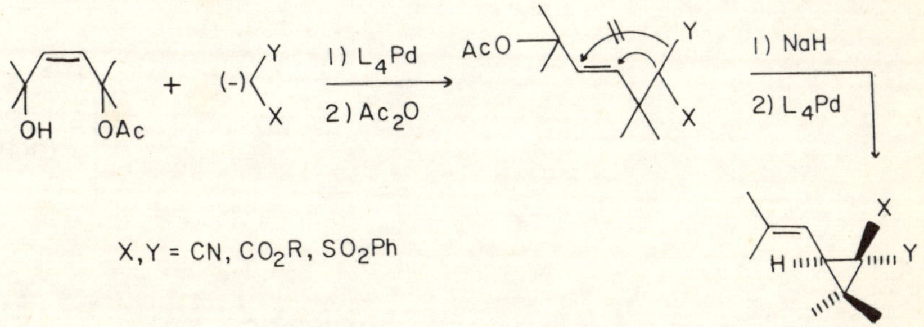

$X,Y = CN, CO_2R, SO_2Ph$

substrates containing other functionality also react cleanly. Vinyl silanes are produced from TMS containing allyl acetates regardless of the initial position of the acetates (equation 102).[114] Cyanohydrin acetates of allylic aldehydes undergo clean S_N2' allylic alkylation (equation 103),[115] as do allyl epoxides (equation 104).[116] Macrocycles are available from intramolecular versions of this process (equations 105-108). Note that with acetate as the leaving group the larger ring always formed, whereas with phenoxide the small ring always formed. This reaction is not restricted to carbon nucleophiles, and intramolecular aminations produce the basic nitrogen ring system of the actinabolamine, ibogamine, and mesembrine alkaloids.[120] This chemistry was used to synthesize ibogamine (equation 109)[121] and the related catharanthine.[122] Spirocyclization of both carbon (equation 110)[123] and nitrogen (equation 111)[124] nucleophiles has also been achieved.

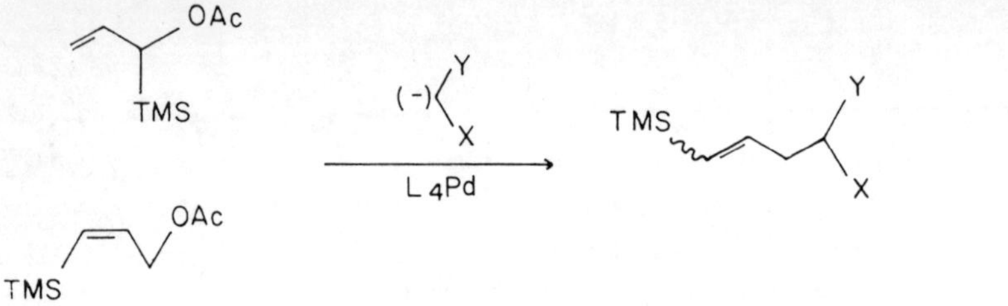

OAc TMS; (-)CH(Y)(X), L_4Pd → TMS∼CH=CH–CH₂–CH(Y)(X) (102)

OAc TMS

(103)

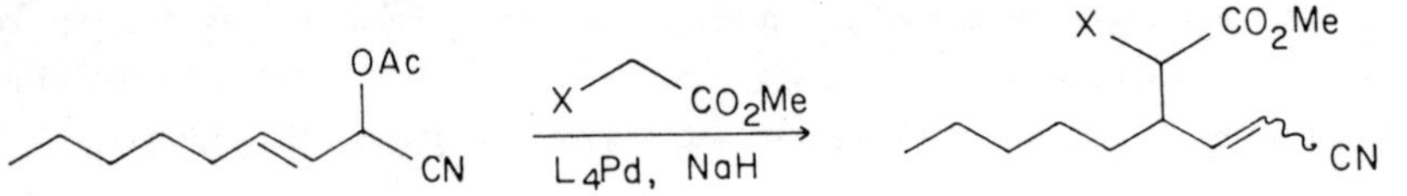

OAc CN; X–CH₂–CO_2Me, L_4Pd, NaH → X, CO_2Me, CN

(104)

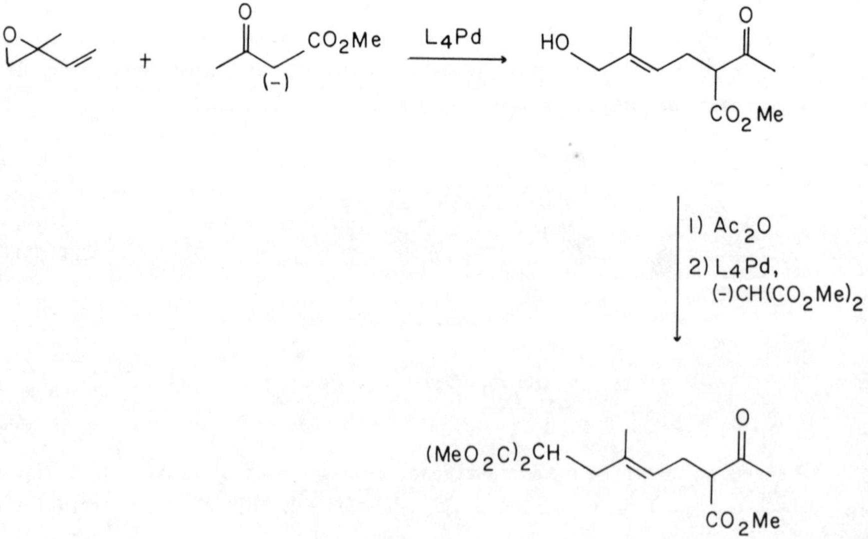

O + O CO_2Me (-) L_4Pd → HO O CO_2Me

1) Ac_2O
2) L_4Pd, (-)$CH(CO_2Me)_2$

$(MeO_2C)_2CH$ O CO_2Me

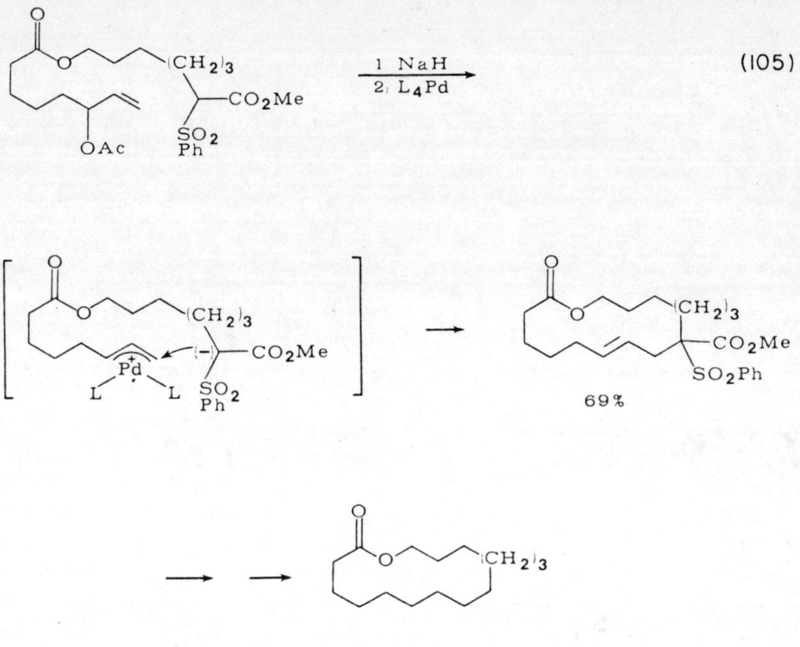

$(CH_2)_3$ CO_2Me OAc SO_2 Ph

1 NaH
2 L_4Pd

(105)

$(CH_2)_3$ CO_2Me Pd   L   L SO_2 Ph

$(CH_2)_3$ CO_2Me SO_2Ph

69%

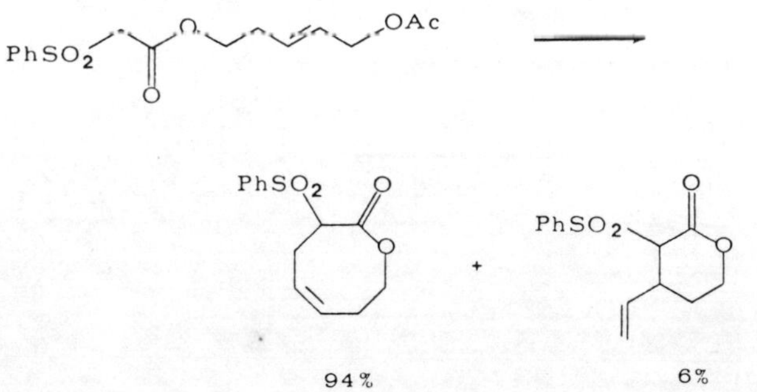

$(CH_2)_3$

$PhSO_2$ OAc

(106)

$PhSO_2$ + $PhSO_2$

94%   6%

(107)

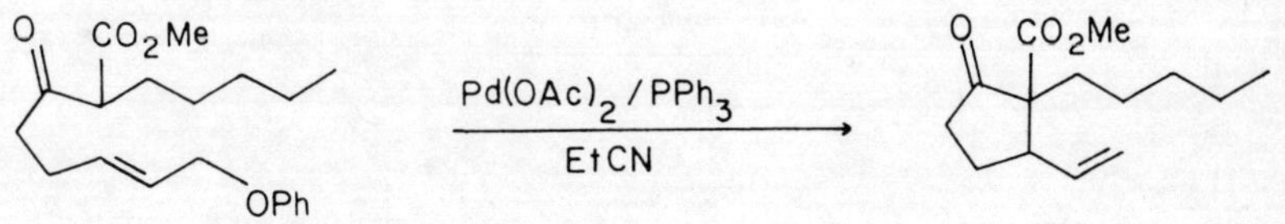

CO_2Me OPh

$Pd(OAc)_2$ / PPh_3
EtCN

CO_2Me

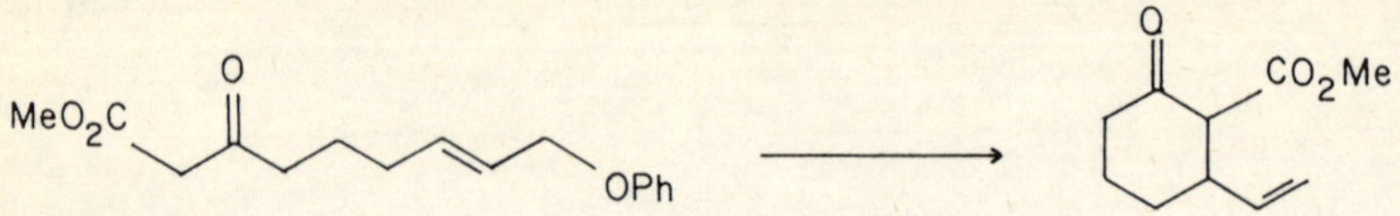

(108)

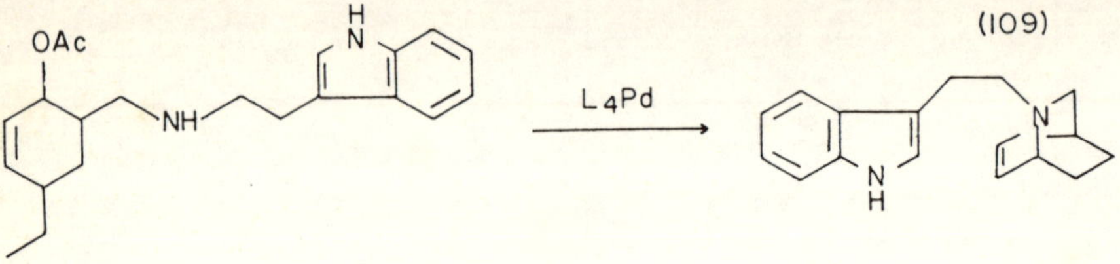

(109)

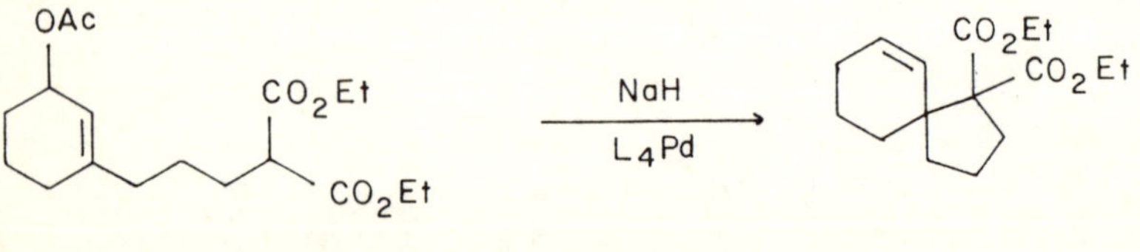

L_4Pd

(110)

NaH

L_4Pd

L_4Pd

Et_3N, MeCN

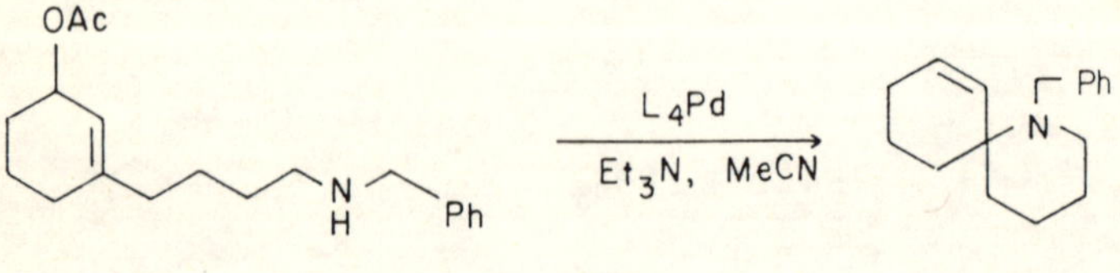

(111)

Control of stereochemistry in acyclic systems is among the most difficult of problems faced in organic synthesis. The stereospecific nature of both the formation of π-allylpalladium complexes and their reactions with carbanions can be used to relay the stereochemistry of one chiral center to a remote position in conformationally nonrigid systems.[125] Thus, organopalladium chemistry was used to transfer the chirality of a vinyl lactone moiety to the remote vinylic carbon, producing an intermediate having two chiral centers in a 1,5 relationship (equation 112). For this process to succeed, formation of the π-allylpalladium complex must proceed only from the conformation shown, the π-allyl complex must maintain its stereochemistry, and the nucleophile must attack regiospecifically at the terminal carbon of the π-allyl system. All of these criteria were met, and the reaction went in 90% yield with greater than 95% stereospecificity. This chemistry was used to synthesize the side chain of Vitamin E.[126] Similar chemistry was used to transfer chirality in the palladium(0) catalyzed rearrangement of vinylogous lactones to cyclopentanones or cycloheptenones (equation 113).[127] The product depends on the catalyst used.

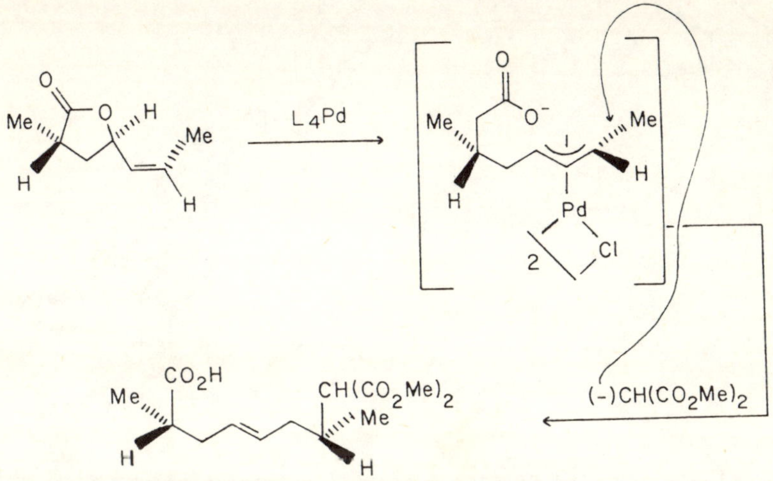

(112)

L_4Pd

$(-)CH(CO_2Me)_2$

(113)

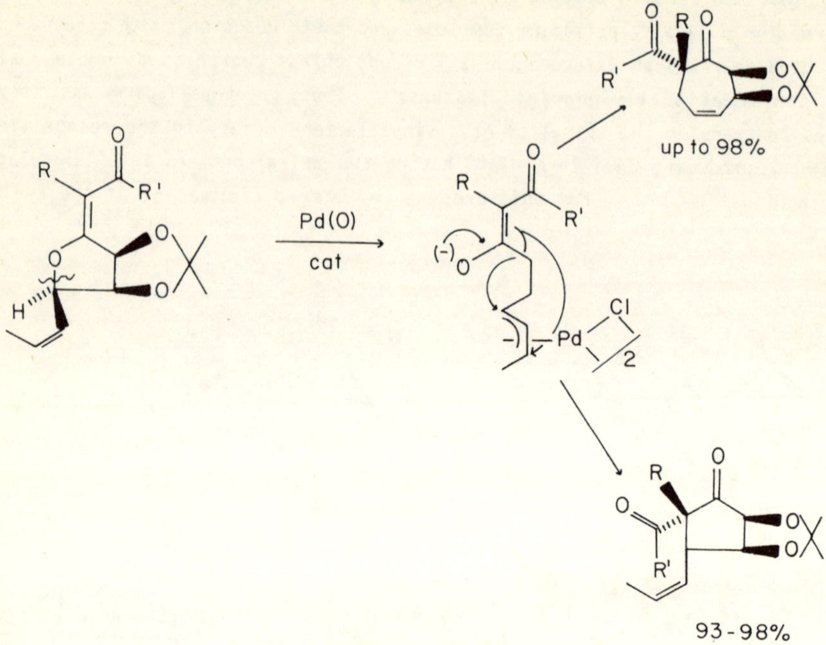

The chemistry reported herein is only illustrative of the synthetic applications of Group VIII transition metals. Although the important classes of reactions and complexes have been presented, many more variations and examples are known. The field is young and rapidly growing. Mechanistic knowledge is lagging far behind synthetic developments, and the ability to predict or even explain the outcome of a particular reaction is low. Thus, a great deal of interesting new chemistry remains to be found.

VII REPRESENTATIVE EXPERIMENTAL PROCEDURES

12-Hydroxy-4,8-Dodecadienoic Acid Lactone[6]

6'-Bromo-4'-hexen-1'-yl 6-bromo-4-hexenoate (119 mg, 0.336 mmol) in N-methylpyrrolidone (2 ml) was slowly added via motor-driven syringe over a 3.5 hr period to nickel carbonyl (CAUTION! VOLATILE, TOXIC) (0.26 ml, 2.02 mmol) in N-methylpyrrolidone (10 ml) at 50°; the initially yellow to yellow-orange solution became green in about one hour. After stirring for 4 hr at 50°C, ether (5 ml) was added and distilled from the reaction mixture 3 times to remove residual nickel carbonyl. The mixture was carefully treated with water (15 ml) and 3N hydrochloric acid (5 ml), and the product was extracted with petroleum ether (5 x 20 ml) and dried (sodium sulfate). The crude product was purified by preparative tlc (10:1 pet.ether-ether, 2 dev.) to ($CDCl_3$), δ yield 49 mg (76%) of lactone possessing a camphor- like odor: IR: 1733(s, C=O), 970 cm^{-1} (s, trans-C=C); nmr ($CDCl_3$) δ 1.79(m, 2, $C\underline{H}_2CH_2O$), 2.09 and 2.33(broad m, 10, CH_2-C=C and CH_2CO), 4.07(t, 2, J=5Hz, CH_2O), 5.32(m, 4, vinyl); mass spectrum; parent m/e 194; tlc R_f 0.53(10:1 pet.ether-ether, 2 dev.). Mol. wt. Calc'd for $C_{12}H_{18}O_2$: 194.1307. Found: 194.1305

Reaction of α,α'-Dibromo Ketones and 1,3-Dienes. Reaction and Workup[10] Conditions

Procedure A: Diiron nonacarbonyl (2.40 mmol) was placed in a 100 ml thick-wall reaction ampule connected to a three-way stopcock equipped with a rubber septum and a nitrogen inlet tube. The system was filled with dry nitrogen by five evacuation-flush with nitrogen cycles. To this was added a mixture of dry benzene (10 ml) a 1,3-diene (20.0 mmol), and an α,α'-dibromo ketone (2.00 mmol) with a syringe and the tube was sealed under nitrogen. The mixture was stirred at 60-90°C for several hours and then cooled to ambient temperature. The resulting precipitates were removed by filtration through a pad of Celite 545 (ca. 2 g) and the filtrate was concentrated by a rotary evaorator to give a yellow oil. Yields were obtained by GLC analysis using an appropriate internal standard. For the isolation of the adducts the residue was directly subjected to preparative TLC followed by distillation using a Kugelrohr apparatus. Otherwise, the mixture was distilled in vacuo and then the distillate was purified by preparative GLC.

Procedure B: In a 300 ml pressure bottle was placed $Fe(CO)_5$ (20.0 mmol). The system was evacuated and flushed with nitrogen. Then a mixture of a 1,3-diene (60.0 mmol) and dry benzene (20.0 ml) was introduced with a syringe. The mixture was irradiated at 60°C for 6-30 hr with a 200-W high-pressure mercury lamp. The bottle was opened and an aliquot of the mixture was analyzed by NMR (dioxane as an internal standard) to confirm the quantitative formation of the diene-iron tricarbonyl complex. To this was added a solution of an α,α'-dibromo ketone (15.0 mmol) in a small amount of benzene. The mixture was again covered with nitrogen and heated at 70-120°C for several hours. To an aliquot of the reaction mixture was added an appropriate internal standard and the mixture was analyzed by GLC using the specified column. The remaining reaction mixture was filtered through a pad of Celite 545 (ca. 2 g) and the filtrate was concentrated by a rotary evaporator to give a pale yellow oil. This oil was dissolved in acetone (20 ml), mixed with solid $CuCl_2;2H_2O$ (10 mmol), and stirred at room temperature for 15 min to decompose the remaining diene-iron tricarbonyl complex. The slurry was then passed through a short column packed with Celite 545 (2 g) and the filtrate was concentrated under reduced pressure. The residue was dissolved in methylene chloride (40-50 ml) and washed with four 20 ml portions of 5% aqueous solution of ethylenediaminetetraacetic acid disodium salt (EDTA disodium salt). The organic layer was dried over anhydrous Na_2SO_4 and the solvent was evaporated by a rotary evaporator to give a yellow oil, distillation of which under reduced pressure using a Kugelrohr apparatus yielded the 1:1 adduct as a colorless oil. Analytical samples were obtained by preparative TLC, GLC, or a combination of both.

Procedure C: Diiron nonacarbonyl (1.30 mmol) was placed in a 50 ml pressure bottle and the air in the vessel was replaced by nitrogen by five evacuation-flush with nitrogen cycles. To this was introduced a mixture of dibromo ketone (1.00 mmol), a 1,3-diene (10.0 mmol), and dry benzene (2.5 ml). The mixture was irradiated at room temperature for 36 hr by a 200-W high-pressure mercury lamp using a 10% aqueous solution of cupric sulfate as a filter. The reaction mixture was worked up as described in procedure B.

<u>2,2,4,5,7,7-Hexamethyl-4-cycloheptenone</u> (Equation 4)[10]

Procedure A: A mixture of $Fe_2(CO)_9$ (1.60 g, 4.40 mmol) 2,4-dimethyl-2,4-dibromopentane-3-one (540 mg, 2.00 mmol), 2,3-dimethylbutadiene (7.28 g, 88.5 mmol), dry benzene (5.0 ml), and a small amount of hydroquinone was heated at 60°C for 38 hr with occasional shaking. Workup gave a yellow oil (290 mg). GLC analysis (column E, 100°C, tetradecane as an internal standard) indicated the production of the cycloadduct in 71% yield. The analytical sample was obtained by preparative GLC (column D, 165°C, retention time (t_R) 20 min). IR: 1685 cm^{-1} (C=O); NMR δ 1.07(s, 4, CH_3), 1.79(s, 2=CCH_3), 2.25(s, 2=CH_2); mass spectrum, m/e 194(M^+), 123(M^+ -71) (base peak), 110 (M^+ -84), 109(M^+ -85). Anal: ($C_{13}H_{22}O$), C, H.

Procedure B: A mixture of $Fe(CO)_5$ (3.92 g, 20.0 mmol), 2,4-dimethyl-2,4-dibromopentane-3-one (4.93 g, 60.0 mmol), and dry benzene (20.0 ml) was irradiated at 55°C for 6.5 hr. To this was added the dibromo ketone (4.08 g, 15.0 mmol) and the mixture was kept at 80°C for 12 hr. The reaction mixture was worked up to give a brown oil (6.20 g), which was distilled under reduced pressure affording (2.90 g, 100%) as a colorless oil, bp 120-130°C (0.02 mm).

The product was homogeneous on GLC and TLC analyses.

Reaction of $Na_2Fe(CO)_4$ with Bromoesters[19]

All reactions were conducted under a nitrogen atmosphere. In a typical preparation 2.16 g (10.3 mmol) of ethyl 5-bromovalerate was added to a solution of 2.37 g (11.1 mmol) of $Na_2Fe(CO)_4$ in 30 ml of N-methyl-2-pyrrolidone (distilled from CaH_2), the reaction stirred for 1 hr, and 2.4 ml (30.0 mmol) of ethyl iodide added. After 24 hr the solution was diluted with ether, washed three times with brine, dried, and filtered, and the ether was removed with a rotary evaporator. The residue was placed on a short column (130 g of silica gel) and washed free of the colored iron by- products with hexane, and the product removed with 50% ether-hexane. Fractional distillation gave 1.41 g (74%) of ethyl 6-ketooctanoate.

The Acylation of Allylic Bromides with Nickel Carbonyl/Organolithium Reagents[22] (equation 12)

To 10 ml of dry, argon-saturated ether in a 100 ml round-bottom flask with sidearm and three-way stopcock opened to an argon filled balloon was added nickel carbonyl (Caution. Toxic) (0.39 ml, 3.0 mmol). The solution was cooled to -50° (Dry Ice--acetonitrile), and the organolithium reagent (2.26 ml of 1.33 *M* methyllithium in ether; 2.50 ml of 1.20 *M* *n*-butyllithium in pentane) was added. The mixture was stirred for 2 hr at -50°. The allylic bromide (2.0 mmol) was then added. The reaction was followed by tlc using phosphomolybdic acid spray for visualization. Upon disappearance of the starting material, the mixture was quenched at the reaction temperature with 10 ml of aqueous ammonium chloride. The mixture was allowed to warm to 25°, and excess nickel carbonyl was decomposed by the addition of iodine in ether (care. carbon monoxide is evolved and the reaction foams) until the brown color of iodine remained for 0.25 hr. The mixture was poured into a separatory funnel containing 50 ml of ether and 50 ml of water and was shaken. The ether layer was washed with 50 ml of aqueous sodium sulfite to decompose excess iodine, two 50 ml portions of water, dried over anhydrous magnesium sulfate, and concentrated. The products were purified by preparative tlc when necessary.

Reaction of 1-Phenethylmagnesium Chloride with (E)-β-Bromostyrene (equation 16)[25]

$PdCl_2$[(S)-(R)-BPPFA] (40 mg, 0.05 mmol) was placed in a 50 ml two-necked flask equipped with a stirring bar, a serum cap, and a three-way stopcock. The reaction vessel was then filled with argon after evacuation and at -78°C charged with 1.83 g (10 mmol) of (E)-β-bromostyrene and 1-phenylethylmagnesium chloride in ether (20 mmol, 17 ml of 1.2 M solution) through the serum cap with a syringe. The reaction mixture was stirred at 0°C for 24 hr and hydrolyzed with 10% hydrochloric acid. Workup as above and distillation under reduced pressure gave a crude product (121-125°C (3 mm Hg)). Purification by preparative GLC gave 1.62 g (78%) of (E)-1,3-diphenyl-1-butene: $[\alpha]^{20}D$ +24.3° (c 3, benzene); NMR ($CDCl_3$) δ 1.45(d, J=7Hz, 3H, CH_3), 3.45-3.78(m, 1H, $CHCH_3$), 6.30-6.45(m, 2H, CH=CH), 7.04-7.45(m, 10H, C_6H_5).

5(E)-Styryl-2'-deoxyuridine Method A (Figure 5)[30]

A solution of 0.1 M lithium tetrachloropalladate (11 ml) was added with stirring to a suspension of 5-chloromercuri-2'-deoxyuridine, (463 mg, 1 mmol) and styrene (310 mg, 3 mmol) in 11 ml of methanol. After refluxing for 12 hr the mixture was saturated with hydrogen sulfide gas and filtered through Celite, and the filtrate evaporated. The mixture was resolved on a silica column (3 x 40 cm) using 0-10% methanol in chloroform as eluent to give 306 mg of white solid. Recrystallization from 1-octanol gave a yield of 175 mg (57%) of compound 3: mp 195-196°C; mass spectrum m/e 330(molecular ion), 214(heterocyclic base), 143, 117; UV (H_2O) γ_{max} 305 (e 17 600), γ_{min} 238 (6000); ^1H NMR (CD_3COOD) δ 8.2(s, 1H, C-6), 7.63-6.92(m, 5H, aromatic), 7.38(d, 1H, J=16.5Hz, PhCH=), 6.86(d, 1H, J=16.5Hz, U-CH=), 6.31(t, 1H, J=6Hz, C-1). Anal: ($C_{17}H_{18}N_2O_5$, 330.33) C, H, N.

5(E)-Styryl-2'-deoxyuridine also was prepared in 47% yield by method A with the exception that a catalytic amount of lithium tetrachloropalladate (26 mg, 0.1 mmol) was used together with cupric chloride (538 mg, 4 mmol).

Cyclization of 2-Iodo-N-(3,3-dimethylallyl)aniline to 3-Isopropylindole (equation 29)[44]

A Fischer-Porter aersol compatibility tube (3.5 x 13 cm) was charged with 0.287 g (1.00 mmol) of the iodoaniline, 2.2 mg (0.01 mmol) of $Pd(OAc)_2$, 0.21 mg (1.5 mmol) of NEt_3, and 6 ml of CH_3CN. The bottle was capped with a rubber serum cap, clamped into place by the pressure bottle head supplied with the unit, and flushed several times with argon. The mixture was heated at 110°C for a total of 72 hr. After 24 hr and 48 hr, the mixture was cooled to room temperature, an additional 2.2 mg of $Pd(OAc)_2$ and 0.07 ml of Et_3N in 1 ml of

CH_3CN was added, and the mixture was reheated to 110°C. After completion, the crude product was purified by medium-pressure liquid chromatography, eluting with 4:1 hexane/ether to give 3-isopropylindole (0.116 g, 73%).

Cyclization of 2-Bromo-4-carbethoxy-N-allylaniline to 3-Methyl-5-carbethoxyindole[44] (Equation 31)

The reaction was run as above except 0.02 equiv of tris(2-tolyl)phosphine (based on substrate) was added in addition to the rest of the components. An additional 0.02 equiv of this phosphine was added with each subsequent addition of $Pd(OAc)_2$. From 0.32 g (1.12 mmol) of the allylaniline was obtained 0.113 g (49%) of 3-methyl-5-carbethoxyindole: NMR (CCl_4) 1.48(t, J=7Hz, 3, $C\underline{H}_3CH_2O$), 2.38(s, 3, CH_3), 4.45(q, J=7Hz, 2, $CH_3C\underline{H}_2O$), 6.95(s, 1, indole 2H), 7.25(d), 7.90(d of d), 8.35(s, 3, ArH), 9.00(br, 1, NH); IR (CCl_4) 3330(NH), 1700(C=0) cm^{-1}.

Synthesis of α-Farnesene (equation 37)[50]

To a solution of trimethylalane (pyrophoric) (2.88 g, 3.84 ml, 40 mmol) and dichlorobis(η^5-cyclopentadienyl)zirconium (2.24 g, 8 mmol) in 1,2-dichloroethane (30 ml) is added at room temperature under nitrogen 1-buten-3-yne (1.04 g, 20 mmol) in xylene. After stirring the reaction mixture for 12 hr at room temperature, geranyl chloride (3.45 g, 20 mmol), tetrakis(tri-phenylphosphine)palladium (1.15 g, 1 mmol), and THF (40 ml) are added at 0°C. The reaction mixture is stirred for 3 hr at room temperature, treated with water (30 ml), and extracted with pentane. The extract is washed with aqueous $NaHCO_3$ and dried over $MgSO_4$. After filtration and evaporation, simple distillation yields 3.50 g (86%) of α-farnesene: bp 30-32°C (0.15 mm Hg); 1.4977; IR (neat): 3080(w); 2960(s); 2900(s); 1664(w); 1635(m); δ 1601(m); 981(m); 883(s) cm^{-1}; ^1H NMR ($CDCl_3$, Me_4Si) δ 1.59(s, 3H); 1.63(s, 3H); 1.66(s, 3H); 1.74(s, 3H); 2.03(m, 4H); 2.82(t, J=6Hz, 2H); ^{13}C NMR ($CDCl_3$, Me_4Si) δ 11.62; 16.07; 17.63; 25.69; 26.89; 27.35; 39.88; 110.37; 122.36; 124.50; 131.10; 131.74; 133.79; 135.55; 141.69; high-resolution MS. Anal. Calc'd for $C_{15}H_{24}$: 204.188. Found: 204.191.

Di-μ-acetato-bis(2-acetaminophenyl-2C,O)dipalladium(II) (2a) (equation 45)[58]

Into a mixture of acetanilide (4.05 g, 30 mmol) and $Pd(OAc)_2$ (3.23 g, 14.4 mmol) was added toluene (50 ml) and the solution was refluxed under a slow stream of nitrogen for 30 min. The precipitate from the solution was quickly filtered and washed with fresh solvent to give greenish yellow 2a: 2.10 g; dec 210-213°C. The filtrate was again refluxed for 90 min, and the deposited 2a (1.24 g; dec 204-205°C) was collected. Further refluxing of the filtrate from the second reaction yielded a third crop of 2a: 0.95 g; dec 205-206°C (total yield of 2a was 4.29 g, 99.3% based on palladium). The final mother liquor of the reaction was concentrated under reduced pressure to give recovered 1a: 1.80 g; IR 3290 (NH); 1625; 1570(NAc); NMR (Me_2SO-d_6) 1.35 and 2.28(OAc, bridged and monodentate, s); 1.90(NAc, s); 6.60-7.10(aromatic); 11.2(NH). Anal. Calc'd for $(C_{10}H_{11}O_3NPd)_2$: C, 40.04; H, 3.70; N, 4.67. Found: C, 40.18; H, 4.24; N, 5.25.

Reaction of 2a with Carbon Monoxide

An ethanolic solution (30 ml) of 2a (900 mg, 1.5 mmol) was placed in a thick-walled tube fitted with a mechanical stirrer. Carbon monoxide (2 atm) was introduced, and the mixture was stirred for 60 min at room temperature. The system changed from a yellow solution to a black suspension. The reaction mixture was allowed to stand overnight. Filtration of the solvent to remove metallic palladium (262 mg, 2.46 mmol) and concentration of the filtrate gave colorless crystals of the desired *o*-ethyl ester: 493 mg (2.38 mmol, 79%); mp 66-68°C; IR 3250(NH); 1710(ester); 1685(NAc).

Reaction of 2a with Methyl Vinyl Ketone

A mixture containing 2a (300 mg, 0.5 mmol), methyl vinyl ketone (300 mg, 4.3 mmol), and Et_3N (350 mg, 3.5 mmol) in toluene (7.5 ml) was refluxed for 45 min. The mixture was filtered and concentrated under reduced pressure. The residue was recrystallized from benzene to yield colorless crystals of the desired conjugated ketone: 168 mg (0.828 mmol, 83%); mp 126-127°C. Anal. Calc'd for $C_{12}H_{13}O_2N$: C, 70.91; H, 6.45; N, 6.89. Found: C, 70.64; H, 6.50; N, 6.90.

General Reaction Procedure for Alkylation of Olefins Without Added HMPA (equation 48)[65]

The $PdCl_2(2CH_3CN)$ (0.76 g, 3.0 mmol) was weighed into a 200 ml, two-necked, round-bottomed flask fitted with a magnetic stir bar, stopcock, and rubber serum cap. The flask was alternately evacuated and filled with argon (four cycles). Addition of THF (100 ml) and stirring for 0.2 hr at room temperature produced an amber suspension. Addition of the olefin (2-4 equiv based on Pd) followed by stirring for 0.2 hr gave a homogeneous solution. The flask was cooled to -78°C.

Triethylamine (2.0 equiv per Pd) was added dropwise slowly with vigorous stirring over 0.2 hr followed by stirring at -78°C for 0.2 hr. The cold bath was allowed to warm to -60°C (ca. 0.25 hr), at which time the anion (1.2-2.0 equiv per Pd) was added as a THF solution (cooled to -78°C before addition) over 0.2 hr via a precooled (dry ice) syringe. The cold bath was maintained at -60°C for 0.5 hr. To obtain saturated (reduced) products the flask was flushed with hydrogen, a hydrogen balloon affixed to the stopcock, and the cold bath removed. The reaction mixture was allowed to warm to room temperature, stirring vigorously, and stirred overnight under a hydrogen atmosphere. The resulting suspension was filtered and the solvent concentrated on a rotary evaporator.

Alternatively, the unsaturated (β-elimination) products were obtained by simply allowing the reaction flask to warm to room temperature under an argon atmosphere. After 2 hr the heterogeneous reaction mixture was filtered and the solvent concentrated on a rotary evaporator. The alkylated products were isolated and purified by standard methods.

General Reaction Procedure for Alkylation in the Presence of HMPA (equation 48)[65]

The $PdCl_2(2CH_3CN)$ (0.76 g, 3.0 mmol) was weighed into a 200 ml, two-necked, round-bottomed flask fitted with a magnetic stir bar, stopcock, and a rubber serum cap. The flask was alternately evacuated and filled with argon (four cycles). Addition of THF (100 ml) and stirring for 0.2 hr at room temperature produced an amber suspension. Addition of the olefin followed by stirring for 0.2 hr gave a homogeneous solution. Addition of HMPA (10-20 equiv per Pd) caused a slight darkening of the amber color. After 0.1 hr the flask was cooled to -78°C. Triethylamine (2.0 equiv per Pd) was added dropwise over 0.2 hr followed by stirring at -78°C for 0.2 hr. The cold bath was allowed to warm to -60°C (ca. 0.25 hr), at which time the anion was added as a THF solution over 0.2 hr via a precooled (dry ice) syringe. The cold bath was maintained at -60°C for 0.5 hr. To obtain saturated (reduced) products the flask was flushed with hydrogen, a hydrogen balloon affixed to the stopcock, and the cold bath removed. The reaciton mixture was allowed to warm to room temperature and stirred overnight under an atmosphere of hydrogen. The resulting suspension was filtered and the solvent concentrated on a rotary evaporator. Alternatively, the unsaturated (β-elimination) products were obtained by simply allowing the reaction flask to warm to room temperature under an argon atmosphere. After 2 hr the heterogeneous reaction mixture was filtered and the solvent concentrated on a rotary evaporator.

The products were separated from the HMPA by the following procedure. The mixture was taken up in ether (100 ml) and washed with water (3 x 60 ml) and brine (1 x 60 ml). The organic phase was dried over $MgSO_4$ and filtered, and the solvent concentrated on a rotary evaporator. The crude products obtained in this manner were virtually free of HMPA and could be purified by standard methods.

<u>General Procedure for the Catalytic Cyclization of 2-Allylanilines</u> (equation 50)[68]

In a one-neck flask were placed $PdCl_2(CH_3CN)_2$ (1-10 mol %), benzoquinone (1 equiv), and LiCl (10 equiv). THF (15 ml/mmol substrate) was then added, and was stirred for 3-5 min. The substrate in THF (3-4 ml/mmol substrate) was added to the flask via syringe and the solution was refluxed for 10 hr. The THF was removed on the rotary evaporator and the residue was taken up in approximately 20 ml of ether and stirred for 20 min with a small amount of decolorizing charcoal. The solution was filtered and the residue purified by preparative layer chromatography.

<u>Preparation of 2-Methylindole</u>

2-(2-Propenyl)aniline (1.0 g, 7.52 mmol), $PdCl_2(CH_3CN)_2$ (0.195 g, 0.75 mmol), benzoquinone (0.812 g, 7.52 mmol), and LiCl (3.158 g, 75.2 mmol) were combined as usual in 95 ml of THF. After 5 hr at reflux, the solvent was removed and the residue was stirred with ether and decolorizing charcoal for approximately 20 min and filtered. The filtrate was washed five times with 50 ml portions of 1 M NaOH. The solvent was removed on the rotary evaporator and the residue was placed on a silica gel column and eluted with 3:1 petroleum ether/ether. 2-Methylindole (0.818 g 86%) was collected as a white, crystalline solid, identical with authentic material.

<u>Preparation of Fp(η^2-5-hexen-2-one)BF_4</u> (**1**) (equation 65)[86]

Fp(isobutene)BF_4 (8.32 g, 26 mmol) was dissolved in 250 ml of methylene chloride in a flask fitted with a septum cap and a reflux condenser.

As the temperature of the solution was slowly raised, 5-hexen-2-one (6 ml, 52 mmol) was added by syringe, and the resulting orange solution was maintained at reflux for 4 hr. Slow addition of ether to the cooled solution precipitated a light orange solid, which was collected and washed thoroughly with ether. Recrystallization from nitromethane-ether afforded 8.21 g (87%) of product as a light yellow crystalline solid: NMR (CD_3NO_2) δ 5.69(s, 5, Cp), 5.13(m, 1, CH=), 3.95(d, 1, J=8Hz, cis, CH_2=), 3.53(d, 1, J=15Hz, trans, CH_2=), 2.75(m, 4, CH_2CH_2), 2.14(s, 3, CH_3); IR (CH_3CN) 2085, 2045, 1715 cm^{-1}. Anal. Calc'd for $C_{13}H_{15}BF_4FeO_3$: C, 43.14; H, 4.18. Found: C, 42.68; H, 4.17.

Preparation of 2-Methyl-5-[Fp(methyl)]-1-pyrroline

Complex **1** (1.8 g, 5 mmol) was dissolved in 45 ml of methylene chloride in a Schlenk flask. Ammonia gas was slowly passed over the organge solution, which turned a deep reddish color while being stirred for 2 hr at room temperature. Ether (50 ml) was then added, and the resulting mixture was filtered through a small volume of Celite and concentrated in vacuo to a viscous red oil. The crude product was chromatographed on basic alumina. Elution with ether-methylene chloride (1:1) afforded 0.93 g (68%) of a red orange oil: NMR (DCl_3) δ 4.79(s, 5, Cp), 3.88(m, 1, CH), 2.6-1.9(m, 4, CH_2CH_2), 1.99(s, 3, CH_3), 1.6-1.1(m, 2, $FpCH_2$); ^{13}C NMR ($CDCl_3$) 217.6 (CO), 171.7(C=N), 85.4(Cp), 79.4(C-N), 39.2, 32.6(CH_2CH_2), 19.9(CH_3), 8.8($FpCH_2$); IR (CH_2Cl_2) 2010, 1937, 1640 cm^{-1}.

Tricarbonyl(4-methoxy-1-methylcyclohexadienylium)iron Hexafluorophosphate (4) (equation 72)[93]

Birch reduction of p-methylanisole gave the unconjugated diene which was treated with a catalytic amount of ρ-toluenesulfonic acid (100 mg/100 g diene) at 80°C under nitrogen for 2 hr, followed by distillation to give the equilibrium mixture of 1,4-diene and 1-methoxy-4-methylcyclohexa-1,3-diene (1:3, 93% recovery). This mixture (93.0 g) was stirred with pentacarbonyliron (250 ml) in di-n-butyl ether (700 ml) under reflux and nitrogen atmosphere for 48 hr (efficient stirring is essential for good yields). The reaction mixture was cooled and filtered through celite, and solvent and excess pentacarbonyliron were removed on the rotary evaporator, to give crude tricarbonyl(1-methoxy-4-methylcyclohexa-1,3-diene)iron (140 g, 58%), which was sufficiently pure for the next step. Treatment with triphenymethylium tetrafluoroborate (190 g) in dichloromethane (1.51) at room temperature for 1 hr, followed by evaporation of most of the solvent and precipitation with wet ether gave tricarbonyl(4-methoxy-1-methylcyclohexadienylium)iron tetrafluoroborate which was converted into the more stable hexafluorophosphate (2) quantitatively by precipitation from a saturanted aqueous solution with a 20% excess of ammonium hexafluorophosphate, yield 184 g (82%).

Tricarbonyl[methyl-1-(η^2-5,4-methoxy-1-methylcyclohexa-2,4-dienyl)-2-oxo-cyclopentanecarboxylate]iron, Diastereoisomers

To a stirred solution of potassium t-butoxide (7.0 g) in dry tetrahydrofuran (300 ml) under nitrogen was added dropwise methyl 2-oxocyclopentanecarboxylate (10.0 g) in tetrahydrofuran (100 ml). The suspension was stirred for 30 min and then cooled to 0°C. With back flushing of nitrogen the flask was opened while tricarbonyl(4-methoxy-1-methylcyclohexa-2,4-dienylium)iron hexafluorophosphate (4) (20.0 g) was added in portions. Complete reaction was evidenced by dissolution of (4) (ca. 15 min), whereupon most of the tetrahydrofuran was removed on the rotary evaporator at 40-50°C. Ether extraction in the usual way afforded the crude partly crystalline mixture of diastereoisomers which was recrystallized from 30% ether in pentane (-40°, overnight) to give yellow crystals of (5) ca. 90% single diastereoisomer (9.0 g, 45%). Further recrystallization afforded pure (5), mp 125-127°, γ_{max} ($CHCl_3$) 2045, 1975, 1750, and 1720 cm^{-1}; δ 5.00(1, H, dd, $J_{2,3}$ 6, $J_{3,5}$ 2.5Hz, 3H). 3.70(3H, s, CO_2Me), 3.63(3H, s, OMe), 3.30(1H, m, 5H), 2.79(1H, dd, J_{gem} 16, $J_{5,6}$ 2.5Hz, exo-6-H), 2.41(1H, d, $J_{2,3}$ 6Hz, 2H), 2.6-1.7(6H, m, 3 x CH_2), 1.50(1H, dd, J_{gem} 16, $J_{5,6}$ 3.5Hz, endo-6-H), 1.09(3H, s, Me); m/e 404(M^+) (Found: C, 53.3; H, 5.1. $C_{18}H_{20}FeO_7$ requires C, 53.5; H 5.0%).

Evaporation of the liquors followed by recrystallization from 10% ether in pentane, gave pure (6) (4.5 g). Repeated evaporation of liquors followed by recrystallization from 5% ether in pentane gave 95% single diastereoisomer (6) (5.5 g) which could be further recrystallized. The total yield of (6) was 9.5 g (48%), mp 78-80°, γ_{max} ($CHCl_3$) 2045, 1975, 1750, 1725 cm^{-1}; δ 4.98(1H, dd, $J_{2,3}$ 6.5, $J_{3,5}$ 2.5Hz), 3.70(3H, s, CO_2Me), 3.64(3H, s, OMe), 3.30(1H, m, 5H), 2.91(1H, d, $J_{2,3}$ 6.5Hz, 2H), 2.50-1.50(7H, m, 3 x CH_2 and exo-6-H), 1.34(1H, dd, J_{gem} 15, $J_{5,6}$ 3.5Hz, endo-6-H), and 1.18(3H, s, Me); m/e 404(M^+) (Found: C, 53.55; H, 5.1%).

dℓ-2,3-Bis(trimethylsilyl)estra-1,3,5(10)-trien-17-one (equation 79)[96]

The cyclopentanone (146 mg, 0.64 mmol) and $CpCo(CO)_2$ (10 μl, 0.08 mmol) in BTMSA (10 ml) were added to $CpCo(CO)_2$ (10 μl, 0.08 mmol) in refluxing BTMSA (25 ml) over a period of 35 hr and then the mixture was refluxed for an additional 6 hr. The solvent was vacuum transferred leaving a reddish brown residue which was chromatographed on silica gel (75 g, ether-petroleum ether (8:92) as eluent) to give two compounds: (A) the benzocyclobutene diastereomers 23 as an oil (142 mg, 56%) [R_f 0.53 (ether-petroleum ether (1:4) as eluent); m/e (rel intensity) 398(M^+, 5.91), 383(4.05), 294(13.28), 259(7.70), 187(19.64), 149(14.36), 73(100); IR (neat) 3110, 2995, 1640, 1250, 1180, 1090, 840, and 760 cm^{-1}; NMR (CCl_4) δ 7.33(br s, 2H), 5.83(m, 1H), 5.06(m, 2H), 3.60-2.43(m, 3H), 2.43-1.10(m, 9H), 0.88(s, 3H), 0.37(s, 18H)] and (B) the bis(trimethylsilyl)-estratrienone (46 mg, 18%) as colorless crystals [mp 155-157°C (recrystallized from petroleum ether); R_f 0.39; m/e (rel intensity) 398(M^+, 27.47), 384(22.97), 383(73.37), 368(16.45), 367(47.75), 147(11.77), 131(12.82), 97(17.25), 81(17.58), 73(100); IR ($CHCl_3$) 2995, 1745, 1440, 1370, 1250, 940, 840, and 755 cm^{-1}; NMR (CCl_4) δ 7.52(brs, 1H), 7.27(brs, 1H), 3.0-1.0(m, 15H), 0.90(s, 3H), 0.37(s, 18H). Anal. Calc'd for $C_{24}H_{38}OSi_2$: C, 72.29; H, 9.60. Found: C, 72.47; H, 9.47.

The benzocyclobutenes 23 (142 mg) were dissolved in degassed decane (35 ml) and refluxed for 20 hr. The solvent was vacuum transferred and the residue was filtered through silica and crystallized from petroleum ether to give more estratrienone 24a (135 mg, 95% conversion; 181 mg total, 71%).

dℓ,Estra 1,3,5(10)-trien-17-one (24b)

The bis(trimethylsilyl)estratrienone 24a (55.1 mg, 0.138 mmol) was dissolved in CCl_4 (1 ml) and CF_3CO_2H (1 ml) and stirred for 20 hr. Standard ethereal aqueous workup gave a colorless solid (34 mg, 97%) identical with an authentic sample of d-estra-1,3,5(10)-trien-17-one: mp 107-109°C (recrystallized from petroleum ether, lit. for d-estratrienone 138-141°C); m/e (rel intensity) 254(M^+, 100); 210(33.72); 198(29.83); 197(37.04); 156(27.58); 143(33.04); 128(42.42); 115(32.42); 97(19.23); 91(25.06); 43(47.36); IR ($CHCl_3$) 2955, 1740, 1495, 1455, 1260, 1050, 740 cm^{-1}; NMR (CCl_4) δ 7.04(m, 4H), 3.1-1.1(m, 15H), 0.89(s, 3H); ^{13}C NMR (C_6D_6) δ 217.66, 140.00, 136.44, 129.17, 127.02, 50.24, 47.61, 44.51, 38.14, 35.48, 32.05, 29.55, 26.59, 25.81, 22.35, 13.58. Anal. Calc'd for $C_{18}H_{22}O$: C, 84.99; H, 8.72. Found: C, 84.77; H, 8.63.

Preparation of π-(2-Methoxyallyl)nickel Bromide (1) (equation 87)[100]

(a) From 2-Methoxyallyl bromide and nickel carbonyl. CAUTION. Nickel carbonyl is an exceedingly toxic, volatile (bp 43°) liquid. All operations involving this compound must be carried out in an efficient draft hood.

Benzene (380 ml) was placed in a 1ℓ three-necked flask, containing a large magnetic bar and fitted with a stopcock, a serum cap, and a cold finger condenser filled with a salt-ice mixture and vented through an empty trap to prevent suck-back, and then a trap filled with concentrated HNO_3 to decompose any nickel carbonyl which escaped from the reaction vessel. The system was thoroughly flushed with argon. With the system open to the trap, nickel carbonyl (38.5 ml) was added via syringe, and the mixture was heated to 50°. The 2-methoxyallyl bromide (70% pure) (16 g, 75 mmol) was added, and the mixture was heated to reflux with rapid stirring. A deep red color rapidly developed, and CO evolution began.

After 0.75 hr at reflux, the mixture was very dark red, and CO evolution had slowed considerably. The solution was cooled to 25° and the condenser replaced with a stopper. The solution was evaporated to dryness under aspirator vacuum through a -78° trap followed by a -196° trap. The red-brown nonvolatile residue was taken up in argon-saturated ether and filtered under a positive pressure of argon, and the residue on the filter was washed with ether until the washes were colorless. The dark red filtrate was concentrated under aspirator vacuum until a brick red precipitate began to form. The solution was then cooled to -78° to complete crystallization, and the supernatant was removed through a syringe needle attached to an aspirator. After drying under vacuum, the yield was 12.9 g (80%) of a brick red solid: mp 96-98° (sealed capillary); IR (KBr), 2980(w), 2940(w), 1720(w), 1480(s), 1470(s), 1460(s), 1432(m), 1332(s), 1260(w), 1186(w), 1060(s), 950(w), 896(m), 820(s) cm^{-1}; NMR (benzene-TMS) AB quartet, δ_A 1.31(2H, protons anti to OCH_3), and δ_B 2.53(2H, protons syn to OCH_3, J_{AB} = 2.5Hz), singlet, δ 3.29(3H, OCH_3); mass spectrum, parent ion cluster m/e = 416, 418, 410, 422, 424, ratio 6:18:5:18:7:1. Anal. Calc'd for $C_8H_{14}Br_2Ni_2O_2$: Ni, 27.99; Br, 38.15. Found: Ni, 27.62; Br, 38.00.

(b) From 2-methoxyallyl bromide and bis(1,5-cyclooctadiene)nickel. Bis(1,5-cyclooctadiene)nickel (2.63 g, 9.60 mmol) was suspended in argon-saturated benzene (40 ml) in a 100 ml two-neck flask fitted with a stopcock and a dropping funnel and was cooled to 4°. The 2-methoxyallyl bromide (1.45 g, 9.6 mmol) in 5 ml of benzene was added dropwise to the stirred slurry. A deep red color developed immediately, and some metallic nickel separated. The mixture was allowed to warm to 25° and was stirred at this temperature for 0.5 hr. It was then filtered under argon and concentrated to 20 ml under aspirator vacuum, and petroleum ether (75 ml) was added. The resulting red slurry was cooled to -20° to complete crystallization, and the supernatant was removed with a syringe. After drying under vacuum the yield was 1.00 g (55%) of a brick red solution.

General Procedure for the Reaction of 1 with Organic Substrates (equation 87)[100]

Reactions were carried out in a 100 ml one-neck flask with a side arm capped with a serum cap containing a magnetic stirring bar and fitted with a stopcock. The reaction flask was flushed with argon and placed in a nitrogen-filled glove bag along with the flask containing complex 1. The desired amount of 1 (1-2 mmol) was transferred into the reaction flask through the side arm (in the glove bag), the side arm was recapped with the serum cap, and the reaction flask was removed from the glove bag. The complex was dissolved in argon-saturated DMF (30 ml solvent/mmol complex) giving a deep red solution. Liquid reactants (1.8-3.6 mmol) were directly added to the reaction flask, while solid reactants were dissolved in a minimum amount of DMF and added as solutions. With organic halides as reactants, the solution turned emerald green upon completion, whereas with ketones and aldehydes, it turned brown-orange. Upon completion, the reaction mixture was poured into a separatory funnel containing 50 ml of aqueous 3% HCl and 50 ml of ether and was thoroughly shaken. The aqueous phase was washed with three 20-ml portions of ether, and the combined ether extracts were washed with three 50 ml portions of aqueous 3% HCl to ensure complete hydrolysis of the enol ether and complete removal of DMF. The organic phase was dried over anhydrous $MgSO_4$ and solvent was removed under vacuum. The crude product, usually more than 90% pure, was purified by silica gel preparative layer chromatography or distillation.

Cyclization of (E)-1-Acetoxy-2-penten-5-yl Benzenesulfonylacetate (equation 105)[117]

Sodium hydride (42.1 g, 1.11 mmol) was added in one portion to a cooled (0°C) solution of the title compound (360 mg, 1.10 mmol) in 2 ml of THF. The mixture was stirred at 0°C for 15 min then at room temperature for 30 min. The resulting clear, yellow solution was diluted to a total volume of 12 ml with THF and added via a syringe pump at a rate of 5 ml/hr to a refluxing solution of tetrakis(triphenylphosphine)palladium(0) (63.5 mg, 0.055 mmol) and 1,2-bis-(diphenylphosphino)ethane (43.8 mg, 0.110 mmol) in 10 ml of THF. The needle used for the addition was fitted with a small glass wool plug. Upon completion of the addition, the cloudy, yellow mixture was cooled, filtered through a small pad of silica gel, and rinsed with ether. The combined organic portions were concentrated in vacuo. The resulting yellow oil was purified via preparative TLC (40% ethyl acetate in hexane) to give 193.4 mg (73%) of the eight membered ring lactone. Recrystallization from absolute ethanol gave short, white needles, mp 87.5-89°C, of pure (Z)-2-benzenesulfonyl-4-hepten-7-olide. Analysis of the chromatographed product mixture by 270 MHz NMR indicated two signals at δ 5.88 and 5.71 assigned as the two olefinic protons. NMR (270 MHz, $CDCl_3$) δ 7.96(d of m, J=8Hz, 2H), 7.69(t of t, J=7.5, 1.5Hz, 1H), 7.58(t of m, J=8Hz, 2H), 5.88(d of t of d, J=11, 7.8, 1.2Hz, 1H), 5.71(d of d of d of d, J=11, 9.3, 7.4, 1.5H, 1H), 4.07(d of d, J=10.7, 4.6Hz, 1H), 3.19(d of d of d of d, J=12.5, 10.5, 9.5, 1Hz, 1H), 2.68(d of d of d, J=12.5, 7.5, 4.8Hz, 1H), 2.61(m, 1H), 2.17(d of d of d of d, J=14.4, 8.2, 3.6, 2.1Hz, 1H). IR (CCl_4): 1762, 1467, 1452, 1358, 1340, 1331, 1250, 1188, 1155, 1085, 1040 cm^{-1}. Anal. Calc'd for $C_{13}H_{14}O_4S$: 266.0613. Found: 266.0607.

VIII REFERENCES

1. For general reviews on the use of metal carbonyls in synthesis see: I. Wender and P. Pino, eds., "Organic Synthesis via Metal Carbonyls," Vol. 1, (1968), Vol. 2, (1977), Wiley-Interscience, New York; M. Ryang, Organometal. Chem. Rev., A., 5, 67 (1970); M. Ryang and S. Tsutsumi, Synthesis, 55 (1971); H. Alper, J. Organometal. Chem. Library, 1, 305 (1976).

2. a) I. D. Webb and G. T. Borcherdt, J. Am. Chem. Soc., 73, 2654 (1951); b) E. J. Corey, M. F. Semmelhack, and L. S. Hegedus, J. Am. Chem. Soc., 90, 2416, 2417 (1968).

3. E. J. Corey and E. Wat, J. Am. Chem. Soc., 89, 2757 (1967).

4. E. J. Corey and E. Hamanaka, J. Am. Chem. Soc., 86, 1641 (1964); 89, 2758 (1967).

5. W. G. Dauben, G. H. Beasely, M. D. Broadhurst, B. Mulber, D. J. Peppard, P. Pesnelle, and C. Sutter, J. Am. Chem. Soc., 96, 4724 (1974).

6. E. J. Corey and H. A. Kirst, J. Am. Chem. Soc., 94, 667 (1972).

7. L. Crombie, G. Kneen, G. Pattenden, and D. Whybrow, J. Chem. Soc., Perkin I, 1711 (1980).

8. R. Noyori, Accts. Chem. Res., 12, 61 (1979).

9. R. Noyori, Y. Hayakawa, H. Takaya, S. Muria, R. Kobayashi, and N. Sonoda, J. Am. Chem. Soc., 100, 1759 (1978).

10. H. Takaya, S Makino, Y. Hayakawa, and R. Noyori, J. Am. Chem. Soc., 100, 1765 (1978).

11. H. Takaya, Y. Hayakawa, S. Makino, and R. Noyori, J. Am. Chem. Soc., 100, 1778 (1978).

12. R. Noyori, T. Sato, and Y. Hawakawa, J. Am. Chem. Soc., 100, 2561 (1978).

13. Y. Hayakawa, Y. Baba, S. Makino, and R. Noyori, J. Am. Chem. Soc., **100**, 1786 (1978).

14. R. Noyori, F. Shimizu, K. Fukuta, H. Takaya, and Y. Hayakawa, J. Am. Chem. Soc., **99**, 5196 (1977).

15. Y. Hayakawa, K. Yokoyama, and R. Noyori, J. Am. Chem. Soc., **100**, 1791 (1978).

16. Y. Hayakawa, K. Yokoyama, and R. Noyori, J. Am. Chem. Soc., **100**, 1799 (1978); Org. Synth., **58**, 56 (1978).

17. a) T. Ishizu, M. Mori, and K. Kanematsu, J. Org. Chem., **46**, 526 (1981); b) T. Ishizu, K. Harano, M. Yasuda, and K. Kanematsu, J. Org. Chem., **46**, 3630 (1981).

18. a) M. P. Cooke, J. Am. Chem. Soc., **92**, 6080 (1970); b) J. P. Collman, Accts. Chem. Res., **8**, 342 (1975).

19. J. P. Collman, R. G. Finke, J. N. Cawse, and J. I. Brauman, J. Am. Chem. Soc., **99**, 2515 (1977); **100**, 4766 (1978) and references therein.

20. R. F. Heck in "Organic Synthesis via Metal Carbonyls," I. Wender and P. Pino, Eds., Wiley Interscience, New York, 1968, Vol. I, pp. 373-404.

21. a) E. J. Corey and L. S. Hegedus, J. Am. Chem. Soc., **91**, 1233 (1969); b) E. J. Corey, H. A. Kirst, and J. A. Katzenellenbogen, J. Am. Chem. Soc., **92**, 6314 (1970).

22. a) L. S. Hegedus, Ph.D. Thesis, Harvard University, 1970, pp. 85-87; b) E. J. Corey and L. S. Hegedus, J. Am. Chem. Soc., **91**, 4926 (1969).

23. a) K. Tamao, K. Sumitani, Y. Kiso, M. Zembayashi, A. Fijioka, S-i Kodama, I. Nakajima, A. Minato, and M. Kumada, Bull. Chem. Soc., Japan, **49**, 1958 (1976); b) M. Kumada, Pure Appl. Chem., **52**, 669 (1980).

24. T. Hayashi, T. Mise, M. Fukushima, M. Kagotani, N. Nagashima, Y. Hamada, A. Matsumoto, S. Kawakami, M. Konishi, K. Yamamoto, and M. Kumada, Bull. Chem. Soc., Japan, **53**, 1138 (1980).

25. T. Hayashi, M. Konishi, M. Fukushima, T. Mise, M. Kagotani, M. Tajika, and M. Kumada, J. Am. Chem. Soc., 104, 180 (1982).

26. a) R. F. Heck, Proc. of the Welch Foundation, XVII, 53 (1973); b) R. C. Larock, Angew. Chem. Int. Ed., 17, 27 (1978); c) J. Organomet. Chem. Libr., 1, 257 (1976).

27. S. F. Dyke and M. J. McCartney, Tetrahedron, 37, 431 (1981).

28. a) D. E. Bergstrom and J. L. Ruth, J. Am. Chem. Soc., 98, 1587 (1976); b) C. F. Bigge, P. Kalarites, and M. P. Mertes, Tetrahedron Lett., 1653 (1979); c) J. L. Ruth and D. E. Bergstrom, J. Org. Chem., 43, 2870 (1978).

29. D. E. Bergstrom, H. Inoue, and P. Anantha Reddy, J. Org. Chem., 47, 2174 (1982).

30. C. F. Bigge, P. Kalarites, J. R. Peck, and M. P. Mertes, J. Am. Chem. Soc., 102, 2033 (1980).

31. D. E. Bergstrom, J. L. Ruth, and P. Warwick, J. Org. Chem., 46, 1432 (1981).

32. a) I. Arai, G. D. Daves, Jr., J. Org. Chem., 43, 4110 (1978); b) I. Arai, G. D. Daves, Jr., J. Am. Chem. Soc., 100, 287 (1978); D. E. Berstron, A. J. Brattesani, M. K. Ogawa, and M. Schweickert, J. Org. Chem., 46, 1423 (1981).

33. R. C. Larock, D. R. Leach, and S. M. Bjorge, Tetrahedron Lett., 23, 715 (1982).

34. R. C. Larock, B. Riefling, and C. A. Fellows, J. Org. Chem., 43, 131 (1978).

35. R. C. Larock and B. Riefling, J. Org. Chem., 43, 1468 (1978).

36. a) H. Yatagai, Bull. Chem. Soc., Japan, 53, 1670 (1980); b) K. Takai, K. Oshima, and H. Nozaki, Tetrahedron Lett., 21, 2531 (1980).

37. A. Cowell and J. K. Stille, Tetrahedron Lett., 133 (1979).

38. M. Mori, K. Chiba, and Y. Ban, J. Org. Chem., **43**, 1684 (1978).

39. M. Mori, K. Chiba, and Y. Ban, Heterocycles, **6**, 1841 (1977).

40. M. Mori, K. Chiba, M. Okita, and Y. Ban, J. C. S. Chem. Commun., 698 (1979).

41. M. Ishikura, M. Mori, T. Ikeda, M. Terashima, and Y. Ban, J. Org. Chem., **47**, 2456 (1982).

42. G. Pandey and K. P. Tiwari, Synth. Comm., **10**, 523 (1980).

43. a) M. Mori, K. Chiba, and Y. Ban, Tetrahedron Lett., 1037 (1977); b) M. Mori and Y. Ban, Tetrahedron Lett., 1803, 1807 (1976); 1133 (1979); M. Mori and Y. Ban, Heterocycles, **9**, 391 (1978).

44. R. Odle, B. Blevins, M. Ratcliff, and L. S. Hegedus, J. Org. Chem., **45**, 2709 (1980).

45. H. Iida, Y. Yuasa, and C. Kibayashi, J. Org. Chem., **45**, 2938 (1980).

46. F. Ziegler, U. Chakrabortz, and R. B. Weisenfeld, Tetrahedron, **37**, 4035 (1981).

47. a) D. Milstein and J. K. Stille, J. Am. Chem. Soc., **101**, 4981; 4992 (1979); b) D. Milstein and J. K. Stille, J. Org. Chem., **44**, 1613 (1979).

48. a) B. M. Trost and E. Keinan, Tetrahedron Lett., **21**, 2591, 2595 (1980); b) J. Godschalx and J. K. Stille, Tetrahedron Lett., **21**, 2599 (1980).

49. a) N. Miyaura, T. Yanagi, and A. Suzuki, Tetrahedron Lett., **22**, 127 (1981); b) R. Rossi, A. Carpita, and M. G. Quirici, Tetrahedron, **37**, 2617 (1981); c) N. Miyaura and A. Suzuki, J. Organomet. Chem., **213**, C53 (1981); d) N. Miyaura, K. Maeda, H. Suginone, and A. Suzuki, J. Org. Chem., **47**, 2117 (1982).

50. H. Matsushida and E-i Negishi, J. Am. Chem. Soc., **103**, 2882 (1981).

51. P. Vinçent, J. P. Beaucourt, and L. Pichat, Tetrahedron Lett., **23**, 63 (1982).

52. M. Kobayashi and E-i Negishi, J. Org. Chem., 45, 5223 (1980).

53. E-i Negishi, Pure Appl. Chem., 53, 2333 (1981).

54. C. E. Russell and L. S. Hegedus, J. Am. Chem. Soc., in press.

55. For reviews on cyclometallation see: a) M. I. Bruce, Angew. Chem. Int. Ed., 16, 73 (1977); b) H. P. Abicht and K. Issleib, Z. Chem., 17, 1 (1977); c) J. Dehand and M. Pfeffer, Coord. Chem. Rev., 18, 327 (1976).

56. J. M. Thompson and R. F. Heck, J. Org. Chem., 40, 2667 (1975).

57. a) R. A. Holton, Tetrahedron Lett., 355 (1977); b) R. A. Holton and R. G. Davis, J. Am. Chem. Soc., 99, 4175 (1977).

58. H. Horino and N. Inoue, J. Org. Chem., 46, 4416 (1981).

59. P. M. Henry, "Palladium Catalyzed Oxidation of Hydrocarbons," D. Reidel, Dordrecht, Holland, 1980.

60. B. Akermark, J. E. Backvall, L. S. Hegedus, K. Siirala-Hansen, K. Sjoberg, and K. Zetterberg, J. Organomet. Chem., 72, 127 (1974).

61. a) L. S. Hegedus, K. Siirala-Hansen, J. Am. Chem. Soc., 97, 1184 (1975); b) L. S. Hegedus, O. P. Anderson, K. Zetterberg, G. Allen, K. Siirala-Hansen, D. J. Olsen, and A. B. Packard, Inorg. Chem., 16, 1887 (1977).

62. J. E. Backvall and E. E. Bjorkman, J. Org. Chem., 45, 2893 (1980).

63. J. E. Backvall, J. C. S. Chem. Commun., 413 (1977).

64. J. E. Backvall, E. E. Bjorkman, S. E. Bystrom, A. Solladie-Cavallo, Tetrahedron Lett, 23, 943 (1982).

65. L. S. Hegedus, R. E. Williams, M. A. McGuire, and T. Hayashi, J. Am. Chem. Soc., 102, 4973 (1980).

66. L. S. Hegedus and W. H. Darlington, J. Am. Chem. Soc., 102, 4980 (1980).

67. T. Hosokawa, S. Miyagi, S-i Murahashi, and A. Sonoda, J. Org. Chem., **43**, 2752 (1978).

68. L. S. Hegedus, G. F. Allen, J. J. Bozell, and E. L. Waterman, J. Am. Chem. Soc., **100**, 5800 (1978).

69. D. E. Korte, L. S. Hegedus, and R. K. Wirth, J. Org. Chem., **42**, 1329 (1977).

70. L. S. Hegedus, P. M. Winton, and S. Varaprath, J. Org. Chem., **46**, 2215 (1981).

71. S. Hatano, M. Sarurvatari, K. Isomura, and H. Taniguchi, Heterocycles, **15**, 747 (1981).

72. T. Hosokawa, M. Hirata, S-i Murahashi, and A. Sonoda, Tetrahedron Lett., 1821 (1976).

73. A. Kasahara, T. Izumi, K. Sato, K. Maemura, and T. Hayasaka, Bull. Chem. Soc., Japan, **50**, 1899 (1977).

74. T. Izumi and A. Kasahara, Bull. Chem. Soc., Japan, **48**, 1673 (1975).

75. K. Maeda, T. Hosokawa, S-i Murahashi, and I. Moritani, Tetrahedron Lett., 5075 (1973).

76. A. Kasahara and T. Saito, Chem. Ind. (London), 745 (1975).

77. A. Kasahara, Chem. Ind. (London), 1032 (1976).

78. A. Kasahara and N. Fukuda, Chem. Ind. (London), 485 (1976).

79. L. S. Hegedus and J. M. McKearin, J. Am. Chem. Soc., **104**, 2444 (1982).

80. L. S. Hegedus, G. F. Allen, and D. J. Olsen, J. Am. Chem. Soc., **102**, 3583 (1980).

81. B. M. Trost and J. M. D. Fortunak, Organometallics, **1**, 7 (1982).

82. A. S. Kende, B. Roth, and P. J. Sanfilippo, *J. Am. Chem. Soc.*, **104**, 1784 (1982).

83. P. Lennon, M. Madhavarao, A. Rosan, and M. Rosenblum, *J. Organomet. Chem.*, **108**, 93 (1976).

84. P. M. Lennon, A. M. Rosan, and M. Rosenblum, *J. Am. Chem. Soc.*, **99**, 8426 (1977).

85. M. Madhavarao, D. F. Marten, and M. Rosenblum, *J. Am. Chem. Soc.*, **99**, 2823 (1977).

86. S. R. Berryhill and M. Rosenblum, *J. Org. Chem.*, **45**, 1984 (1980).

87. a) T. C. T. Chang, M. Rosenblum, and S. B. Samuels, *J. Am. Chem. Soc.*, **102**, 5930 (1980); b) T. C. T. Chang and M. Rosenblum, *J. Org. Chem.*, **46**, 4103 (1981).

88. T. C. T. Chang and M. Rosenblum, *J. Org. Chem.*, **46**, 4626 (1981).

89. a) A. J. Pearson, *Accts. Chem. Res.*, **13**, 463 (1980); *Transition Met. Chem.*, **6**, 67 (1981); b) A. J. Birch and I. D. Jenkins in H. Alper, Ed., *Transition Metal Organics on Organic Synthesis*, Vol. 1, Academic Press, New York, 1976, pp. 1-75.

90. For a review of this area see: L. S. Hegedus, "Nucleophilic Attack on Transition Metal Organometallic Compounds", in "Chemistry of the Metal-Carbon Bond," S. Patai, F. R. Hartley Eds., Wiley, New York, in press.

91. L. F. Kelly, P. Dahler, A. S. Narula, and A. J. Birch, *Tetrahedron Lett.*, **22**, 1433 (1981).

92. A. J. Pearson and P. R. Rathby, *J. C. S. Perkin I*, 395 (1980).

93. A. J. Pearson and C. W. Ong, *J. Am. Chem. Soc.*, **103**, 6686 (1981).

94. A. J. Pearson and D. C. Rees, *J. Am. Chem. Soc.*, **104**, 1118 (1982).

95. K. P. C. Vollhardt, *Accts. Chem. Res.*, **10**, 1 (1977) and references therein.

96. R. L. Funk and K. P. C. Vollhard, J. Am. Chem. Soc., **102**, 5253 (1980).

97. D. J. Brien, A. Naiman, and K. P. C. Vollhardt, J. C. S. Chem. Commun., 133 (1982).

98. For reivews see: a) M. F. Semmelhack, Org. React., **19**, 115 (1972); b) R. Baker, Chem. Rev., **73**, 487 (1973); c) L. S. Hegedus, J. Organomet. Chem. Libr., **1**, 329 (1976).

99. E. J. Corey and M. F. Semmelhack, J. Am. Chem. Soc., **89**, 2755 (1967).

100. L. S. Hegedus and R. K. Stiverson, J. Am. Chem. Soc., **96**, 3250 (1974).

101. a) K. Saito, S. Inoue, and K. Sato, J. C. S. Perkin I, 2289 (1973); b) S. Inoue, R. Yamaguchi, K. Saito, and K. Sato, Bull. Chem. Soc., Japan, **47**, 3098 (1974).

102. L. S. Hegedus, S. D. Wagner, E. L. Waterman, and K. Siirala-Hansen, J. Org. Chem., **40**, 593 (1975).

103. L. S. Hegedus and S. Varaprath, Organometallics, **1**, 259 (1982).

104. B. M. Trost, P. E. Strege, L. Weber, T. J. Fullerton, and T. J. Dietsch, J. Am. Chem. Soc., **100**, 3407 (1977).

105. B. M. Trost and P. J. Metzner, J. Am. Chem. Soc., **102**, 3572 (1980).

106. W. T. Dent, R. Long, and A. J. Wilkinson, J. Chem. Soc., 1585 (1964).

107. B. Akermark, A. Ljungquist, and M. Panunzio, Tetrahedron Lett., **22**, 1055 (1981).

108. D. J. Collins, W. R. Jackson, and R. N. Timms, Tetrahedron Lett., 495 (1976).

109. P. S. Marchand, H. S. Wong, and J. F. Blount, J. Org. Chem., **43**, 4769 (1978).

110. J. S. Temple, M. Riediker, and J. Schwartz, J. Am. Chem. Soc., **104**, 1310 (1982).

111. J. McMurray, J. R. Matz, K. L. Kees, and P. A. Bock, Tetrahedron Lett., **23**, 1777 (1982).

112. For reviews see: a) B. M. Trost, Accts. Chem. Res., **13**, 385 (1980); b) B. M. Trost, Tetrahedron, **33**, 2615 (1977); c) B. M. Trost, Pure and Appl. Chem., **53**, 2357 (1981); d) J. Tsuji, Pure and Appl. Chem., **54**, 0000 (1982); e) J. Tsuji, Pure and Appl. Chem., **53**, 2371 (1981); f) J. Tsuji, Topics Current Chem., **91**, 30 (1980); g) J. Tsuji, "Organic Synthesis with Palladium Compounds," Springer-Verlag, Berlin, 1980.

113. J. P. Genet and F. Piau, J. Org. Chem., **46**, 2414 (1981); Tetrahedron Lett., **23**, 331 (1982).

114. T. Hirao, J. Enda, Y. Ohshiro, and T. Agawa, Tetrahedron Lett., **22**, 3079 (1981).

115. J. Tsuji, H. Ueno, Y. Kobayashi, and H. Okumoto, Tetrahedron Lett., **22**, 2573 (1981).

116. J. Tsuji, H. Kataoka, and Y. Kobayashi, Tetrahedron Lett., **22**, 2575 (1981).

117. B. M. Trost and T. R. Verhoeven, J. Am. Chem. Soc., **99**, 3867 (1977); **102**, 4743 (1980).

118. B. M. Trost and T. R. Verhoeven, J. Am. Chem. Soc., **101**, 1595 (1979).

119. J. Tsuji, Y. Kobayashi, H. Kataoka, and T. Takahashi, Tetrahedron Lett., **21**, 1475 (1980).

120. B. M. Trost and J. P. Genet, J. Am. Chem. Soc., **98**, 8516 (1976).

121. B. M. Trost, S. A. Godleski, and J. P. Genet, J. Am. Chem. Soc., **100**, 3930 (1978).

122. B. M. Trost, S. Godleski, and J. Belletire, J. Org. Chem., **44**, 2052 (1979).

123. S. A. Godleski and R. S. Valpez, J. Org. Chem., **47**, 381 (1982).

124. S. A. Godleski, J. D. Meinhart, D. J. Miller, and S. VanWallendael, Tetrahedron Lett., **22**, 2247 (1981).

125. B. M. Trost and T. P. Klun, J. Am. Chem. Soc., **101**, 6756 (1979).

126. B. M. Trost and T. P. Klun, J. Am. Chem. Soc., **103**, 1864 (1981).

127. B. M. Trost and T. Runge, J. Am. Chem. Soc., **103**, 2485 (1981).

ORGANOCOPPER AND ORGANOMANGANOUS REAGENTS

Professor Jean François Normant
Laboratoire de Chimie des Organo-Eléments
tour 44, 4 place Jussieu
75230 Paris Cédex 05, France

Part 1 PREPARATION OF CONJUGATED DIENES AND ENYNES VIA ORGANO COPPER REAGENTS

Professor Jean François Normant and Dr. Alexandre Alexakis
Laboratoire de Chimie des Organo-Eléments
tour 44, 4 place Jussieu
75230 Paris Cédex 05, France

4-12 1-Ethylthio 1Z, 3E-Hexadiene

4-13 1-Ethoxy-4-Butyl-2(E), 4-pentadiene

1 - OCCURENCE and INTEREST of CONJUGATED DIENES

The conjugated diene unit is extremely widespread among natural products ; although it rapidly caught the attention of chemists and led to a detailed study of electron delocalization, cycloadditions, etc... The preparation of isomerically pure dienes, without tedious purifications, has been an important challenge for years. The goal is to prepare

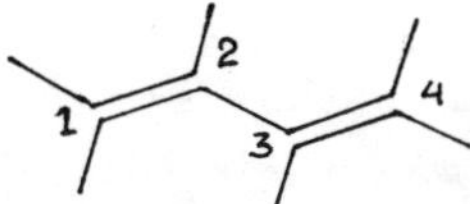

mono- to hexa-substituted butadiene showing a given geometry, namely EE, E-Z, Z-E, Z-Z. Compounds of first type are generaly the most stable, and may be formed unduly during attempted synthesis of the other three types. However, they are the more useful for Diels Alder reactions, since the s-cis conformation may be adopted freely. E-Z isomers have been used however, more recently, in such 2+4 additions[1] and appear to be frequently encountered in sexual pheromones of lepidopterae. Z-Z isomers also exist in nature, but are less frequent. In the following pages, attention will be focussed on the EZ 1-4 disubstituted butadienes, and on tri, tetra ... substituted ones.

2 - VARIOUS SYNTHETIC METHODS LEADING To (mainly E,Z) DIENES

The most well known strategies to prepare these structures are the following :

2-1 Wittig Reaction

From an α, β ethylenic aldehyde (purely E if kept some time at room temperature), or Z, if prepared according to[2] with a Wittig reagent, either in the presence or absence of lithium salts[3,4] it is theoretically possible to prepare all four types of 1-4 disubstituted 1-3 butadienes. However high purity is found only in the preparation of the E,E isomers

$$R\text{-}CH{=}PPh_3 + R'\text{-}CH{=}CH\text{-}CHO \xrightarrow{\text{salt}} R'\text{-}CH{=}CH\text{-}CH{=}CH\text{-}R \ (E,E)$$
$$\xrightarrow{\text{salt free}} R'\text{-}CH{=}CH\text{-}CH{=}CH\text{-}R \ (E,Z)$$

2-2 From Z-conjugated enynes

prepared from Z vinyl copper derivatives (v.i) and 1-halo-1-alkynes[5,6] or from an 1-alkyne[7] or an acetylenic Grignard[7] and a Z 1-halo-1-alkene in the presence of Pd°. A subsequent antihydrogenation of the triple bond leads to the desired Z,E diene.

2-3 From E-conjugated enynes

These may be prepared via boron chemistry, by reaction of E vinyl boranes with either 1-halo alkynes in the presence of Pd° [8,9]

$$Bu\text{-}CH{=}CH\text{-}BSia_2 \ + \ Hex\text{-}C{\equiv}C\text{-}Br \xrightarrow{1\%\ Pd(PPh_3)_4} Hex\text{-}C{\equiv}C\text{-}CH{=}CH\text{-}Bu$$

or with lithium acetylides, followed by carbon migration in the ate complex thus formed[10,11]. Use of bicyclohexyl borane

$$R^1\text{-}CH{=}CH\text{-}B\,cHex_2 \xrightarrow[2/\ I_2\ \ NaOAc]{1/\ LiC{\equiv}C\text{-}R^2} R_1\text{-}CH{=}CH\text{-}C{\equiv}C\text{-}R_2$$

proved particularly efficient[4].

An other excellent approach is the reaction of acetylenic Zinc chlorides with E-1-iodo-1-alkenes in the presence of Palladium[7]

$$R\text{-}C{\equiv}C\text{-}ZnCl \ + \ Bu\text{-}CH{=}CH\text{-}I \xrightarrow{Pd°P(Ph_3)_4} R\text{-}C{\equiv}C\text{-}CH{=}CH\text{-}Bu$$

These enynes are further reduced in a Syn way by Lindlar's semi hydrogenation[13] or by bicyclohexyl[14] or disiamyl boranes[10,15] or Zinc in propanol in the presence of potassium cyanide[16]. Hydro-alanation of terminal enynes is also possible[17,18]

2-4 From diynes

It should be raised, that the Lindlar's hydrogenation of such substrates may lead to a mixture of isomers, e.g. :[19]

Pent-≡-≡-CH_2OH $\xrightarrow[\text{Lindlar}]{H_2}$ Pent–(Z,Z)–CH_2OH + Pent–(Z,E)–CH_2OH + Pent–(E,Z)–CH_2OH + Pent–(E,E)–CH_2OH

75/8/16/ 1%

but that dicyclohexyl borane has led to pure Z,Z dienes[20]

2-5 Miscellaneous methods

Specific methods can be employed when a given functionality in allylic position is hoped for

$\xrightarrow{OH^-}$ R–diene–CH_2N< (21)

Pent–allene–COOEt $\xrightarrow{Al_2O_3}$ Pent–diene–COOEt (22)

OAc $\xrightarrow{Pd^{II}}$ OAc (23)

Pent $\xrightarrow[\text{2, F B(OMe)}_2 \quad \text{3, [O]}]{\text{1, nBuLi,KOtBu}}$ Pent–diene–CH_2OH (19)

2-6 From coupling of vinyl organometallics with vinylic halides

The palladium catalyzed coupling reaction of a Z vinyl Grignard with an E-1-iodo-1 alkene gives good yields of Z-E diene, but the isomeric Z-1-iodo-1-alkene leads to a mixture of isomers.[24]

One of the most useful approach is the use of organo-vinyl Zirconium[25] or -Aluminium[26] prepared by hydrometallation of a terminal, or internal alkyne. These reagents were shown by Negishi[27,28] to couple with 1-iodo-1 alkenes under palladium or nickel catalysis

Pent-≡ —dibal→ Pent⁄⁄AliBu$_2$ —(Bu⁄⁄I, Pd°L$_4$ cat.)→ Pent⁄⁄⁄Bu

in certain cases, it may be appropriate to switch from the alane compound, to the Zinc derivative, which couples even more efficiently[27,28]

Vinyl boranes either E or Z (the latter being less stable), also couple with E or Z 1-halo-1-alkenes, opening a way to all classes of isomeric dienes[9] under palladium catalysis (Suzuki)[8-15]

Bu⁄⁄B(catechol) —(Ph⁄⁄Br, Pd°L$_4$ cat.)→ Bu⁄⁄⁄Ph 86% (99% ZE)

Excellent, and more detailed reviews have appeared[4,15,29,32]

3 - USE OF ORGANO COPPER REAGENTS FOR THE SYNTHESIS OF CONJUGATED DIENES AND ENYNES

3-1 Introduction

From these various examples, the last method (vinyl-vinyl coupling) is the most straight forward since retention of both vinylic partners seems to be a general pattern, through the several substitutions which occur during the palladium catalysis :

R-=-I (E) → R-=-PdIL$_2$ (E) —R'-=-M (Z)→ R-=-Pd-=-R' (E, Z) → R-=-=R' (E Z) + PdL$_4$

However, we are left with the problem of preparing isomerically pure reagents, and the syn hydrometallation of alkynes is best suited for the access to E units

R-≡ —H-M→ R⁄⁄M
R'-≡ —1/H-M, 2/ I$_2$→ R'⁄⁄I
} —"Pd°"→ R⁄⁄⁄R'

Here is the point where organo copper chemistry may be of some help, since carbocupration of acetylenes leads, in high yields and high purity, to Z reagents, or to β,β disubstituted vinyl copper reagents of given geometry

$$R_2CuLi \; + \; 2\; HC{\equiv}CH \longrightarrow (R\text{-}CH{=}CH)_2CuLi$$

$$RCu,\, MgX_2 \; + \; R'C{\equiv}CH \longrightarrow R'(R)C{=}CH\text{-}Cu,\, MgX_2$$

These can be iodinolyzed with total retention of configuration to the corresponding 1-iodo-1-alkenes in high yield.[33]

Also worth of note, is the fact that hydro alumination of alkynes bearing oxygenated functions is almost inoperative,[34] while hydrozirconation operates regioselectively, whatever the extra functions borne by the acetylenic substrate.[27]

In the case of carbocupration of such functionalized alkynes, a high control of the regioselectivity may occur, according to the nature, and proximity of these functions (v.i), the nature of the solvent, and of the main group metal salts.[33]

3-2 Preparation of the vinyl copper reagents[33]

Z,β-Mono substituted vinyl copper reagents are best prepared from the corresponding lithium reagent, either converted to a homocuprate and added to two equivalents of acetylene in ether

$$2\; RLi \; + \; 1\; CuI \longrightarrow R_2CuLi \xrightarrow[\text{ether}]{2\; HC{\equiv}CH} (R\text{-}CH{=}CH)_2CuLi$$

(or 1 $CuBr.Me_2S$)

or converted to a copper reagent or to an unsymetrical cuprate and added to one equivalent of acetylene.

$$RLi \; + \; 1\; CuI \longrightarrow RCu,LiI \xrightarrow{HC{\equiv}CH} R\text{-}CH{=}CH\text{-}Cu,LiI$$

$$'' \; + \; 1\; CuOtBu \longrightarrow RCuOtBuLi \xrightarrow{''} R\text{-}CH{=}CH\text{-}CuOtBu\; Li$$

$$'' \; + \; 1\; CuSPh \longrightarrow RCuSPhLi \xrightarrow{''} R\text{-}CH{=}CH\text{-}CuSPh\; Li$$

$$'' \; + \; 1\; CuC{\equiv}CR' \longrightarrow RCuC{\equiv}CR'Li \xrightarrow{''} R\text{-}CH{=}CH\text{-}CuC{\equiv}CR'\; Li$$

β,β disubstituted vinyl copper reagents are best prepared from the corresponding Grignard reagent, in ether or in THF

$$\text{R-MgX} + \text{CuBr} \longrightarrow \text{RCu, MgXBr} \xrightarrow[\text{ether or THF}]{\text{R'C}\equiv\text{CH}} \text{R(R')C=CH-Cu,MgXBr}$$

(or CuBr, 2 LiCl } soluble
or $CuBr,Me_2S$ } in THF)

and added to a terminal alkyne. THF is best suited for secondary and tertiary alkyl copper reagents. $CuBr,Me_2S$ in a Me_2S-ether mixture leads to more stable species and avoids thermal decomposition to copper zero and the unwanted symmetrical diene R(R')C=CH-CH=CR(R').[35] THF is best suited for the preparation of magnesium homo cuprates[36]

$$2\ \text{RMgX} + 1\ \text{CuX} \longrightarrow \text{R}_2\text{CUMgX} \xrightarrow[\text{THF}]{2\ \text{R'C}\equiv\text{CH}} \text{R(R')C=CH-Cu-R MgX}$$

or unsymmetrical cuprates

$$\text{RMgX} + \text{CuZ} \longrightarrow \text{R-CuZ MgX} \xrightarrow{\text{R'C}\equiv\text{CH}} \text{R(R')C=C-CuZ MgX}$$

Z= OtBu, SPh, C≡C-R"

but the symmetrical homocuprates insert only one mole equivalent of alkyne, so that a vinyl alkyl cuprate results, in which the Csp_3-Cu bond is more reactive than the Csp_2-Cu bond, towards usual electrophiles, and the synthetic use of them is accordingly troublesome. On the other hand, the saturated alkyl-alkoxy cuprates present no extra stabilization, as compared with the corresponding alkyl copper reagents.

3-3 Addition of alkenyl cuprates to alkynes

One of the easiest way to construct a conjugated diene system would be the addition of an alkenyl copper or cuprate derivative to the triple bond of an appropriate alkyne, leading to the formation of a dienic copper or cuprate reagent, which may react further with an electrophile :

$R_2CuLi \xrightarrow{HC\equiv CH}$ (R-CH=CH)$_2$CuLi $\xrightarrow{R'-C\equiv CH}$ (R-CH=CH-C(R')=CH)$_2$CuLi $\xrightarrow{E}$ R-CH=CH-C(R')=CH-E

However, among the organo-copper or cuprate reagents, the methyl-,aryl-,and alkenyl-copper are the less reactive ones as compared to their alkyl counterparts, for the carbocupration reaction.[33] Thus, only reactive alkynes are suitable substrates. In fact, many alkynes may react under appropriate conditions and only the non-functionalized mono- or bisubstituted alkynes are totally inert towards alkenyl copper reagents.

3-3-1 Acetylenic esters

Acetylenes bearing a conjugated carbonyl function, such as esters, ketones or aldehydes are, of course, very prone to add a cuprate reagent. This reaction may also be viewed as a conjugate addition to the ynone system ; the examples reported in the litterature include reactions with mono or bisubstituted esters :

(CH$_2$=CH)$_2$CuLi + nPent-CH(OSiMe$_3$)-C≡C-COOMe $\xrightarrow[-78°C,3h]{THF}$ nPent-CH(OSiMe$_3$)-C(CH=CH$_2$)=CH-COOMe 90% 37

(nPent-CH=CH)$_2$CuLi + HC≡C-COOEt $\xrightarrow[-15°C,\ 10min]{Et_2O}$ nPent-CH=CH-CH=CH-COOEt 90% 38

(EtO-CH=CH)$_2$CuLi + HC≡C-COOMe $\xrightarrow[-10°C,5h]{THF}$ EtO-CH=CH-CH=CH-COOMe 65% 39

Me-CH=C(CH(OEt)$_2$)-CuSPh Li + HC≡C-COOEt $\xrightarrow[-60°C,30min]{Et_2O}$ Me-CH=C(CH(OEt)$_2$)-CH=CH-COOEt 82% 40

THPO-CH$_2$-CH=CH-Cu,LiX + Me-C≡C-COOMe $\xrightarrow[-25°C,\ 4h]{Et_2O}$ THPOCH$_2$-CH=CH-C(Me)=CH-COOMe 44% 41

In these examples only half of the cuprate reagent was used. However, with the particularly reactive propiolic ester, both alkenyl groups may add :

$$\text{nPent}_2\text{CuLi} \xrightarrow{2\ \text{HC}\equiv\text{CH}} (\text{nPent-CH=CH})_2\text{CuLi} \xrightarrow{2\ \text{HC}\equiv\text{C-COOEt}} 2\ \text{nPent-CH=CH-CH=CH-COOEt} \quad 78\%$$ [42]

2(E)-4(Z)-Ethyl decadienoate, which is obtained in a 95% E-Z purity, is the major component of Bartlett's Pear aroma.[43] The stereochemical purity of the conjugated dienes obtained by this way is high enough but care should be taken to the experimental conditions, because of the facile isomerisation via the corresponding allenolate.[44]

$$\text{RR'C=C("M")COOEt} \longrightarrow \text{RR'C=C=C(O"M")OEt} \longrightarrow \text{RR'C=C(COOEt)"M"}$$

The best results are obtained in tetrahedrofuran (THF) solvent. An alternative is also to use the acid instead of the ester :

$$(\text{nPent-CH=CH})_2\text{CuLi} + \text{HC}\equiv\text{C-COOH} \longrightarrow \text{nPent-CH=CH-CH=CH-COOH} \quad 80\%$$ [45]

$$(\text{cyclohexenyl})_2\text{CuLi} + \text{HC}\equiv\text{C-COOH} \longrightarrow \text{cyclohexenyl-CH=CH-COOH} \quad 86\%$$ [46]

3-3-2 Acetylenic acetals

Acetylenic acetals and ketals, although less reactive than the corresponding esters are also prone to add alkenyl cuprates. However, only mono-substituted alkynes of this type can be used :

$$(\text{nPent-CH=CH})_2\text{CuLi} + \text{HC}\equiv\text{C-CH(OEt)}_2 \xrightarrow[-30°\text{C, 1h}]{\text{Et}_2\text{O}} \text{nPent-CH=CH-CH=CH-CH(OEt)}_2 \quad 76\%$$ [42]

nPr–CH=CH)$_2$CuLi + 2 HC≡C-CH(OEt)$_2$ $\xrightarrow[-30°C,3h]{Et_2O}$ 2 nPr–CH=CH–CH=CH–CH(OEt)$_2$ 47 54%

(CH$_2$=C(nBu))$_2$CuLi + 2 HC≡C-CH(OEt)$_2$ $\xrightarrow[-30°C,2h]{Et_2O}$ CH$_2$=C(Bu)–CH=CH–CH(OEt)$_2$ 40,47 80%

β(Z)-Monosubstituted alkenyl cuprates transfer well one of the two alkenyl groups, the second one being incompletely transferred. On the other hand α-mono-substituted cuprates (which are more reactive) transfer their two alkenyl groups.

In contrast to the esters, there isno possibility to isomerise the dienyl cuprate in these cases, and the stereochemical purity of the dienes is almost 100%.

3-3-3 Heterosubstituted alkynes

Acetylenes bearing an heteroatom directly linked to the sp carbon atom may react with one or whith the two alkenyl groups of the cuprate, depending on their degree of substitution :

nBu_2CuLi $\xrightarrow{HC≡CH}$ (Bu–CH=CH)$_2$CuLi $\xrightarrow[2/\ H_3O^+]{1/\ 2\ HC≡C-SEt}$ 2 nBu–CH=CH–CH=CH–SEt 42 86%

nBu_2CuLi $\xrightarrow[2/\ 2\ HC≡C-OEt]{1/\ 2\ HC≡CH}$ (nBu–CH=CH–C(OEt)=CH)$_2$CuLi $\xrightarrow{H_3O^+}$ 2 nBu–CH=CH–C(=O)–Me 42 75%

nBu_2CuLi $\xrightarrow[2/\ 1\ Me-C≡C-SPh]{1/\ 2\ HC≡CH}$ nBu–CH=CH–C(Me)=CH–SPh 42 98%

In this last case, the bisubstitution of the alkyne moiety allows only one alkenyl group to react. The opposite regio-selectivity observed, according to the heteroatom positionned on the starting acetylene, leads to the formation of an alkenyl carbenoïd species (use of sulfur) or an E-β alkoxy alkenyl cuprate (in the case of oxygen) which are respectively equivalents of acyl anions or enolates of ethylenic ketones :

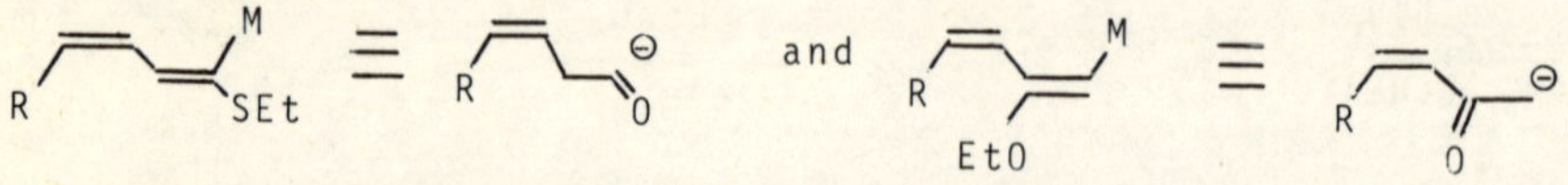

These heterosubstituted dienes are also useful building blocks of Diels-Alder reaction[48], where the stereochemistry of the diene system determines the substitution pattern of the cyclohexene ring.

3-3-4 Acetylene

Acetylene itself is also reactive enough to allow carbocupration. It is also the more versatile for synthetic purposes. In the normal carbocupration of acetylene with dialkyl cuprates[49] we observed that under particular conditions the following bis-carbocupration proceeds to some limited extent

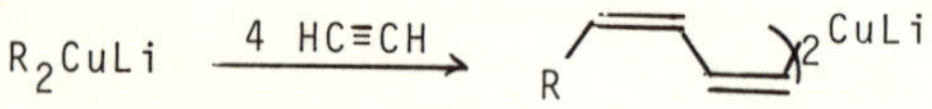

$$R_2CuLi \xrightarrow{4\ HC\equiv CH} (R\text{–CH=CH–CH=CH})_2CuLi$$

We were thus led to a systematic study of the addition of alkenyl cuprates to acetylene, according to their structural features, namely the degree of substitution of the sp_2 carbon atoms :

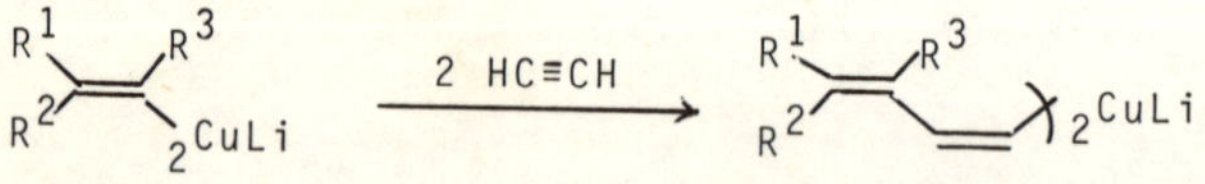

$$(R^1R^2C{=}CR^3)_2CuLi \xrightarrow{2\ HC\equiv CH} (R^1R^2C{=}CR^3\text{–CH=CH})_2CuLi$$

Since the temperature of addition is now higher than that which is required in the case of dialkyl cuprates addition, the results are highly depending upon the thermal stability of the starting alkenyl cuprate and the resulting dienyl cuprate. A notable side reaction which is also observed in some cases is the metallation of acetylene due to its two acidic hydrogens :

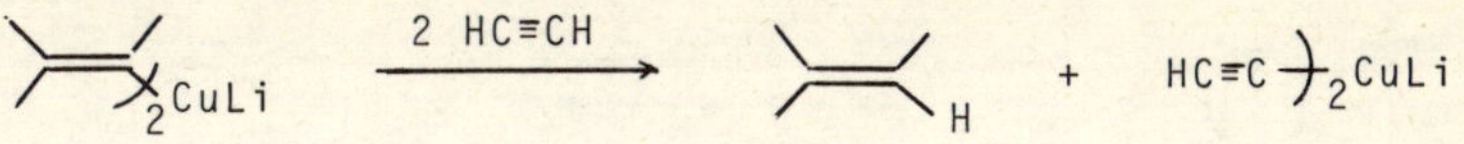

It turns out that the presence of the R^3 group, geminated to copper is essential for the success of the carbocupration, so that the reagents of type :

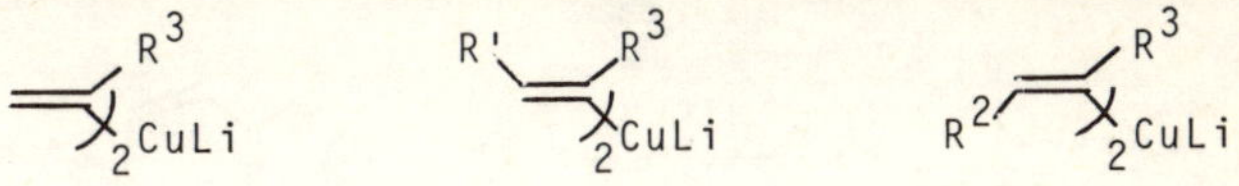

do insert smoothly two equivalents of acetylene, leading respectively to reagents of type :

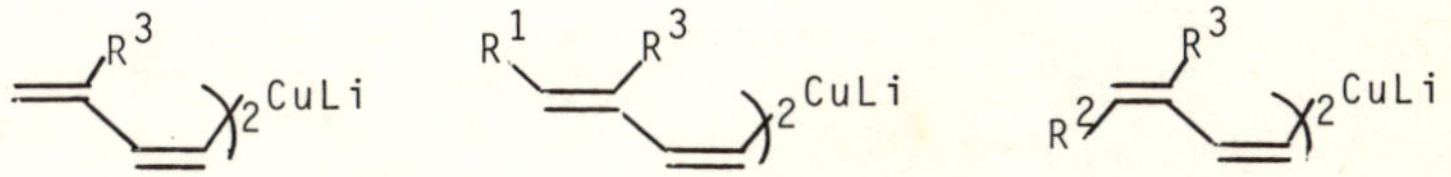

It is worthy of note that no further addition is observed, this fact may be tentatively assigned to an internal chelation of the copper atom by the former C=C bond. Reagents which do not bear the R^3 group, such as the following :

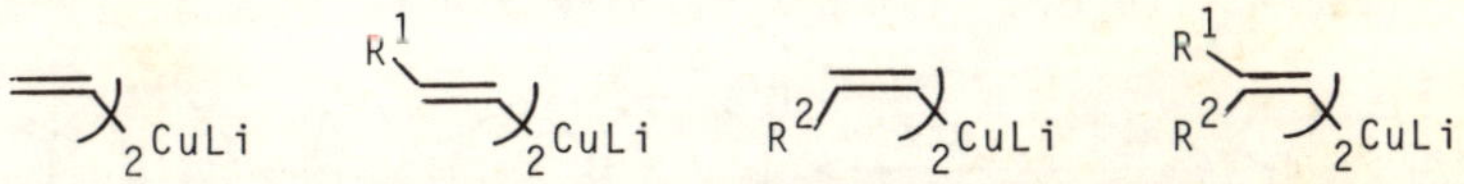

do also add acetylene, but the yields are, at present, disappointingly low (10-50%).

The dienyl cuprates obtained in this manner may be further used for chain elongation.[46] They behave, in fact, as normal alkenyl cuprates in their reactions with electrophiles. Thus, they have been carbonated or alkylated or aminomethylated as shown below[46] :

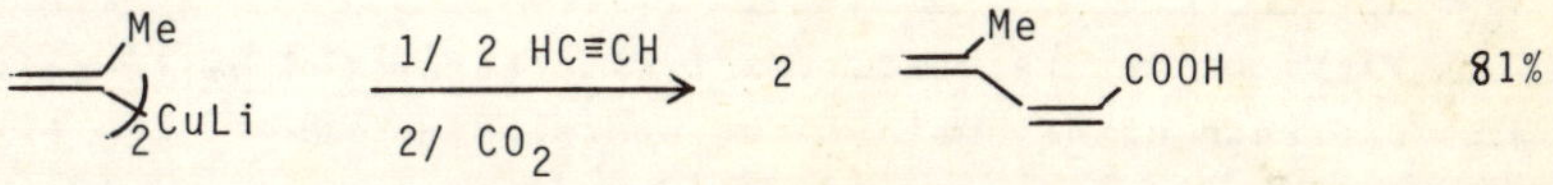

81%

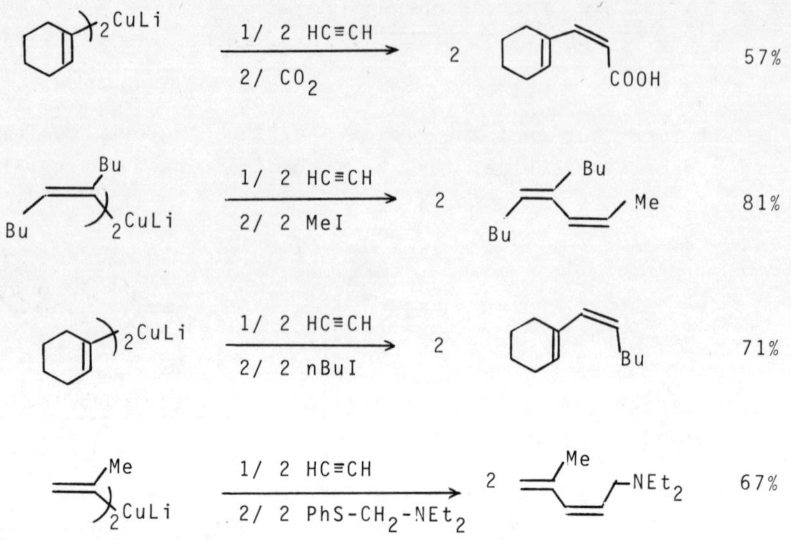

These dienyl cuprates may be used also in reactions with other electrophiles such as enones or epoxides for various synthetic purposes. Particularly interesting are the cyclo alkenyl cuprates which add efficiently to acetylene, and which may lead, after condensation, with an electrophilic reagent, to cyclisations such as the following :

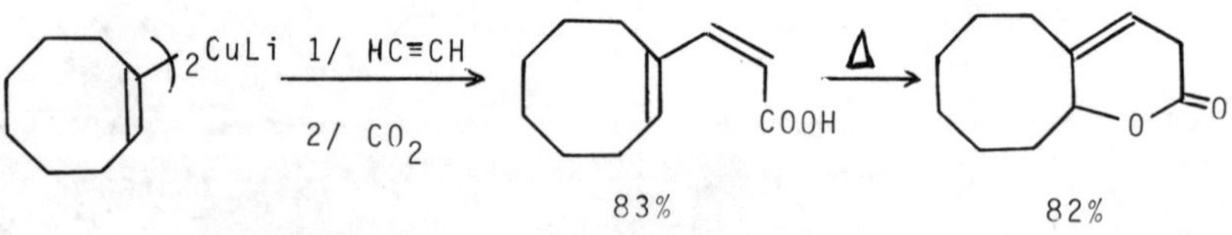

In summary, alkenyl cuprates derivatives, obtained via carbometallation (or not), are good precursors of conjugated dienes via their addition to acetylene or other alkynes, affording dienyl cuprates of given geometry which may introduce directly the dienyl synthon in more sophisticated structures.

3-4 Coupling between vinyl copper reagents and vinylic halides

Vinyl copper reagents react with saturated halides in the presence of hexamethyl phosphotriamide (HMPT) and triethyl phosphite, but are inert towards vinylic halides.[33] On the other hand, lithium vinyl cuprates react very sluggishly with the latter substrates (yield 5-10%). As a general rule for the use of such organometallics, a low temperature is recommended, to avoid decomposition, and the desired coupling reaction must be fast enough so that the temperature may not exceed +5, +15°.

3-4-1 Starting from vinyl copper-magnesium halide reagents

We have shown that under palladium zero catalysis, vinyl copper reagents couple with 1-iodo-1-alkenes, in the presence of magnesium salts[50,51]

$$R^1R^2C{=}C{-}Cu,MgX + R^3R^4C{=}C{-}I \xrightarrow{5\%\,Pd^0L_4} R^1R^2C{=}C{-}C{=}CR^4R^3$$

Table 1 shows that the 1-3 butadienes thus obtained, may be 1-mono- or 1,1-disubstituted with a large array of alkyl groups (primary, secondary, tertiary) or aryl groups. 1,1,4 tri- or 1,1,4,4-tetra-substituted ones may also be obtained with an excellent purity. Commercially available β-bromo styrene (90% E, 10% Z) may even be used stereoselectively; since Z-1-halo-1-alkenes react much slower than their E isomers, the corresponding diene is found 99.0% pure.

The use of hetero substituted vinyl cuprates present no synthetic advantage, as compared with the ViCu,MgX_2 reagent

$$Me(iPr)C{=}C{-}Cu\,Z\,MgX \xrightarrow[5\%\,Pd^0L_4]{I{-}CH{=}CH{-}Pent} Me(iPr)C{=}C{-}CH{=}CH{-}Pent$$

Z=X	Yield	64%	Purity	99.8 % Z,Z
Z=OtBu		60%		99.8 % Z,Z
Z=SPh		50%		99.5 % Z,Z

and the symmetrical magnesium divinyl cuprates cannot be obtained via carbocupration (as said above). However, these latter ones may be prepared (with some loss of isomeric purity) from the corresponding Grignard reagents.

Table 1 - Synthesis of dienes by the reaction of haloalkenes with $\text{Me(Et)C=C(Cu,MgClBr)}$ in the presence of 5% $Pd(PPh_3)_4$

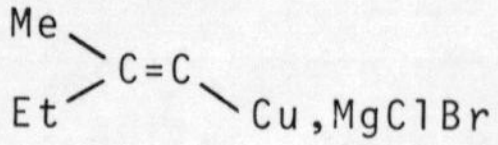

1-halo alkene	Product	Yield %	Purity
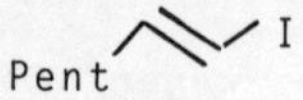 Pent–CH=CH–I	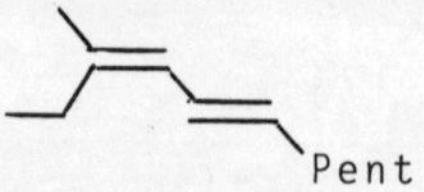 (diene)–Pent	78	99.5 Z,E
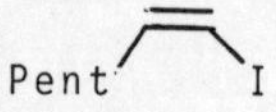 Pent–CH=CH–I (Z)	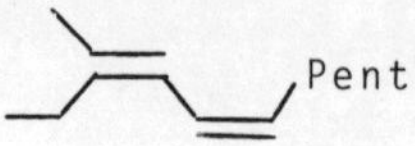 (diene)–Pent	70	99.8 Z,Z
 Ph–CH=CH–Br	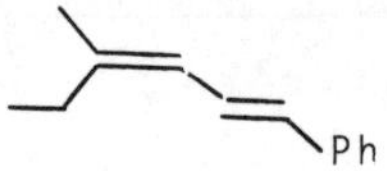 (diene)–Ph	74	99.0 Z,E
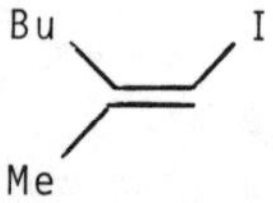 Bu, Me, I (alkene)	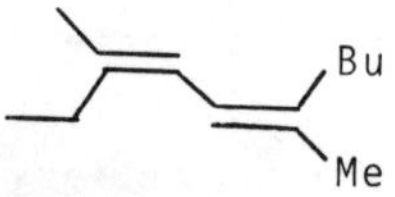 (diene) Bu, Me	55	99.8 Z,Z

Their palladium catalyzed reaction with 1-iodo-1-alkenes is extremely slow, but becomes rapid in the presence of one equivalent (or less) of Zinc halides.

2 (alkenyl)MgBr —1 CuBr→ (alkenyl)$_2$CuMgBr —2 I–CH=CH–Pent, 5%Pd°L_4→ 2 (diene)–Pent

without $ZnBr_2$: yield 24%

with $ZnBr_2$: yield 94%

(Note that both vinylic groups have been converted to diene). The use of an ate complex derivatized from a magnesium acetylide proved unsuccessful :

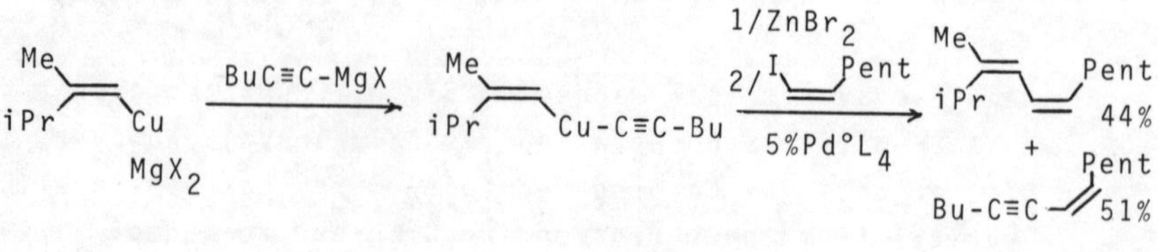

Me(iPr)C=C–Cu MgX_2 —BuC≡C-MgX→ Me(iPr)C=C–Cu-C≡C-Bu —1/$ZnBr_2$, 2/ I–CH=CH–Pent, 5%Pd°L_4→ Me(iPr)(diene)–Pent 44% + Bu-C≡C–CH=CH–Pent 51%

since transfer of the alkynyl moiety competes largely.

3-4-2 Starting from lithium cuprates[52]

This strategy is followed when 1-3 butadienes, mono substituted on carbon 1 are the target molecules.

Lithium divinyl cuprates react with 1-iodo-1-alkenes in the presence of Ni° or Pd° catalysts but yields and isomeric purity are low

$$(\text{CH}_3\text{CH=CH})_2\text{CuLi} + \text{I-CH=CH-Pent} \xrightarrow[\text{solvent, } -10 \text{ to } +20°]{\text{catalyst}} \text{CH}_3\text{CH=CH-CH=CH-Pent} \quad (Z + E)$$

Solvent	Catalyst	Time (h)	Yield %	Ratio Z/E
ether or THF	none	72	trace	
ether/THF	3%$NiBr_2(PPh_3)_2$	4	51	80/20
ether	5%$Pd(PPh_3)_4$	24	50	91/ 9
"	"	0.5	94[a]	97/ 3

a : inverse addition

except in the case of inverse addition of the cuprate to the haloalkene. However, in the presence of zinc halides a rapid reaction takes place and leads to high yields of diene, more than 99% pure. The reason must lie in the fact that the metal-metal exchange:

$$R_2CuLi + ZnX_2 \longrightarrow RCu + RZnX \xrightarrow{ZnX_2}\!\!\!\!\!\!\!\times\;\; 2\ RZnX$$

leads to a reactive Zinc species (The reverse reaction

$$Me_2Zn + CuX \longrightarrow MeZnX + MeCu$$

had been already disclosed by Thiele[53]).

Some examples are presented in table II.

Table II - Coupling between lithium vinyl cuprates and 1-halo-1-alkenes in the presence of 5% $Pd(PPh_3)_4$ and 1 equivalent of $ZnBr_2$

Cuprate	Vinyl-X (purity)	Product	Yield %	Purity %
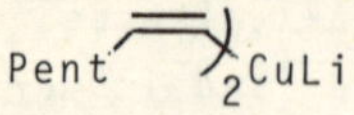 (Pent-CH=CH)$_2$CuLi	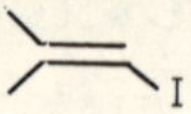 Me_2C=CH-I	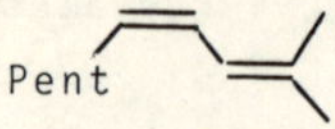 Pent-CH=CH-CH=CMe_2	94	99.8
"	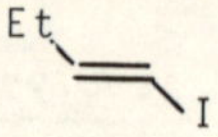 Et-CH=CH-I	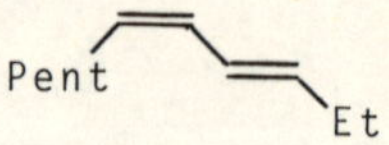 Pent-CH=CH-CH=CH-Et	80	98.9
"	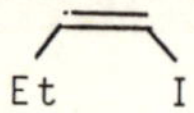 Et-CH=CH-I (Z)	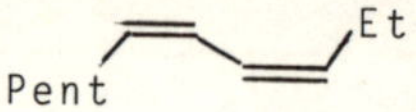 Pent-CH=CH-CH=CH-Et	86	99.5
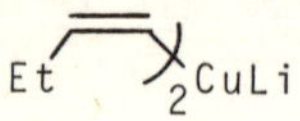 (Et-CH=CH)$_2$CuLi	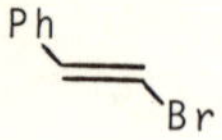 Ph-CH=CH-Br	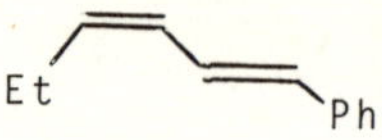 Et-CH=CH-CH=CH-Ph	85	97.6
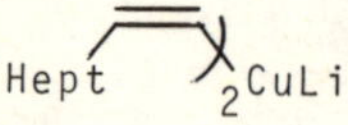 (Hept-CH=CH)$_2$CuLi	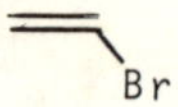 CH_2=CH-Br	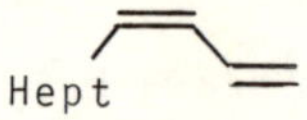 Hept-CH=CH-CH=CH_2	91	99.8[a]

a : in the presence of HMPT, and of excess of bromo ethylene.

Table III - (Et-CH=CH)$_2$CuLi → Et-CH=CH-C$_6$H$_4$-Y

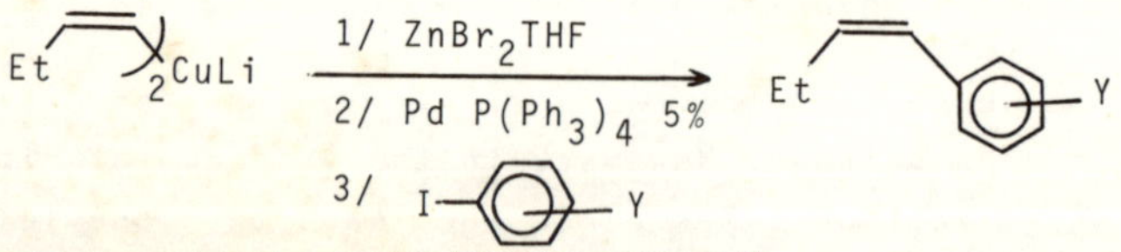

1/ $ZnBr_2$THF
2/ Pd $P(Ph_3)_4$ 5%
3/ I-C$_6$H$_4$-Y

Y	Yield %
p.NO_2	80
p.OMe	65
H	67
p.Br	71
o.COOMe	72
p.COOMe	81

An extension of this study has shown that iodo arenes behave similarly : in the presence of zinc halides and a catalytic amount of $Pd°L_4$, they couple efficiently with lithium divinyl cuprates to give β-substituted styrenes of Z geometry. The aryl ring may bear extra functionalities, like alkoxy, bromo, nitro, carbomethoxy groups, the electron with drawing substituents giving the fastest reaction (table III)[54]. Heterocyclic substrates, as functionalized cuprates, behave as well[54]

$$Et_2CuLi + Me{-}{\equiv}{-}CH(OMe)_2 \longrightarrow (Me)(Et)C{=}C(CH(OMe)_2)\}_2CuLi$$

$$(Me)(Et)C{=}C(CH(OMe)_2)\}_2CuLi + \text{2-iodothiophene} \xrightarrow[ZnBr_2]{5\%\ Pd°L_4} (Me)(Et)C{=}C(CH(OMe)_2)(\text{2-thienyl}) \quad 85\%$$

The amount of zinc halides may be less than one equivalent (0.5, 0.25 and 0.05 equivalents have been used) but the velocity decreases accordingly.

In these reactions leading to dienes or styrenes, it should be pointed out that the yields were calculated from the cuprate reagent (or the haloalkene in a 1/1 ratio) : only one vinyl moiety is transferred. This may be inconvenient when a more sophisticated cuprate is used. The metal-metal exchange, leading from the lithium cuprate to a reactive vinyl zinc halide, and an unreactive vinyl copper – lithium halide reagent, the latter remains unchanged. However we have seen (3-4-1) that the vinyl copper – magnesium halide reagent is indeed reactive towards 1-iodo-1-alkenes. It turns out that vinyl copper – lithium halide reagents are also converted to reactive species if magnesium halides are added to the reaction mixture. Following these lines, it is then possible to use both vinyl moieties of a lithium cuprate, if (i)MgX_2 salts and (ii)ZnX_2 salts are added prior to the halo alkene and catalyst :

$$(Et{-}CH{=}CH)_2CuLi + 2\ Pent{-}CH{=}CH{-}I \xrightarrow{5\%Pd°L_4} 2\ Et{-}CH{=}CH{-}CH{=}CH{-}Pent$$

with $ZnBr_2$ alone, the yield is 84% for one butenyl moiety transferred, whereas with $ZnBr_2$ and $MgCl_2$ the yield raises to 84% for two transferred butenyl moieties.

3-5 Coupling between vinylic copper species and 1-halo-1-alkynes

3-5-1 Starting from a vinyl copper-magnesium halide reagent

We have shown that the direct coupling between vinylic copper reagents and 1-halo-1 alkynes in ether may lead to a variety of products :

ViCu + RC≡C-X ⟶ Vi-C≡C-R + Vi-Vi + RC≡C-C≡C-R ...

resulting from metal/halogen exchanges. However the presence of THF and of tetramethylethylen diamine (1,2 to 2 equivalents) suppresses these side reactions so that the desired Z enyne is obtained in good yields[55,56]

Table IV - R-C≡C-X + ViCu ⟶ R-C≡C-Vi

ViCu	R-C≡CX	Product	Yield %
Bu(Et)C=CH-Cu	n.Bu-C≡C-I	Bu(Et)C=CH-C≡C-Bu	77
	n.Bu-C≡C-Br	"	78
	n.Bu-C≡C-Cl	"	8
"	$Me_3SiC≡C-I$	Bu(Et)C=CH-C≡C-$SiMe_3$	82
"	THPO-CH_2-C≡C-Br	Bu(Et)C=CH-C≡C-CH_2OH	80[a]
Me(Et)C=CH-Cu	EtOCO-C≡C-Br	Me(Et)C=CH-C≡C-COOEt	78
"	EtS-C≡C-Br	Me(Et)C=CH-C≡C-SEt	82

a : yield in alcohol after acidic treatment

1-bromo- and 1-iodo-1-alkynes behave similarly but 1-chloro give poor results.

As shown in table IV, the acetylenic moiety may bear a variety of functions, and use of trimethyl silyl halo-acetylene allows the preparation of terminal enynes after basic treatment.

3-5-2 Starting from lithium cuprates[57]

The access to Z-β-monosubstituted vinyl copper species being quantitative from the lithium cuprates and acetylene, we choose this way to prepare Z enynes

$$R_2CuLi \xrightarrow{2\ HC{\equiv}CH} (R\text{-}CH{=}CH)_2CuLi \xrightarrow{R'C{\equiv}C\text{-}X} \text{metal halogen exchange}$$

but only the undesired metal-halogen exchange took place.

It is possible however to add one equivalent of copper halide to the divinyl cuprate :

$$(R\text{-}CH{=}CH)_2CuLi + CuI \longrightarrow 2\ R\text{-}CH{=}CH\text{-}Cu,\ LiI$$

and to couple the vinyl copper reagent, thus formed, in the presence of TMEDA with 1-halo-1-alkynes : the reaction is so fast that even halo alkynes bearing a free hydroxyl group may be used :

$$(n.Pr\text{-}CH{=}CH)_2CuLi \xrightarrow[\substack{2/\ BrC{\equiv}C\text{-}CH_2\text{-}CH_2\text{-}OZ \\ THF\ 1,2\ eq.TMEDA \\ 3/\ H_3O^+}]{1/\ CuI} n.Pr\text{-}CH{=}CH\text{-}C{\equiv}C\text{-}CH_2\text{-}CH_2OH$$

Z=SiMe$_3$ Yield 83%

Z=H " 75%

As an example, we prepared the pheromone of the processionary moth Thaumetopoea pityocampa according to this procedure [58]

$$Et_2CuLi \xrightarrow[2/\ I\ CuI]{1/\ 2\ HC{\equiv}CH} 2\ Et\text{-}CH{=}CH\text{-}Cu,LiI \xrightarrow[2/\ AcO(CH_3)_{10}C{\equiv}CI]{1/\ TMEDA} Et\text{-}CH{=}CH\text{-}C{\equiv}C\text{-}(CH_2)_{10}\ OAc \quad 76\%$$

3-5-3 Transformation of enynes to dienes

The reduction of the propargylic en-yn-ols (prepared according to the above procedure) by lithium aluminium hydride in boiling THF only affects the triple bond and represents an easy way to 2(Z), 4(E) alkadien-1-ols[55]

$$nPr\text{-}CH{=}CH\text{-}C{\equiv}C\text{-}CH_2OH \xrightarrow{LAH} nPr\text{-}CH{=}CH\text{-}CH{=}CH\text{-}CH_2OH \quad 91\%$$

(E,Z 99%)

It is also possible to add alkyl copper reagents to conjugated enynes : the addition is exclusively of 1,2 type, and leads to alkylated butadienes[59,60] :

$$\text{CH}_2{=}\text{CH}{-}\text{C}{\equiv}\text{CH} + Bu_2CuMgCl \rightarrow \text{(dienyl)}CuBuMgCl \xrightarrow[\text{HMPT, } 0.1\ P(OEt)_3]{CO_2} \text{(dienyl, Bu)}COOH \quad 70\%$$

Hetero substituted enynes may be used as substrates, with the usual regioselectivity :

$$\text{(enyne)}OEt \xrightarrow[2/\ H_3O^+]{1/\ Bu_2CuMgCl} \text{(Bu-diene)}OEt \quad 55\%$$

or the "reverse" regioselectivity :

$$\text{(enyne)}SEt \xrightarrow[2/\ H_3O^+]{1/\ BuCu,MgClBr} Bu\text{(diene)}SEt \quad 80\%$$

CONCLUSION -

Organo copper reagents may be considered as good candidates for the elaboration of conjugated 1-3 butadienes, bearing substituents on carbon 1,2,3 or 4.

The vinyl-vinyl coupling presents a large scope since a large variety of vinyl copper may be derivatized from acetylene or terminal alkynes. The Z 1-iodo-1 alkenes are easily prepared from vinyl cuprates by iodonolysis;the E ones derive from other sources (hydroalanation, boration ...). The other main pathway, namely the addition of a vinyl copper species to an alkyne, is even more promising since a dienyl metal reagent is thus obtained, which may undergo further transformations. This latter aspect is actively studied in our laboratory.

This survey should help to keep in mind, how important is the selection of the various kinds of organo copper reagents, the selection of the main group metal salts, and the nature of the solvent.

ACKNOLEDGEMENTS -

The authors thank all the research staff of the Laboratory of Chimie des Organoéléments whose work is reported here, and for financial help from the C.N.R.S. The manuscript was typed by Mrs J. Mathé who is warmly thanked for her skill and her patience.

4 - EXPERIMENTAL PART

4-1 Preparation of alkenyl copper and cuprates reagents

Alkenyl cuprates which are not prepared by carbocupration of an alkyne, are prepared by addition of the corresponding alkenyl-lithium reagent (2 eq.) to a suspension of $CuBr.Me_2S$ complex (1 eq.) in ether solvent, at -30°C for an hour. A yellow greenish solution is obtained ready for further use.

Lithium (Z) dialkenyl cuprates are prepared as follows :

to a suspension of purified copper iodide (28 mmol) in 70 ml ether, at -40°C, is added an ethereal solution of alkyl lithium reagent (50 mmol). After 30 min a blue solution of dialkyl cuprate is obtained. Acetylene gas (1200 ml, 55 mmol) (the proper volume is measured in a water gasometer) is bubbled rapidly into the solution of dialkyl cuprate at -50°C. The solution becomes green, while the temperature is raised at -25°C for 30 min. (Z) dialkenyl cuprates are ready for further use.

$\beta\beta$-Bisubstituted magnesium alkenyl copper reagents are prepared as follows:

To a suspension of purified copper (I) bromide (55 mol) or copper bromide dimethyl sulfide complex, in 80 ml ether, is added a solution of the appropriate alkyl magnesium halide in ether solvent, at -35°C. After stirring for 30 min a yellow precipitate of alkyl copper reagent is formed. An appropiate alkyne is added (50 mmol) and the temperature of the mixture is raised to -15°C for 2 h. A deep green solution of alkenyl copper reagent is obtained and is ready for further use.

4-2 Ethyl (2-E), (4-Z) Decadienoate

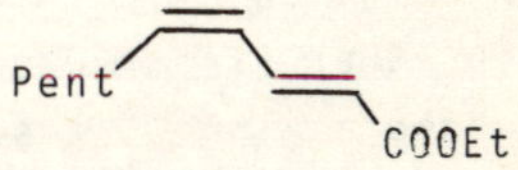

Pent COOEt

To a solution of 25 mmol of (Z) diheptenyl cuprate (prepared from 50 mmol n.pentyl lithium reagent) are added 50 ml of THF at -50°C with cooling. Ethyl propiolate (25 mmol) is slowly added in solution in THF (15 ml) at -50°C (very exothermic). The reaction mixture which turns red, is stirred 15 min at -30°C, then hydrolysed with 100 ml NH_4Cl sat. solution. 100 ml of hexane are added and the precipitate filtered off. The organic phase is washed one or two times with NH_4Cl sat. solution, dried over $MgSO_4$ and evaporated. The residue is distilled through a 15 cm Vigreux column to afford 4.4 g of pure Pear Ester. B.p. 81-82°/0.1 mmHg ; n_D^{20} 1.4881 ; IR (neat) cm^{-1} 1715 (COOEt), 1640, 1620 (C=C-C=C) ; ^1H NMR (CCl_4,δ) 7.22 (CH=,dd,1), 5.45-5.95 (CH=,m,3), 3.98 (CH_2-O,q,2), 2.13 (CH_2-C=q,2).

4-3 1,1 - Diethoxy4Butyl-2 E,4-pentadiene

Bu

$CH(OEt)_2$

To a solution of lithium di-1-hexen-2-yl cuprate (prepared from 50 mmol of alkenyl lithium) in 100 ml ether, at -40°C, is added an ethereal solution (15 ml) of 1,1-diethoxy propyne (50 mmol). The solution is stirred at -30°C for three hours, then hydrolysed with 80 ml NH_4Cl sat. solution and 20 ml 17% aqueous ammonia. The mixture is stirred at room temperature untill all precipitate dissolves. The layers are separated, the organic phase washed once with NH_4Cl sat solution, dried over K_2CO_3 and evaporated. The residue is distilled through a 15 cm Vigreux column. B.p. 55-66°/0.05 mmHg ; n_D^{20} 1.4563. IR (neat) cm^{-1} 3015, 1635, 940, 900. ^1H NMR (CCl_4,δ) 6.34 (CH-,d,1), 5.0-5.8 (CH=,m,3), 4.52 (CH_2-O,m,4), 2.22 (CH_2-C=,t,2).

4-4 1-Ethylthio-1(E),3(Z)-Hexadiene

CH_3-CH_2 $S-CH_2-CH_3$

A solution of 25 mmol of (Z) dibutenyl cuprate is prepared in ether (from 50 mmol ethyl lithium). 80 ml of THF are added at -50°C, then 4.3 g (50 mmol) ethylthioacetylene in 20 ml THF. The mixture is stirred two hours at -5°C, then hydrolysed with 80 ml NH_4Cl sat. solution and 20 ml 20% aqueous HCl. After filtration, the organic phase is washed twice with NH_4Cl sat. solution, then dried over $MgSO_4$ and concentrated in vacuo. The residue is distilled to afford 6.12 g (86%) of the title compound. B.p. 82-84°/14 mmHg ;

n_D^{20} 1.5373 ; IR (neat) cm^{-1} 3010, 1635, 1570 (HC=CH) ; ^1H NMR (CCl_4,δ) 6.35 (CH,dd,1), 6.05 (CH=,d,1), 5.86 (CH=, dd,1), 5.18 (CH=,d,1), 2.66 (S-CH_2,q,2), 2.13 (CH_2-C=,m,2).

4-5 3(Z)-1(1-cyclohexenyl) acrylic acid

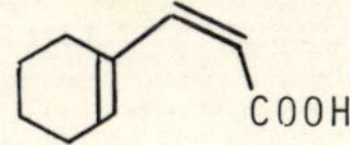

An ethereal solution of cyclohexenyl lithium (prepared according to Braude[62]) (30 mmol) is added to a suspension of CuBr,Me_2S (16 mmol) in 70 ml ether at -30°C. After stirring for 30 min a pale green solution is obtained. The temperature is raised to 0°C and acetylene gas (675 ml, 33 mmol) is bubbled into the solution. The gas is absorbed and stirring is continued for 30 min at +20°C. The dark solution is cooled to -20°C, THF is added (50 ml), then HMPT (5.4 g, 30 mmol) and 100 mg $P(OEt)_3$ (as catalyst). Dry CO_2 is bubbled into the reaction mixture, which is allowed to reach room temperature. After stirring for three hours, the mixture is hydrolysed with 70 ml HCl 5N. After filtration the organic phase is washed twice with HCl 5N, then with dilute aqueous KOH. This basic aqueous phase is separated and acidified with dilute H_2SO_4. The acid is extracted twice with 50 ml portions of ether, dried over $MgSO_4$ and concentrated. The crude acid melted at 59°C and amounted to 2.6 g (57% yied). IR cm^{-1} 3050 (COOH) 1690, 1615 (C=C COOH) 920, 855, 820 (C=C-C=C) ; ^1H NMR ($CDCl_3$,δ) 6.61 (=CH,d,1) 6.23 (=CH,t,1) 5.76 (=CH,d,1) J_{cis} : 13 Hz ; ^{13}C NMR ($CDCl_3$,δ) 172.7 (COOH) 147.4, 137.2, 135.5, 115.0 (=CH).

4-6 3 (E)-(1-Cyclohexenyl) acrylic acid

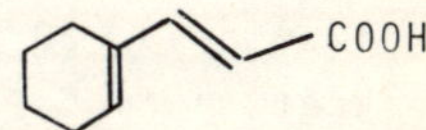

Lithium dicyclohexenyl cuprate is prepared as above (from 50 mmol) of lithium reagent and 27 mmol CuBr,Me_2S). Propiolic acid (25 mmol) is added in ether at -45°C. The mixture is stirred 30 min. at 0°C, then hydrolysed and worked up as above. 3.3 g of crude acid are obtained (86%) which melted at 109°. ^1H NMR ($CDCl_3$,δ) 7.62 (=CH,d,1) 6.44 (=CH,t,1)

5.96 (=CH,d,1) J_{trans} : 17 Hz ; ^{13}C NMR ($CDCl_3$,δ) 173.5 (COOH) 150.7, 140.5, 135.0, 113.6 (=CH).

4-7 1-Phenyl, 4-Methyl, 1(E)-3(Z) Hexadiene

Me
Et
Ph

30 mmol of magnesium alkenyl copper reagent are prepared as described above. To this solution are added 50 ml THF at -20°C, then a mixture of 25 mmol β-bromo styrene (E/Z : 91/9 and 1.25 mmol of $Pd(PPh_3)_4$ in 20 ml THF. The reaction mixture is allowed to reach room temperature, stirred for 30 min., then hydrolysed with 80 ml NH_4Cl sat. solution. The organic phase is separated and concentrated in vacuo without drying. The crude product is diluted with 150 ml pentane to precipitate all organic salts, then filtered and washed once with NH_4Cl sat. sol. and aqueous NH_4OH, dried over $MgSO_4$ and the solvent evaporated. Distillation through a 10 cm Vigreux column afforded 3.2 g of pure product (74%). B.p. 73°/0.01 mmHg ; n_D^{20} 1.5985 ; IR (neat) cm^{-1} 3030, 3015, 1640, 1600, 960, 750, 690. ^1H NMR (CCl_4,δ) 7.09 (=CH,d,1) 6.47 (=CH,d,1) 6.01 (=CH,dd,1) 1.84 (CH_3,s,3).

4-8 7 (E), 9(Z)-Dodecadien 1-yl acetate

Et
$(CH_2)_6$-OAc

To a solution of (Z) dibutenyl cuprate, (prepared from 6 mmol of EtLi in 30 ml ether, are added at -40°C 20 ml THF, then a solution of 700 mg $ZnBr_2$ in 10 ml THF. After stirring 30 min at -20°C, a mixture of 3 mmol 1-iodo, 1(E)-octen-8-yl acetate (94% E purity) and 0.15 mmol $Pd(PPh_3)_4$ in 10 ml THF is added. The reaction mixture is slowly warmed to +10°C and after 30 min at this temperature, 40 ml NH_4Cl sat. solution are added. The organic phase is concentrated in vacuo and 50 ml of pentane are added to precipitate the inorganic salts which are filtered off. The organic solution is washed twice with 20 ml NH_4Cl sat. solution, dried over $MgSO_4$

and the solvent removed in vacuo. The crude residue is purified by preparative T.L.C. to afford a 78% yield of pheromone (97% EZ purity). ^{1}H NMR (CCl_4,δ) 6.74 (dt,1) 6.25 (dd;1) 6.02 (dd,1) 5.38 (dt,1). ^{13}C NMR ($CDCl_3$,δ) 179.0 (-COO-) 134.3, 131.6, 128.1, 125.7 (=CH)

4-9 5(Z),7(E)-Tridecadien-1-yl-acetate Pent$\diagdown$=$\diagdown$=$\diagup$$(CH_2)_4$OAc

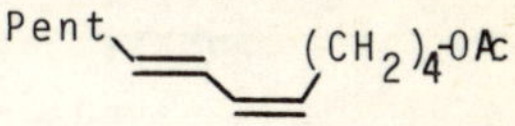

To a solution of 15 mmol of di-6-ethoxyethyl 1-(Z) hexenyl cuprate (prepared from 30 mmol of 4-ethoxyethyl butyl lithium) in 100 ml of ether, are added 30 ml THF at -30°C. Then, a solution of 30mmol $MgCl_2$ in 50 ml THF is added at the same temperature, followed (after 10 min) by a solution of 30 mmol $ZnBr_2$ in 20 ml THF. After stirring for 15 min a mixture of 30 mmol 1-iodo-1(E) heptene and 1.25 mmol $Pd(PPh_3)_4$ in 50 ml THF is added at -15°C. The reaction mixture is warmed slowly (30 min) to +15°C maintained one hour at this temperature, then hydrolysed with 80 ml NH_4Cl and worked up as described in 4-7. The crude product is acetylated by pouring into 30 ml of a mixture of AcCl in AcOH (1:10) at 40°C for twelve hours. After addition of 10ml pentane and washing with 30 ml $NaHCO_3$ sat. solution, the organic solution is dried on $MgSO_4$ and the solvents evaporated. The residue is distilled through a 10 cm Vigreux column to afford 3.7 g of the title product (52% yield). B.p. 85-87°/0.01 mmHg ; n_D^{20} : 1.4701 ; IR (neat) cm^{-1} : 3020 (C=C) 1745 (-COO-) 1645, 985, 950, 730 (C=C) ; ^{1}H NMR ($CDCl_3$,δ) 6.40 (=CH,dd,1) 6.04 (=CH, dd,1) 5.72 (=CH,dt,1) 5.38 (=CH,dt,1) ; ^{13}C NMR ($CDCl_3$,δ) 170.8(-COO-) 135.0, 129.5, 128.9, 125.6 (=CH).

4-10 5-Methyl-4-Hepten-2-yn-1-ol

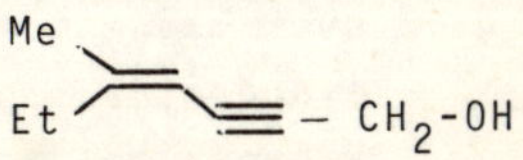

Me, Et >C=CH–C≡C – CH_2-OH

2-Methyl-1(Z) Butenyl copper, magnesium halide, is prepared (50 mmol) in 150 ml ether as described previously (see 4.1). 80 ml of THF and 100 mmol tetramethylethylene diamine (TMEDA) are added at -20°C and then a solution of 1-bromo-3-trimethylsiloxy-1-propyne in 20 ml THF. The green solution is discoloured in about 15-20 min and a white precipitate appears. After stirring for one hour at -10°C the mixture is hydrolysed with 80 ml H_2SO_4(2N), the salts are filtered off and the organic phase is diluted with 150 ml pentane. This phase is washed once again with 50 ml H_2SO_4(2N), then with 80 ml $NaHCO_3$ sat. solution, dried over $MgSO_4$ and the solvents are removed in vacuo. Distillation of the residue affords the title alcohol in 82% yield. B.p. 57°C/0.4 mmHg ; n_D^{20} : 1.5022 ; IR (neat) cm^{-1} 3320 (-OH) 2210 (C≡C) 1625 (C=C) 1015 (C-O) ; ^{1}H NMR (CCl_4,δ) 5.21 (=CH,s,1) 4.32 (-CH-O,s,2) 1.76 (CH_3,s,3).

4-11 13-Hexadecen-11-yn-1-yl acetate   Et–CH=CH–C≡C–$(CH_2)_{10}$–OAc

A solution of lithium di 1-(Z) Butenyl cuprate (50 mmol) is prepared in 180 ml ether as described previously (see 4-1) (from 100 mmol EtLi). To this reagent are successively added at -30°C, CuI (60 mmol), 100 ml THF and 18.5 ml TMEDA, whereupon all CuI was dissolved. A solution of 12-iodo-11-decyn-1-yl acetate in 30 ml THF is added dropwise and the stirred mixture is warmed to 0°C for 30 min. The mixture is then hydrolysed with 50 ml of ammonium chloride solution and 50 ml 2N hydrochloric acid, filtered on celite and partitioned. The organic phase to which 200 ml of hexane is added is washed successively with ammonium chloride solution, with 5% sodium thiosulfate, with 10% ammonium hydroxide, and with sat. sol. ammonium chloride again, then dried over magnesium sulfate and concentrated on a rotatory evaporator. The residue is distilled through a 15 cm Vigreux column to afford 12.6 g of pure pheromone (76%) yield. B.p. 105-107°/0.01 mmHg ; n_D^{20} : 1.4701 ; IR (neat) cm^{-1} 3020 (HC=CH) 2210 (C≡C) 1740 (OAc)

730 ($HC\overset{Z}{=}CH$) ; ^{1}H NMR (CCl_4,δ) 5.88 (=CH,dt,1) 5.47 (=CH,d,1) 4.12 (CH_2-O,t,2) ; ^{13}C NMR ($CDCl_3$,δ) 170.9 ($C\lesssim^{O}_{O}$) 143.7, 109.0 (C=C) 94.4, 77.3 (C≡C) 64.6 (CH_2-O)

4-12 1-Ethylthio,1(Z)-3(E)-Hexadiene CH_3-CH_2 ... SCH_2-CH_3

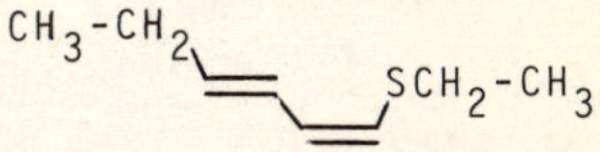

50 mmol of ethyl copper are prepared in 100 ml ether from copper bromide and ethyl magnesium bromide at -35°C. To this yellow suspension are added 50 mmol of 1-ethylthio-1-buten-3-yne (prepared according to ref. 63) in 20 ml ether, at -40°C. The reaction mixture is stirred at -20°C for two hours, then hydrolysed with 50 ml NH_4Cl and 30 ml HCl 5N, filtered and washed twice with NH_4Cl sat. solution. The organic phase is dried over $MgSO_4$ and concentrated in vacuo. The residue is distilled to afford 5.8 g of product (82%). B.p. 87°/15 mmHg ; n_D^{20} : 1.5331 ; IR (neat) cm^{-1} 3010, 1640, 1570, 970 ; ^{1}H NMR (CCl_4,δ) 6.29 (ddt,1) 6.04 (t,1) 5.77 (d,1) 5.64 (dt,1).

4-13 1-Ethoxy-4-n butyl-2(E),4-pentadiene Bu ... CH_2OEt

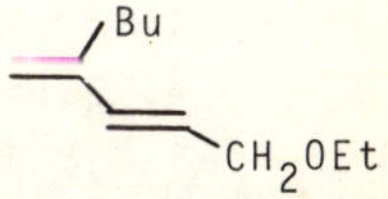

To a suspension of 25 mmol CuBr in 50 ml THF are added 50 mmol of a THF solution of butyl magnesium bromide, at -50°C. After stirring for two hours at this temperature, 1-ethoxy-2(E)-penten-4-yne (25 mmol) in 20 ml THF is added and the reaction mixture is kept at -40°C for two hours, then hydrolysed with 80 ml HCl 5N. 100 ml hexane or pentane are added and the organic phase is washed twice with NH_4Cl sat. solution, dried over $MgSO_4$ and concentrated in vacuo. The residue is distilled to afford 2.3 g (55%) of pure product. B.p. 42°/0.05 mmHg ; n_D^{20} : 1.4607 ; IR (neat) cm^{-1} 1610, 1105, 970, 890 ; ^{1}H NMR (CCl_4,δ) 6.2 (=CH,d,1) 5.71 (=CH,dt,1) 4.94 (=CH_2,s,2).

REFERENCES and NOTES -

1. a/ H.O. House, T.H. Cronin, J. Org. Chem. 30 1061 (1965)

   b/ S.G. Pyne, M.J. Hensel, S.R. Byrn, A.T. Mc Kenzie, P.L. Fuchs, J. Amer. Chem. Soc. 102 5960, (1980)

   c/ R.K. Boeckman, T.R. Alessi, ibid 104, 3216 (1982)

2. H.J. Bestmann, K. Roth, M. Ettlinger, Ber. 115, 161 (1982)

3. M. Schlosser, K.F. Christmann, Liebigs Ann. Chem. 708, 1 (1967)

4. H.J. Bestmann, O. Vostrowsky, Chemistry and Physics of Lipids 24, 335 (1979)

5. J.F. Normant, A. Commerçon, J. Villieras, Tetrahedron Lett. 1465 (1975)

6. Or from a vinyl boronate, through a vinyl copper species : H.C. Brown, G.A. Molander, J. Org. Chem. 46, 647, (1981)

7. a/ K. Sonogashira, Y. Fonda, N. Hagihara, Tetrahedron Lett. 4467 (1975)
   V. Ratovelomanana, G. Linstrumelle, Synthetic Comm. 11 917 (1981)

   b/ A.O. King, N. Okukado, E. Negishi, J. Chem. Soc. Chem. Comm. 683 (1977)

8. N. Miyaura, K. Yamada, A. Suzuki, Tetrahedron Lett. 3437(1979)

9. A. Suzuki Acc. Chem. Res. 15 178 (1982)

10. E.I. Negishi, G. Lew, T. Yoshida, J. Chem. Soc. Chem. Comm. 874 (1973)

11. G. Zweifel, S.J. Backlund, J. Organomet. Chem. 156, 159 (1978)

12. E. Negishi, A. Abramovitch Tetrahedron Lett. 411 (1977)

13. D. Holme, E.R.H. Jones, M.C. Whiting Chem. and Ind.928 (1956)

14. G. Zweifel, N.L. Polston J. Amer. Chem. Soc. 92 4068 (1970)

15. R. Rossi, A. Carpita, M. Grazia Quirici, Tetrahedron, 37 2616 (1981)

16. F. Näf, P. Decorzant, W.T. Thommen, B. Willhalm, G. Ohloff, Helv. Chim. Acta 58, 1016 (1975)

17. G. Zweifel, J.T. Snow, C.C. Whitney, J. Amer. Chem. Soc. 90, 7139 (1968)

18. A.P. Kozikowski, Y. Kitigawa, Tetrahedron Lett., 23,2087(1982)

19. H. Bosshardt, M. Schlosser, Helv. Chim. Acta 63, 2393(1980)

20. P.I. Svirskaya, C.C. Leznoff, W.L. Roelofs, Synthetic Comm. 10, 391 (1980)

21. G. Decodts, G. Dressaire, Y. Langlois, Synthesis, 510,(1979)

22. S. Tsuboi, T. Masuda, H. Makino, A. Takeda, Tetrahedron Lett. 23, 209 (1982)

23. B.T. Golding, G. Pierpoint, R. Aneja, J. Chem. Soc. Chem. Comm., 1030,(1981)

24. H.P. Dang, G. Linstrumelle, Tetrahedron Lett. 191 (1978)

25. J. Schwartz, J. Organometal. Chem. Library, 1, 461 (1976)

26. G. Wilke, H. Müller, Chem. Ber. 89, 444, (1956)

27. E.I. Negishi, Aspects of mechanism and organometallic Chemistry, J.H. Brewster, Ed. Plenum Press, New York 285(1978)

28. E.I. Negishi, Pure and Applied Chem. 53, 2333 (1981)

29. C.A. Henrick, Tetrahedron 33, 1845,(1977)

30. K. Mori Aspects of mechanism and organometallic Chemistry, J.H. Brewster Ed. Plenum Press, New York 285 (1978)

31. K. Mori, Pure and Applied Chem. 53, 2333 (1981)

32. R. Rossi, Synthesis 817 (1977)

33. J.F. Normant, A. Alexakis, Synthesis 841 (1981)

34. N. Okukado, D.E. Van Horn, W.L. Klima, E.I. Negishi, Tetrahedron Lett. 1027, 1978

35. A. Marfat, P.R. Mac Guirk, P. Cramer, P. Helquist, J. Amer. Chem. Soc. 99, 253 (1977)

36. H. Westmijze, J. Meijer, T.J.T. Bos, P. Vermer, Recl. Trav. Chim. Pays Bas, 95 299 and 304 (1976)

37. E.J. Corey, C.U. Kim, R.H.K. Chen, M. Takeda, J. Am. Chem. Soc. 94 4395 (1972)

38. F. Näf, P. Degen, Helv. Chim. Acta 54 1939 (1971)

39. R.H. Wollemberg, K.F. Albizati, R. Peries, J. Am. Chem. Soc. 99 7365 (1977)

40. A. Alexakis, A. Commerçon, J. Villieras, J.F. Normant, Tetrahedron Lett. 2313 (1976)

41. R.J. Sundberg, B.C. Pearce, J. Org. Chem. 47 725 (1982)

42. A. Alexakis, G. Cahiez, J.F. Normant, Tetrahedron 36 1961 (1980)

43. W.G. Jennings, R.K. Creveling, D.E. Heinz, J. Food Sci. 29 730 (1964)

44. J. Klein, R. Levene, J. Chem. Soc., Perkin II 1971 (1973)

45. N.Jabri, A. Alexakis, J.F. Normant, unpublished result

46. A. Alexakis, J.F. Normant, Tetrahedron Lett. 0000 (1982)

47. A. Alexakis, J.F. Normant, unpublished results

48. M. Petrzilka, J.I. Grayson, Synthesis 753 (1981)

49. A. Alexakis, G. Cahiez, J.F. Normant, J. Organomet. Chem. 177, 293 (1979)

50. N. Jabri, A. Alexakis, J.F. Normant, Tetrahedron Lett. 23, 1589 (1982)

51. N. Jabri Thèse de Spécialité Université P. et M. Curie(1982)

52. N. Jabri, A. Alexakis, J.F. Normant, Tetrahedron Lett. 22, 959 (1981)

53. K.H. Thiele, J. Köhler, J. Organomet. Chem. 12, 225 (1968)

54. N. Jabri, A. Alexakis, J.F. Normant, Tetrahedron Lett. 22, 3851 (1981)

55. J.F. Normant, A. Commerçon, J. Villieras, Tetrahedron Lett. 1465 (1975)

56. A. Commerçon, J.F. Normant, J. Villieras, Tetrahedron 36 1215 (1980)

57. A. Alexakis, G. Cahiez, J.F. Normant, Synthesis 826 (1979)

58. M. Gardette, A. Alexakis, J.F. Normant, J. Chem. Ecol. (1983) 0000

59. H. Westmijzc, H. Klein, J. Meijer, P. Vermeer, Tetrahedron Lett. 869 (1977)

60. F. Scott, G. Cahiez, J.F. Normant, J. Villieras, J. Organomet. Chem. 144, 13 (1978)

61. A. Alexakis, J.F. Normant, J. Villieras, J. Organomet. Chem. 96, 471 (1975)

62. E.A. Braude, J.A. Coles, J. Chem. Soc. 2014 (1950)

63. L. Brandsma, Preparative acetylenic Chemistry, Elsevier Ed. 1971

Part 2 ORGANOMANGANOUS REAGENTS: THEIR USE IN ORGANIC SYNTHESIS

Professor Jean-François Normant and Dr. Gerard Cahiez
Laboratoire de Chimie des Organo-Eléments
tour 44, 4 place Jussieu
75230 Paris Cédex 05, France

1 - INTRODUCTION

Organomanganous reagents are among the less expensive organo transition metal compounds, due to the low cost of manganese metal.

However, contrary to the well-known manganese carbonyl derivatives, the more simple organomanganese (II) compounds have been almost ignored until these last seven years. Thus the first organomanganous compounds, diphenyl manganese and phenyl manganous iodide have been prepared by Gilman and Bailie[1] in 1937. Since this work, only a few reports about the preparation of organomanganous derivatives are found in the literature before 1976. The major descriptions concern :

- methylmanganous iodide, dimethylmanganese and lithium trimethylmanganate by Beermann and Clauss[2] ;
- lithium triethyl, butyl, phenyl, para-anisyl and para-tolylmanganates by Riemschneider et al[3] ;
- potassium, sodium or barium tetraalkynyl manganates (alkynyl = ethynyl, propynyl and phenylethynyl) and also potassium hexaethynyl manganate by Nast and Grieshammer[4] ;
- dineopentylmanganese (and dineophyl or bis trimethylsilymethylmanganese) by Wilkinson et al[5]. These authors reported that attempts to isolate the lithium trialkylmanganates always led to the dilithium tetraalkylmanganate complexes.

This survey must be completed by a study on the mechanism of the decomposition of dialkylmanganese reagents effected by Kochi[6].

Recently the use of organomanganous derivatives in organic synthesis has given rise to a large extension of their preparations, and a large number of alkyl, alkenyl, alkynyl and aryl manganous iodides or chlorides (as also lithium and magnesium manganates reagents) has been described since 1976 [7 to 17](section 2).

The synthetic use of organomanganous compounds is even less exemplified before 1976, except for :

- the trimerization of alkynes by diphenylmanganese studied by Tsutsui and Zeiss[18] ;

- The attempts to alkylate some organic halides by lithium trimethyl- and tributylmanganates reported by Corey and Posner[19].

In fact, all the synthetic methods using organomanganous reagents have been described these last seven years.

The subject of this review is restricted to the reactions of organomanganous reagents with carbonyl derivatives. Nevertheless these reagents can be also used for the regiospecific methylation of allylic halides($S_N2 \geqslant 99.5\%$)[20] and they are the effective intermediates for two reactions which are catalyzed by manganous salts :

- The reduction of ethylenic and aromatic halides by isopropylmagnesium chloride[7].
- The stereospecific coupling of ethylenic iodides to conjugated dienes by butyl lithium[8].

2 - VARIOUS TYPES AND PREPARATION OF ORGANOMANGANOUS REAGENTS

2-1 General considerations about their preparation

These reagents are easily obtained by action of organolithium or magnesium derivatives on the manganese halides (chloride, bromide or iodide).

According to the ratio organometallic/manganese halide, it is possible to prepare three types of organomanganous reagents :

- the organomanganous halides

$$\underset{(M=Li,MgX')}{RM} + MnX_2 \longrightarrow \underset{(X=Cl,Br,I)}{RMnX}$$

- the dialkymanganese

$$\underset{(M=Li,MgX')}{2\ RM} + MnX_2 \longrightarrow R_2Mn$$

- the organomanganates

$$\underset{(M=Li,MgX')}{3\ RM} + MnX_2 \longrightarrow R_3MnM$$

The preparation is generally effected in ether or THF. The major differences between these two solvents concern :

- the stabilization of the organomanganous reagents (section 2-4)
- the choice of the starting manganese halide, which is selected according to its solubility in either or both solvents (sections 2-2 and 2-3)

These preparations can be effected directly from the cheap commercial anhydrous manganese chloride or bromide but not from the commercial manganese iodide which is too impure. Fortunately this compound is quantitatively and easily prepared by the simple reaction of iodine on manganese in ether at room temperature, the anhydrous manganous iodide so obtained is employed without further treatment (see experimental part).

For synthetic purposes it is important to note the very large array of R groups which can be used.

Thus, all types of t,s or n-alkyl, alkenyl, alkynyl, allyl and aryl manganous derivatives can be prepared and the only limitations originate from the main group organometallics used.

These reagents are always employed in situ for subsequent synthetic applications.

Some results about the structure of organomanganates species which are obtained from isolated complexes are reported in the literature. These results do not concern directly the present subject but they are discussed farther (section 2-5) in connection with their potential for mechanistical considerations.

2-2 Preparation in ether

In ether the organomanganous reagents are rapidly and quantitatively obtained from manganese iodide. On the other hand the slightly soluble manganese bromide or the insoluble manganese chloride cannot be employed quantitatively.

a/ Organomanganous iodides :

They are quantitatively prepared by action of organolithium or magnesium derivatives on manganese iodide. The reaction of metal-metal exchange is performed between 0 and 20°C (R=n-alkyl, aryl, alkenyl, alkynyl) or below when the thermal stability of the final reagent is lower (generally -35°C for R=s or t-alkyl) :

$$RM + MnI_2 \xrightarrow{\text{ether}} RMnI + MI$$

(M=Li or MgX)

These reagents are readily obtained as a beige to brownish suspension or red-brownish solutions in ether (when R=vinyl or aryl, a greenish coloration is often observed).

b/ Dialkylmanganese

These reagents are more unstable than the other organomanganous derivatives. They are prepared from organolithium or magnesium derivatives and manganous iodide in ether at -30°C (section 2-4).

$$2\ RM + MnI_2 \xrightarrow{\text{ether}} R_2Mn + 2\ MI$$

(M=Li or MgX)

The dialkylmanganese are obtained as a brownish ethereal suspension (sometimes a greenish coloration is observed when R=vinyl or aryl).

c/ Lithium organomanganates

They are quantitatively prepared by action of organolithium derivatives on manganous iodide between 0 and 20°C

$$3\ RLi + MnI_2 \xrightarrow{\text{ether}} R_3MnLi + 2\ LiI$$

These reagents are the most stable among the organomanganous derivatives which are prepared in ether. They are generally obtained as (red-) brownish solutions.

It is important to note that magnesium organomanganates reagents cannot be obtained from organomagnesium derivatives in ether.

Sometimes the preparation of lithium organomanganate reagents can be effected from manganese bromide instead of manganese iodide but the reaction time increases .

2-3 Preparation in THF

Organomanganous reagents are more stable in THF than in ether (section 2-4). They are quantitatively prepared from manganese chloride or bromide. The manganese iodide can be used but it is less convenient since it requires a longer reaction time.

It is possible to dissolve manganous chloride (or bromide) by complex formation with lithium chloride (or bromide). Thus a yellow solution of "ate" complex is easily prepared by addition of lithium chloride to a manganese chloride suspension in THF (see experimental part).

$$MnCl_2 + 2\ LiCl \xrightarrow{THF} "MnCl_4Li_2"$$

The cheapest manganous chloride is generally employed for the following preparations :

a/ Organomanganous chlorides

They are readily prepared by action of organomagnesium derivatives on the soluble "ate" complex of manganese chloride (v.s). The latter is used in order to obtain soluble organomanganous chloride reagents :

$$RMgX + MnCl_4Li_2 \xrightarrow{THF} RMnCl + 2\ LiCl$$

(X=Br,Cl)

when organolithium derivatives are used, it is possible to start from a manganese chloride suspension in THF. In fact, the lithium salt which results of the rapid metal-metal exchange reaction is able to dissolve the organomanganese chloride reagents by "ate" complex formation :

$$RLi + MnCl_2 \xrightarrow{THF} RMnCl + LiCl$$

All these reactions are performed at room temperature ; they lead to a yellow to brownish solution of organomanganous chlorides. When the THF is pure and particularly free from oxygen a clear yellow solution is always obtained. Organomanganous chloride reagents can be also obtained as a suspension from organomagnesium derivatives and manganese chloride instead of its "ate" complex.

b/ Dialkylmanganese

They are prepared at room temperature from organomagnesium or organolithium derivatives and a manganese chloride suspension (the use of the soluble "ate" complex is not necessary)

$$2\ RMgX + MnCl_4Li_2 \xrightarrow{THF} R_2Mn + 2\ LiCl + 2\ MgXCl$$

(X=Br,Cl)

$$2\ RLi + MnCl_2 \xrightarrow{THF} R_2Mn + 2\ LiCl$$

A yellow to brownish solution is obtained. The influence of the purity of THF is the same as for organomanganous chlorides (v.s).

c/ Lithium and magnesium organomanganates

They are readily prepared by action of organolithium or magnesium derivatives on manganese chloride at room temperature :

$$3\ RLi + MnCl_2 \xrightarrow{THF} R_3MnLi + 2\ LiCl$$

$$3\ RMgX + MnCl_2 \xrightarrow{THF} R_3MnMgX + 2\ MgXCl$$

(X=Cl,Br)

These manganate species are soluble in THF, thus it is not necessary to use a soluble "ate" complex of manganese chloride even with organomagnesium derivatives. A yellow to brownish solution is obtained. As previously, a pure THF, free of oxygen, gives a clear yellow solution.

Lastly it is important to note that magnesium organomanganates which cannot be obtained in ether are prepared easily in THF.

2-4 Stability of the reagents

The stability of organomanganous reagents is predominantly influenced by the three following factors :

- The type of organomanganous compounds

  A decreasing stability is observed according to :

  R_3MnLi (or MgX) $>$ RMnX $\gg$ R_2Mn

  As an example, with R=n-Bu, the corresponding reagents must not be used above :

Bu_2Mn	BuMnI	Bu_3MnLi
-30°C	+20°C	well above +35°C

- The nature of R groups

The number of available β-hydrogens is the most important factor related to the stability of organomanganous reagents on account to their mechanism of decomposition which proceeds principally by a β-manganese hydride elimination[6]. Such an influence is well-known in the field of organic derivatives of transition metals.

Thus, for the following organomanganous iodides (RMnI) in ether, the temperature must not exceed :

t-Bu	i-Pr	n-Bu	Me,Me_2C=CH,Ph
-35°C	-30°C	+20°C	above +35°C

- The influence of solvents or ligands

This influence is clearly illustrated by the temperature threshold, necessary to avoid any decomposition of the following organomanganous halides

	t-BuMnX	i-PrMnX	n-BuMnX	PhMnX
Et_2O(X=I) :	-35°C	-30°C	+20°C	above +35°C

THF(X=Cl) : All reagents are stable above +66°C

A similar stabilization is observed with organomanganous iodides (in ether) when THF is used as ligand.

Sometimes this complexation by THF can be very useful for synthetic applications. Thus it is possible to increase the yield of ketones obtained from t or s-alkylmanganous iodides in ether by simple addition of five molar equivalents of THF (section 5-5, application 1).

With regard to the stabilization of organomanganous reagents, THF or more generally oxygenated ligands are superior to most other common ligands such as trialkyl phosphites or phosphines and dialkyl sulfides. For instance isopropylmanganous iodide which decomposes above -30°C in ether is stabilized by five molar equivalents of ligand, up to the following temperatures :

THF	THT	THP	Me_2S	Ph_3P
-5°C	-10°C	-20°C	-30°C	-30°C

(THP = tetrahydropyran ; THT = tetrahydrothiophene)

Lastly the influence of dimethylformamide (DMF) is very important : isopropyl manganous iodide prepared in ether is not decomposed at +35°C in the presence of five equivalents of DMF.

2-5 Structure of the reagents

The RMnX or R_2Mn formulation of organomanganous halides or dialkyl manganese is generally accepted contrary to the structure of manganate species which has given rise to some controversy[5]. In fact the trialkylmanganate reagents R_3MnLi have never been isolated and the solid tetraalkylmanganate derivatives are always obtained instead[4,5,6c]. Nevertheless it is impossible to conclude on the exact nature of the species present in solution.

In absence of a suitable study on this problem the formulation R_3MnLi or R_3MnMgX will be used according to the ratio of organolithium or magnesium/manganese halide.

3 - REACTIONS WITH CARBONYL COMPOUNDS

3-1 Esters, amides

Organomanganous reagents do not react with amides at room

temperature but they form very stable complexes with these compounds (section 2-4). Thus dimethylformamide can be used as cosolvent leading to a very stable organomanganous complex at the expense of its reactivity[16].

Esters do not react at room temperature with organomanganous derivatives[12] except for the formiate derivatives. These latter lead (at room temperature) to the corresponding alcohols with good yields[14] :

$$\text{BuMnI} + \text{HCOOMe} \xrightarrow[20°\text{C}]{\text{ether}} 1/2\ \text{Bu}_2\text{CHOH} \qquad 98\%$$

For a comparison, under the same conditions ethyl pentanoate gives no reaction after 6 h[12] :

$$\text{BuMnI} + \text{BuCOOEt} \xrightarrow[20°\text{C, 6 h}]{\text{ether}} \!\!\!\!\!\!\times$$

Even more, ethyl acetate can be used without drawback as cosolvent for the preparation of ketones from acyl halides and organomanganous iodide reagents[16] (section 5-5).

3-2 Aldehydes - ketones ; comparison

Organomanganous derivatives react with ketones and aldehydes at room temperature in ether to give the corresponding alcohols with good yields[12] :

$$\text{BuMnI} + \text{BuCHO} \xrightarrow[20°\text{C,15min}]{\text{ether}} \text{Bu}_2\text{CHOH} \qquad 89\%$$

$$'' + \text{Et}_2\text{C=O} \longrightarrow \text{Bu(Et)}_2\text{C-OH} \qquad 86\%$$

$$'' + \text{i-Pr}_2\text{C=O} \longrightarrow \text{Bu(i-Pr)}_2\text{C-OH} \qquad 88\%$$

These reactions are fast at room temperature ; however when the temperature is lowered the reaction time increases more rapidly for ketones than for aldehydes. Thus at -50°C an almost selective attack of the aldehyde moiety is possible during competitive experiments[12] :

$$\text{RMnI} + [\text{n-BuCHO} + \text{MeCOEt}] \xrightarrow[-50°\text{C}]{\text{ether}} \text{RCHOHBu}$$

R	Bu	Ph	Pr-C≡C	$Me_2C=CH$
yield %	90	86	82	68

These yields are obtained by direct addition but the last yield can be improved to 89% by addition of $Me_2C=CHMnI$ into the mixture of aldehyde and ketone.

This discrimination between ketones and aldehydes has been used for the preparation of a keto-alcohol from a keto-aldehyde[12] :

$$BuMnI + MeCO(CH_2)_3CHO \xrightarrow[-50°C]{ether} MeCO(CH_2)_3CHOHBu \quad 78\%$$

In other instances the selectivity of attack of a keto-ester by an organomanganous iodide reagent has been used for the solid phase synthesis of tertiary hydroxyesters via the corresponding keto esters[21].

Lastly the transformation of a multifunctional aldehyde into the corresponding allylic alcohol has been performed more easily and more selectively with vinyl manganous iodide (-20°C ; 92%) than with vinylmagnesium bromide (-78°C ; 79%)[22].

4 - REACTIONS WITH CO_2, CS_2, RN=C=O, SO_2

4-1 Reactions with CO_2[14]

Organomanganous halides, dialkylmanganese and lithium or magnesium organomanganate reagents react with carbon dioxide at room temperature leading, after hydrolysis, to carboxylic acids :

$$HeptM + CO_2 \xrightarrow{20°C} \xrightarrow{H_3O^+} HeptCOOH$$

HeptM	Solvent	Yield%	HeptM	Solvent	Yield%
HeptMnCl	THF	50	$Hept_3MnLi$	Et_2O	81
HeptMnI	Et_2O	88	$Hept_3MnLi$	THF	83
$Hept_2Mn$	Et_2O	86	$Hept_3MnMgCl$	THF	72

The scope of this reaction is very large :

$$R_3MnLi \text{ (or MgCl)} \xrightarrow[THF, 20°C]{CO_2} \xrightarrow{H_3O^+} 3\ RCOOH$$

R	Yield%	R	Yield%
Bu	86	$Me_2C=CH$	96
Hept	83	$BuC\equiv C$	83
s-Bu	77	$PhC\equiv C$	75
t-Pent	50	$Pr_2C=CHCH_2$	71
Ph	89		

Carboxylic acids are obtained with good yields.

The manganese reagents present an interesting advantage since their carbonation can be effected smoothly at room temperature contrary to organo-magnesium or lithium reagents which require generally the use of low temperatures.

Actually the temperature exerts no influence upon this reaction since the manganous carboxylates which are formed before hydrolysis do not react with the lithium or magnesium organomanganate reagents.

$$(RCOO)_2Mn + R'_3MnLi \xrightarrow[20°C]{\text{ether or THF}} \not\rightarrow$$

Allylic manganate reagents are carboxylated exclusively at the most substituted carbon :

$$(Pr_2C=CHCH_2)_3MnMgCl \xrightarrow[THF, 20°C]{CO_2} \xrightarrow{H_3O^+} 3\ Pr_2C(COOH)-CH_2CH=CH_2 \quad 71\%$$

Finally, the carbonation of an alkenyl manganate reagent is performed with retention of the starting configuration :

$$(Bu\text{-}CH=CH)_3MnLi \xrightarrow[\text{ether}, 20°C]{CO_2} 3\ Bu\text{-}CH=CH\text{-}COOH \quad 80\% \; (Z \geqslant 99.5\%)$$

4-2 Reactions with CS_2, RN=C=O

The organomanganate reagents react on carbon disulfide at low temperature. However the resulting manganous dithiocarboxylates give the corresponding dithioesters with low yields[14] :

$$Hept_3MnLi \xrightarrow[THF,-70°C]{CS_2} \xrightarrow[20°C,4h]{MeI} 3\ HeptCS_2Me \qquad 34\%$$

Isocyanates give the corresponding amides with excellent yields by reaction with the organomanganous halides or organomanganate reagents[12,14] :

$$BuMnI + PhN{=}C{=}O \xrightarrow[-15° to\ 20°C]{ether} PhNHCOBu \qquad 93\%$$

$$Me_3MnLi + 3\ PhN{=}C{=}O \xrightarrow[-50° to\ 20°C]{THF} 3\ PhNHCOMe \qquad 98\%$$

4-3 Reactions with SO_2

Sulfinic acids are obtained with good yields by action of sulfur dioxide on organomanganate reagents at low temperature[14] :

$$Hept_3MnLi \xrightarrow[THF,-60°C]{SO_2} 3\ HeptSO_2H \qquad 90\%$$

5 - PREPARATION OF KETONES

5-1 Introduction

One of the most useful applications of organomanganous reagents is the preparation of ketones from trivalent functionalized molecules. Esters and amides do not react (section 3-1), but acid halides and anhydrides are excellent candidates. In fact the challenging aspect of manganese reagents has to be compared to that of a large variety of organometallic reagents, well known for their ability to yield ketones from acid halides : Boron[23], Rhodium[24], Silicon [25], Tin[26], Zirconium[27], Mercury[28], Aluminium[23b,29] species are used, but less frequently, due to the access to the organometallic reagent, cost, or toxicity. Cadmium[30-33] and Zinc[31-34] reagents usually lead to fair yields of ketones and their preparation may be troublesome because it sometimes requires lengthy heating, or handling insoluble reagents. Copper[35,36] reagents have been often preferred to the latter ones since cuprates give excellent yields of ketones, but require low temperature conditions, and very often a large excess of the reagent. Magnesium compounds[32,37] may be used at very low temperatures (-78°) a condition which meets with problems when scaling up and with alkyl Grignards it is necessary to use a large excess (2/1) of acid halides. Lithium[38] derivatives are the more useful in their reaction with lithium carboxylates but satisfactory yields are obtained only with methyl- or phenyl-lithium.

5-2 Saturated and aromatic ketones

5-2-1 from acid halides

Alkyl manganous halides, prepared in ether, react with acid chlorides : RMnI compounds give the best yields.[9,10] Acid bromides can be used.

$$\text{BuMnX} + \text{HeptCOCl} \xrightarrow{\text{ether}} \text{Bu-CO-Hept}$$

X =	
Cl	73%
Br	78%
I	91%

$$\text{HeptMnI} + \text{MeCOBr} \longrightarrow \text{HeptCOMe} \quad 93\%$$

Some examples are given in the following table :

$$\text{RM} + \text{MnI}_2 \xrightarrow{\text{ether}} \text{RMnI} \xrightarrow{\text{R'COCl}} \text{R'COR (M=Li or MgBr)}$$

R	R'	Yield% in isolated ketone
Me	Hept	83
Bu	Hept	91
Hept	i-Pr	85
Hept	t-Bu	80
Bu	C-Hex	86
i-Pr	Hept	78 **
t-Bu	Hept	60 *

* with 1.5 equivalent of tBuMnI at -30°C

** with 1.5 eq. of iPrMnI and 5 eq. of THF (section 5-5)

Aromatic ketones may be prepared along the same lines via :

$$BuMnI + Ph\text{-}COCl \xrightarrow{ether} Bu\text{-}CO\text{-}Ph \quad 92\%$$

or

$$PhMnI + Bu\text{-}COCl \longrightarrow \quad " \quad 80\%$$

dialkyl manganese also react and transfer both alkyl groups

$$Bu_2Mn + 2\ HeptCOCl \xrightarrow{ether} 2\ Bu\text{-}CO\text{-}Hept \quad 85\%$$

but are less easy to handle (section 2-4), as compared with manganates (which consume less manganese salts). These latter, however, lead to slightly lower yields, but all three organic moieties are transferred.

$$Bu_3MnLi + 3\ HeptCOCl \longrightarrow 3\ Bu\text{-}CO\text{-}Hept \quad 80\%$$

5-2-2 from acid anhydrides

a/Symmetrical acid anhydrides

The cheap manganese dichloride is best used in THF (section 2) but behaves as a good Lewis acid towards ring opening of the solvent in the presence of acid halides. It was therefore needed to turn to other trivalent functions.

Symmetrical acid anhydrides react with RMnI in ether or RMnCl in THF[13] less rapidly than with acid halides, but yields are excellent especially in ether.

$$HeptMnX + (EtCO)_2O \longrightarrow Et\text{-}CO\text{-}Hept$$

X = I, solvent ether : 96%
X = Cl, solvent THF : 85%

a limitation is observed when tertiary organometallics are used.

$$RM + MnI_2 \xrightarrow{ether} RMnI \xrightarrow{(R'CO)_2O} R'COR \quad (M=Li\ or\ MgBr)$$

R	R'	Yield% of isolated ketone
Hept	Et	96
Hept	t-Bu	89
t-Bu	i-Bu	40-50
Ph	Et	92

$$RM + MnCl_2 \xrightarrow{THF} RMnCl \xrightarrow{(R'CO)_2O} R'COR \ (M=Li,\ MgCl)$$

R	R'	Yield% of isolated ketone
Hept	Et	85
Hept	i-Pr	81
Hept	t-Bu	81
i-Pr	Hept	15
t-Bu	i-Bu	25
Ph	i-Bu	83

Manganates also react in ether or THF with acid anhydrides. Yields are slightly lower than those observed with RMnI

$$Hept_3MnM + 3\ (EtCO)_2O \longrightarrow 3\ HeptCO\text{-}Et$$

M= Li (ether) 80%

M= MgCl (THF) 74%

b/ Unsymmetrical acid anhydrides

In order to use the totality of the interesting R group, various anhydrides of type R-COOCOR' have been tried, the most interesting ones are mixed pivalic anhydrides.

$$HeptMnI + PentCOOCOtBu \xrightarrow{ether} Hept\text{-}CO\text{-}Pent \quad 72\%$$

5-2-3 from mixed acid anhydrides

a/ Carboxylic-carbonic anhydrides : RCOOCOOEt

The easily accessible carboxylic-carbonic anhydrides may be used. They are conveniently prepared according to :

$$RCOOH + ClCOOEt \xrightarrow[Et_2O \text{ or } CH_2Cl_2]{Et_3N} R\text{-}COOCOOEt \quad \text{quant.}$$

and used without further purification after removal of the solvent (do not delay). They give excellent yields of ketone in ether.

$$BuMnI + Ph\text{-}COOCOOEt \xrightarrow{ether} Bu\text{-}CO\text{-}Ph \qquad 90\%$$

but in THF reproducibility of the reaction is highly dependent on the purity of the manganous halides (for still unknown reasons).(section 5-5, application 3).

b/ Other mixed acid anhydrides

Other unsymmetrical reagents have been tried, of type R-COOCOZR' (ZR'=OMe,OPh,SEt), but are less stable and reagents R-CO-O-P(O)(OMe)$_2$ give poor results. Pyridyl-thio-esters are also reactive.

$$Py\text{-}S\text{-}COBu + HeptMnX \longrightarrow HeptCOBu$$

(Py = 2-pyridyl)	X=I, ether	80%
	X=Cl, THF	77%

but R-CO-SPh fail to react.

5-2-4 Mechanistical considerations

As far as mechanism is concerned, it may be surprising that the ketones prepared according to the above methodology are never spoiled by the corresponding tertiary alkohols, since already at -50°C ketones are found to be reactive with the organometallic reagent. Two explanations may be put forward :

$$RMnX + R'C(=O)X \rightarrow \begin{cases} RR'C(OMnX)X \quad (\underset{\sim}{A}) \\ RR'C{=}O,MnX_2 \quad (\underset{\sim}{B}) \end{cases} \xrightarrow[\times]{RMnX} RR'RC\text{-}OMnX$$

Various results point to the conclusion that complex intermediate $\underset{\sim}{B}$ should be favoured over the hemiketalate form $\underset{\sim}{A}$. This property is indeed a key point for the preference of the manganese reagent over its precursors (Li,MgX) and since also low temperatures are no more needed.

However, in the presence of an excess of the manganous reagent, and at room temperature, the very slow formation of the carbinol may be observed :

$$2\ BuMnI + 1\ BuCOCl \xrightarrow[20°C]{Ether} Bu\text{-}CO\text{-}Bu + Bu_3C\text{-}OH$$

after 24 h	61%	25%
after 36 h	20%	63%

This process has been used efficiently for the synthesis of polymer bound carbinols according to :[21]

$$\text{⌇-}Ph\text{-}CH_2\text{-}OOC\text{-}(CH_2)_4\text{-}COCl \xrightarrow{PhMnI} \text{⌇-}Ph\text{-}CH_2\text{-}OOC\text{-}(CH_2)_4C(OH)Ph_2$$

85%

5-3 Ethylenic and acetylenic ketones[10,17]

5-3-1 from acid halides

Vinyl and alkynyl manganous iodides in ether react with a variety of acid halides to give good yields of conjugated ketones.

The acetylenic reagents are less reactive than the vinylic ones.

$$RMnI + R'COCl \xrightarrow{ether} RCOR'$$

R	R'	Yield %
$Me_2C{=}CH$	Hept	92
$Me_2C{=}CH$	Ph	85
$Me_2C{=}CH$	$Me_2C{=}CH$	78
$BuC{\equiv}C$	n-Bu	89
$Bu\text{-}C{\equiv}C$	t-Bu	75

It is also possible to start with α-ethylenic acid chlorides, but so far, only β,β-substituted ones ($R_2C{=}CH\text{-}COCl$) have been used successfully, β-mono substituted ones lead to polymerization. However we know of no other methods for attaining these α,α'-biethylenic ketones in good yields. Allyl manganous iodide does not allow the preparation of ketones according to this procedure. The geometry of the vinyl manganous iodide is preserved during these reactions. (Care must be taken that Z,E isomerisation does not occur: for this purpose, manganous salts are quantitatively removed by treatment with sodium carbonate.

Bu⌒MnI + BuCOCl $\xrightarrow{\text{ether}}$ Bu⌒CO—Bu 80% (Z : 95%)

Me(Et)C=CH-MnI + BuCOCl $\xrightarrow{\text{ether}}$ Me(Et)C=CH-CO-Bu 82% (Z : 97%)

Bu⌒MnI + $ClCO-(CH_2)_4COOEt$ ⟶ Bu⌒$CO-(CH_2)_4COOEt$ 70% (Z≥95%)

The starting reagent is made from the corresponding 1-iodo-1-alkene[39,40]

R⌒I $\xrightarrow{\text{BuLi}}$ R⌒Li $\xrightarrow{MnI_2}$ R⌒MnI

the elaboration of a pure Z or E ethylenic acid chloride is difficult and isomerisation takes place[17]

Me(Et)C=CH-COOH + ClCOCOCl ⟶ Me(Et)C=CH∿COCl Z/E=85/15

5-3-2 from acid anhydrides

Vinyl manganous halides give excellent yields of ketones when treated by acid anhydrides in ether

$Me_2C=CHMnI$ + $(iPrCO)_2O$ $\xrightarrow{\text{ether}}$ $Me_2C=CHCO-iPr$ 85%

$Me_2C=CHMnI$ + Bu-COOCOOEt $\xrightarrow{\text{ether}}$ $Me_2C=CHCOBu$ 90%

but yields are lower in THF

$Me_2C=CHMnCl$ + $(BuCO)_2O$ $\xrightarrow{\text{THF}}$ $Me_2C=CHCOBu$ 40%

5-3-3 Purity of the E or Z ethylenic ketones

Starting from α-ethylenic carboxylic acids, the pure mixed anhydrides are obtained without isomerisation[17]

$$\underset{Et}{\overset{Me}{>}}=\!\!\diagdown_{COOH} \xrightarrow[Et_3N \text{ or } CH_2Cl_2]{ClCOOEt} \underset{Et}{\overset{Me}{>}}=\!\!\diagdown_{COOCOOEt} \quad \text{quant.}$$

but unfortunately the reaction of the latter with butyl manganous iodide gives the desired ketone in both isomeric forms (E/Z=60/40).

Alkynyl manganous iodides react with symmetrical anhydrides much better in THF than in ether :

$$BuC{\equiv}CMnI \ + \ (iPrCO)_2O \longrightarrow Bu\text{-}C{\equiv}C\text{-}CO\text{-}iPr$$

in ether : 35%
in THF : 75%

whereas in the above reaction, a mixed carbonic carboxylic anhydride allows only fair yields

$$BuC{\equiv}C\text{-}MnI \ + \ BuCOOCOOEt \xrightarrow{ether} BuC{\equiv}C\text{-}COBu \qquad 63\%$$

5-4 Preparation of functionalized ketones

Some extra functionalities may be present on the acid halide moiety, which are not attacked by the manganous reagent.

5-4-1 Halogenated ketones

Their access via organolithium[41,42], magnesium[37-d,43], cadmium[43,44,45], Zinc[43] reagents or cuprates[36] is still a problem specially if the halogen is in α-position, and organo manganous halides seem best suited for such a strategy. For example if an equimolar mixture of acid halide and allyl chloride is reacted with one equivalent of alkyl manganous iodide, only the former reagent is attacked

$$1\ BuCOCl \ + \ 1\ CH_2{=}CHCH_2Cl \xrightarrow{1\ BuMnI} Bu\text{-}CO\text{-}Bu \qquad 92\%$$

α-halo acid halides may be employed (the corresponding mixed anhydrides are very labile and cannot be used)

$$RMnI + Cl_3COCl \xrightarrow{ether} R\text{-}CO\text{-}CCl_3$$

R = Hept 75%
R = BuC≡C 55%

This scheme applies also to trifluoromethyl ketones, starting with the anhydride

$$RMnI + (CF_3CO)_2O \longrightarrow R\text{-}CO\text{-}CF_3$$

$n\text{-}C_{10}H_{21}$	70%
$Bu_2C{=}CH$	89%
$HexC{\equiv}C\text{-}$	45%
P-MeO-Ph	84%

α,α-dichloroketones are among the most challenging ketones to obtain by an organometallic route.

They have been obtained from the RMnCl reagent prepared in the minimum quantity of THF (30 ml for 50 mmole) and condensed in ether with the α,α-dihalo acid chloride

$$BuLi + MnCl_2 \xrightarrow{(THF)} BuMnCl \xrightarrow[CHCl_2COCl]{Ether} CHCl_2\text{-}CO\text{-}Bu \quad 69\%$$

If RMnI is used directly in ether a 50% yield is obtained. α-monochloroketones are obtained by the same procedure

$$HeptLi + MnCl_2 \xrightarrow{(THF)} HeptMnCl \xrightarrow[CH_2ClCOCl]{ether} CH_2Cl\text{-}CO\text{-}Hept \quad 66\%$$

β-haloketones cannot be obtained from β-halo acid halides since elimination of HCl is too fast.

$\gamma \ldots \omega$ haloketones are prepared in excellent yield from RMnI in ether

$$RMnI + Cl\text{-}(CH_2)_3\text{-}COCl \longrightarrow R\text{-}CO\text{-}(CH_2)_3Cl$$

R=Hept	86%
$Me_2C{=}CH$	70%
$PrC{\equiv}C$	88%

p.chloro benzoyl chloride reacts as well (93%).

5-4-2 Alkoxy-, alkylthio-, heterocyclic-ketones

These ketones are prepared in a straightforward manner from RMnI and acid halides in ether. Alcoholic functions may also be blocked by ester formation.

$$HeptMnI + EtO\text{-}CH_2COCl \longrightarrow EtO\text{-}CH_2\text{-}CO\text{-}Hept \quad 84\%$$

$$BuMnI + p\text{-}MeO\text{-}PhCOCl \longrightarrow p\text{-}MeO\text{-}Ph\text{-}COBu \quad \text{quant.}$$

$$BuMnI + PhS\text{-}CH_2\text{-}COCl \longrightarrow PhS\text{-}CH_2\text{-}CO\text{-}Bu \quad 80\%$$

$$BuC{\equiv}CMnI + AcO\text{-}\underset{Me}{CH}\text{-}COCl \longrightarrow AcO\text{-}\underset{Me}{CH}\text{-}COC{\equiv}CBu \quad 68\%$$

$$Me_2C{=}CHMnI + AcO\text{-}(CH_2)_5COCl \longrightarrow Me_2C{=}CHCO(CH_2)_5OAc \quad 89\%$$

For the isomeric acetoxy benzoic acid chlorides,the para isomer reacts much more easily than the ortho one,where the proximity of both functions promotes attack of the acetate: in similar cases organo cadmium halides proved disappointing[31-33] but a nearly 60% yield is observed with the manganese reagent.

$$MeO\text{-}Ph\text{-}COCl + BuMnI \longrightarrow MeO\text{-}Ph\text{-}COBu$$

ortho 58%
para 96%

Furyl or thienyl acid halides react normally with RMnI

$$RMnI \quad (\text{Y-ring})\text{-}COCl \xrightarrow{\text{ether}} (\text{Y-ring})\text{-}COR$$

Y	R	Position	Yield% of isolated ketone
O	Bu	3	91
O	Bu	2	87 *
O	$Me_2C{=}CH$	2	72
O	$BuC{\equiv}C$	2	43
S	Bu	2	96 **

* the reaction from the mixed ethyl carbonic carboxylic anhydride gives 79%

** the reaction from the mixed ethyl carbonic carboxylic anhydride gives 92%

5-4-3 Amino-ketones

As a general rule, the presence of amines on the carboxylic substrate, or in the medium (solvents ...) must be avoided with organomanganous reagents : yields are generally low

BuMnI + (3-pyridyl)-COOCOOEt $\xrightarrow{\text{ether}}$ (3-pyridyl)-COBu 38%

5-4-4 Keto-esters, keto-nitriles

Nitriles, like esters are inert toward RMnX so that cyano- or carboalkoxy acid halides may be used.

BuMnI + $NC-(CH_2)_4-COCl$ $\xrightarrow{\text{ether}}$ $NC-(CH_2)_4-CO-Bu$ 86%

BuMnI + (3-CN-C_6H_4)-COCl $\xrightarrow{\text{ether}}$ (3-CN-C_6H_4)-COBu 84%

BuMnI + $MeOOC-(CH_2)_2-COCl$ $\longrightarrow$ $BuCO(CH_2)_2COOMe$ 87%

$Me_2C=CHMnI$ + $EtOOC-(CH_2)_6-COCl$ $\longrightarrow$ $Me_2C=CHCO(CH_2)_6COOEt$ 92%

Mixed acid anhydrides may also be used. Ethoxy oxalyl chloride fails to yield the α keto ester.

5-4-5 Keto-acids

The reaction of RMnI with cyclic anhydrides can be used in some specific examples, but yields are variable.

RMnI + $(CH_2)_n(CO)_2O$ $\longrightarrow$ $R-CO-(CH_2)_n-COOH$

R	n	Yield %
Hept	2	77
Bu	3	55
Ph	3	33

5-4-6 Diketones

These ketones are readily available from the corresponding keto acid halides, since the keto moiety is much less reactive towards RMnI

EtMnI + Bu-CO-$(CH_2)_4$-COCl ⟶ Et-CO-$(CH_2)_4$CO-Bu 90%

BuMnI + 2-oxocyclopentyl-$(CH_2)_3$COCl ⟶ 2-oxocyclopentyl-$(CH_2)_3$COBu 77%

The reaction could not be extended to γ-oxo or δ-oxo acid halides which cyclise readily during their preparation, but mixed carbonic-levulinic anhydride may be used in the presence of hexane (section 5-5)

Me-CO-$(CH_2)_2$COOCOOEt + EtCH=CHCH$_2$CH$_2$MnCl ⟶ MeCO$(CH_2)_2$CO$(CH_2)_2$CH=CHEt 50%

5-4-7 Multifunctionalized ketones[15]

The following examples, in the field of steroïds, show several interesting chemoselectivities

AcO ... OAc ... COOH —1/ $SOCl_2$-C_6H_6; 2/ iBuMnI, Et_2O CH_2Cl_2⟶ AcO ... OAc ... O=

93%

HCOO ... OOCH ... COOH —1/$(COCl)_2$-C_6H_6; 2/ RMnI, Et_2O-CH_2Cl_2⟶ HCOO ... OOCH ... C(=O)R

a/	R= iBu-	89%
b/	R= Me_2C=CH-	78,5%
c/	R= Pr-C≡C-	73,5%

The mild formylation-deformylation of hydroxy functions is tolerated by the reagent, since the acid chloride reacts much faster than the formiate esters.

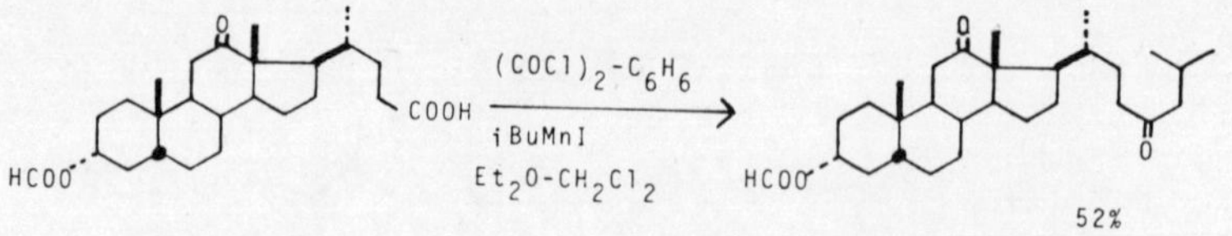

$$\text{HCOO-steroid-COOH} \xrightarrow[\text{iBuMnI, } Et_2O\text{-}CH_2Cl_2]{(COCl)_2\text{-}C_6H_6} \text{HCOO-steroid ketone} \quad 52\%$$

Only in the case of this 12-oxo steroid, did we observe a competition of carbonyl versus chlorocarbonyl attack. This might be attributed to the allowed vicinity of the two functions during complexation of the manganous reagent. Such a competition was not observed in the case of

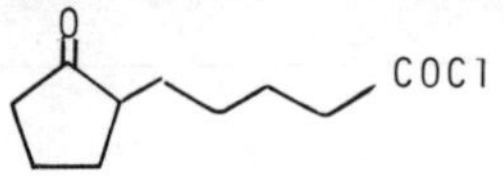

2-(COCl-alkyl)cyclopentanone (section 5-4-6) where a chemioselective reaction was observed

5-5 Influence of the nature of the solvents and of the ligands

a/ Solvents

These acylation reactions can be performed in the presence of a large variety of solvents :

$$\text{HeptMnCl} \xrightarrow[\text{THF + cosolvent (1.5/1)}]{(EtCO)_2O} \text{Hept-CO-Et}$$

Cosolvent	0	AcOEt	CO_3Et_2	CH_3CN
Yield %	85	75	75	78

dichloromethane may be used, and was the best cosolvent in the reaction mixture, able to dissolve steroidal acid halides (section 5-4-7). All these cosolvents are particularly unusual for the corresponding main group organometallics reactions. Dimethyl formamide and dioxane slow down dramatically the reaction, pyridin (and amines) must be avoided.

b/ ligands

Some ligands of manganese have been tried in a stoechiometric ratio (5 equivalents by Mn atom)

$$BuMnI + BuCOCl \xrightarrow[\text{ligand}]{\text{ether}} BuCOBu$$

Ligand	0	THF	THP	DME	Ph_3P	Me_2S	Et_3N
Yield %	92	93	92	79	84	98	0

(THP : tetrahydropyran, DME : dimethoxyethane). The influence of these small amounts of THF and THP is exemplified in application n° 1 (stabilization), that of Me_2S is to solubilize the RMnI reagent.

- application 1

The fact that five equivalents of THF may be used, without ring opening by the acid halide, finds a useful application in the case of the secondary and tertiary reagents, which decompose by -30°C in ether (section 2) so that an excess of reagent must be used. If 5 equivalents of THF are now used,

$$1.5\ iPrMnI + HeptCOCl \xrightarrow{\text{ether}} iPr\text{-}CO\text{-}Hept \qquad 65\%$$

$$1.5\ iPrMnI + HeptCOCl \xrightarrow[\text{5 THF}]{\text{ether}} \quad " \qquad 78\%$$

the stabilization of the reagent is improved (dec. by -10° C instead of -30°) and the yield of ketone too.

- application 2

RMnCl derivatives are not prepared quantitatively in ether, although their use would be more economical than use of RMnI. If THF is added in low proportions, the reagent is now obtained in much better yields, and acid halides do not lead to troublesome ring openings.

$$MnCl_2 + 2\ LiCl \xrightarrow[\text{THF}]{\text{minimum}} MnCl_4Li_2 \xrightarrow{\text{BuM/ether}} BuMnCl \xrightarrow{\text{BuCOCl}} BuCOBu$$

M = Li 72%

M = MgCl 71%

The ratio is 30 ml of THF for 50 mmole of $MnCl_4Li_2$. THP may be used in even higher concentrations since the ring opening occurs much less rapidly

$$BuLi + MnCl_2 \xrightarrow{\text{ether, THP(1/1)}} BuMnCl \xrightarrow{\text{BuCOCl}} Bu\text{-}CO\text{-}Bu \qquad 80\%$$

but DME leads to poor yields (54%).

- application 3

The reaction of RMnCl species with mixed anhydrides, for example:

$$\text{PentMnCl} + \text{PentCOOCOOEt} \xrightarrow[\text{S}]{\text{THF}} \text{PentCOPent} + \text{PentCOOEt}$$

may lead to the formation of undesired ethyl carboxylates, responsible for the drop of the yield. This is the major drawback mentionned above, and related to the quality of $MnCl_2$. It can be circumvented however when an adequate, less polar, medium is used. The following ratios of solvents have been tried :

Solvents (S)	Proportions vol/vol*	Yield % of ketone
THF		35
THF/ether	60/40	42
THF/ether	20/80	75
THF/ether/Hexane	15/15/70	89
THF/Hexane	30/70	88

The organomanganous reagent is prepared in high concentration (25 mmolesin 30 ml THF) (heterogeneous) hexane is then added (two phase system) and the mixed anhydride added under good stirring. This process is however limited to small scale operations.

In conclusion, the use of organomanganous halides may be recommanded for the preparation of a large array of ketones starting from acid halides or acid anhydrides.

The most interesting features are : the absence of alcohol in the product, the fact that low temperatures may be avoided, the access to ethylenic ketones with a given geometry, and to acetylenic ketones. α-haloketones may also be obtained and several polyfunctionalized ketones as well.

---

* in ml for 25 mmoles

6 - REACTIONS WITH $ClCOOEt$, $COCl_2$

a/ ClCOOEt

Organomanganous iodides, dialkylmanganese and organomanganate reagents react at room temperature with ethylchlorocarbonate to give the corresponding esters in good yields.[11]

$$\text{HeptMnI} + \text{ClCOOEt} \xrightarrow[20°C]{\text{ether}} \text{HeptCOOEt} \qquad 83\%$$

(70% from $Hept_2Mn$ and 65% from $Hept_3MnLi$).

Aryl or vinyl manganous iodides behave similarly :

$Me_2C{=}CHCOOEt$ (70%)   PhCOOEt (77%).

However the chloroformates are not as reactive as acid chlorides.

Thus the alkynylmanganous iodides which react generally as their alkyl or alkenyl analogues show for the first time a discrepancy since they cannot lead to the corresponding esters.

b/ $COCl_2$

Symmetrical ketones can be obtained from organomanganous iodides and phosgene.[11]

$$\text{HeptMnI} + COCl_2 \xrightarrow[-40°C \text{ to } 20°C]{\text{ether}} \underset{55\%}{\text{HeptCOHept}} + \underset{35\%}{\text{Hept-Hept}}$$

This reaction is very exothermic and the yield of ketone is lowered by the side formation of a coupling product which results from the oxydation of manganous reagent by the phosgene.

CONCLUSION

Organomanganous reagents have shown several advantages in comparison with other organometallic derivatives :

- the metal is cheap,
- they are prepared in usual solvents at room temperature or nearly,
- they react generally in very mild conditions, without the use of low temperature or expensive additives,
- they lead usually to high yields,
- they are especially efficient for selective attack of multifunctional compounds,
- they show a great tolerance towards unusual solvents in organometallic chemistry,
- they are particularly suited for the preparation of various simple or functional ketones.

In this latter case they may be considered as the most versatile organometallic reagents.

Acknowledgements

The authors acknowledge for the active participation of several coworkers, and particularly Dr. Bernard and Dr. Friour. Financial support comes from the C.N.R.S. Mrs Mathé is gratefully acknowledged for typing this manuscript.

7 - EXPERIMENTAL

7-1 Preparation of RMnX, R_2Mn and R_3MnLi (or MgX) in ether and THF

a/ Preparation of manganese iodide : MnI_2

Under a nitrogen atmosphere, 1.01 moles (256.5 g) of iodine (in solid form) are added in several portions of 30-50 g to 1 mole (55 g) of manganese chips (crushed sheets of manganese ; electrolytical grade $\geqslant$ 99.9%) in 400 ml of anhydrous ether with a good stirring at room temperature. Additions of iodine are effected at intervals of about 30 mn. On a smaller scale until 0.4 mole, iodine is added at once. When the reaction starts after the first addition of iodine, the temperature of the mixture raises slowly and occasional coolings with a cold water bath are used in order to keep the temperature below 30°C. After all iodine has been added, stirring is continued at room temperature until the total consumption of manganese chips is observed (5-8 h). The beige precipitate of manganous iodide is filtered on a sintered glass funnel, then washed twice with anhydrous ether (2 x 100 ml). The resulting powder is kept 2 h under reduced pressure (10^{-2} torr) in order to remove the residual ether. 280 g (or more) of pure manganese iodide are thus obtained (yield $\geqslant$ 90%).

This product is hygroscopic and all the preceding manipulations must be performed quickly or under a nitrogen atmosphere.

Manganese iodide must be kept dry in the dark.

b/ Preparation of the soluble "ate" complex of manganese chloride : $MnCl_4Li_2$ (this formulation only refects the ratio of the two components, smaller amounts of LiCl may be used but solubilization is then slower)

Under a nitrogen atmosphere, 100 mmol (8.5 g) of anhydrous lithium chloride powder are added, at room temperature to a suspension of 50 mmol (6.3 g) of anhydrous manganese chloride in anhydrous THF (50 ml). After stirring for about 15 mn a clear yellow solution is obtained, which is generally used without delay.

c/ Preparation of organomanganous iodides : RMnI

Under a nitrogen atmosphere, the solution of organolithium or magnesium reagent (50 mmol) is added with stirring, at 0°C to a suspension of 52 mmol (16.1 g) of manganese iodide in anhydrous ether (80 ml). After 10 mn, the mixture is allowed to warm to room temperature and stirred for 30 mn (for the less stable s- or t-alkyl manganous iodides (section 2-4) the addition is performed at -35°C with further stirring for 1 to 2 h at -35°C). The organomanganous iodide is thus obtained as a red-brownish solution or a beige to brownish suspension (when R = vinyl or aryl group a greenish coloration is often observed).

d/ Preparation of lithium organomanganates in ether : R_3MnLi

They are prepared, in 80 ml ether, from 16.7 mmol of manganese iodide and 50 mmol of organolithium reagent.

The procedure is the same as above, generally they form brownish solutions.

e/ Preparation of organomanganous chloride : RMnCl

- from organolithium reagents :

Under a nitrogen atmosphere, the solution of organolithium reagent (50 mmol) is added with stirring, at 0°C, to a suspension of 52 mmol (6.55 g) of anhydrous manganese chloride in anhydrous THF (80 ml). The mixture is allowed to warm to room temperature and to stirr for 15 mn ;

- from organomagnesium reagents :

Same general procedure except that a solution of 52 mmol of the "ate" complex $MnCl_4Li_2$ in 80 ml of anhydrous THF (prepared as above) is used instead of the manganese chloride suspension.

In both cases, a yellow to brownish solution of organomanganous chloride is thus obtained (yellow if THF is free from oxygen) ;

f/ Preparation of lithium or magnesium organomanganates in THF : R_3MnLi, R_3MnMgX

These reagents are prepared, in THF, from 16.7 mmol (2.1 g) of anhydrous manganese chloride and 50 mmol of organolithium or magnesium reagents according to the procedure used for the preparation of RMnCl from RLi. Yellow to brown solutions are obtained.

7-2 Preparation of n-butyl diisopropyl carbinol[12]

$$BuMnI + i\text{-}Pr_2C{=}O \longrightarrow i\text{-}Pr_2(Bu)C\text{-}OH$$

Under a nitrogen atmosphere, a solution of diisopropylketone (50 mmol ; 5.7 g) in anhydrous ether (30 ml) is added, at -10°C, to a stirred ethereal suspension of butyl manganous iodide (52 mmol, section 7-1). The mixture is allowed to warm to room temperature, stirred for 30 mn, and then hydrolyzed at 0°C with 80 ml of a 0.5N HCl solution. The aqueous layer is decanted, extracted twice with hexane or ether (2 x 50ml) then the combined organic layers are washed with a saturated sodium carbonate solution (40 ml) and dried ($MgSO_4$). Solvents are evaporated in vacuo and the crude product is distilled to give 7.6 g (88%) of n-Butyl diisopropyl carbinol.

B.p. 93-94°C/14 torr ; ^{1}H NMR (100Mhz, CCl_4, TMS) : 0.95 and 1.00 (t and d, 15 H, CH_3) ; 1.25-1.70 (m, 7H, CH_2 and OH) ; 1.97 (m, 2H, CH).

7-3 Selective attack of 5-oxo hexanal, by n-butyl manganous iodide[12]

$$BuMnI + MeCO(CH_2)_3CHO \longrightarrow MeCO(CH_2)_3CHOHBu$$

Under a nitrogen atmosphere, a solution of 5-oxo hexanal (50 mmol ; 5.7 g) in anhydrous ether (30 ml) is added quickly to a stirred ethereal suspension of butyl manganous iodide (50 mmol, section 7-1), so that the temperature be kept below -50°C. After stirring for 30 mn at -50°C, the mixture is hydrolyzed at -20°C with 80 ml of a 0.5N HCl solution. The aqueous layer is decanted, extracted twice with hexane or ether (2 x 50 ml) then the combined organic layers are washed with a saturated sodium hydrogen carbonate solution (40 ml) and dried ($MgSO_4$). Solvents are removed in vacuo and the crude product is distilled to give 6.7 g (78%) of 6-hydroxy decan-3-one.

B.p. 58-59°C/2 torr ; ^{1}H NMR (100 MHz, CCl_4, TMS) : 0.94 (m, 3 H, $C\underline{H}_3CH_2$) ; 1.20-1.60 (m, 10 H, CH_2) ; 1.68 (s, 3 H, $CH_3C{=}O$) ; 1.98 (m, 2 H, $CH_2C{=}O$) ; 3.75 (m, 1 H, $C\underline{H}OH$) ; 4.37 (s, 1 H, OH).

7-4 Carbonation of magnesium tris 3-propyl hex-2-en-1-yl manganate[14]

$$(Pr_2C{=}CHCH_2)_3MnMgCl \xrightarrow{CO_2} 3\ Pr_2\underset{\displaystyle COOH}{\underset{|}{C}}{-}CH{=}CH_2$$

Under a nitrogen atmosphere, a stream of dry CO_2 is introduced into the flask which contains a stirred solution of the above magnesium allylic manganate (16.7 mmol) in THF (80 ml) at 0°C (ice bath) until the end of the exothermic effect. The mixture is then stirred for 1 h at room temperature and hydrolyzed at 0°C with 80 ml of a 1N HCl solution. The aqueous layer is decanted, extracted once with ether (50 ml) then twice with hexane (2 x 50 ml). The combined organic layers are dried ($MgSO_4$) and solvents are evaporated in vacuo. The product is isolated by distillation. 6 g (70.5%) of 2,2 dipropyl 3-butenoic acid are thus obtained.

B.p. 91°C/0.1 torr ; n_D^{20}=1.4485 ; ^{1}H NMR (100 MHz, CCl_4, TMS) : 0.92 (t, 6H, CH_3) ; 1.28 (m, 2 H, $CH_3CH_2C\underline{H}_2$) and 1.70 (m, 2 H, $CH_3CH_2C\underline{H}_2$) ; 5.15 (d, 1 H, trans C-C=C$\underline{H}$) ; 5.20 (d, 1 H, cis C-C=C$\underline{H}$) ; 6.08 (dd, 1 H, C-C$\underline{H}$=C) ; J_{cis}= 12 Hz ; J_{trans}= 18 Hz ^{13}C-NMR (22.5 MHz, $CDCl_3$, TMS) : 52.6 ($\underline{C}$-COOH) 114.8 ($\underline{C}H_2$=) ; 140.1 ($\underline{C}$H=) ; 183.1 (COOH).

7-5 Sulfinylation of lithium tri n-heptyl manganate

$$\text{Hept}_3\text{MnLi} \xrightarrow{SO_2} 3\ \text{HeptSO}_2\text{H}$$

Under a nitrogen atmosphere, a stream of SO_2 is introduced into the flask which contains a stirred solution of the lithium triheptyl manganate (16.7 mmol, section 7-1) in THF (80 ml) at -50°C, until the end of the exothermic effect. The rate of introduction is adjusted so that the temperature does not exceed -40°C. The mixture is then allowed to warm to room temperature, stirred for 30 mn and hydrolyzed at 0°C with 80 ml of a 1N HCl solution. The aqueous layer is decanted, extracted twice with hexane or ether (2 x 50 ml) then the combined organic layers are dried ($MgSO_4$). Solvents are removed in vacuo to give 7.4 g (90%) of heptane sulfinic acid.

^{1}H NMR (100 MHz, CCl_4, TMS) : 0.94 (t, 3 H, CH_3) ; 1.37 (m, 8 H, $CH_3(C\underline{H}_2)_4$) ; 1.70 (m, 2 H, $C\underline{H}_2CH_2SO_2H$) ; 1.84 (t, 2 H, CH_2SO_2) ; 11.35 (s, 1 H, SO_2H).

Characterization is performed from the crude product by addition of methyl vinyl ketone.

$$HeptSO_2H + CH_2{=}CHCOMe \longrightarrow HeptSO_2CH_2CH_2COMe$$

After one recrystallization of the crude adduct in pentane/ethanol (9/1) ; 7.9 g (75% from crude sulfinic acid) of pure Heptyl, oxo-3-butyl sulfone are obtained, m.p. 70°C (lit. 70°C).[46]

7-6 Preparation of 5-oxo decan-1-ol acetate

$$BuMnI + MeCOO(CH_2)_5COCl \longrightarrow MeCOO(CH_2)_5COBu$$

Under a nitrogen atmosphere, a solution of 6-acetoxy hexanoyl chloride (50 mmol, 9.6 g) in anhydrous ether (30 ml) is added, at -10°C, to the stirred suspension of butyl manganous iodide (52 mmol, section 7-1) in ether (80 ml). After 15 mn the mixture is allowed to warm to room temperature, stirred for 2 h and then hydrolyzed at 0°C with 80 ml of 0.5N HCl. The aqueous layer is decanted, extracted twice with hexane (2 x 50 ml) then the combined organic layers are washed with a sodium thiosulfate solution (40 ml) and with a saturated sodium hydrogen carbonate solution (40 ml) before drying ($MgSO_4$). Solvents are evaporated in vacuo and the crude product is distilled to give 10.2 g (95%) of 6-oxo decan-1-ol acetate.

B.p. 113°C/0.8 torr ; n_D^{20}=1.4380 ; ^1H NMR (100 MHz, CCl_4, TMS) : 0.90 (t, 3 H, $C\underline{H}_3CH_2$) ; 1.96 (s, 3 H, CH_3COO) ; 2.37 (t, 4 H, CH_2COCH_2) ; 3.98 (t, 2 H, CH_2O) ; ^{13}C NMR (22.5 MHz, $CDCl_3$, TMS) : 13.85 ($\underline{C}H_3CH_2$) ; 20.85 ($\underline{C}H_3COO$) ; 42.45 ($\underline{C}H_2COCH_2$) ; 64.20 (CH_2O) ; 170.75 (COO) ; 210.55 (C=O).

7-7 Preparation of 2-methyl decan-3-one (THF as ligand)

$$MnI_2 \xrightarrow[\text{1/ 5 eq.THF; 2/ i-PrMgBr}]{\text{ether}} \text{i-PrMnI ; 5 THF} \xrightarrow{\text{HeptCOCl}} \text{i-PrCOHept}$$

Under a nitrogen atmosphere, 5.4 g THF (5 eq.) are added at 20°C to the suspension of 75 mmol of manganous iodide in

70 ml ether and 75 mmol of isopropyl magnesium bromide in ether are added at -40°C (then stirred for 1 h at -35°C).

To this stirred suspension is added at -40°C a solution of heptanoyl chloride (50 mmol, 7.4 g) in anhydrous ether (30 ml). The mixture is stirred for 3 h at -25°C, then hydrolyzed at this temperature with 80 ml of 0.5N HCl. The aqueous layer is decanted, extracted twice with hexane or ether (2 x 50 ml), then combined organic layers are washed with a sodium thiosulfate solution (40 ml) and with a sodium hydrogen carbonate solution (40 ml). After drying ($MgSO_4$), solvents are removed in vacuo and the product is isolated by distillation, 6.65 g (78%) of isopropyl heptyl ketone are thus obtained.

B.p. 101°C/13 torr ; n_D^{20} = 1,4260 ; ^1H NMR (100 MHz, CCl_4, TMS) : 0.90 (t, 3H, $C\underline{H}_3CH_2$) ; 1.06 (d, 6 H, $(C\underline{H}_3)_2CH$) ; 2.38 (t, 2 H, $CH_2C{=}O$) ; 2.50 (m, 1 H, CHC=O).

7-8 Preparation of 2,6-dimethyl-2,5-heptadien-4-one

$$Me_2C{=}CHMnI\ ;\ 5\ THF + Me_2C{=}CHCOCl \xrightarrow{ether} Me_2C{=}CHCOCH{=}CMe_2$$

Under a nitrogen atmosphere, a solution of 3-methyl prop-2-enoyl chloride (50 mmol, 5.9 g) in anhydrous ether (30 ml) is added at -50°C to a stirred suspension of 2-methyl-1-propenyl manganous iodide (52 mmol) complexed with 5 equivalents of THF (this is not compulsory) in 80 ml of ether (see section 7-7). After 30 mn at -50°C, the reaction mixture is allowed to warm to room temperature, stirred for 2 h, then hydrolyzed at 0°C with 80 ml of a saturated sodium hydrogen carbonate solution. The resulting mixture is stirred for 20 mn at room temperature then the precipitate of manganous carbonate is filtered off and washed with ether (80 ml). The aqueous layer is decanted, extracted twice with hexane (2 x 50 ml) then combined organic layers are washed once with a sodium thiosulfate solution (40 ml) and once with a sodium hydrogen carbonate solution (40 ml). After drying ($MgSO_4$), the solvents are evaporated in vacuo and the crude product is distilled to give 5.4 g (78%) of 2,6-dimethyl-2,5-heptadien-4-one.

B.p. 81°C/12 torr ; ^{1}H NMR (100 MHz, CCl_4, TMS) : 1.85 (s, 6H, trans-CH_3C=C-C=O) ; 2.13 (s, 6H, cis-CH_3C=C-C=O) ; 5.95 (m, 2H, C=CH).

Important remark :
Hydrolysis with saturated sodium hydrogen carbonate solution <u>must be used</u> for the preparation of pure Z or E conjugated ketones but it is not compulsory when no stereochemistry is expected (this case).

7-9 Preparation of 1-chloro nonan-5-yn-4-one

$$PrC{\equiv}CMnI \quad + \quad Cl(CH_2)_3COCl \longrightarrow PrC{\equiv}CCO(CH_2)_3Cl$$

Under a nitrogen atmosphere, a solution of 4-chlorobutyryl chloride (50 mmol, 7.05 g) in anhydrous ether (30 ml) is added, at -10°C, to a stirred suspension of 1-pentynyl manganous iodide (52 mmol) (prepared from pentynyl lithium which is obtained by metallation of pentyne with butyl lithium in ether), in ether (80 ml). Then the mixture is allowed to warm to room temperature, stirred for 4 h and hydrolyzed at 0°C with 80 ml of 0.5 N HCl. The aqueous layer is decanted, extracted twice with hexane (2 x 50 ml) then the combined organic layers are washed once with a sodium thiosulfate solution (40 ml) and once with a saturated sodium hydrogen carbonate solution (40 ml). After drying ($MgSO_4$), the solvents are removed in vacuo and the crude product is distilled to give 7.6 g (88%) of 1-chloro nonan-5-yn-4-one.

B.p. 70-72°C/0.1 torr ; n_D^{20} = 1,4750 ; ^{1}H NMR (100 MHz, CCl_4, TMS) : 1.04 (t, 3H, CH_3) ; 1.65 (m, 2H, $CH_3\underline{CH_2}$) ; 2.08 (m, 2H, $\underline{CH_2}CH_2Cl$) ; 2.38 (t, 2H, $C{\equiv}CCH_2$) ; 2.72 (t, 2H, CH_2C=O) ; 3.60 (t, 2H, CH_2Cl) ; ^{13}C NMR (22.5 MHz, $CDCl_3$, TMS) : 13.45 (CH_3) ; 44.05 (CH_2Cl) ; 80.85 (C=<u>C</u>-C=O) ; 94.50 (<u>C</u>≡C-C=O) ; 186.30 (C=O).

7-10 Preparation of 4-butyl-1,1,1-trifluoro oct-3-en-2-one

$$Bu_2C{=}CHMnI \quad + \quad (CF_3CO)_2O \longrightarrow Bu_2C{=}CHCOCF_3$$

Under a nitrogen atmosphere, a solution of trifluoroacetic

anhydride (50 mmol, 10.5 g) in anhydrous ether (30 ml) is added at -20°C to a stirred suspension of 2-butyl-1-hexenyl manganous iodide* (52 mmol) in ether (80 ml). Then the mixture is allowed to warm to room temperature, stirred for 2 h and hydrolyzed at 0°C with 80 ml of 0.5N HCl. The aqueous layer is decanted, extracted twice with hexane (2 x 50 ml) then combined organic layers are washed once with a saturated potassium hydrogen carbonate solution (60 ml) and once with a sodium thiosulfate solution (40 ml). After drying ($MgSO_4$), the solvents are removed in vacuo and the crude product is distilled to give 10.5 g (89%) of 4-butyl-1,1,1- trifluoro oct-3-en-2-one.

B.p. 90°C/12 torr ; n_D^{20}= 1.4225 ; ^1H NMR (100 MHz, CCl_4,TMS): 0.97 (t, 6H, CH_3) ; 1.20-1.75 (m, 8H, $CH_3\underline{CH_2}\underline{CH_2}$) ; 2.35 (t, 2H, trans-$CH_2C$=C) ; 2.70 (t, 2H, cis-$CH_2C$=C) ; 6.32 (s, 1H, C=CH) ; ^{13}C NMR (22.5 MHz, $CDCl_3$, TMS) : 13.85 (CH_3) ; 34.15 (cis-$\underline{C}$-C=C) ; 39.85 (trans-$\underline{C}$-C=C) ; 114.65 (C=$\underline{C}$H) ; 116.65 (CF_3) ; 177.25 ($\underline{C}$=CH) ; 179.25 (C=O) ; J_{C-F}=291.3 Hz (CF_3) ; 33 Hz ($\underline{C}OCF_3$).

7-11 Preparation of 1,1-dichloro hexan-2-one

$$BuMnCl \text{ (minimum of THF)} + Cl_2CHCOCl \xrightarrow{ether} BuCOCHCl_2$$

BuMnCl is prepared from an ethereal solution of BuLi (52mmol) and a suspension of $MnCl_2$ (55 mmol) in 30 ml of anhydrous THF. The addition is performed at -30°C. After 10 mn the temperature is allowed to raise to room temperature, the mixture is stirred for 30 mn and cooled to -70°C. 80 ml of anhydrous ether are added and then, after 10 mn, a solution of dichloroacetyl chloride (50 mmol, 7.4 g) in anhydrous ether (30 ml). After 15 mn, the mixture is allowed to warm to room temperature, stirred for 1 h 30 and hydrolyzed at 0°C with 80 ml of diluted sulfuric acid. The aqueous layer

---

* The corresponding alcenyl lithium can be prepared in ether by metal-halogen exchange according to[47] :

$$Bu_2C{=}CHI \quad + \quad RLi \xrightarrow{-60°C,\ 15\ mn} Bu_2C{=}CHLi \quad + \quad RI \ (R = Et, Bu)$$

Alcenyl manganous iodide is obtained by addition of MnI_2 at -50°C then 30 mn of stirring at -5°C.

is quickly decanted, extracted twice with pentane and the combined organic layers are washed with a diluted sodium thiosulfate solution (only if iodine is present) and with 30 ml of H_2O. After drying ($CaCl_2$) the solvents are removed in vacuo. Distillation of the crude product affords 5.85 g (69%) of 1,1-dichloro hexan-2-one.

B.p. 69-70°C/12 torr ; n_D^{20} = 1.4510 ; ^1H NMR (100 MHz, CCl_4, TMS) : 0.96 (t, 3H, CH_3) ; 1.25-1.82 (m, 4H, $CH_3C\underline{H}_2C\underline{H}_2$) ; 2.82 (t, 2H, $CH_2C{=}O$) ; 5.87 (s, 1H, $CHCl_2$) ; ^{13}C NMR (22.5 MHz, $CDCl_3$, TMS) : 13.85 (CH_3) ; 22.15 (CH_3-$\underline{C}H_2$) ; 25.90 ($CH_3CH_2\underline{C}H_2$) ; 34.85 ($\underline{C}H_2C{=}O$) ; 70.05 ($CHCl_2$) ; 197.00 (C=O)

7-12 Preparation of dodecan-3,8-dione

$$EtMnI + BuCO(CH_2)_4COCl \longrightarrow BuCO(CH_2)_4COEt$$

Under a nitrogen atmosphere, a solution of 6-oxo decanoyl chloride (50 mmol, 10.3 g) in anhydrous ether (30 ml) is added, at -70°C, to a stirred suspension of ethyl manganous iodide (52 mmol) in ether (80 ml). After 20 mn at this temperature the mixture is allowed to warm to room temperature and stirred for 2 h, then hydrolyzed at 0°C with 80 ml of diluted HCl (0.5N). The aqueous layer is decanted, extracted twice with ether (2 x 50 ml) and organic layers are washed with a sodium thiosulfate solution (40 ml) and with a saturated sodium carbonate solution (40 ml) then dried ($MgSO_4$). Solvents are evaporated in vacuo and the crude product is distilled to give 8.95 g (90%) of dodecan-3,8-dione.

B.p. 93°C/0.1 torr ; m.p. 48°C (pentane) ; ^1H NMR (100 MHz, CCl_4, TMS) : 0.90 (t, 3H, $C\underline{H}_3CH_2CH_2$) ; 1.02 (t, 3H, $C\underline{H}_3CH_2C{=}O$) 1.15-1.80 (m, 8H, $CH_3C\underline{H}_2C\underline{H}_2$ and $CH_3C\underline{H}_2$) ; 2.25-2.60 (m, 8H, $CH_2C{=}O$) ; ^{13}C NMR (22.5 MHz, $CDCl_3$, TMS) : 7.80 ($\underline{C}H_3CH_2C{=}O$); 13.85 ($\underline{C}H_3CH_2CH_2$) ; 35.60 ($CH_3\underline{C}H_2C{=}O$) ; 42.05 ($CH_3CH_2CO\underline{C}H_2$); 42.45 ($CH_3CH_2CH_2\underline{C}H_2CO\underline{C}H_2$) ; 210.65 (Et$\underline{C}$=O) ; 210.95 (Bu$\underline{C}$=O).

REFERENCES

1. a/ H. Gilman, J.C. Bailie
   J. Org. Chem. 2, 84 (1937)
   b/ H. Gilman, R. Kirby
   J. Am. Chem. Soc. 63, 2046 (1941) ; see also ref. 2
2. a/ C. Beermann, K. Clauss
   Angew. Chem. 71, 627 (1959)
   b/ K. Clauss, H. Bestian
   Liebigs Ann. Chem. 654, 8 (1962) ; see also ref. 3
3. R. Riemschneider, H.G. Kassahn, W. Schneider
   Z. Naturforsch. 15b, 547 (1960)

   see also for a manganate derived from dilithium diphenylether :
   H. Drevs
   Z. Chem. 15, 451 (1975)
4. R. Nast, H. Griesshammer
   Chem. Ber. 90, 1315 (1957)
5. R.A. Andersen, E. Carmona-Guzman, J.F. Gibson, G. Wilkinson
   J. Chem. Soc. Dalton 1976, 2204
6. a/ M. Tamura, J. Kochi
   J. Organometal. Chem. 29, 111 (1971)
   b/ M. Tamura, J. Kochi
   Bull. Chem. Soc. Japan 44, 3063, 1971

   see also :
   c/ R.P.A. Sneeden, H.H. Zeiss
   J. Organometal. Chem. 22, 713 (1970)
7. G. Cahiez, D. Bernard, J.F. Normant
   J. Organometal. Chem. 113, 99 (1976)
8. G. Cahiez, D. Bernard, J.F. Normant
   J. Organometal. Chem. 113, 107 (1976)
9. G. Cahiez, A. Masuda, D. Bernard, J.F. Normant
   Tetrahedron Letters 36, 3155 (1976)
10. G. Cahiez, D. Bernard, J.F. Normant
    Synthesis 1977, 130
11. G. Cahiez, J.F. Normant
    Bull. Soc. Chim. France 1977, 570

12. G. Cahiez, J.F. Normant
Tetrahedron Letters 1977, 3383

13. G. Cahiez, A. Alexakis, J.F. Normant
Synthetic Comm. 1979, 639

14. G. Friour, A. Alexakis, G. Cahiez, J.F. Normant
Bull. Soc. Chim. France II 1979, 515

15. G. Cahiez
Tetrahedron Letters 1981, 1239

16. G. Friour, A. Alexakis, G. Cahiez, J.F. Normant
Tetrahedron, in press

17. G. Friour, G. Cahiez, J.F. Normant
Preparation of various functionnalized ketones from organomanganous reagents : to be published

18. a/ M. Tsutsui, H. Zeiss
J. Am. Chem. Soc. 83, 825 (1961)

see also :
E.O. Fischer, M.W. Schmidt
Chem. Ber. 100, 3782 (1967)

19. E.J. Corey, G.H. Posner
Tetrahedron Letters 1970, 315

20. A. Fischer, G. Cahiez, J.F. Normant
Selective methylation of allylic halides : to be published

21. C.C. Leznoff, V. Yedidia
Can. J. Chem. 58, 287 (1980)

22. R.E. Ireland, W.J. Thompson
J. Org. Chem. 44, 3044 (1979)

23. a/ E. Negishi, K. Chiu, T. Yosida
J. Organomet. Chem. 40, 1676 (1975)

b/ E. Negishi
Organometallics in organic Synthesis, Ed. J. Wiley, New York 1980

24. a/ L.S. Hegedus, P.M. Kendal, S.M. Lo, J.R. Sheats
J. Am. Chem. Soc. 97, 5448 (1975)

b/ C.U. Pittman Jr, R.M. Hanes
J. Org. Chem. 42, 1194 (1977)

25. a/ E. Frainnet, R. Calas, P. Gerval
C.R. Acad. Sci. 261, 1329 (1965)

b/ T.H. Chan, E. Chang, E. Vinokur
Tetrahedron Letters 1970, 1137

c/ T.H. Chan, I. Fleming
Synthesis 1979, 761

26. D. Milstein, J.K. Stille
J. Am. Chem. Soc. 100, 3636 (1978)

27. D.W. Harts, J. Schwartz
J. Am. Chem. Soc. 96, 8115 (1974)

28. a/ A.L. Kurts, I.P. Beletskaya, I.A. Savchenko, O.A. Reutov
J. Organomet. Chem. 17, P 21 (1969)

b/ R.C. Larock,
Angew. Chem. Int. Ed. Engl. 17, 27 (1978)

29. H. Adkins, C. Stanley
J. Am. Chem. Soc. 73, 2854 (1961)

30. a/ H. Gilman, J.F. Nelson
Rec. Trav. Chim. des Pays-Bas, 55, 518 (1936)

b/ P.L. De Benneville
J. Org. Chem. 6, 462 (1940)

c/ J. Cason
Chemical Reviews 40, 15 (1947)

31. N.I. Sheverdina, K.A. Kocheshkov
The Organic compounds of Zinc and Cadmium
North Holland 1967 Amsterdam

32. D.A. Shirley
Organic Reactions 8, 28

33. a/ F. Wingler
Houben Weyl. Methoden der Organischen Chemie 7/2a, 548,
Georg. Thieme Verlag 1973 Stuttgart

b/ P.R. Jones, P.J. Desio
Chemical Reviews 78, 491 (1978)

34. a/ E.E. Blaise, M. Maire
C.R. Acad. Sci. 142, 215 (1906)

b/ E.E. Blaise, A. Koehler
C.R. Acad. Sci. 148, 489 (1909)
Bull. Soc. Chim. Fr. 5, 681 (1909)
Bull. Soc. Chim. Fr. 7, 215 (1910)

c/ R.G. Jones
J. Am. Chem. Soc. 69, 2350 (1947)

d/ F.L. Breush, F. Baykut
Chem. Ber. 86, 684 (1953)

35. a/ C. Jallabert, NGOC-Tuyet Luong-Thi, H. Riviere
Bull. Soc. Chim. Fr. 1970, 797

b/ J.F. Normant, M. Bourgain
Tetrahedron Letters 1970, 2659

c/ G.H. Posner, C.E. Whitten
Tetrahedron Letters 1970, 4647

d/ G.H. Posner, C.E. Whitten, P.E. Mc Farland
J. Am. Chem. Soc. 94, 5106 (1972)

e/ M. Bourgain, J. Villieras, J.F. Normant
C.R. Acad. Sci (c) 276, 1477 (1973)

36. G.H. Posner
Organic Reactions 22, 284

37. a/ M.S. Newman, W.T. Booth Jr
J. Am. Chem. Soc. 67, 154 (1945)

b/ M.S. Newman, A.S. Smith
J. Org. Chem. 13, 592 (1948)

c/ F. Sato, M. Inoue, K. Oguro, M. Sato
Tetrahedron Letters 1979, 4303

d/ M.K. Eberle, G.G. Kahle
Tetrahedron Letters 1980, 2303

38. a/ H. Gilman, P.R. Van Hesse
J. Am. Chem. Soc. 55, 1258 (1933)

b/ E.A. Evans
J. Chem. Soc. 1956, 4691

c/ M.J. Jorgenson
Organic Reactions 18, 1

d/ R. Levine, M.J. Karten, W.M. Kadunce
J. Org. Chem. 40, 1770 (1975)

e/ H. Hauptman, M. Mader
Synthesis 1978, 307

39. a/ J.F. Normant, G. Cahiez, C. Chuit
J. Organomet. Chem. 77, 269 (1974)

b/ A. Alexakis, G. Cahiez, J.F. Normant
J. Organomet. Chem. 177, 293 (1979)

40. G. Cahiez, D. Bernard, J.F. Normant
Synthesis 1976, 245

41. T.F. Mc Grath, R. Levine
J. Am. Chem. Soc. 77, 3656 (1955)

42. H.F. Bluhm, H.V. Donn, H.D. Zook
J. Am. Chem. Soc. 77, 4406 (1955)

43. R.G. Jones
J. Am. Chem. Soc. 70, 143 (1948)

44. J.F. Bunnet, D.S. Tarbel
J. Am. Chem. Soc. 67, 1944 (1945)

45. K.N. Campbell, J.O. Knobloch, B.K. Campbell
J. Am. Chem. Soc. 72, 4380 (1950)

46. G. Cahiez, D. Bernard, J.F. Normant, J. Villieras
J. Organometal. Chem. 121, 123 (1976)

47. G. Cahiez, D. Bernard, J.F. Normant
Synthesis 1976, 245

TITANIUM AND ZIRCONIUM DERIVATES IN ORGANIC SYNTHESIS

A Review with Procedures

by Professor Dieter Seebach, Dr. Beat Weidmann and Leo Widler

Laboratorium für Organische Chemie der Eidgenössischen Technischen Hochschule, ETH-Zentrum, Universitätstrasse 16, CH - 8092 Zürich (Switzerland)

List of Contents

I INTRODUCTION

Derivatives of titanium and zirconium have become increasingly important in organic synthesis during the past decade. This is documented in several review articles covering different facets. These reviews will be referred to in the appropriate sections of the present treatise, which emphasizes the practical and experimental aspects. With few exceptions, we also restrict the discussion to non-catalytic uses of derivatives of these metals. Thus, their industrially important role as components of the catalytic mixtures used for *Ziegler-Natta* polymerization [1)], for metathesis [2)], and for nitrogen fixation [3)] will not be covered here.

The laboratory use of titanium and zirconium derivatives has focused on selectivity of known transformations rather than on novel reactivity. The applications can be divided into two big groups; one includes carbon carbon bond forming processes, the other one functional group transformations. After some comments on the availability, with procedures describing the preparation and purification of starting materials, section II, these two groups of applications will be the main subject, sections III and IV.

II REAGENTS AND STARTING MATERIALS

In almost all applications simple, readily available derivatives of tetravalent titanium and zirconium are employed. The actual reagent is often generated *in situ* and may be derived from lower-valent titanium or zirconium. The original reagents are either titanocene or zirconocene derivatives 1-3 or they are halo-, alkoxy-, dialkylamino- and organometallic compounds of type MX_4 (4-13) or $XM(Y)_3$ (14-21) without π-ligands on the central atom. Many of these compounds are commercially available, see Table 1; all of them are easily prepared on large scale, see the procedures below. Unlike most other transition metal reagents, the titanium and zirconium derivatives can be employed stoichiometrically, not only because they are cheap, but also because they are non-toxic. This is especially true of the second group of reagents 4-21: on contact with water they hydrolyze to give oxide-hydrates and eventually TiO_2 or ZrO_2, highly insoluble and, if desired, readily recovered materials. In contrast, the Cp_2-derivatives are very stable to hydrolysis to the inorganic oxides. After aqueous workup, Cp_2MCl_2

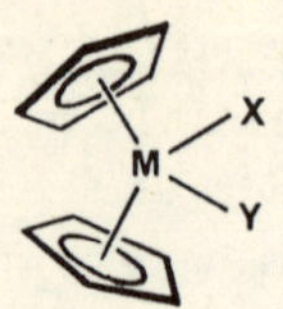

a: X = Y = Cl

b: X = Y = Br

c: X = Cl, Y = H

1: M = Ti

2: M = Zr

Cp CH$_3$
CH$_2$
Ti Al
Cl
Cp CH$_3$

3

M(Halogen)$_4$:	4:	$TiCl_4$	6:	$ZrCl_4$
	5:	$TiBr_4$		
M(OR)$_4$:	7:	$Ti(OEt)_4$	9:	$Zr(OProp)_4$
	8:	$Ti(OCHMe_2)_4$	10:	$Zr(OBut)_4$
			11:	$Zr(OC_6H_5)_4 \cdot THF$
M(NR$_2$)$_4$:	12:	$Ti(NMe_2)_4$		
	13:	$Ti(NEt_2)_4$		
ClM(OR)$_3$:	14:	$ClTi(OCHMe_2)_3$	16:	$ClZr(OBut)_3$
	15:	$ClTi(OC_6H_5)_3$	17:	$ClZr(OC_6H_5)_3$
XTi(NR$_2$)$_3$:	18:	$ClTi(NMe_2)_3$		
	19:	$BrTi(NEt_2)_3$		
RTi(OCHMe$_2$)$_3$:	20a:	$H_3C\text{-}Ti(OCHMe_2)_3$		
	20g:	$H_5C_6Ti(OCHMe_2)_3$		

Table 1. Approximate prices of some commercially available titanium and zirconium derivatives [January 1983; see fine chemical catalogs and offers by Dynamit Nobel (Germany), DuPont (USA), Titanium Intermediates Ltd. (England); 1 \$ ∿ 2 Sfr. ∿ 2.3 DM]

Compound	Mol. Wt.	Prices in \$	
1a	249	15/10 g; 60/50 g	300.--/mol
2a	292	70/25 g; 200/100 g	580.--/mol
2c	258	50/5 g; 180/25 g	1,850.--/mol
4	190	15/liter	1.60/mol
5	367	65/100 g	240.--/mol
6	233	10/250 g; 30/kg	7.--/mol
7	228	30/500 ml	14.--/mol
8	284	25/500 g; 10/kg (50 kg order)	2.80/mol
9	328	10/kg	3.30/mol
10	458	50/kg; 20/kg (50 kg order)	9.20/mol
12	224	70/5 g; 260/25 g	2,330.--/mol

can be regenerated with hydrochloric acid. Otherwise, the hydrated titanocene or zirconocene can be separated from products by chromatography (see section III, F).

Some representative procedures are given on the following pages. These can also be used for the preparation of analogous starting materials.

Dicyclopentadienyl titanium and zirconium dichlorides 1a and 2a [4)]

Cl_4M (4 or 6) + 2 C_5H_5Na → Cp_2MCl_2 + NaCl

A suspension of 1 mol of sodium cyclopentadienide in 1 L of dimethoxyethane is added under argon from a pressure equalized dropping funnel to a mechanically stirred mixture of 200 ml benzene and 0.5 mol MCl_4 (2 L three-necked flask, reflux condenser) at such a rate that the reaction does not become too vigorous. After the reaction mixture has cooled to room temperature, stirring is continued for 1 hour and the solvents are removed evaporatively (water aspirator vacuum). The residue is quickly transferred into a Soxhlet thimble and extracted with ca. 500 ml $CHCl_3$, from which the product crystallizes and is filtered after cooling. Yield at least 50%. 1a: red crystals of m.p. 290°C; 2a: colorless crystals of m.p. 248°C (dec.).

Chloro-bis(cyclopentadienyl)-hydrido-zirconium (2c) [5)]

$$2\ Cp_2ZrCl_2 + NaAlH_2(OCH_2CH_2OCH_3)_2 \rightarrow 2\ Cp_2ZrClH$$

To a ca. 0.2 M solution of the dichloride 2a in THF is added under argon a 0.5 mol equivalent of the reducing reagent (Vitride; as a benzene solution). The hydride 2c precipitates and is always contaminated with some NaCl, which does not interfere with subsequent reactions.

As reducing reagents, $LiAlH_4$ and $LiAl(O\text{-}t\text{-}C_4H_9)_3H$ have been recommended. The latter reagent produces analytically pure 2c after filtration and washing with THF.

Tebbe-Reagent (3) [6,7)]

$$Cp_2TiCl_2 + 2\ Al(CH_3)_2 \rightarrow Cp_2Ti\langle{}^{CH_2}_{Cl}\rangle Al(CH_3)_2$$

A solution of 62 g of Cp_2TiCl_2 and 48 ml of Me_3Al in 250 ml of toluene is allowed to stand 60 hours at room temperature. The nonvolatile products are recrystallized from toluene to produce 35 g of crude 3 (80-90% pure). Pure samples can be obtained by recrystallization from a solution of Me_3Al in toluene and from pentane.

Purification of tetrabutoxy-zirconium (10) [8)]

The commercial material contains ca. 1 equivalent of butanol per mol of 10. For many purposes, pure 10 is necessary as a starting material. Removal of the alcohol is accomplished in the following way: ca. 350 g of alcohol-containing material is heated in a 1 L flask (Aldrich Kugelrohr, Cat. No. Z 10,46-3) at 100°C/oil pump vacuum. After removal of most of the butanol, the gelatinous residue is distilled under high vacuum (200-250°C air bath temperature/0.001 Torr). The distillate has the consistency of melting glass and is dissolved in twice the volume of ether to give a stock solution for further use.

The tetrapropoxy derivative 9 is purified similarly.

Tetrakis(diethylamino)titanium (13) [9,10)]

$$TiCl_4 + 4\ LiNEt_2 \rightarrow Ti(NEt_2)_4 + 4\ LiCl$$

A 2 L three necked flask equipped with a pressure equalized dropping funnel, a thermometer, and a mechanical stirrer and immersed in an ice/NaCl cooling mixture is charged under argon with 530 ml 1.55 N butyllithium in hexane

(0.82 mol). Anhydrous diethyl amine (60 g, 0.82 mol) is added at such a rate that the internal temperature is kept at -10°C. The cooling bath is removed and stirring continued at room temperature for 30 min. In the dropping funnel, a solution of 32.4 g 4 (0.17 mol) in 200 ml toluene is prepared. This red solution is added with cooling and stirring to the white $LiNEt_2$ suspension at such a rate that the temperature of the reaction mixture is maintained at ca. 10°C. The black-brown mixture is stirred at room temperature for 12 hours. The precipitate is allowed to settle, and the supernatant solution is decanted by suction through a sintered funnel under argon, using teflon tubing. The filtrate is freed of solvent under exclusion of air, and the residue distilled *in vacuo*. Yield of 13 37.5 g (66%), yellow-red liquid, b.p. 115°C/0.15 Torr, stable under an inert atmosphere.

The dimethylamino-derivative 12 is prepared similarly; a ca. 4 M, cold solution of dimethylamine in hexane is used instead of the neat diethylamine.

Chloro-triisopropoxy-titanium (14) [11,12,13]

$$3\ Ti(OCHMe_2)_4 + TiCl_4 \rightarrow 4\ ClTi(OCHMe_2)_3$$

Neat 4 (47.5 g, 0.25 mol) is added dropwise to a stirred solution of 213 g (0.75 mol) 8 in 250 ml hexane at 0°C. After warming to room temperature, the solvent is removed *in vacuo*, and the residue is distilled. Yield 245 g (94%), b.p. 60-61°C/0.2 Torr. All operations must be carried out with exclusion of moisture. The chloride 14 is highly viscous and may crystallize. With pure starting materials and accurate stoichiometry, undistilled 14 can be employed equally well for reactions. The components 4 and 8 can even be combined neat to give 14. For titanations it is useful to keep stock solutions of 14, for instance in hexane (250 g 14 diluted to 500 ml). These can easily be handled with syringes and added to organolithium or *Grignard* solutions for transmetallations.

Other halo-trialkoxy-titanium and -zirconium derivatives can be prepared in the same way. - Cf. also preparation of 19 below.

Chloro-triphenoxy-titanium (15) [14]

$$ClTi(OCHMe_2)_3 + 3\ C_6H_5OH \rightarrow ClTi(OC_6H_5)_3 + 3\ Me_2CHOH$$

To a solution of 14 (71.9 g, 276 mmol) in 500 ml toluene is added 77.9 g (828 mmol) phenol. The deep red solution is concentrated by distillation through a 10 cm *Vigreux* column which removes most of the isopropanol azeotropically. The residue is distilled in a Kugelrohr (see purification of 10, above) at 250°C/0.001 Torr to give 95.8 g (96%) 15 which crystallizes upon cooling. A 0.26 M stock solution in THF can be kept under exclusion of air.

This method of exchanging OR-groups is generally useful, for instance also for *in situ* preparation of derivatives with chiral OR-groups (see below).

Bromo-tris(diethylamino)titanium (19) [10,15]

$TiBr_4 + 3\ Ti(NEt_2)_4 \rightarrow 4\ BrTi(NEt_2)_3$

A 250 ml three-necked flask equipped with dropping funnel, thermometer, reflux condenser, and magnetic stirring bar (mechanical stirring in larger scale preparations!) is charged with 100 ml absol. hexane and 24.4 g (0.073 mol) 13. The yellow-red solution is cooled to -30°C under argon with stirring, and combined dropwise with a solution of 8.9 g (0.024 mol) 5 in 60 ml hexane. A dark suspension is formed which is allowed to warm to room temperature and then heated at reflux for 2.5 hours, whereupon most of the precipitate dissolves. The reflux condenser is quickly replaced by a *Liebig* condenser, and the hexane distilled off at normal pressure. The liquid residue is purified by short path distillation (b.p. 101-102°C/2×10^{-5} Torr) to give an orange-red viscous liquid which crystallizes occasionally (m.p. ca. 50°C), yield 29.6 g (89%).

Other halo-tris(dialkylamino)titanium derivatives can be prepared likewise.

Methyl-triisopropoxy-titanium (20a) [16]

$CH_3Li + ClTi(OCHMe_2)_3 \rightarrow H_3C\text{-}Ti(OCHMe_2)_3 + LiCl$

In a 2 L three-necked flask (dropping funnel, thermometer, Ar-inlet, magnetic stirrer) a solution of 635 ml 0.84 M CH_3Li in ether is added at -50°C within ca. 1.5 hour to a stirred solution of 125 ml (0.53 mol) 14 in 400 ml of the same solvent. After 2 hours stirring is stopped, and the cold supernatant solution is decanted from the LiCl precipitate. This is done by forcing the solution with argon pressure through teflon tubing into another flask. The solvent is removed by condensing the solvent from the magnetically stirred solution under reduced pressure into a cold trap. The residue is distilled *in vacuo* through a short path distillation apparatus. Yield of 20a 122 g (95%), bright yellow liquid of b.p. 50°C/0.01 Torr, m.p. 10°C. Samples of 20a have been kept without decomposition under an inert atmosphere in a refrigerator (ca. +5°C) for many months.

The solid phenyl derivative 20g [17] is prepared similarly. The crude product containing some LiCl is purified by extraction with pentane, rather than by distillation; 20g can be recrystallized by cooling pentane solutions to -80°C. It is a nearly colorless substance which can be stored under an inert atmosphere.

Other triisopropoxy organotitanium compounds are prepared *in situ* from 14 and the corresponding Li- or Mg-derivatives by the same type of reaction as the two derivatives described here.

III C,C-BOND-FORMING PROCESSES

There are certain selectivity problems with all classical C,C bond-forming organo reactions, see Scheme 1. Thus, the simple additions of nucleophilic organometallic reagents to carbonyl groups is complicated by the fact that aldehydes, ketones and esters are not well differentiated, that other electrophilic functional groups such as cyano, nitro, halo, trialkylstannyl may interfere, and that proton abstraction rather than addition occurs, see A, B, and C in Scheme 1. With α.β-unsaturated carbonyl substrates 1.2- and 1.4-additions compete and carbonyl derivatives with asymmetric carbon atoms may lead to diastereomeric mixtures, see D and E, respectively. In alkylations with halides or sulfonates β-elimination is often competing, see F. Finally, ambident nucleophiles such as metal enolates and unsymmetrically substituted allylic organometallic compounds can give rise to two constitutional isomers non-selectively, see G and H, respectively.

A) Nucleophilic Additions to Carbonyl Groups

We have covered the general aspects of this subject in a recent *Angewandte Chemie* review article entitled "Organometallic Compounds of Titanium and Zirconium as Selective Nucleophilic Reagents in Organic Synthesis" [18], and another survey has appeared last year [19]. We can therefore just briefly summarize the results here.

The different types of reagents which have thus far been employed for simple nucleophilic additions are shown in the Formulae 20-27, with the R-groups being specified in Table 2; for derivatives with chiral OR*-groups see section III A4, for allyl-metal reagents III F2. The titanocene and zirconocene derivatives (see 1 and 2) with alkyl and aryl groups attached to the metal are not useful for carbonyl additions. Cp_2MXR-Derivatives are used for carbonylations, for olefinations and for transfer of allylic groups, see below.

By far, most of the reagents 20-27 have been generated *in situ* by simply adding the corresponding transition metal derivatives 7, 8, 12, 13, 14-19 to the conventionally prepared organolithium or *Grignard* reagents. This trans-

Scheme 1

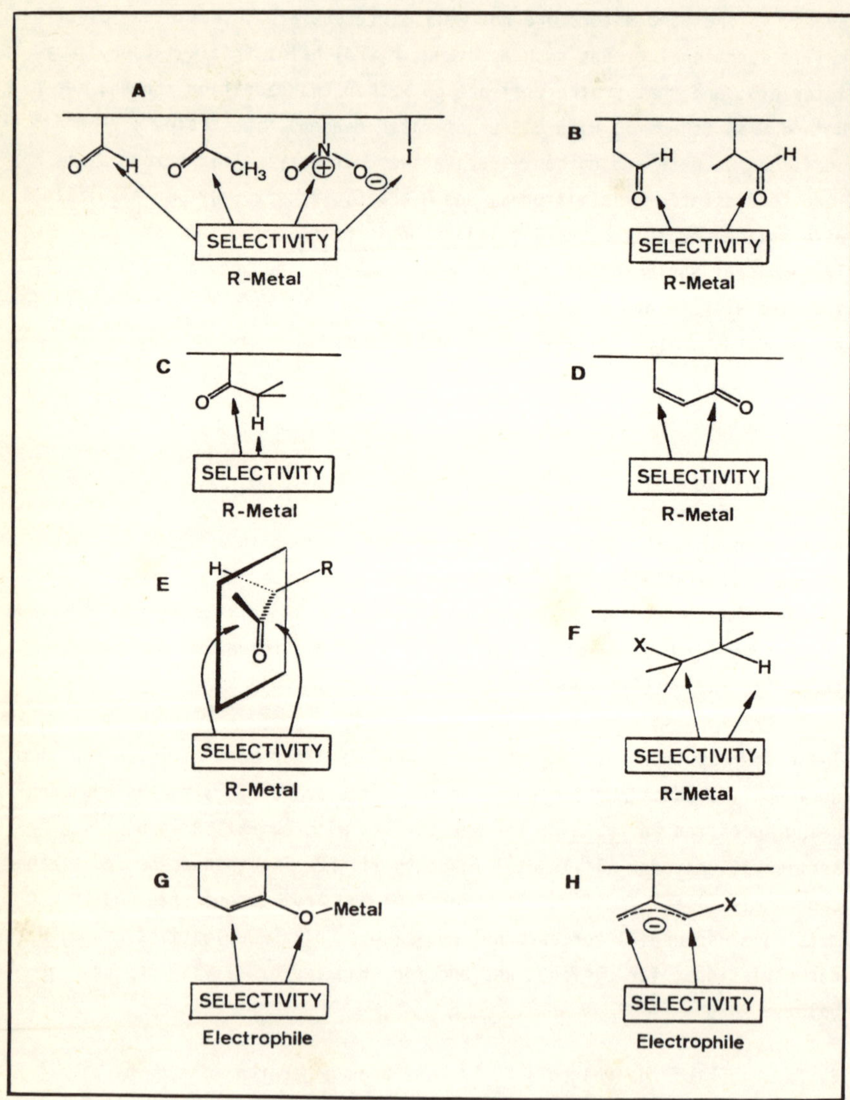

A
O
H
O
CH3
N
O
O
I
SELECTIVITY
R-Metal
B
H
O
H
O
SELECTIVITY
R-Metal
C
O
H
SELECTIVITY
R-Metal
D
O
SELECTIVITY
R-Metal
E
H
R
O
SELECTIVITY
R-Metal
F
X
H
SELECTIVITY
R-Metal
G
O
Metal
SELECTIVITY
Electrophile
H
X
SELECTIVITY
Electrophile

$R^1Ti(OR^2)_3$ $\underline{20}$ $R^1Zr(OR^2)_3$ $\underline{21}$ $R^1Ti(NR^2_2)_3$ $\underline{22}$

$R^1_2Ti(X)_2$ $\underline{23}$ R^1TiCl_3 $\underline{24}$ R^1_4M $\underline{25}$ (M = Ti,Zr)

$[R^1Ti(OR^2)_4]^{\ominus}M^{\oplus}$ $\underline{26}$ (M = Li or MgX) $[R^1Ti(NR^2_2)_4]^{\ominus}M^{\oplus}$ $\underline{27}$

metallation reaction is essentially instantaneous with titanium, even at low temperatures; with zirconium longer reaction times and higher temperatures up to 0°C may be required [8]. The reagents 20-27 are surprisingly stable. The R^1-groups attached to the metal may be varied widely (Table 2): primary, secondary and even tertiary alkyl, allyl, propargyl, vinyl, aryl, α- and β-heterosubstituted alkyl and aryl. Some simple rules about stability and reactivity are:

(a) The reactivity of the Ti-derivatives has been found to increase in the order $(R'_2N)_3TiR < (R'O)_3TiR < Cl_3TiR \ll (RO)_2TiR_2$, Cl_2TiR_2.

(b) The stability decreases in the same order: the aminoderivatives 22 with β-hydrogens on the R^1-groups can be distilled at temperatures up to 100°C, while the trichloro-analogues are unstable even in solution at low temperatures.

(c) The zirconium derivatives $(R'O)_3ZrR$ are more stable with respect to all processes which occur with reduction of the metal, such as β-hydrogen elimination, reductive elimination and one-electron transfer to electrophiles.

Table 2. List of organotitanium and -zirconium reagents of types 20-27 which have been used for C,C bond forming reactions. Allylic derivatives of this type are collected separately (Table 5), see section F3.

No.	R^1	R^2 (X, M)	Ref.
20a	CH_3	$CHMe_2$	20,21)
20b	CH_3	C_6H_5	19)
20c	C_2H_5	$CHMe_2$	19)
20d	$n\text{-}C_4H_9$	$CHMe_2$	22)
20e	$CH(CH_3)_2$	$CHMe_2$	12)
20f	[structure]	$CHMe_2$	23)
20g	C_6H_5	$CHMe_2$	17,20,21)
20h	[structure]	$CHMe_2$	23)
20i	[structure: F]	$CHMe_2$	22)
20j	C_6F_5	$CHMe_2$	19,24)
20k	$C_6H_5\text{-S-}CH_2$	$CHMe_2$	22)
20l	[structure: S, S]	$CHMe_2$	22)
20m	$(Me_3Si)_3C$	$CHMe_2$	22)
20n	[structure: Br]	$CHMe_2$	23)
20o	$(C_6H_5)_2AsCH_2$	$CHMe_2$	25)
20p	$(C_6H_5)_2SbCH_2$	$CHMe_2$	25)
21a	CH_3	C_4H_9	8)
21b	CH_3	C_3H_7	26)

No.	R^1	R^2 (X, M)	Ref.
21c	CH_3	C_6H_5	27)
21d	C_4H_9	C_4H_9	8)
21e	*t*-C_4H_9	C_4H_9	8)
21f	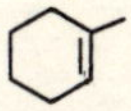	C_4H_9	8)
21g	C_6H_5	C_4H_9	8)
21h	C_6H_5	C_6H_5	27)
21i	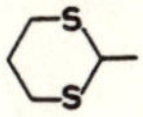	C_4H_9	8)
22a	CH_3	C_2H_5	19,28)
22b	CH_3	$-(CH_2)_5^-$	28)
22c	C_4H_9	C_2H_5	28)
23a	CH_3	(X = $OCHMe_2$)	29)
23b	CH_3	(X = Cl)	21,30,31,32)
23c	C_2H_5	(X = Cl)	32)
23d	C_4H_9	(X = Cl)	32)
23e	C_6H_5	(X = Cl)	21)
24a	CH_3	---	21,30,32,33)
24b	C_6H_5	---	21)
24c	$(H_3C)_3Si-CH_2$	---	34)
25a	CH_3	(M = Ti)	26)
25b	CH_3	(M = Zr)	26)
26a	CH_3	$CHMe_2$ (M = Li)	10,19)
26b	CH_3	$CHMe_2$ (M = MgI)	19)
27	CH_3	C_2H_5 (M = Li)	10)

1) Selectivity of Addition

The most characteristic difference between the reagents $(R'O)_3MR$ (20, 21) and the classical nucleophilic reagents RLi and RMgX is the selectivity of the transition metal derivatives. Thus, $CH_3Ti(OCHMe_2)_3$ (20a) adds to benzaldehyde within a few hours at -60°C, while acetophenone requires room

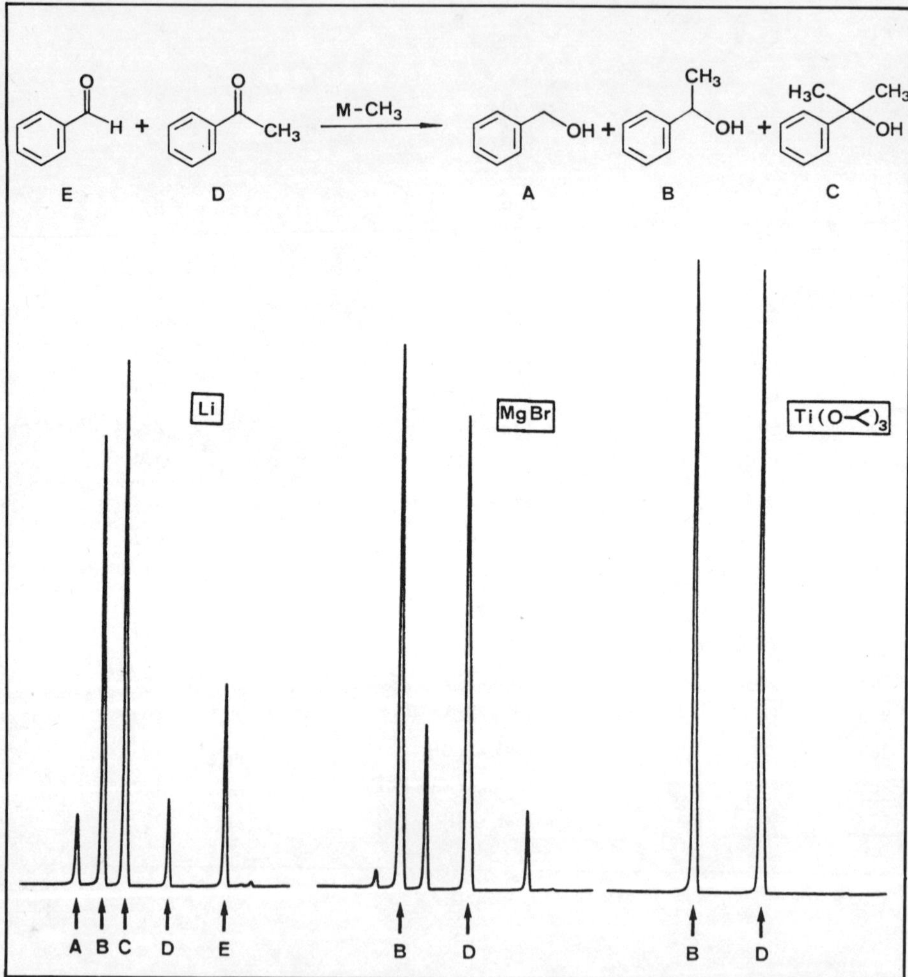

Fig. 1. Comparison of the 1:1:1 reactions of methyl-metal derivatives with benzaldehyde and acetophenone in ether at room temperature. The titanium reagent distinguishes perfectly between aldehyde and ketone! (The reactions with CH_3Li and CH_3MgBr are further complicated by the occurence of aldol reactions between the carbonyl compounds.)

temperature to give the adduct within a similar period of time. Addition of 20a to a 1:1 mixture of benzaldehyde and acetophenone in THF at room temperature produces only the adduct to the aldehyde, see Figure 1 and Figure 2 [20]. As is evident from Table 3, branched and unbranched aldehydes, cyclic and acyclic ketones, terminal and non-terminal ketones, and saturated and α.β-unsaturated ketones can also be distinguished with excellent selectivity. The examples in Scheme 2 show that other functional groups, such as COOR, $CONR_2$, CN, $C\text{-}NO_2$, C-halogen, and C-O-C do also not interfere with the addition to aldehyde groups.

Epoxides are also inert. The functional group selectivity of these new nucleophilic reagents $(R^2O)_3MR^1$ allows their use in solvents, such as acetonitrile, carbon tetrachloride, chloroform, and methylene chloride, which are not compatible with Li- and MgX-derivatives. To avoid hazardous ether solvents is attractive for industrial applications; for this purpose, the organotitanium compounds can even be generated directly in these solvents, using dialkylzinc or trialkylaluminium instead of RLi or RMgX [35].

The halogeno-titaniumorganic compounds (23,24), the tris(dialkylamino)-(22) and the ate-complexes (26,27) do not appear to be suitable for simple nucleophilic additions of alkyl and aryl groups to aldehydes or ketones; for their applications see later sections.

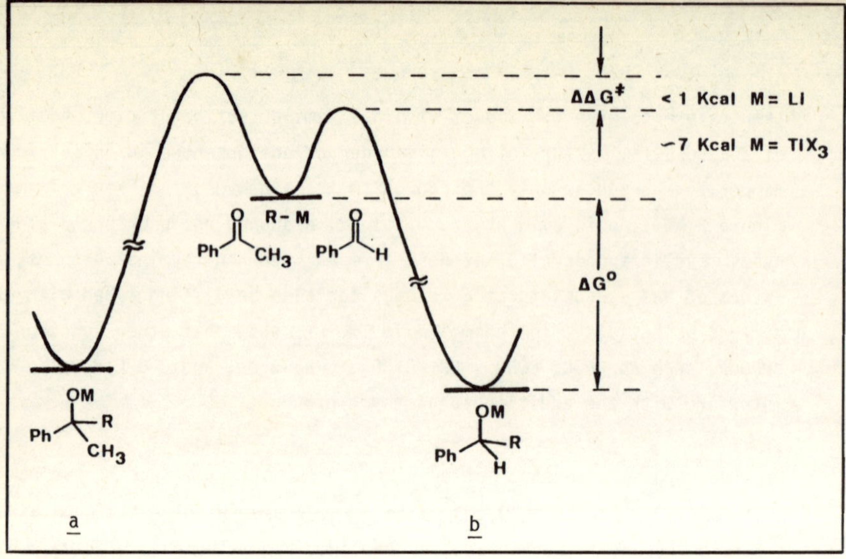

Fig. 2. Selectivity of organolithium and organotitanium reagents in reactions with 1:1 mixture of benzaldehyde and acetophenone. Reaction energy profile and examples.

R- in R-Li and R-Ti(O*i*Pr)$_3$	Li-Reagent [^{o}C]	a / b	Ti-Reagent [^{o}C]	a / b
H_3C-	20	40 / 60	20	>98 / 2
H_5C_6-	0	40 / 60	0	>98 / 2
		--	-95 → r.t.	>98 / 2
O		--	-80 → r.t.	>98 / 2
SLi	-80	33 / 67	-80 → r.t.	>95 / 5
$H_5C_6SCH_2$-	-50	67 / 33	20	>98 / 2
S S	-70	67 / 33	20	>98 / 2
H Br H		--	-95 → r.t.	>98 / 2

Table 3. Selectivities in intermolecular competitions of CH_3-addition with $CH_3Ti(OCHMe_2)_3$ (20a) to pairs of carbonyl compounds in THF [19,20].

substrate pair: more reactive	substrate pair: less reactive	% selectivity
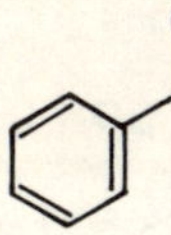 O	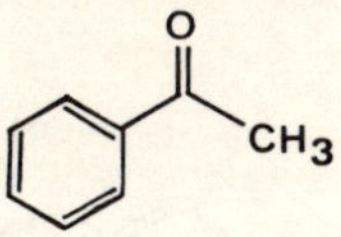 O, CH_3	>99.9
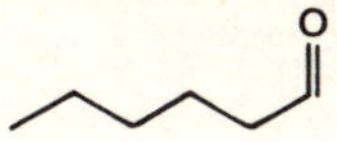 O	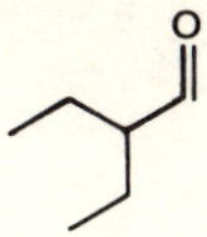 O	92
 O	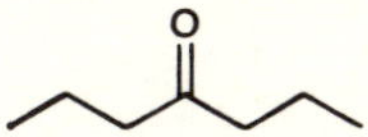 O	>97
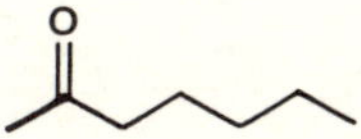 O	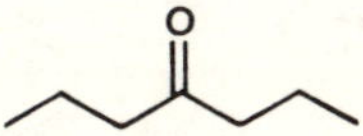 O	85
 O	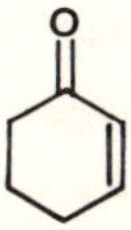 O	94
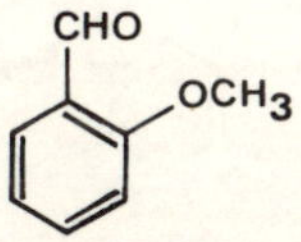 CHO, OCH_3	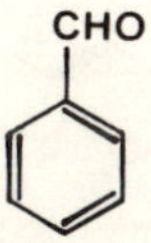 CHO	87

Of the following two typical procedures, the first one demonstrates the stability and the small tendency of trialkoxy-organotitanium compounds to do one-electron transfer. In the second procedure, the butyltitanium 20d is added to iodo-benzaldehyde to give the carbonyl adduct, which is not formed with butyllithium because of attack at iodine.

Scheme 2. Adducts of organotitanium reagents of type R-Ti($OCHMe_2)_3$ to multifunctional aromatic aldehydes. The newly formed bonds are indicated by heavy lines. All yields are high [19,20].

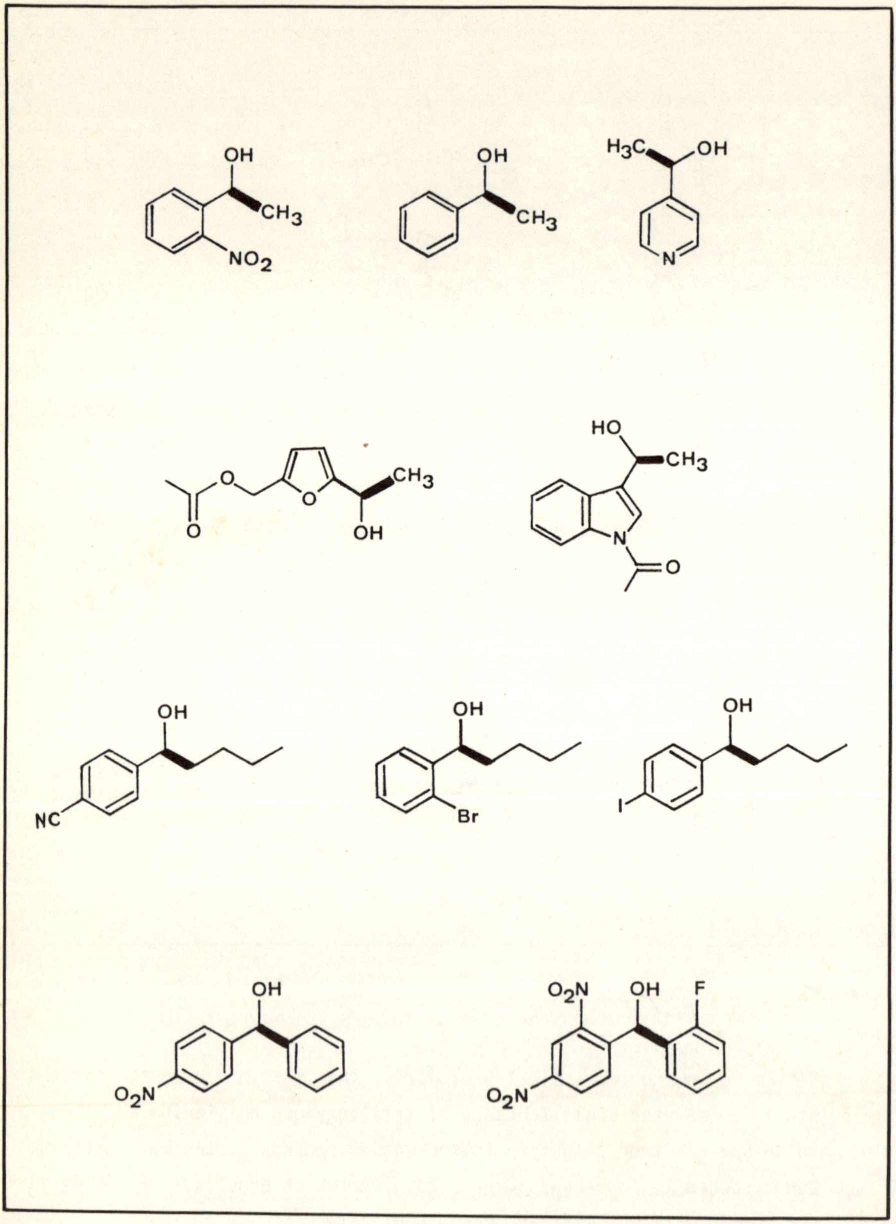

(o-Fluorophenyl)-triisopropoxy-titanium (20i) and its addition to 2.4-dinitrobenzaldehyde to give 2-fluoro-2'.4'-dinitro-benzhydrol [22)]

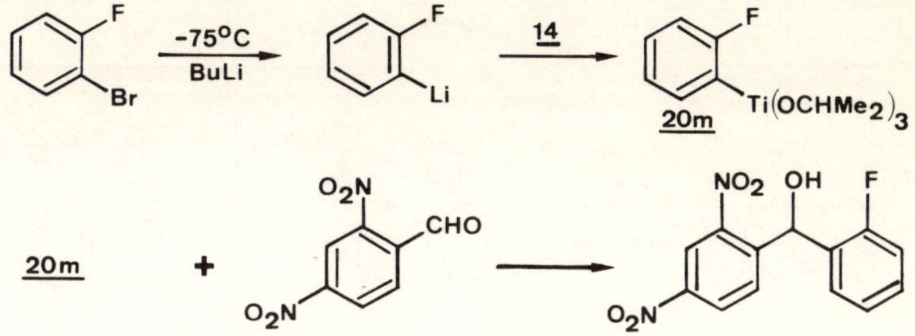

A solution of 1.08 ml (10 mmol) 2-fluoro-brombenzene in 20 ml THF is treated with butyllithium in hexane (6.3 ml, 10 mmol) at -100°C. After 10 min. $ClTi(OCHMe_2)_3$ (14) (4.4 ml, 2.27 M in hexane) is added dropwise and the solution allowed to reach room temperature. (20i may be isolated similarly to 20g). After the addition of 2.4-dinitrobenzaldehyde (1.57 g, 8 mmol, in 5 ml THF) stirring is continued for 1 hour. Extractive workup (1 M HCl/ether) provides 2.28 g (98% yield) of the desired carbinol, m.p. 94-95 C^{o}(ether/pentane).

1-p-Iodophenyl-1-pentanol [22)]

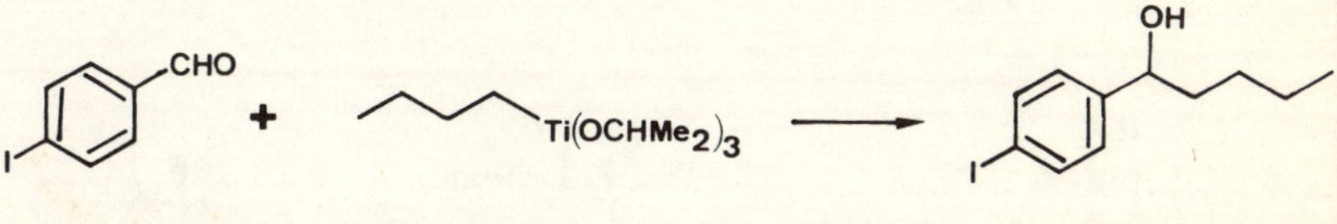

At -15°C butyllithium in hexane (6.33 ml, 10 mmol) is added to 2.2 ml (5 mmol) of 14 (2.27 M in hexane), diluted with 10 ml anhydrous ether. After the solution has been stirred for 1 hour at that temperature, p-iodobenzaldehyde dissolved in 10 ml THF is added and stirring is continued for 15 hours at room temperature. Extractive workup (1 M HCl/ether) affords 1.19 g (82%) spectroscopically pure 1-p-iodophenyl-1-pentanol.

2) Nucleophilicity vs. Basicity

The problem of enolate instead of adduct formation from certain ketones can not always be avoided with $(RO)_3Ti$-derivatives. The nucleophilicity can, however, be enhanced by two modifications, see Scheme 3 and Scheme 4. One is to switch for instance to the dimethyl-titanium [19] derivative 23a,

Scheme 3

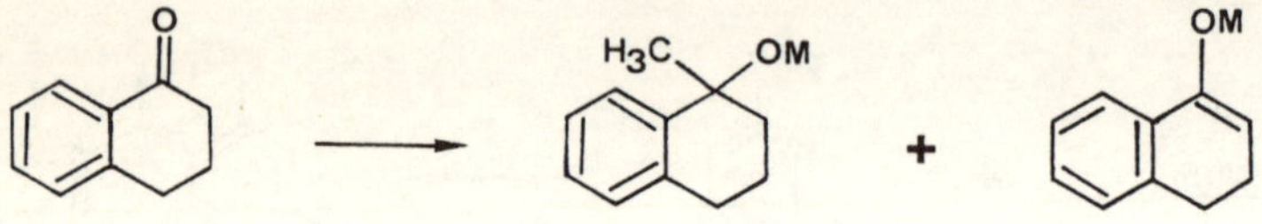

$CH_3-Ti(OCHMe_2)_3$	(20a)	50 : 50
$(CH_3)_2Ti(OCHMe_2)_2$	(23a)	90 : 10
$CH_3Zr(OBut)_3$	(21a)	90 : 10

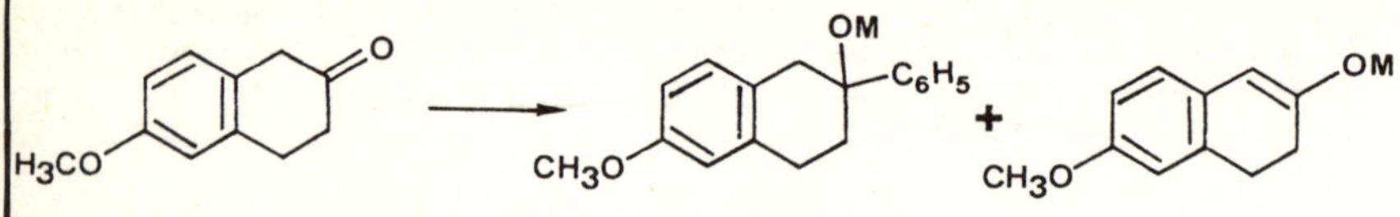

$C_6H_5Ti(OCHMe_2)_3$	(20g)	25 : 75
$C_6H_5Zr(OBut)_3$	(21g)	90 : 10

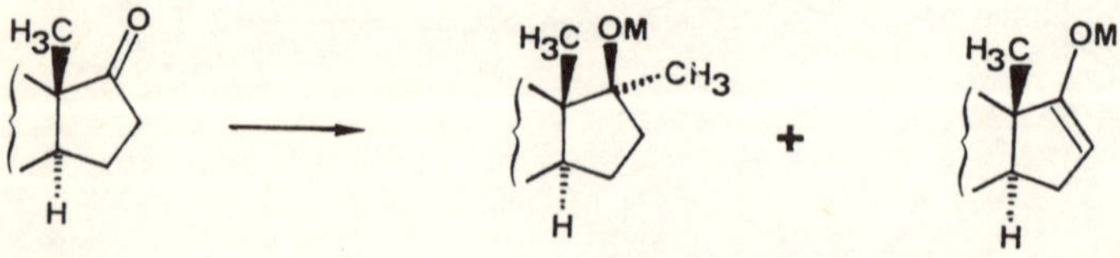

(pregnenolone)	20a	5 : 95
	21a	80 : 20

Scheme 4. Adducts obtained from ketones and $RZr(OBut)_3$ (21). The newly formed bonds are indicated by heavy lines. % ds gives the diastereoselectivity[37] of addition[8,12].

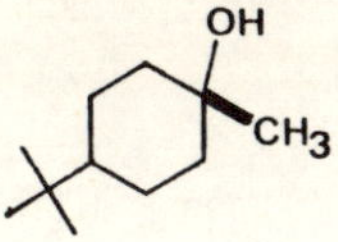

95% (80% ds)	78% (85% ds)	90% (>95% ds)
84%	80%	60%
78%	72%	74% (one Diastereomer)

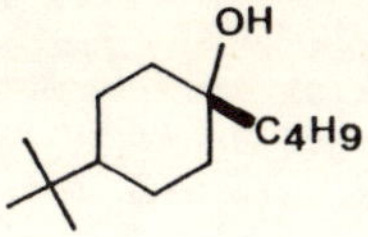

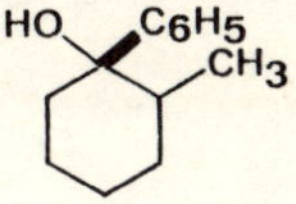

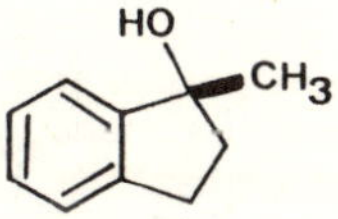

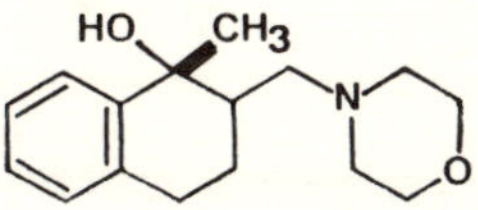

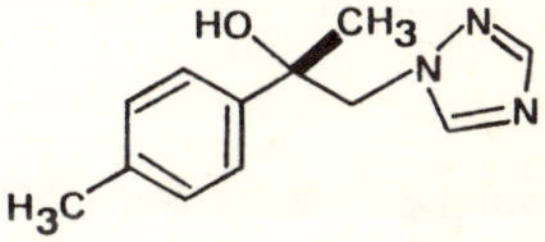

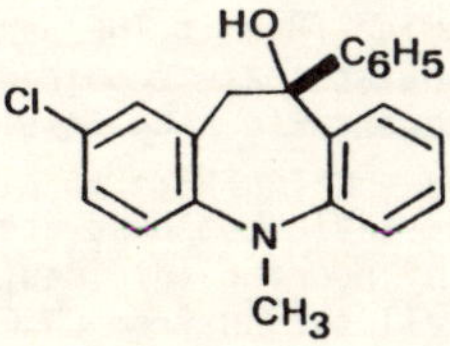

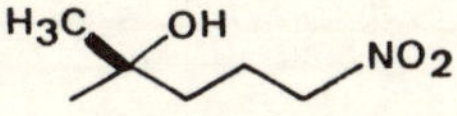

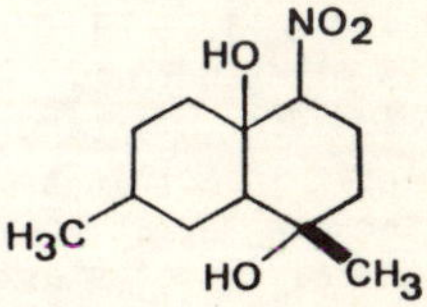

with the obvious disadvantage that only half of the CH_3-groups of the reagent are transferred. The other, more general and more useful method is the employment of zirconium derivatives of type 21 which are much more nucleophilic than the titanium analogues. Their drawback is that they may transfer hydride from the α-OCH-position, see section IV 1. For two applications of organozirconium reagents see the accompanying procedures, one of which also demonstrates the functional-group selectivity and diastereoselectivity.

The ate-complexes 26 appear to be especially basic: Upon addition of acetophenone to $[CH_3Ti(OCHMe_2)_4]Li$, vigorous methane gas evolution ensues, and the ketone is recovered after workup [10].

6-Methoxy-2-phenyl-1.2.3.4-tetrahydro-2-naphthol [12]

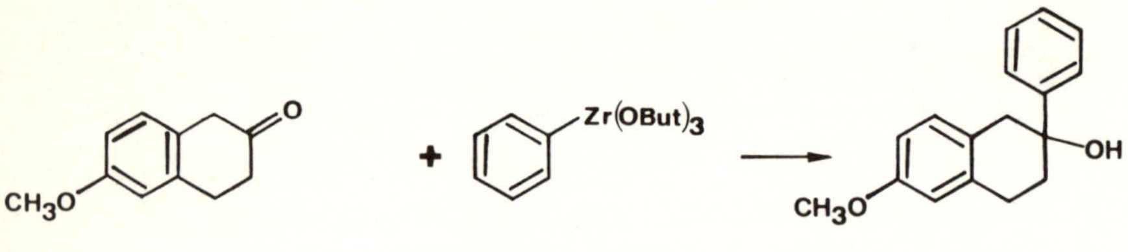

(21g)

At $-20^{o}C$, 1.25 ml (2.5 mmol) phenylmagnesiumbromide (2 M in THF) and 1.25 ml (2.55 mmol) $ClZr(OBu)_3$ (16) (2.04 M in hexane) are combined to yield after stirring for 1 hour at $-10^{o}C$ the reagent 21g. The resulting solution is then treated with 300 mg (1.7 mmol) 6-methoxy-2-tetralone [38]. After one day at room temperature, 330 mg of the title compound (together with ca. 10% of starting material) is isolated by pouring into neutral 20% KF-solution and extracting with ether. Recrystallization from CH_2Cl_2/pentane affords colorless crystals of m.p. 88-89°C.

5.8-Dimethyl-2-nitro-bicyclo[4.4.0]decan-1.5-diol [8)]

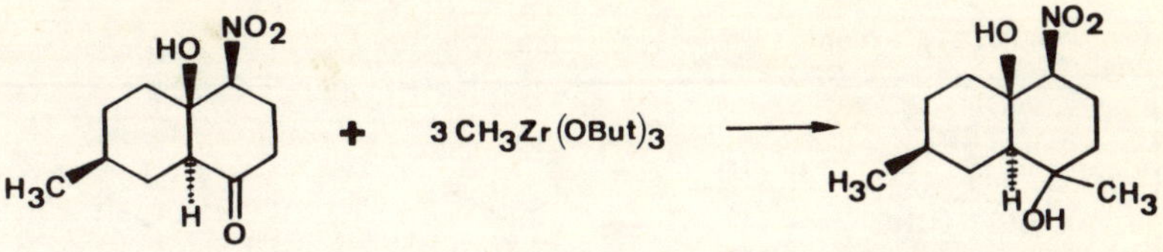

(21a)

A solution of 1.5 ml (3.1 mmol) $ClZr(OBu)_3$ (2.04 M in hexane) is diluted with 6 ml dry ether and treated with 1.75 ml (3 mmol) methyllithium (1.71 M in ether) at -20°C. After stirring 1 hour at -10°C, 223 mg (1 mmol) of 1-hydroxy-2-methyl-2-nitro-bicyclo[4.4.0]decan-5-on [39)] dissolved in 2 ml CH_2Cl_2 is added. Progress of the reaction is followed by TLC. After 15 hours, the reaction mixture is hydrolyzed with 20% neutral KF-solution and the product extracted with ether. Removal of the solvent is followed by purification with prep. TLC (silicagel, CH_2Cl/EtOAc 10:1). The product obtained (178 mg, 74%) is a highly viscous, slightly yellow oil, which was found to consist of a single diastereomer (^{13}C-NMR), the configuration of which was not determined.

3) Diastereoselectivity

As might be expected, the $(RO)_3M$-derivatives are also more diastereoselective than their Li- and XMg-counterparts. An example is the addition of the methylzirconium 21a to a decalone in the preceding procedure, which furnishes a single diastereomer of as yet unknown configuration, cf. also some of the products from organozirconium derivatives listed in Scheme 3, above. For 1.2-, 1.3-, and 1.4-asymmetric inductions [40)] observed in additions of organotitanium compounds, see the reactions in Scheme 5 and Scheme 6. Diastereoselective additions of crotyl and other allylic and also of propargylic Ti- and Zr-reagents are dicussed in section III F.

Scheme 5. Asymmetric inductions in the additions of methyl- and butyl-titanium derivatives to aldehydes and ketones containing asymmetric centers [19,26]. The relative topicities are specified *lk* (*Cram*'s rule), except in the last case.

Ph–CH(CH$_3$)–CHO —(lk-1.2)→ Ph–CH(CH$_3$)–CH(OH)–R + Ph–CH(R)–CH(OH)–R

CH_3Li or CH_3MgCl		66 : 34
$CH_3Ti(OC_6H_5)_3$	(20b)	93 : 7
n-$C_4H_9Ti(OCHMe_2)_3$	(20d)	89 : 11

pregnenolone → (H$_3$C, CD$_3$, OH) + (H$_3$C, OH, CD$_3$)

$D_3CTi(OCHMe_2)_3$	96 : 4

Ph–CH(OH)–CHO —(lk-1.2)→ Ph–CH(OH)–C(Ph)(H$_3$C)(OH) + Ph–CH(OH)–C(Ph)(HO)(CH$_3$)

CH_3MgI		96 : 4
$CH_3Ti(OCHMe_2)_3$	(20a)	>99 : 1

(OBz, O-isopropylidene furanose aldehyde, O=C–H) → (HO, H, CH$_3$; OBz) + (HO, CH$_3$, H; OBz)

CH_3MgX	(*lk*-1.2)	88 : 12
20a	(*ul*-1.2)	<2 : >98

Scheme 6 [19,20)]

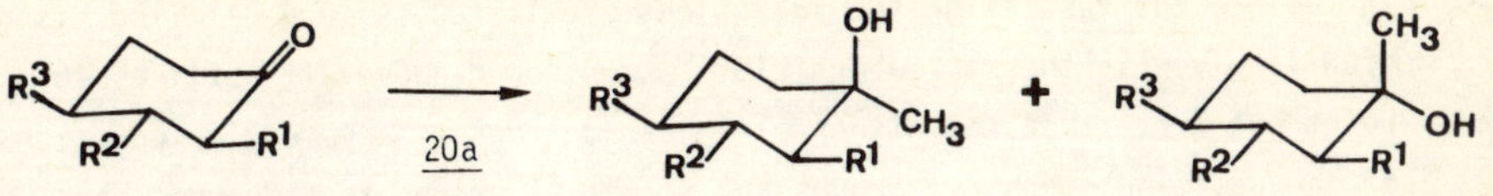

CH$_3$Ti(OCHMe$_2$)$_3$ (20a)

$R^1 = CH_3$, $R^2 = R^3 = H$	94 : 6
$R^2 = CH_3$, $R^1 = R^3 = H$	89 : 11
$R^3 = CH_3$, $R^1 = R^2 = H$	88 : 12
$R^3 = t\text{-}C_4H_9$, $R^1 = R^2 = H$	94 : 6

4) Enantioselectivity

There have been numerous attempts [41-44)] to render organolithium and -magnesium compounds enantioselective by complexation with chiral amines, ethers, and aminoethers, see 28. Only special combinations of RLi and R'CHO gave inductions which are high enough for practical purposes. The organotitanium derivatives of type 29 are expected to give higher enantioface-differentiations because their chiral OR*-groups are bonded less loosely to

28 **29**

(see Scheme 7 for specification of R*)

the metal. Furthermore, a rational design of ligands for titanium appears to be easier. Scheme 7 contains all the chiral reagents tested so far, and in Table 4 the results of the inductions are listed. A more detailed discussion is given in the review article [18] mentioned above. It suffices to point out here, that the hitherto highest degrees of asymmetric inductions of the addition of phenyl groups and of methyl groups to aldehydes have been achieved with this new method; enantiomeric excesses above 50% are not rare (Table 4). Two procedures describe the method of attaching the chiral OR*-groups to titanium and of carrying out the additions.

(S)-(-)-p-Methyl-benzhydrol of 88% e.e. [12]

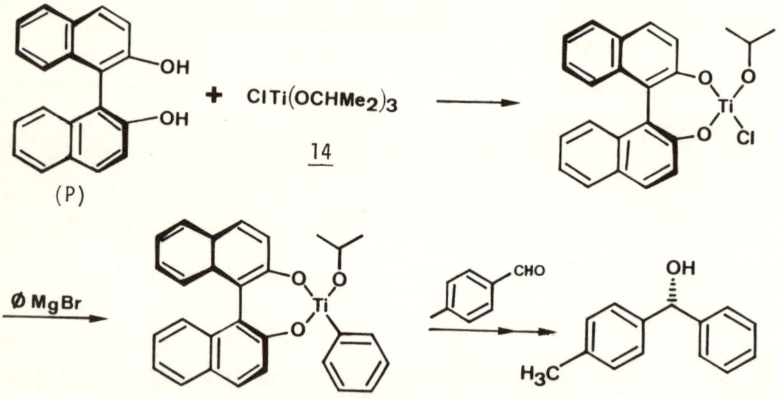

OH OH (P) + $ClTi(OCHMe_2)_3$ 14 → O O Ti O Cl

Ø MgBr → O O Ti O Ph ; CHO → OH H_3C

m (see Scheme 7)

To a solution of (S)-(-)-binaphthol (750 mg; 2.62 mmol) in 100 ml benzene is added an equimolar amount of 14. The solvent is carefully removed by fractionating through a 10 cm *Vigreux* column. The gradually rising boiling temperature indicates the change from benzene/2-propanol azeotrope (b.p. 71°C) to pure benzene. The residue which may retain some benzene is dissolved in 30 ml THF and treated at -20°C with 1.3 ml (2.60 mmol) phenylmagnesiumbromide (2 M in THF). After stirring for 1 hour at -20°C, p-tolylaldehyde (203 mg, 1.7 mmol) is added. The solution is allowed to reach room temperature overnight and is hydrolyzed with 20% neutral KF-solution. The combined ethereal extracts are washed three times with 1 N NaOH to remove the freed chiral ligand (which can be recovered in 70-80% yield by acidification of the aqueous layer with conc. phosphoric acid). After distillation (b.p. 120°C/0.05 Torr) (S)-(-)-p-methyl-benzhydrol (250 mg) is obtained in 74% yield and 88% e.e. $[\alpha]^{25}_{D} = -8.84^{\circ}$ (c = 4.8 in benzene).

Scheme 7. Different types of chiral, non-racemic Ti-derivatives which have been tested so far in asymmetric additions to aldehydes and a ketone. For applications see Table 4.

R-Ti(OR*)$_3$

	R	OR*
a	CH_3	(S)-2-methylbutoxy
b	CH_3	(R)-2-butoxy
c	CH_3	(1R,2S,5R)-menthyloxy
d	C_6H_5	(1R,2S,5R)-menthyloxy
e	1-naphthyl	(1R,2S,5R)-menthyloxy

R-Ti(O')(OR*)$_2$

	R^1	R^2	R^3
f	CH_3	CH_3	CH_3
g	C_6H_5	CH_3	CH_3
h	CH_3	H	$C(CH_3)_3$

R-Ti(OR')$_2$OR*

n: H_3C–Ti($OCHMe_2$)$_2$–O–menthyl

o: Ti($OCHMe_2$)$_3$, CH_3, Ph (bornyl-type alkoxide)

	R	R^1	R^2
i	CH_3	CH_2CH_3	H
j	CH_3	$CHMe_2$	H
k	CH_3	CMe_3	H
l	CH_3	$CHMe_2$	CH_3
m	C_6H_5	$CHMe_2$	H

Table 4. Asymmetric additions of chiral organotitanium derivatives a - o (specification see in Scheme 7) to aldehydes.

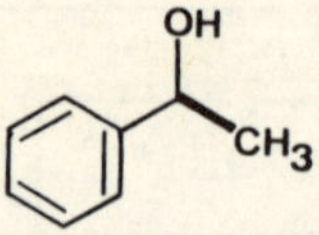

e.e.	Config.	with
8% e.e.	(S)	with a
12%	(S)	b
23%	(S)	c
20%	(R)	f
70%	(S)	g
32%	(R)	h
28%	(S)	i
59%	(S)	j
66%	(S)	k
54%	(S)	l
13%	(S)	n

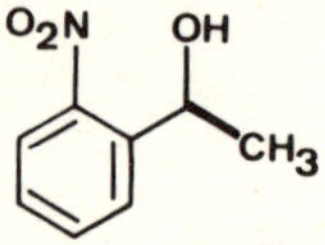

e.e.	Config.	with
58% e.e.	(S)	with c
25%	(R)	f
33%	(S)	g
46%	(S)	j
40%	(S)	l

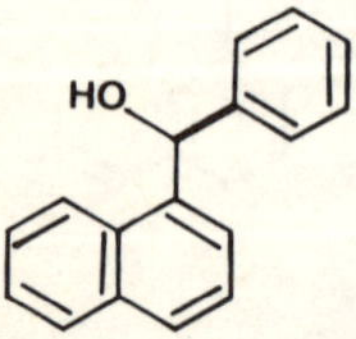

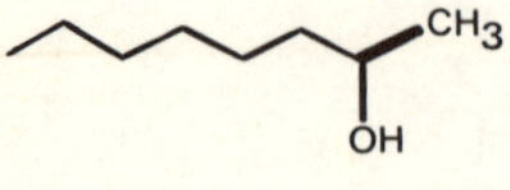

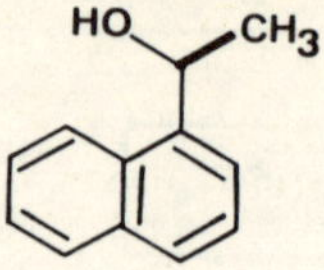

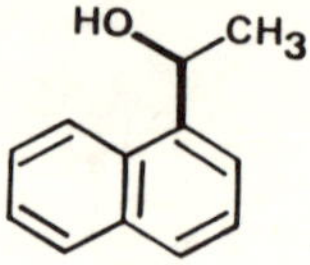

5% e.e. (R) with e

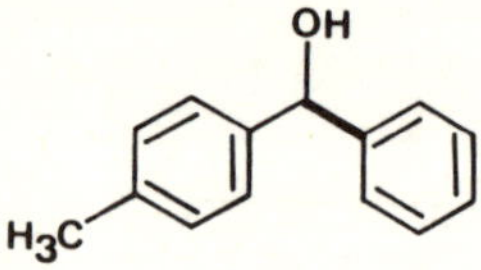

76% e.e. (S) with c

e.e.	Config.	with
29% e.e.	(R)	with d
88%	(S)	m
67%	(R)	h

>80% e.e. with m

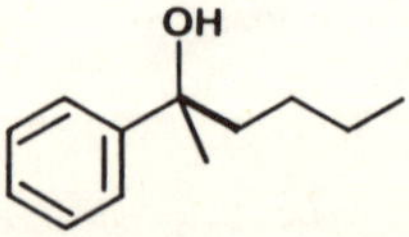

~40% e.e. with j

52% e.e. (S) with g

(S)-(-)-1-Phenyl-ethanol of 70% e.e. [45]

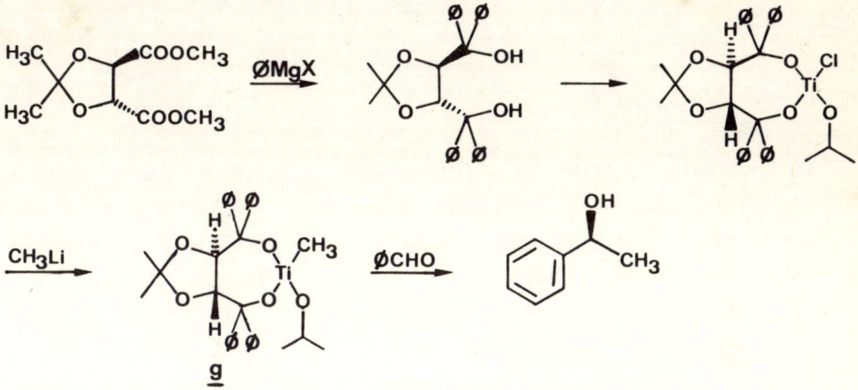

In a 1 L three-necked flask, equipped with a reflux condenser and a dropping funnel, is placed 9.72 g (0.40 g atoms) of magnesium turnings in an argon atmosphere. Ether (40 ml) is added to cover the magnesium and the reaction started through addition of ca. 2 ml bromobenzene and heating to reflux. The remaining bromobenzene (42 ml, 0.40 mol, total) dissolved in 80 ml of ether is added at such a rate that a moderate reflux is maintained. After the addition has been completed (ca. 1.5 hours), the solution is refluxed for 1 hour. The brown *Grignard* solution is then cooled to room temperature and diluted with 200 ml of ether. To this mixture the tartaric ester derivative[46] (dissolved in 100 ml of ether) is added from the dropping funnel. A vigorous reaction is observed and a yellow solid precipitates. After the addition is complete (ca. 90 min.), the reaction mixture is refluxed for 2 hours and then stirred for an additional 4 hours at room temperature. The flask is then placed in an ice-bath and the mixture hydrolyzed by the slow addition of 300 ml of a saturated NH_4Cl solution. The two layers are separated and the aqueous phase is extracted four times with ether. The combined etheral extract is washed twice with saturated aqueous NaCl-solution, dried and concentrated. The crude product is suspended in 100 ml of pentane and stirred for 1 hour at room temperature. The product is collected on a *Büchner* funnel and washed with 100 ml of pentane. Repeating this purification procedure yields 17.15 g (74% yield) of (2R,3R)-1.1.4.4-tetraphenyl--2.3-(2-propylidenedioxy)-butan-1.4-diol; m.p. 190-192°C. $[\alpha]_D = -59.2^0$ (c = 1, $CHCl_3$).

A solution of 2.80 g (6.00 mmol) of the diol in 160 ml of anhydrous toluene is treated with 4.8 ml (6.00 mmol) of 14 (1.25 M in hexane). Through azeotropic distillation, 2-propanol is removed from the mixture. The residue, a brown oily liquid, is dissolved in 50 ml of ether and cooled to -15°C. After addition of 4 ml (5.92 mmol) of MeLi (1.48 M in ether), stirring is continued for 1 hour. The brown suspension is cooled to ca. -20°C and 0.5 ml (4.95 mmol) of benzaldehyde added. The reaction mixture is allowed to come to room temperature overnight, is hydrolyzed with 20% neutral KF-solution and extracted with ether. The crude product is purified by flash-chromatography, employing ether/pentane 1:2 as eluant. The chiral auxiliary, which is eluted first, is recovered in nearly quantitative yield. 0.58 g (4.75 mmol) of methyl-phenyl-carbinol is obtained corresponding to a yield of 96%; $[\alpha]_D = -31.8^o$ (c = 5.1 in ethanol); e.e. = 70%.

B) *Michael* Additions

In the previous section, it was shown that the reagents of type $R^1M(OR^2)_3$ (20,21) are most useful for simple, highly selective additions to aldehydes and ketones. These same reagents show no tendency to add in a 1.4-fashion to α.β-unsaturated aldehydes and ketones [20,47]. Together with the fact that E/Z isomeric vinylzirconium derivatives react without loss of configurational purity, see formation of 30 [12], this may be taken as evidence [48] that no one-electron transfer is involved in their reactions. In

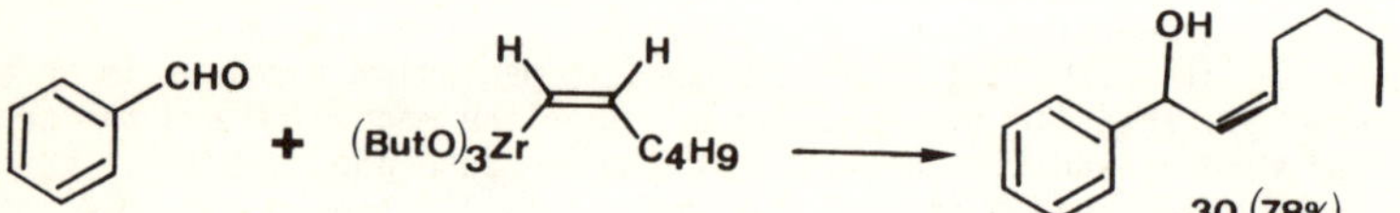

contrast, the vinyl Cp_2ZrCl-derivatives can be caused to add to the 3-position of enones exclusively in a nickel-catalyzed process [49]. The reactive species is probably an organonickel compound. The reaction furnishes *trans*--2.3-disubstituted ketones 31 from α-substitued cyclic enones. No *Michael* addition occurs with the alkyl analogues under these conditions. Since the vinylic reagents are made by hydrozirconation of acetylenes [50] (section IV D), an overall stereoselective reductive *Michael* addition of acetylenes to enones can be achieved.

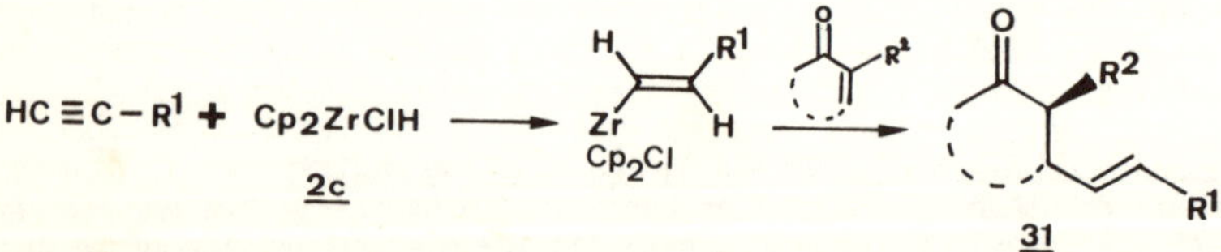

Titanium mediated *Michael* additions of silyl enolethers will be mentioned in section III E3.

3-((E)-3.3-Dimethyl-1-butenyl)-cyclohexanone [51)]

$ZrCp_2Cl$ + (2-cyclohexenone) $\xrightarrow{Ni(AcAc)_2}$

A solution of the vinylzirconium complex (1.92 g, 5.67 mmol, from 2c and *t*-butylacetylene) and 2-cyclohexenone (0.62 g, 6.48 mmol) in 30 ml THF is stirred under argon at 0^oC. After addition of 0.15 g (0.59 mmol) $Ni(AcAc)_2$ stirring is continued in an ice-bath for 6.5 hours. Workup after quenching with saturated NH_4Cl by extraction with ether, washing of the ether solution with $NaHCO_3$ and brine, drying and evaporation of solvents gives after column chromatography 0.744 g (73%) of the $\gamma.\delta$-unsaturated ketone.

C) Olefination and Other Transformations of Carbonyl Derivatives with Removal of the Carbonyl Oxygen

There is a family of transformations by which the oxygen atom of carbonyl derivatives is replaced, Scheme 8. The methods of achieving such transformations are often elaborate, require several steps or are not applicable to certain types of carbonyl substrates. Thus, the geminal dimethylation is normally carried out by the three-step sequence olefination, cyclopropanation, hydrogenolysis, Scheme 9 (top). Another example is the conversion of carboxylic acid derivatives to heterosubstituted olefins, formally a *Wittig* reacton; since acylation of the *Wittig* reagents occurs instead of olefination, a conversion to a methylketone and its subsequent regioselective transformation are necessary, Scheme 8 (bottom part). Again, there are recent titanium-based methods by which this and other processes such as aminative alkylation can be achieved in one-pot reactions.

Scheme 8.

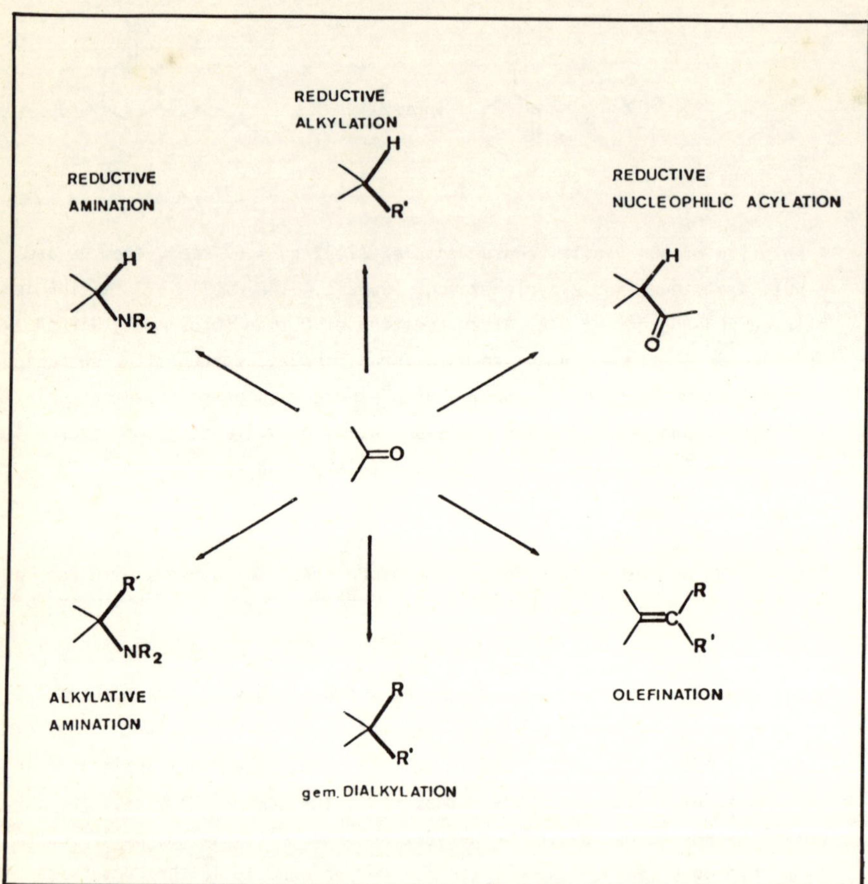
REDUCTIVE
ALKYLATION
H
R'
REDUCTIVE
AMINATION
H
NR2
REDUCTIVE
NUCLEOPHILIC ACYLATION
H
O
O
R'
NR2
ALKYLATIVE
AMINATION
R
R'
gem. DIALKYLATION
R
C
R'
OLEFINATION

Scheme 9. Multistep conversion of a ketone to a gem. dimethyl derivative and of an ester to an enol ether by conventional methods.

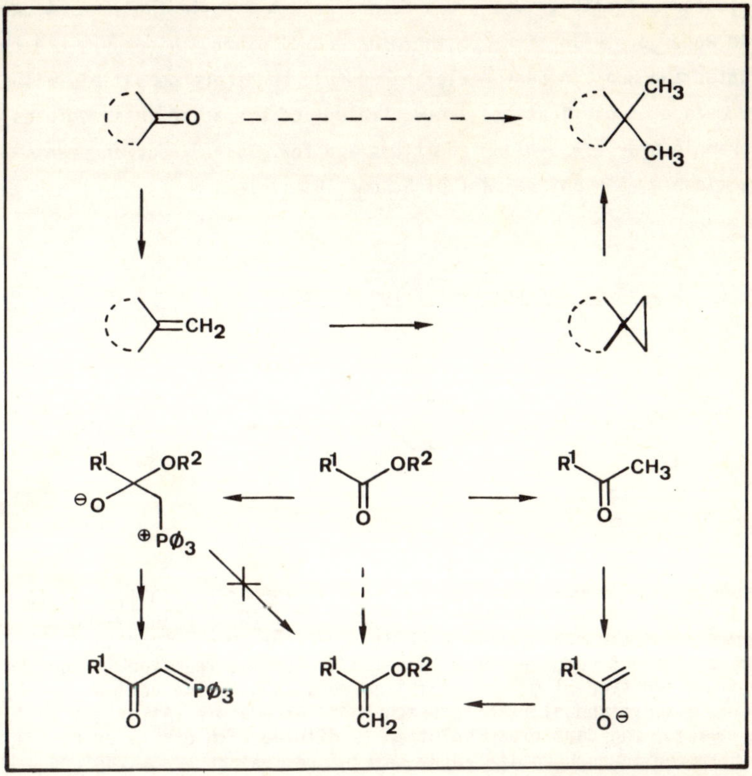

1) Olefinations with the *Tebbe* Reagent (3)

The methylene complex 3 was first prepared and characterized by *Tebbe* [6] who noticed that it smoothly converts cyclohexanone to methylene-cyclohexane, a reaction which can be carried out with *Wittig's* less fancy phosphorous ylid reagent. In the meantime, simple procedures for the preparation of 3 have been elaborated, and it is likely that 3 will be commercially available, shortly. Substituted methylene complexes of this type, with CHR instead of CH_2, have become accessible just recently [52]. The increased interest is not only due to numerous reports about the use of 3 in metatheses [53] but especially because it was found [54] that esters and lactones are directly converted to enol ethers by this reagent. The process is shown in Scheme 10, together with a number of examples. The reaction is not ester-selective, i.e. aldehyde and keto groups in the same molecule also undergo olefination. On the other hand, ether, thioether, and conjugated and non-conjugated olefinic double bonds are compatible with this new method of olefination. A new strategy of preparing intermediates for intramolecular *Diels-Alder* additions and for *Claisen* rearrangements is thus practicable, see bottom part of Scheme 10.

General procedure for transformation of esters into vinyl ethers [54]

$$\text{RO-C(=O)} + Cp_2Ti(\mu\text{-}CH_2)(\mu\text{-}Cl)AlMe_2 \longrightarrow \text{RO-C(=CH}_2\text{)}$$

A solution of 1 mmol of ester in 2 ml of toluene/THF (3:1) and 10 µl of pyridine is cooled to -40°C. 1.1 mmol of 3 (0.55 M in toluene) is added dropwise and the reaction is kept at -40°C for 30 min. and then allowed to warm to ambient temperature within 1.5 hours. The reaction is quenched by dropwise addition of 0.3 ml of 15% aqueous NaOH to the cooled (-10°C) reaction mixture and allowed to reach room temperature. After gas evolution has ceased, the dark green solution is diluted with ether, dried (Na_2SO_4) nd filtered through Celite. Upon solvent removal *in vacuo* the crude vinyl ether may be purified by filtration through a short column of alumina (neutral, activity III), using hexane as eluent to give the pure methylenation product in high yield.

Aldehyde-selective *Peterson* olefinations can be achieved with $Me_3SiCH_2TiCl_3$ (24c) [34]. This reaction also demonstrates the great influence of the X-groups in $RTiX_3$-reagents: the corresponding triisopropoxy-derivative does not react similarly [23] (see also next section). For further examples of olefinations, with allylic silylated nucleophiles, see section III F4; compare also olefinations with low-valent Ti-species/CH_2Br_2 in section III C1.

Scheme 10. Olefinations of esters and lactones with the *Tebbe* reagent. The dotted line shown in the olefins indicates the site of the CO double bond in the substrate [54,55].

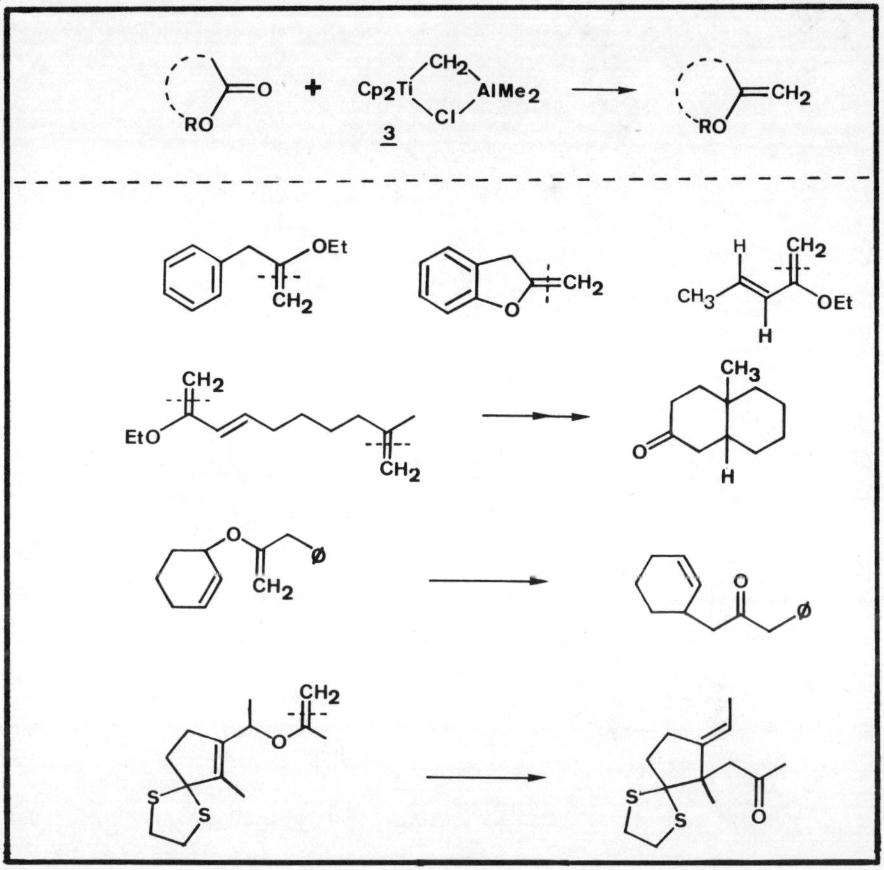

2) Geminal Dimethylation and Amination/Alkylation of Carbonyl Derivatives

The stability of $RTiX_3$ derivatives increases with increasing donor ability of the X-groups, i.e. in the order X = CH_3, Cl, OR, NR_2. In the same order, the *Lewis* acidity and the reactivity decrease. Thus, CH_3TiCl_3 (24a) and $(CH_3)_2Ti(OCHMe_2)_2$ (23a) are both more reactive than methyl-triisopropoxy--titanium. It is likely, that the higher homologues of CH_3TiCl_3, of $(CH_3)_2TiCl_2$, and of $(CH_3)_2Ti(OR)_2$ are not stable enough to become synthetically useful. However, the CH_3-derivatives can be used for unique one step carbon-carbon bond-forming reactions, the detailed mechanisms of which are by no means clear. Thus, trichloro-methyl-titanium causes Br/CH_3 exchange at a bridgehead position (→ 32) [30], while tertiary alcohols undergo an overall OH/CH_3 exchange with dichloro-dimethyl-titanium (→ 33) [31]. Not surprisingly, the same reagent effects a geminal dimethylation of ketones (→ 34). In this last mentioned reaction, the substrate structure is such that intermediate carbocations would rearrange.

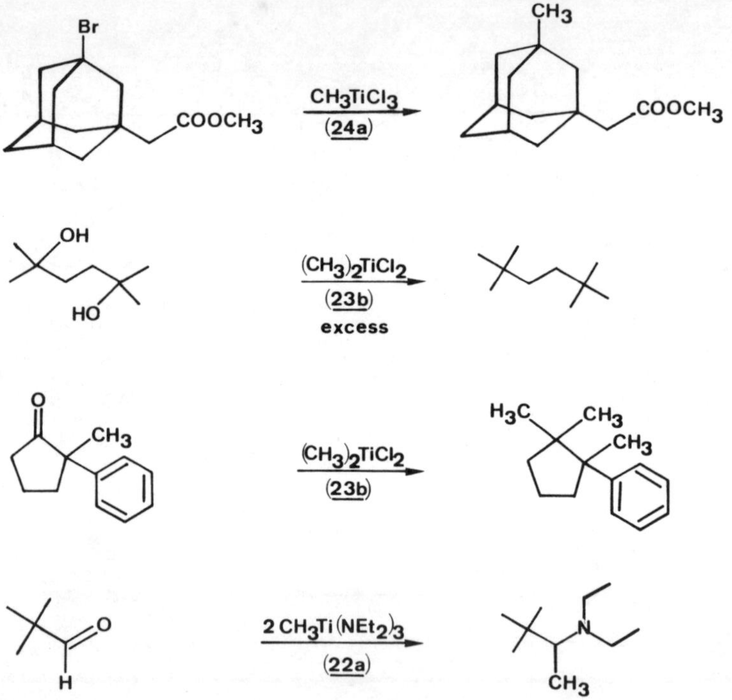

CH_3TiCl_3 (24a) → 32

$(CH_3)_2TiCl_2$ (23b) excess → 33

$(CH_3)_2TiCl_2$ (23b) → 34

2 $CH_3Ti(NEt_2)_3$ (22a) → 35

On the other end of the reactivity scale, the dialkylamino derivative 22 converts non-enolizable aldehydes to tertiary amines (→ 35) [28]. This new method for the one-step alkylative amination has been described in full detail. The yields are not as yet satisfactory; improvements seem possible. The accompanying Table 5 and Scheme 11 contain lists of examples of tertiary alkylations and of conversions of aldehydes to tertiary amines with simultaneous alkylation, respectively. Two typical procedures follow.

Table 5. Some further examples of the preparation of compounds with quaternary carbon atoms produced with methyl titanium reagents[31,32].

Starting material	Product	Method[a] or Reagent	Yield [%]
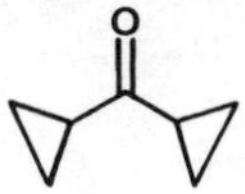 dicyclopropyl ketone	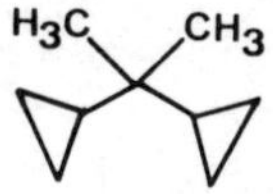 2,2-dicyclopropylpropane	B	84
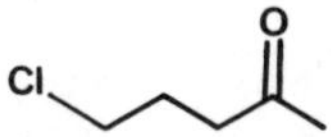 5-chloro-2-pentanone	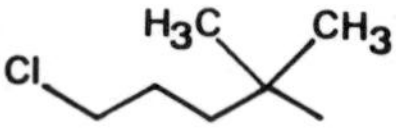 5-chloro-2,2-dimethylpentane	B	82
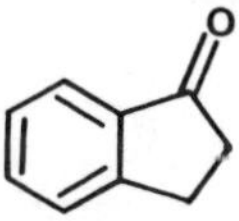 1-indanone	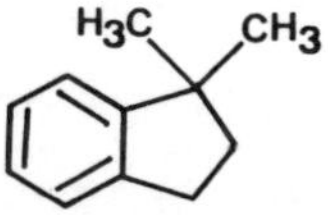 1,1-dimethylindane	A	86
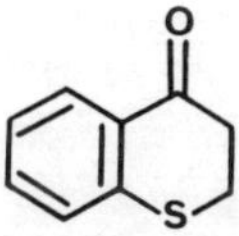 thiochroman-4-one	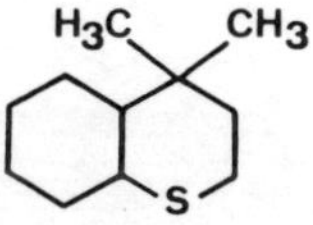 4,4-dimethylthiochroman	A	77
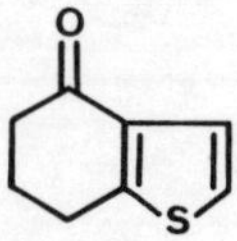 6,7-dihydrobenzo[b]thiophen-4(5H)-one	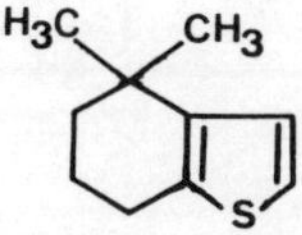 4,4-dimethyl-4,5,6,7-tetrahydrobenzo[b]thiophene	B	74
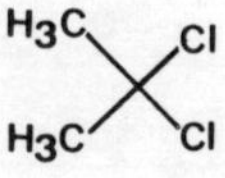 $(H_3C)_2CCl_2$	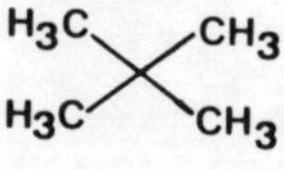 $(H_3C)_2C(CH_3)_2$	$MeTiCl_3$	95

a) See experimental details below.

Scheme 11. Products from aldehydes and alkyl-tris(dialkylamino)titanium derivatives [28]. The newly formed C.C and C.N bonds are indicated by heavy lines.

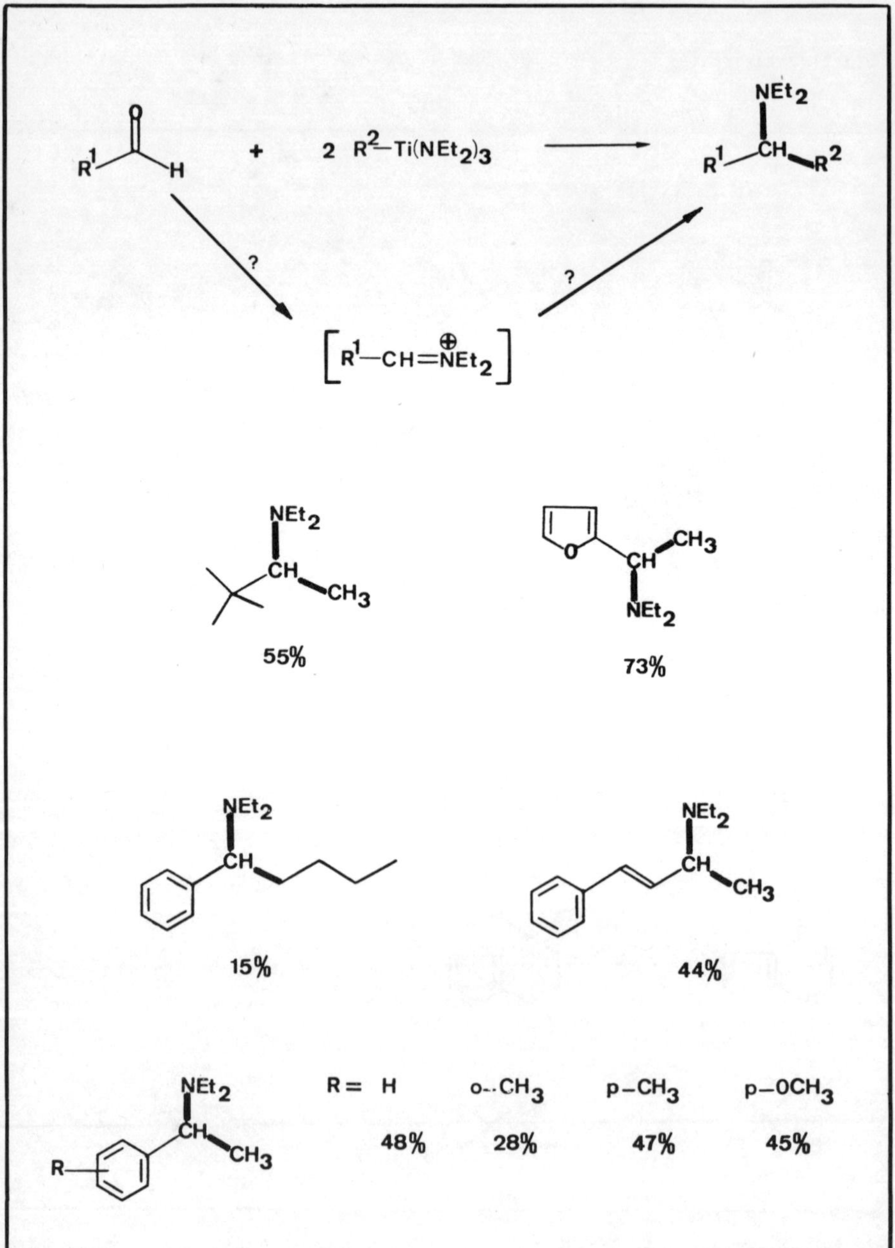

R^1CHO + 2 R^2–Ti(NEt$_2$)$_3$ → R^1–CH(NEt_2)–R^2

? [R^1–CH=$\overset{\oplus}{N}Et_2$] ?

55%

73%

15%

44%

R= H 48%; o-CH_3 28%; p-CH_3 47%; p-OCH_3 45%

Procedures for direct geminal dimethylation of carbonyl compounds [31)]

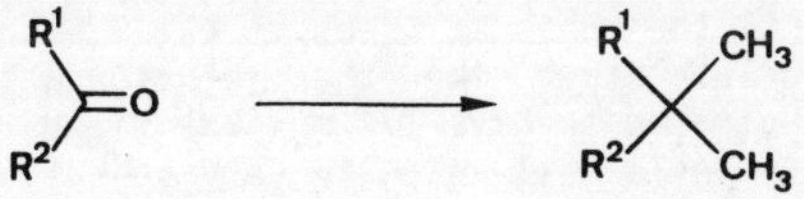

Method A: At -30°C the ketone (1 mol-equiv.) is treated with dimethyltitanium dichloride (23b) (2 mol-equiv.) in CH_2Cl_2 and the reaction mixture allowed to reach room temperature.

Method B: The ketone is combined with Me_2TiCl_2 (23b) at -30°C and after ca. 20 min. titaniumtetrachloride and dimethylzinc (1 mol-equiv. each) are added in rapid succession. The mixture is then allowed to warm to room temperature. This two-step procedure yields the desired products in a comparable yield to the method A; in some cases it proves to be even more successful.

Methylation of tertiary alcohols [31)]

Titaniumtetrachloride (4) (1 mol-equiv.) is combined with the alcohol (1 mol-equiv.) in CH_2Cl_2 at ca. -30°C. After 5-10 min. the reaction mixture is treated with Me_2Zn (1 mol-equiv.) and the temperature maintained for 1 hour before workup. Alternatively Me_2TiCl_2 can be used equally well.

Diethyl-[1-(2'-furyl)ethyl]amine [28)]

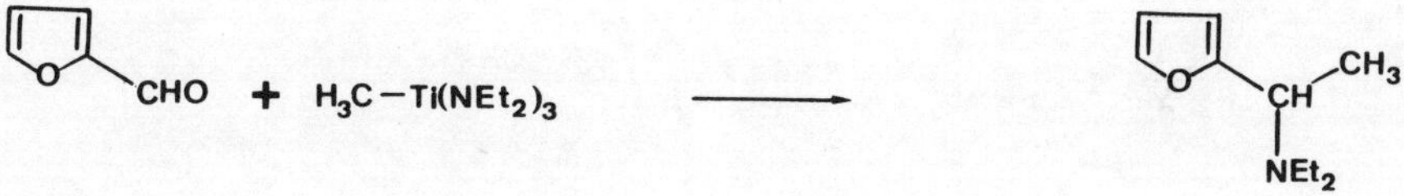

Methyllithium (3.2 mmol, 1.78 M in ether) is added slowly to a stirred solution of 1.116 g (3.2 mmol) of $BrTi(NEt_2)_3$ in 30 ml of ether at -30°C. After 30 min. at that temperature and 1 hour at room temperature, the yellow solution of 22a is combined at -60°C with 124 µl (1.49 mmol) of furfural and then stirred at room temperature for 18 hours. The reaction mixture, containing a white precipitate, is quenched with H_2O and the aqueous phase extracted three times with ether. After washing the combined etheral extracts with 2 N aqueous HCl the pH is adjusted to ca. 9 by addition of Na_2CO_3. The amine is extracted into ether, the ether solution washed with H_2O, dried (Na_2SO_4) and concentrated to give 174 mg (73% yield) of diethyl-[1-(2'-furyl]ethyl)amine as a yellow oil which is very sensitive to air.

D) Carbonylation and Reactions of Isonitriles

A special case of addition to C,X multiple bonds is carbonylation, i.e. reactions with carbon monoxide and additions to isonitriles. These can also be considered as additions to carbene carbon atoms. A process of great practical importance involving carbon monoxide and transition metals is the hydroformylation [56]. Isonitriles are also important industrial starting materials [57]. Carbonylation leads to acyl metal complexes 36 which are also formed by alkylation of anionic metal carbonyl complexes (cf. of iron [58]), less often by direct metalation of acid halides [58,59] (cf. also the formation of carbene complexes 37).

$$RM + CO \longrightarrow R{-}C(=O){-}M \;(\underline{36}) \longleftarrow RCOX + 2M$$

$$RX + [M(CO)_n]^{\ominus} \longrightarrow \underline{36}$$

$$M(CO)_n + RM'/RX' \longrightarrow (CO)_{n-1}\ M{=}C(OR')R \qquad \underline{37}$$

$$R_3C{-}N{=}C + R'M \longrightarrow R'{-}C(={N{-}CR_3}){-}M \qquad \underline{38}$$

In contrast to acyl alkali and earth alkali derivatives of type 36, which are rather unstable [60], the imino analogues 38 are stable at least in solution with M = Li or MgX [61] and are useful nucleophilic acylating reagents (d^1-reactivity [62]). These can be alkylated (→ ketones) and hydroxylalkylated (→ α-hydroxyketones).

There are two interesting applications of titanium and zirconium compounds with carbonylation and additions to isonitriles: the hydrozirconation products (see section IV D) of olefins (39) and of acetylenes (40) react smoothly with CO to give acylzirconocene derivatives 41 and 42, respectively. These in turn can be converted (→ 43,44 [63]) to aldehydes, carboxylic acids or acyl halides, but they do not combine with carbon electrophiles such as aldehydes, ketones, and alkyl halides.

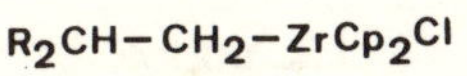

$R_2CH{-}CH_2{-}ZrCp_2Cl$ **39**

R–CH=CH–$ZrCp_2Cl$ **40**

$R_2CH{-}CH_2{-}C(=O){-}ZrCp_2Cl$ **41**

R–CH=CH–C(=O)–$ZrCp_2Cl$ **42**

$R_2CH{-}CH_2{-}C(=O){-}X$ **43**

R–CH=CH–C(=O)–X **44**

X= H,OH, OR,Br

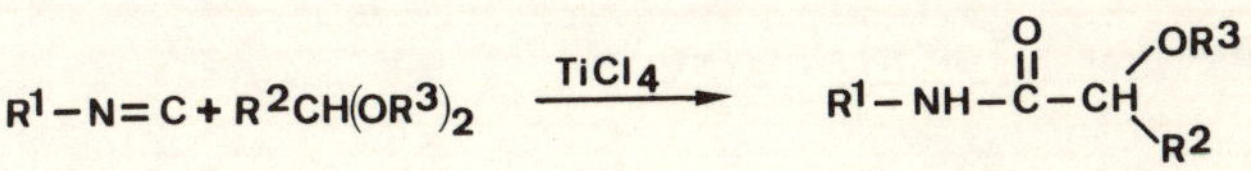

$$R^1{-}N{\equiv}C + R^2CH(OR^3)_2 \xrightarrow{TiCl_4} R^1{-}NH{-}C(=O){-}CH(OR^3)R^2 \quad \mathbf{45}$$

$$R^1CHO + H_3C{-}N{=}C(Cl)TiX_3 \ (\mathbf{46}) \longrightarrow R^1CH(OH){-}C(=O){-}NHCH_3 \ (\mathbf{47})$$

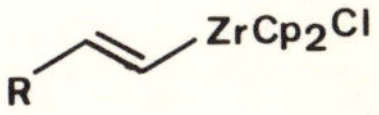

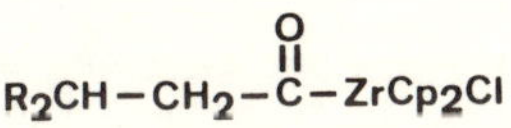

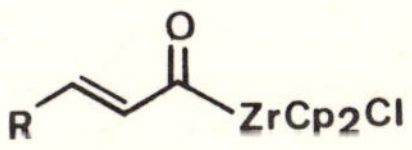

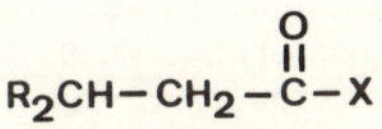

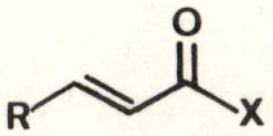

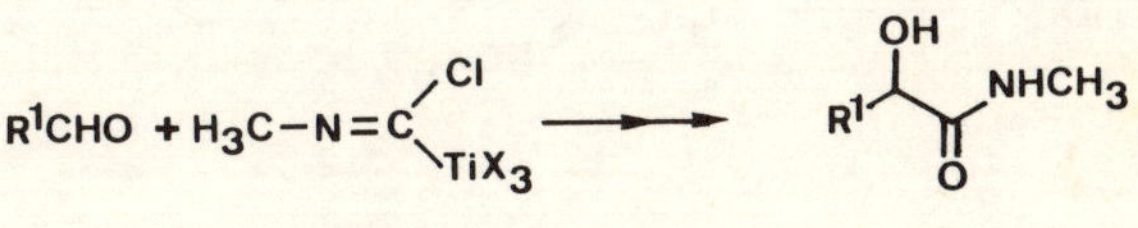

Two new titanium-mediated versions of the *Passerini* reaction [64] lead to α-alkoxyamides 45 and to amides 46 of aliphatic amines. In the first case, it is not clear whether an intermediate with a Ti-C bond is involved or whether $TiCl_4$ merely acts as a *Lewis* acid [65]. In the other reaction, one of the known adducts of $TiCl_4$ to isonitriles, for which structure 46 has been proposed [66], adds to aldehydes [10]. If structure 46 is correct, it is a case in which an organotitanium reagent is formed directly from an organic compound and TiX_4 rather than by transmetalation (see Table 2).

Carbonylation of organozirconium derivatives [63]

$$Cp_2Zr(Cl)R \xrightarrow{CO} Cp_2Zr(Cl)C(=O)R \rightarrow \begin{cases} \xrightarrow{H^{\oplus}} RCHO \\ \xrightarrow{[O]} RCOOH \\ \xrightarrow{NBS,\ MeOH} RCOOMe \end{cases}$$

A solution of 2 mmol of the organozirconium compound 39 (cf. section IV, D) in 5 ml of benzene is stirred under 1.4 atm. of CO at room temperature. Carbonyl insertion to yield the acyl complex 41 is complete within some hours. 41 may be converted to aldehydes by addition of aqueous HCl in benzene solution at room temperature. Carboxylic acids are produced on treatment with aqueous NaOH followed by 30% H_2O_2 and acidification, whereas slow addition of NBS solution in methanol at room temperature results in formation of methyl esters.

2-Hydroxy-N-methyl-heptanoic amide [10]

$$CH_3{-}N{=}C \ + \ TiCl_4 \longrightarrow Cl_3Ti{-}C(Cl){=}NCH_3 \xrightarrow{C_5H_{11}CHO} CH_3(CH_2)_4CH(OH)C(=O)NHCH_3$$

A solution of 1.1 ml (10 mmol) of titaniumtetrachloride in 40 ml CH_2Cl_2 is cooled to ca. -5°C. After slow addition of 0.57 ml (10 mmol) of methylisocyanide the reaction mixture is warmed to room temperature. From the yellow solution a white solid precipitates. Stirring is continued for ca. 2.5 hours. Cooling to -60°C is followed by the addition of 1.17 ml (9.5 mmol) of hexanal. The mixture is warmed slowly to ca. -5°C. The precipitate begins to dissolve at -15°C and a clear yellow solution is obtained. Hydrolysis is achieved by treatment first with 25 ml 2 N aqueous HCl, and after the initially formed white solid is completely dissolved, an additional 20 ml of water is added. The green-yellow colour disappears during vigorous stirring for 1.5 hours. The layers are separated and the aqueous phase is extracted twice with CH_2Cl_2. After washing the combined organic extracts with saturated NaCl solution and drying ($MgSO_4$), the product, a white powder, is obtained on concentration in vacuo. Yield: 1.45 g (9.10 mmol; 96%). m.p. 109.5-110°C (CH_2Cl_2).

E) α- and γ-Alkylations of Carbonyl Derivatives

Replacement of hydrogen by trialkyl silyl groups has led to a "new generation" of carbon nucleophiles for carbon-carbon bond formations. Although vinylsilanes and arylsilanes have also gained considerable importance, the most useful representatives of this class of compounds are undoubtedly the silyl enol ethers 48 [67-71] and the allylic silanes 49 [71,72]. The reactions of these species with electrophiles require catalytic or stoichiometric

48 (OSiMe3) + $TiCl_4$ ⇌ (?) (OTiCl3) + $ClSiMe_3$

49 (SiMe3) + $TiCl_4$ ⇌ (TiCl3) + $ClSiMe_3$

$$\geq C-X + TiCl_4 \rightleftharpoons \geq C-\overset{\oplus}{X}-\overset{\ominus}{Ti}Cl_4 \;(\mathbf{50}) \overset{?}{\rightleftharpoons} \geq C^{\oplus} \;\; XTiCl_4^{\ominus}$$

"activation". For many purposes, the most powerful activation is provided by a stoichiometric amount of $TiCl_4$; with sensitive substrates this activation can be modulated [73] by using mixtures of $TiCl_4$ and $Ti(OR)_4$, i.e. $Cl_2Ti(OR)_2$ in the case of a 1:1 ratio. Although the mechanism is still subject to discussions, it is conceivable [74] that with silyl enol ethers 48 the actual reagent is a titanium enolate, and that likewise an allylic organotitanium derivative is formed from allylsilanes 49. Titanium derivatives of this kind have been generated deliberately from lithium enolates [75] and allylic *Grignard* reagents [14,76,77], respectively, and have been shown to be extremely reactive even at temperatures as low as -110°C, see next section (F). Of course, the titanium tetrachloride will also activate the electrophiles such as halogen and oxygen derivatives (→50), but many applications suggest that no free carbenium ions are formed. Maybe a simultaneous activation of the nucleophiles (→45, 49) and of the electrophiles (→50) is accomplished by Ti(IV) species. For extensive discussions, the reader is referred to some excellent recent review articles [67-72].

Whatever the detailed mechanism might be, the following most useful transformations can be achieved under these conditions. In the presence of $TiCl_4$ silyl enol ethers are converted to α-alkylated carbonyl derivatives with high regioselectivity and in good yields when treated with *t*-alkyl halides, *t*-alkyl ethers or *t*-alkyl esters, with benzylic halides, with α-chloro-thioethers, and with acetals of aldehydes and ketones. O-Alkylation and poor yields because of competing elimination and other side reactions are typical when lithium enolates are treated with the same "S_N1-type electrophiles". Also, aldol additions even to ketones occur with $TiCl_4$/silyl enol ethers. Thus, this kind of silyl enol ether chemistry is not alternative, but complementary to classical enolate chemistry.

It is remarkable that silyl enol derivatives can also undergo remote reactions, i.e. *Michael* additions with enones and d^4-reactions [62] when silyl dienol ethers are employed.

Same characteristic examples of $TiCl_4$-mediated transformations of silyl enol ethers are collected in Table 6, and several procedures follow.

2-*t*-Butyl-cyclopentanone [78)]

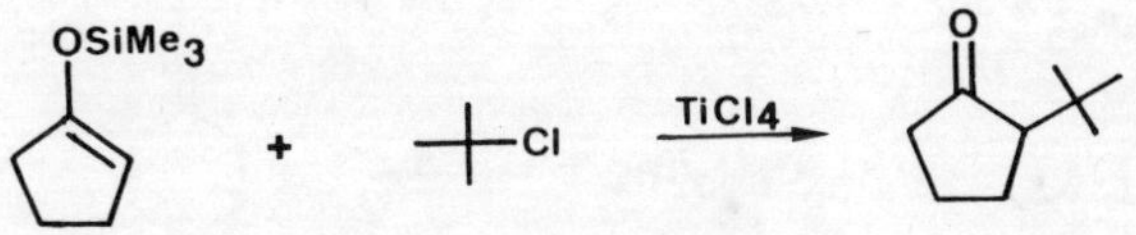

Under inert atmosphere a solution of 15.6 g (100 mmol) of 1-trimethylsilyl-oxy cyclopentene and 10.0 g (108 mmol) of *tert*-butylchloride is treated at -45°C with 20 g (105 mmol) of titanium tetrachloride dissolved in 30 ml of cold (-45°C) CH_2Cl_2. The reaction mixture is stirred for 30 min. and then diluted with 100 ml of CH_2Cl_2. After aqueous workup the crude product is purified by fractional distillation to yield 9.2 g (66% yield) of 2-*t*-butyl-cyclopentanone. b.p. 95°C/45 Torr.

α-(1-Methoxy-cyclohexyl)-α-phenyl-aceton [79)]

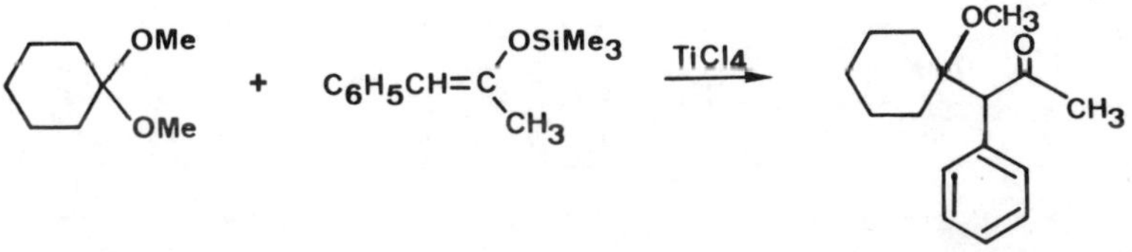

To a solution of 2.6 mmol $TiCl_4$ in 5 ml CH_2Cl_2 stirred under argon at -78°C is first added a methylene chloride solution of the dimethyl acetal of cyclo-hexanone (2.5 mmol), immediately followed by a solution of 1-phenyl-(2-tri-methylsilyloxy)-1-propene in the same solvent. After stirring for 3 hours at -78 C the mixture is hydrolyzed and the aqueous layer extracted with ether. After drying, the solvents are evaporated and the residue is chromatographed on silicagel to give a 91% yield of the β-methoxy-ketone.

2-(Hydroxy-phenyl-methyl)-cyclohexanone [80)]

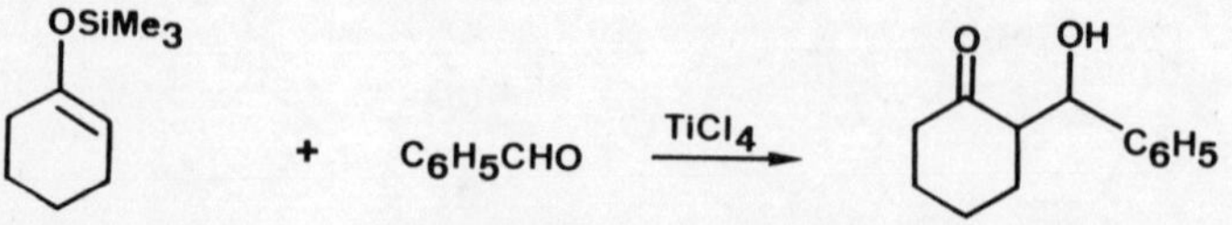

A methylene chloride solution (10 ml) of the silyl enol ether (0.43 g, 2.5 mmol) is added dropwise to a mixture obtained from 20 ml CH_2Cl_2, 0.29 g (2.75 mmol) benzaldehyde and 0.55 g (2.75 mmol) $TiCl_4$ and stirred at -78°C under argon. After 1 hour, hydrolysis at low temperature, extraction of the inorganic layer with ether, washing of the organic layers with water, drying (Na_2SO_4), and evaporating the solvents furnishes a residue which is chromatographed on silicagel with CH_2Cl_2 as eluant. First, 115 g (23%) of the *l*-diastereomer (m.p. 103°C), and then 346 mg (69%) of the *u*-diastereomer (m.p. 74°C) is eluted.

3-(2-Phenyl-2-oxoethyl)-cyclohexanone [81)]

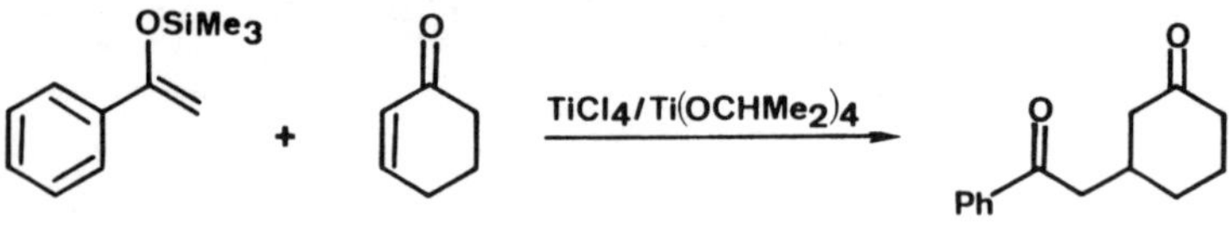

To a methylene chloride solution (4 ml) of $TiCl_4$ (1 mmol) and $Ti(OCHMe_2)_4$ (0.4 mmol) is added a mixture of α-trimethylsilyloxystyrene (1 mmol) and 2-cyclohexen-1-one (1 mmol) in methylenechloride (6 ml) at -78°C. After being stirred for 30 min., the mixture is hydrolyzed with 12 ml aqueous K_2CO_3 (0.75 g) and extracted with ether. The extract is washed with brine, dried (Na_2SO_4) and concentrated under reduced pressure. The residue is chromatographed on silica gel, yielding 70% product of m.p. 75-76°C.

Table 6. Examples of $TiCl_4$-mediated reactions of silyl enol ethers with "S_N1--type" electrophiles. Equimolar amounts of $TiCl_4$ are usually employed, see procedures.

Silyl enol ethers	Electrophile	Product	Yield %	[ref.]
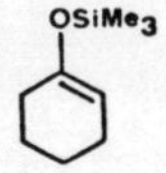 $OSiMe_3$	t-C_4H_9-Cl	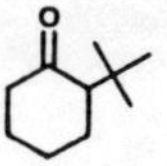	90	78)
	t-C_4H_9-OMe		30-55	67)
	t-C_4H_9-OAc			
	$HC(OCH_3)_3$	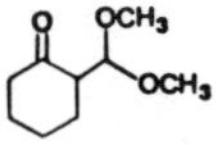 OCH_3 OCH_3	71	79)
	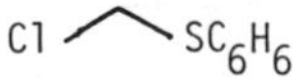 Cl ⁀ SC_6H_6	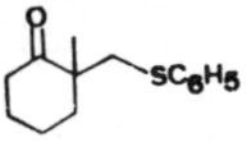 SC_6H_5	71	82)
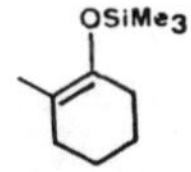 $OSiMe_3$	t-C_4H_9Cl	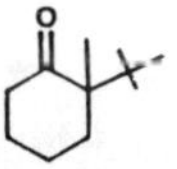	86	78)
	$C_6H_5CHC_4H_9$ (Cl)	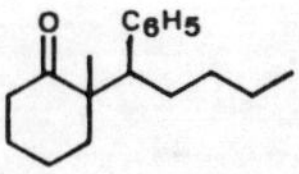 C_6H_5	82	67,83)
	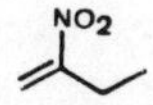 NO_2	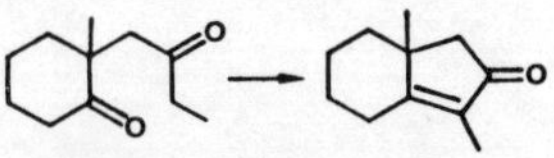 82% → 87%		84)
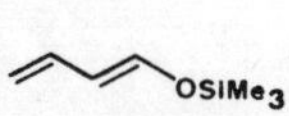 $OSiMe_3$	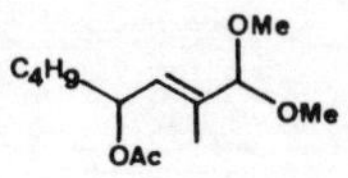 C_4H_9, OMe, OMe, OAc	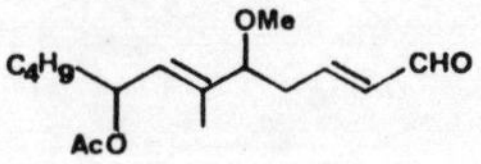 C_4H_9, OMe, CHO, AcO	85	85)
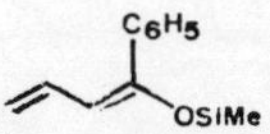 C_6H_5, OSiMe	$Me_2C(OMe)_2$	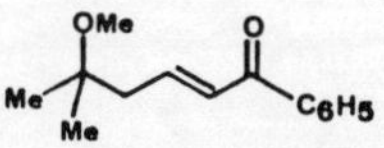 OMe, Me, Me, C_6H_5	78	86)

F) Aldol Additions and Related Processes

There is no synthetic reaction about which so many papers and review articles appeared within a few years as about the diastereoselective aldol addition. The most comprehensive, competent and up-to-date survey is that by *C.H. Heathcock* and is entitled "The Aldol Addition Reaction", to be published in the forthcoming edition of *J.D. Morrison*'s book "Asymmetric Synthesis, Vol. 2". A complete list of the reviews with titles is given in references [87-97]. The interest in acyclic stereoselection arose when macrocyclic and open chain ionophores, antibiotics and other highly functionalized, physiologically active compounds became the target molecules of natural product synthesis; see for instance the most recent efforts towards the total synthesis of palitoxin [104].

Like other processes [101,102] in which two trigonal, two-dimensionally chiral centers combine to give two new asymmetric carbon atoms, the aldol addition can furnish two diastereomeric products. It is shown in Scheme 12

Scheme 12

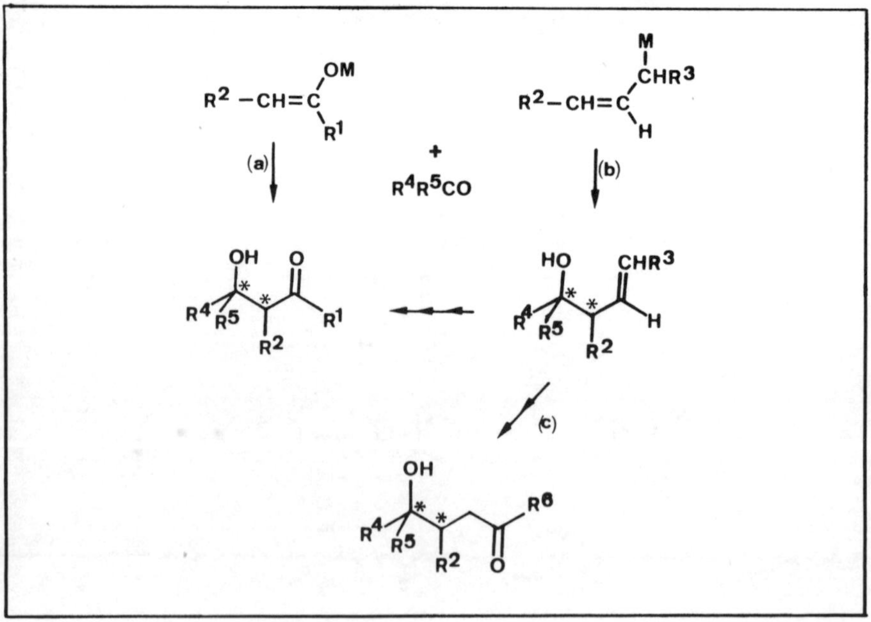

that the same β-hydroxy-carbonyl compounds, which are formed by aldol addition, route (a), can also be obtained from allylic organometallic compounds through homoallylic alcohols, route (b). This latter route requires more steps to the aldol-type products, but it is at the same time more flexible, because conversions of the olefinic double bond other than oxidative cleavage can produce a variety of structures, cf. the γ-hydroxy-carbonyl derivatives, route (c) in Scheme 12. If R^3 of the allylic metal derivative is a heteroatom, step (c) consists of a simple hydrolysis. For these reasons, it is not surprising that the generation and the regio- as well as diastereoselective carbonyl additions of allylic nucleophiles has also attracted much recent interest, as demonstrated by several review articles [96-100].

The stereochemical course of diastereoselective C,C bond formation by combination of two trigonal centers can be discussed with the *Newman* projec-

Scheme 13

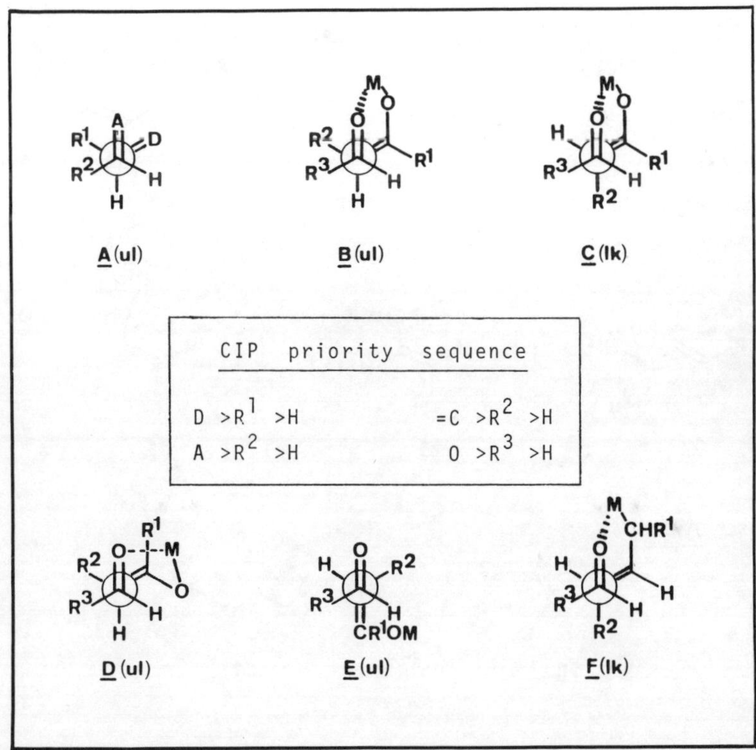

tions in Scheme 13. The relative topicities of approach of the donor and acceptor trigonal centers are specified [40] assuming the "normal" order of priority [105] of the groups as indicated in the center of Scheme 13. As a general rule, the *ul*-approach A is preferred [102]: (i) if no geometrical isomers exist of either the donor or the acceptor component, they combine as shown in A; (ii) the rate of reaction and the degree of selectivity are higher with *Z*-donors, B, than with *E*-donors, C; (iii) under many conditions, the aldol process occurs with *ul*-preference, independant of enolate configuration; whether this is due to a switch from the chair-like arrangement C to a boat-like arrangement D or to a so-called open approach E, and why this is so, is still subject to discussion [87,102,106]. The present authors think [102] that the open model E is highly unfavorable in aprotic solvents and in the absence of ions, when the temporary partial charge separation which must develop along the reaction coordinate is neither stabilized by solvation nor by chelation, nor by counterions.

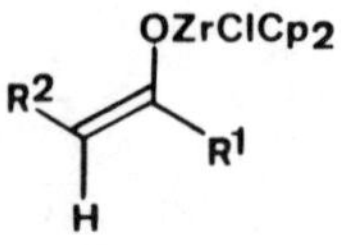

$OZrClCp_2$ · $OZrClCp_2$ · $OTiX_3$ · $OTiX_3$

Z−**51** 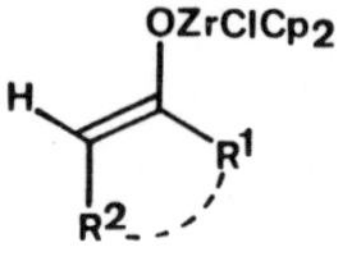 E−**51** 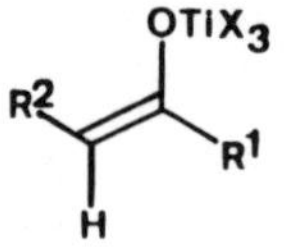 Z−**52** 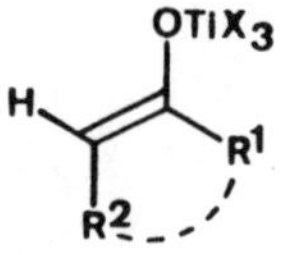 E−**52**

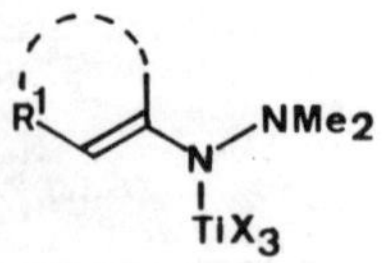

NMe_2, TiX_3 · $CHMX_3$, H (R^3) · $CH{-}TiX_4^{\ominus}M^{\oplus}$

53 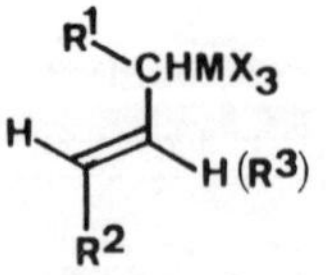 **54** 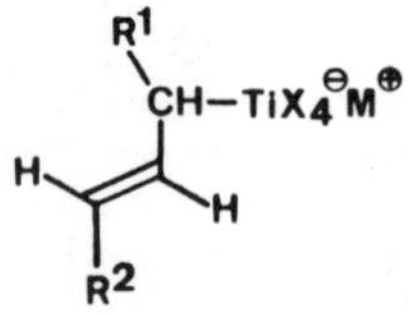 **55**

Cp_2Ti^{III} · XCp_2Ti^{IV}

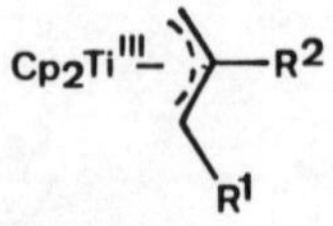

56 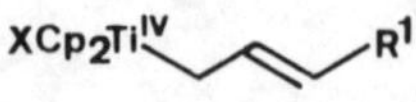 **57**

Returning to the subject matter, it is a fact that the metal enolates 51 and 52 and the enhydrazinolates 53 of titanium and zirconium are among the reagents with *ul*-preference, independant of the enolate geometry. In contrast, both titanocene and titanate derivatives with crotyl or higher or heterosubstituted allylic ligands (54, 55) show a strong *lk*-preference, see F in Scheme 13.

A collection of nucleophiles of type 51 - 57 is specified in Table 7.

Table 7. List of enolates, enhydrazinolates and allylic derivatives ot titanium and zirconium which were used for regio- and diastereoselective C,C bond formation.

	R^1	R^2	X (M)	Ref.
51a	C_2H_5	CH_3		107)
51b	C_6H_5	CH_3		108)
51c	$-(CH_2)_3-$			107)
51d	$-(CH_2)_4-$			107)
51e	$N(CHMe_2)_3$	CH_3		108)
51f	$N(CH_2)_3CH_2$	CH_3		108)
51g, 51h	A (see Scheme 14)	CH_3		109)
51i	OCH_3	CH_3		108)
51j	t-C_4H_9O	CH_3		108)
51k	t-C_4H_9S	CH_3		108)
52a	C_2H_5	CH_3	$OCHMe_2$	75)
52b	C_2H_5	CH_3	$N(C_2H_5)_2$	75)
52c	t-C_4H_9	CH_3	$OCHMe_2$	75)
52d	C_6H_5	CH_3	$OCHMe_2$	75)

(Table 7 continued)

	R^1	R^2	X (M)	Ref.
52e	$C_6H_2(CH_3)_3$	CH_3	$OCHMe_2$	75)
52f	$-(CH_2)_3-$		$N(C_2H_5)_2$	75)
52g	$-(CH_2)_4-$		$OCHMe_2$	75)
52h	$-(CH_2)_4-$		$N(C_2H_5)_2$	75)
52i	$-(CH_2)_5-$		$OCHMe_2$	75)
52j	$-(CH_2)_5-$		$N(C_2H_5)_2$	75)
52k	OC_2H_5	CH_3	$OCHMe_2$	19)
52l	$-O-(CH_2)_2-$		$N(C_2H_5)_2$	19)
52m	$CH_3N-(CH_2)_2-$		$N(C_2H_5)_2$	19)
52n	$CH_3N-(CH_2)_4-$		$N(C_2H_5)_2$	19)
53a	CH_3		$N(C_2H_5)_2$	110)
53b	CH_3		$OCHMe_2$	110)
53c	$CH(CH_3)_2$		$OCHMe_2$	110)
53d	C_6H_5		$OCHMe_2$	110)
53e	C_6H_5		$N(C_2H_5)_2$	110)
54a	H	H	$OCHMe_2$ (Ti)	21)
54b	H	CH_3	OC_6H_5 [a] (Ti)	14)
54c	H	CH_3	OC_6H_5 (Zr)	111)
54d	H	C_4H_9	OC_6H_5 (Ti)	77)
54e	H	$CH(CH_3)_2$	OC_6H_5 (Ti)	77)
54f	LiS	H	$OCHMe_2$ (Ti)	22,112)
54g	LiS	H ($R^3 = CH_3$)	$OCHMe_2$ (Ti)	112,113)
54h	H	SCH_3	$OCHMe_2$ (Ti)	76,112,113)
54i	H	SC_6H_5	$OCHMe_2$ (Ti)	76,112,113)
54j	H	H	NEt_2 (Ti)	19)

(Table 7 continued)

	R^1	R^2	X (M)	Ref.
54k	H	CH_3	NEt_2 (Ti)	19)
54l	$OCON^iPr$	CH_3	NEt_2 (Ti)	114)
55a	H	H	$OCHMe_2$ (MgCl)	19,115)
55b	H	CH_3	$OCHMe_2$ (MgCl)	19,115)
55c	H	CH_3	OC_6H_5 (MgCl)	19,115)
55d	H	$Si(CH_3)_3$	$OCHMe_2$ (Li)	19,115)
55e	H	H	NMe_2 (MgCl)	19,115)
55f	H	H	NMe_2 (Li)	19,115)
55g	H	H	NEt_2 (Li)	19,115)
55h	H	CH_3	NEt_2 (MgCl)	19,115)
56a	H	H		116)
56b	CH_3	H		116,117)
56c	CH_3	CH_3		117)
56d	C_2H_5	H		117)
56e	$(CH_3)_3Si$	H		99)
57a	CH_3		X = Cl	118)
57b			X = Br	118)
57c			X = I	118)

a) The crotyl titanium reagents were also tested with other OR-groups. The phenoxy derivatives gave the best diastereoselectivities in the standard reaction, i.e. addition to benzaldehyde. Therefore, all other experiments were conducted with the triphenoxy compounds.

1) Ti- and Zr-Enolates for Aldol Additions

The trialkoxytitanium and tris(diethylamino)titanium enolates 52 and analogous derivatives 53 of aldehyde dimethylhydrazones have been shown to combine with aldehydes with relative topicity *ul*[75,110]. The selectivity was found to be highest with the hydrazone derivatives (53b-53d) when X = $OCHMe_2$, yielding aldol-type products 58 of two different aldehydes with

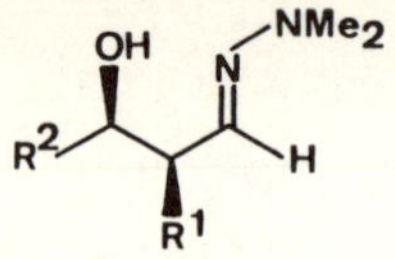

58 R^1: Me, $CHMe_2$, C_6H_5
R^2: Me, $CHMe_2$, CMe_3, C_6H_5

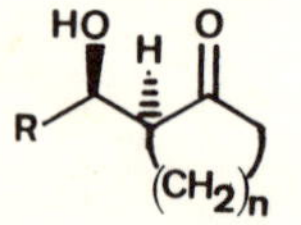

59 n = 2,3,4
R = $CHMe_2$, C_6H_5

diastereoselectivities of up to and above 98:2; with $R^1 = R^2 = CH_3$ the *ul/l*-ratio[40] (R*,S*/R*,R*) is 95:5. This is in contrast to direct aldol reactions of aldehyde enolates, which are poor reactions with respect to both chemical yields and diastereoselectivities[87]. On the other hand, hydrazones of type 58 are not easy to hydrolyze, and oxidative methods of cleavage have the additional disadvantage of producing hazardous dimethylnitrosamine[119]. The titanium enolates 52f-52j derived from cyclic ketones give aldols 59 (>90% ds) with best selectivities, while the open chain counterparts give poorer diastereoselectivities[75].

Most spectacular results are obtained with Cp_2ZrCl-enolates 51, which are not so selective when derived from ketones[107] and esters and more selective from thiolesters and amides[108], but which show a very high enantioselectivity when derived from certain chiral amides[91,109], see 51g and 51h in Table 7. The results of addition of these two Zr-enolates to pentanal, 2-methylpropanal and benzaldehyde are shown in Scheme 14. The proline-derived amide enolate produces almost exclusively the (2S,3R)-β-hydroxy amide, while the valine-derived analogue yields, again with "almost perfect" selectivity, the (2R,3S)-diastereomer. Hydrolytic cleavage of the amides gives practically enantiomerically pure samples of the free acids of opposite chirality [109]:

All the enolates 51 - 53 of titanium or zirconium are best generated *in situ* from the corresponding Li-enolates. It is not necessary to start from configurationally pure enolates. Some comments on the practical execution of the reaction of chiral Cp_2ClZr-enolates follow.

Scheme 14

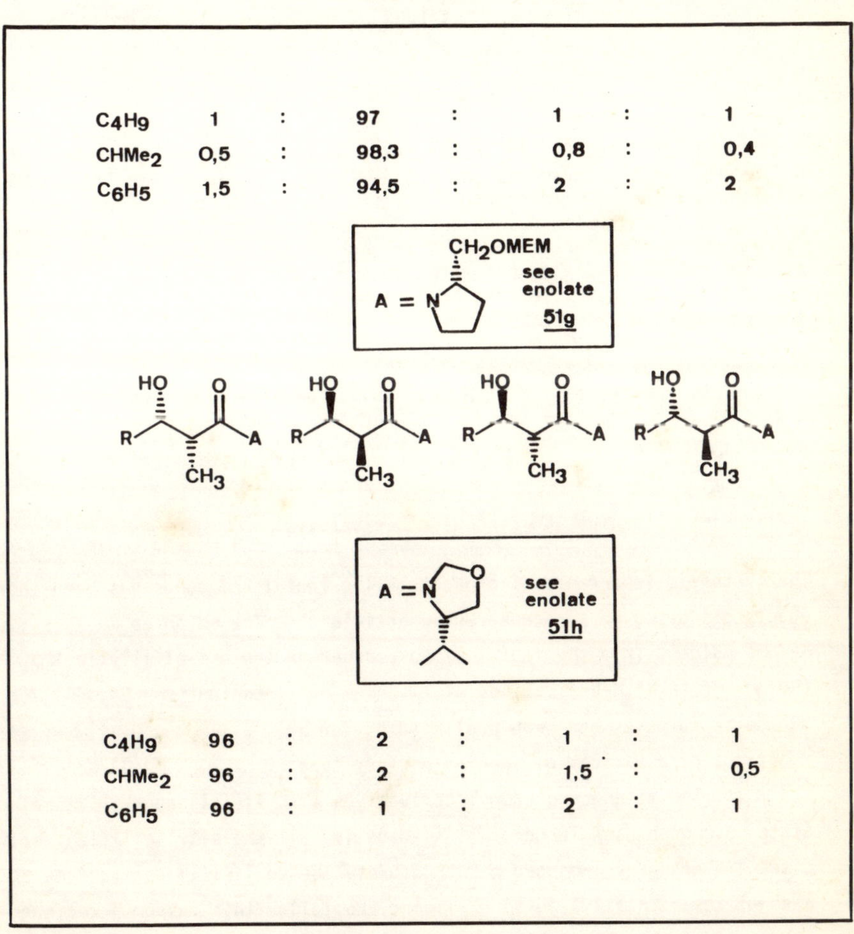

Enantioselective Aldol Additions of Zirconium Enolates. General Procedure[108,109]

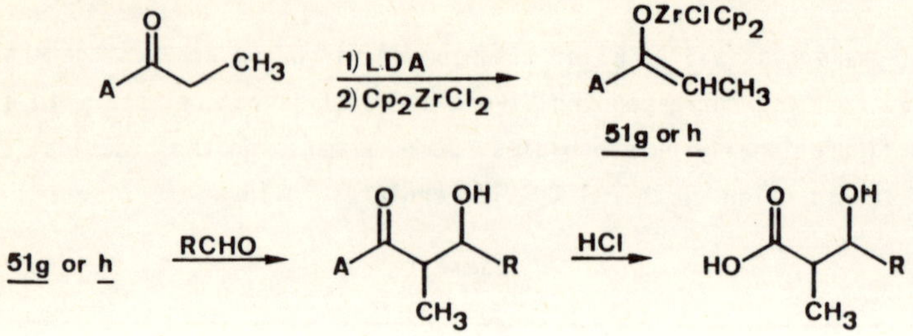

The lithium enolates 51 (Li instead of $ZrCp_2Cl$), which are generated from the respective propionamides and LDA (THF, 0°C, 30 min.), are transformed into the corresponding zirconium enolates 51g or 51h by treatment with 1.1 equiv. of Cp_2ZrCl_2 (0.16 M in THF). After cooling to -78°C, 0.9 to 1.1 equiv. of aldehyde is added, stirring is continued for 1 hour, and the reaction is quenched with saturated aqueous ammonium chloride. After filtration, the product is extracted into methylene chloride, dried (Na_2SO_4) and concentrated *in vacuo*.

Treatment of the aldol adducts with 5% HCl (10 equiv. $H^{\oplus}$, 100°C, 2 hours) followed by neutralization with aqueous sodium bicarbonate and stirring for 5 min. at 25°C liberates the desired carboxylic acids.

2) 2-Alken-1-yl Titanium and Zirconium Derivatives

The synthetic importance of organometallic crotyl compounds has been thoroughly discussed in a recent review article[98]. The advances in this field become evident from the fact that no account on the use of allylic $RMX_4^{\ominus}$-, $(RO)_3M$- or $(R_2N)_3M$-derivatives of titanium or zirconium (see 54, 55) for synthetic purposes had been published a year ago, while now well over 50 reactions of a variety of species of this type are known[14,19,40,77,114,120,121]. The titanocene crotyl derivatives with Ti(III) and Ti(IV) valence state have been used longer[98,99]. Usually, all the allylic Ti(IV) derivatives are drawn as σ-bonded η^1-structures, the $Cp_2Ti(III)$ derivatives as π-complexes with allyl as an η^3-ligand. We follow this custom here (see Scheme 12, Scheme 13, Table 7, and *Formulae* 54 - 57), although the structures of the species which are generated in solution or which actually react with electrophiles are not known. The structures of some products, and with it the relative topicity of their formation have, however, been proved by comparison with compounds of known configuration. In most cases, the configuration was tentatively proposed by analogy. The parent Ti- and Zr-allyl reagents [18,19] are only interesting for functional group selective transformations. Two types of constitutional isomers can be formed with

Scheme 15

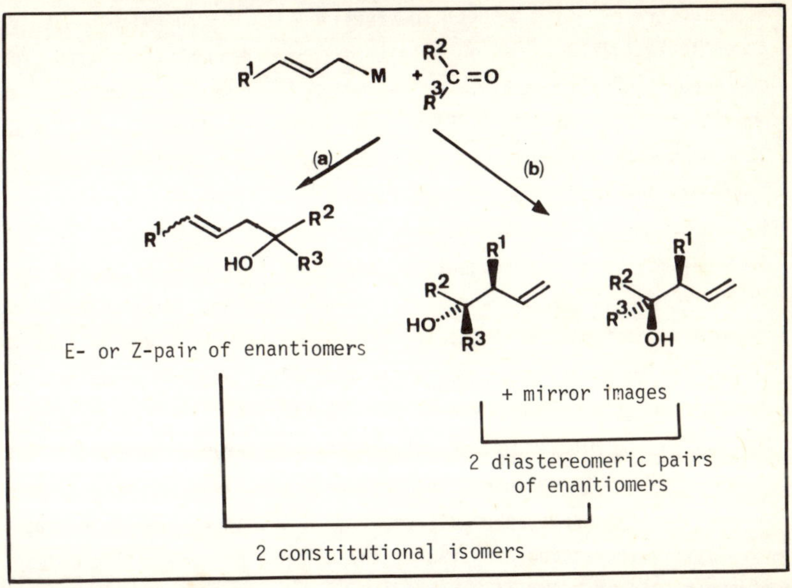

E- or Z-pair of enantiomers

+ mirror images

2 diastereomeric pairs of enantiomers

2 constitutional isomers

substituted allylic organometallics of this type, see Scheme 15. As with other metals, 2-alken-1-yl titanium or zirconium compounds give rise to branched products exclusively, which result from attack of the electrophile at the higher substitued carbon atom, route (b) in Scheme 15; with hetero-substituted systems, R_1 = X, route (a) may be followed (see section F3).

In contrast to alkenyl *Grignard* reagents, the corresponding titanium, but not so much the zirconium analogues (see 54b-54e) add to aldehydes [14,111] with high diastereoselectivity. Even ketones, with the naturally smaller size differences between the two groups bonded to the carbonyl carbon as compaired to aldehydes, give products with surprisingly high preference of one diastereomer [77]: the optimized selectivity of the crotyltitanium 54b with benzaldehyde is 85% ds, with acetophenone 88% ds. Possibly, the smaller difference in size is counterbalanced by the lower reactivity of the ketone (later transition state?). (For aldehyde/ketone selectivity reversal see section IV, B2, below). With chiral, non-racemic OR*-groups on titanium or zirconium, no great enantiomeric excesses could be detected in aldehyde adducts of these supposedly S_E1-type reactions [111].

After numerous RO-groups in crotyl-$Ti(OR)_3$ were examined with benzaldehyde as a substrate [14], the "best" derivative with R = phenyl was used for all other investigations [14,-7]. Allylic *Grignard* reagents, prepared from the corresponding halide and *Rieke* magnesium [122], were combined with an equiv. amount of the triphenoxytitanium chloride (see section II above) and then at -100°C with carbonyl compounds. The results of α-methallylations of aldehydes are outlined in Scheme 16, those of additions of unsymmetrical ketones in Scheme 17. Finally, Scheme 18 contains a comparison of crotyl with other alkenyl titanium reagents in additions to aldehydes and ketones. The following points are noteworthy: (i) α-Branched and unbranched aliphatic aldehydes react most selectively. (ii) Benzaldehydes give >90% ds if substituted with donating groups, while the selectivity drops below 90% with acceptors. (iii) Ketones give useful selectivities (>80%) only when the difference in size of the groups attached to the carbonyl carbon is large enough. (iv) In three cases, the diastereoselectivity is equally good or even better with acetophenone (C_6H_5 *vs.* CH_3) than with benzaldehyde (C_6H_5 *vs.* H). (v) Increasing the size of the R-group on the allylic moiety, i.e. of the steric crowding of R between R_L and R_S in the transition state as indicated by the *Newman* projection in Scheme 18, hardly alters the diastereoselectivity of addition.

Further investigations are necessary in order to understand the reasons for the facts described above under (i), (ii), (iv), and (v), while the effect mentioned under (iii) is expected from simple steric considerations.

Scheme 16. Diastereoselectivities (% ds) of additions of crotyl-triphenoxy-titanium to aldehydes [14]. The numbers given underneath each aldehyde *formula* are the diastereoselectivities, i.e. the preference of relative topicity *lk* over *ul* [40] in percent.

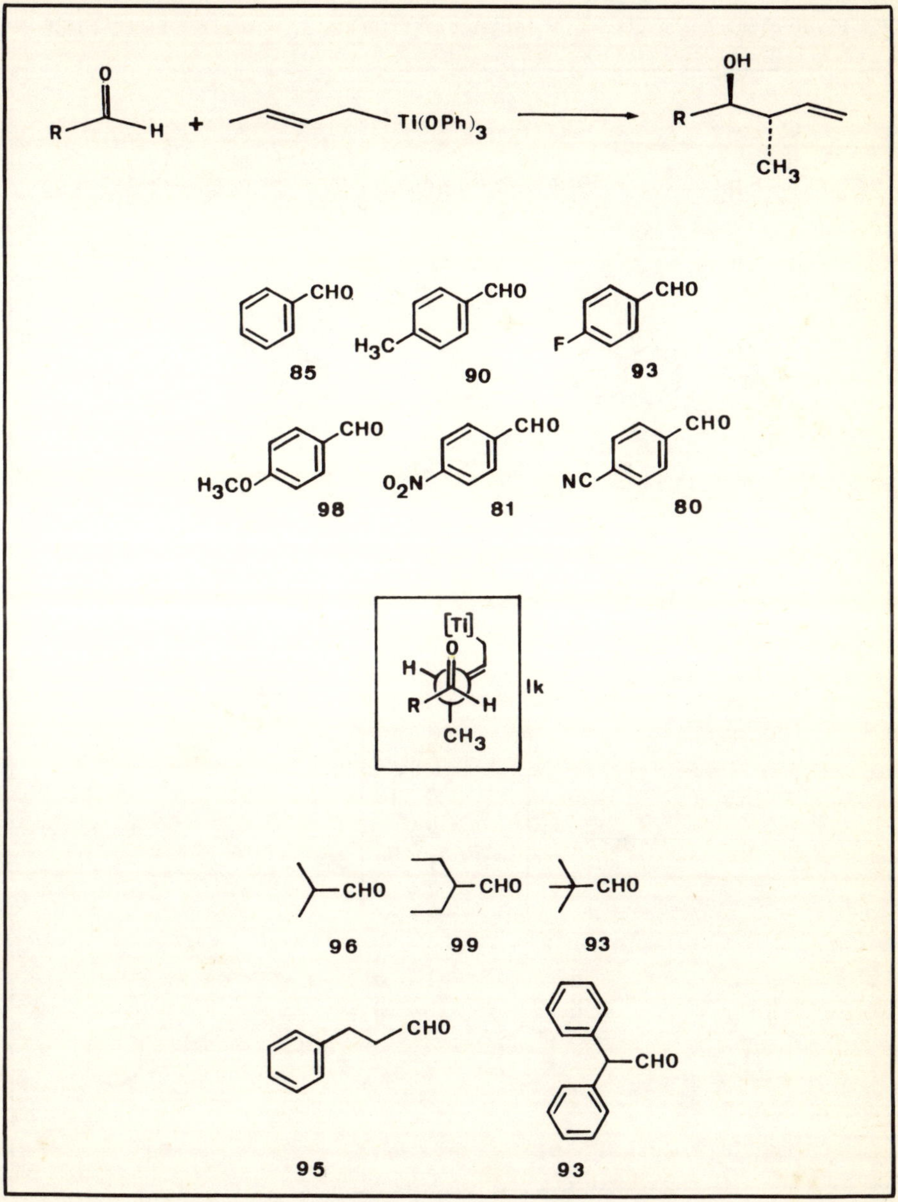

R–CHO + crotyl–Ti(OPh)$_3$ → R–CH(OH)–CH(CH$_3$)–CH=CH$_2$

Aldehyde	% ds
C$_6$H$_5$CHO	85
4-H$_3$C-C$_6$H$_4$CHO	90
4-F-C$_6$H$_4$CHO	93
4-H$_3$CO-C$_6$H$_4$CHO	98
4-O$_2$N-C$_6$H$_4$CHO	81
4-NC-C$_6$H$_4$CHO	80
(CH$_3$)$_2$CH-CHO	96
(C$_2$H$_5$)$_2$CH-CHO	99
(CH$_3$)$_3$C-CHO	93
C$_6$H$_5$CH$_2$CH$_2$CHO	95
(C$_6$H$_5$)$_2$CH-CHO	93

[Ti], O, H, R, H, CH$_3$ — lk

Scheme 17. Diastereoselectivities (% ds) of additions of crotyl-triphenoxytitanium to ketones[77]. The relative topicity of combination of the two trigonal centers is probably as indicated in the *Newman* projection of the approach projectory. However, proof by correlation was supplied only for the acetophenone and pivalophenone adducts. R_L = larger substituent; R_S = smaller substituent.

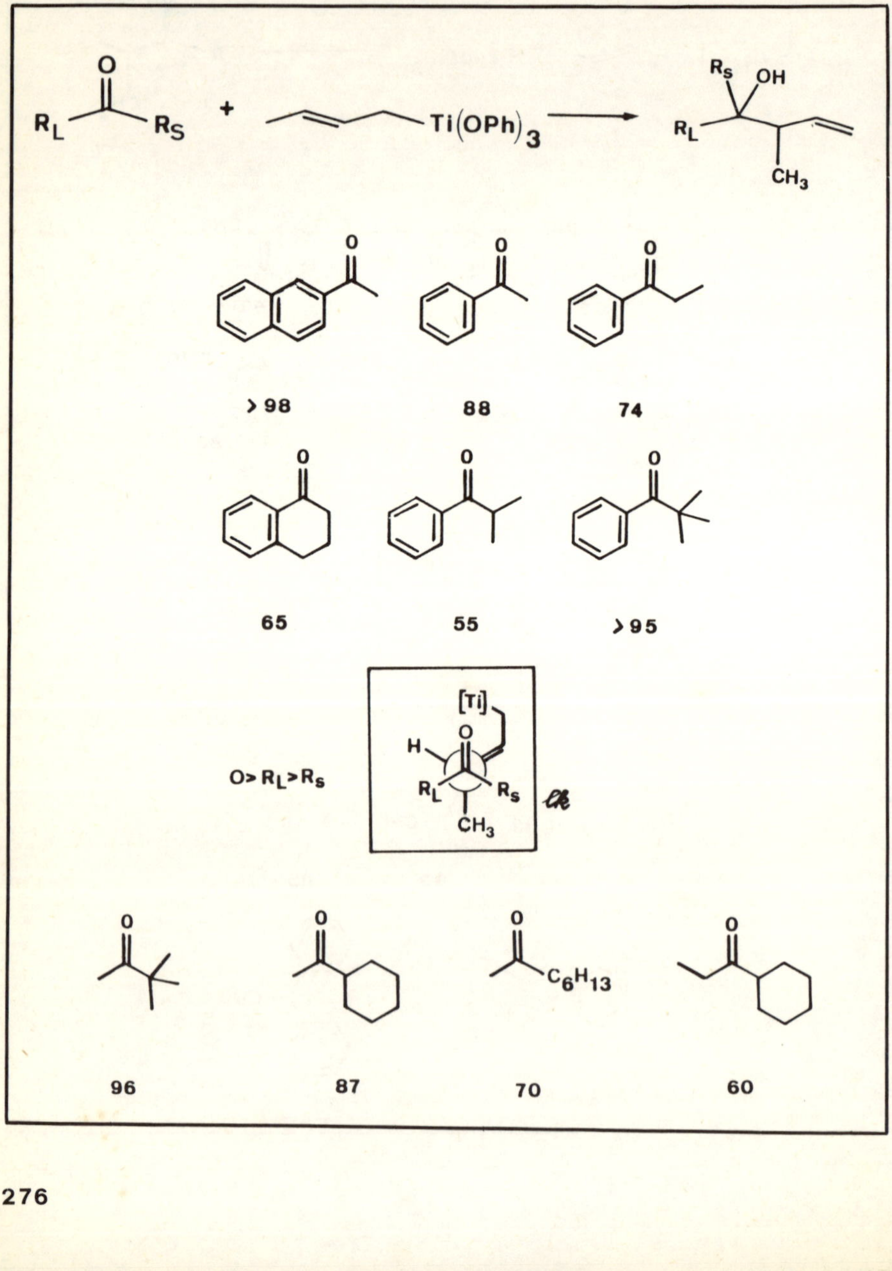

R_L–CO–R_S + crotyl–Ti(OPh)$_3$ → R_L, R_S, OH, CH_3

Ketone	% ds
2-acetonaphthone	>98
acetophenone	88
propiophenone	74
1-tetralone	65
isobutyrophenone	55
pivalophenone	>95
pinacolone	96
cyclohexyl methyl ketone	87
$CH_3COC_6H_{13}$	70
cyclohexyl ethyl ketone	60

O > R_L > R_S

[Ti], O, H, R_L, R_S, CH_3 lk

Scheme 18. Comparison of 2-buten-1-yl-, 2-hepten-1-yl-, and 4-methyl-2-penten-1-yl-triphenoxy-titanium in additions to ketones and aldehydes.

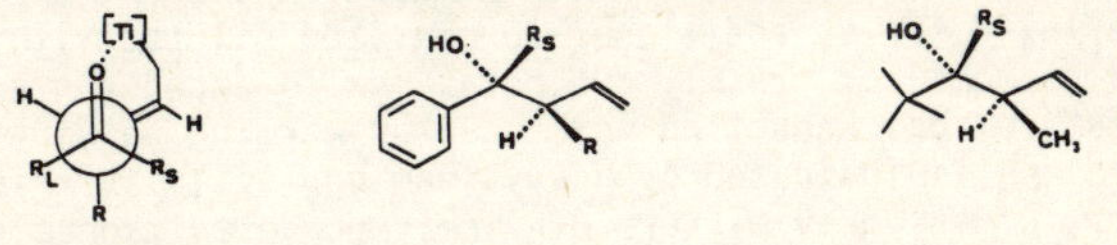

lk ($R_L > R_S$)
ul ($R_S > R_L$)

R_L	R_S	% ds WITH R		
		CH_3	C_4H_9	$CH(CH_3)_2$
Ø	H	85	93	98
Ø	CH_3	88	87	87
Ø	$C{\equiv}C{-}CH_3$	72	77	77
Me_3C	H	>98		
Me_3C	CH_3	>98		
Me_3C	Ø	96		

A drastic increase in selectivity was observed when going from Li, Mg, or Zn to titanium ate-complexes in the addition of propargyl metal derivatives to aldehydes. The results can be explained by assuming a chelated transition state with relative topicity *lk*, see 60 → 61 [123].

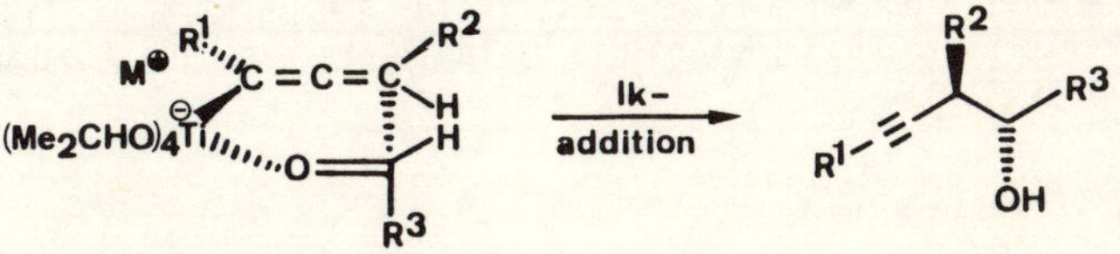

lk-
addition

60

61

Preparation of (2-butenyl)magnesiumhalogenide [14)]

Preparation of *Rieke* magnesium: Under Ar-atmosphere 2.12 g (87.2 g atoms) of magnesium turnings are covered with 20 ml THF, and 7.5 ml (87.0 mmol) of 1.2-dibromo ethane (in 30 ml THF) are added at such a rate, that reflux is maintained without heating. After complete addition, the reaction mixture is refluxed for an additional 30 min. and diluted with 50 ml THF. 6.4 g (164 g atoms) of freshly cut potassium is added to the hot solution of magnesiumbromide. Reduction commences on warming up cautiously to reflux temperature and is indicated by heavy foaming. After stirring at reflux for 2 - 3 hours, a dark grey mixture of finely powdered magnesium results.

Preparation of the crotyl *Grignard* reagent: To the cold suspension 4.2 ml (431 mmol) of 2-butenylchloride is slowly added (over 5 - 10 min.) with vigorous stirring. After sedimentation overnight of the excess of magnesium an aliquot of the supernatant yellow (2-butenyl)magnesiumhalide solution is titrated, using N-phenyl-1-naphthylamine as indicator [124)].

(*u*)-4-Methyl-1-phenyl-5-hexen-3-ol [14)]

$$\text{PhCH}_2\text{CH}_2\text{CHO} + \text{CH}_3\text{CH=CHCH}_2\text{Ti(OC}_6\text{H}_5)_3 \longrightarrow \text{PhCH}_2\text{CH}_2\text{CH(OH)CH(CH}_3)\text{CH=CH}_2$$

A solution of 18 ml (4.77 mmol) chloro-triphenoxy-titanium (0.265 M in THF, see section II) is cooled to -80°C. Ten ml (4.50 mmol) 2-butenyl-magnesiumchloride (0.45 M in THF) is slowly added by syringe. On addition of the first few drops the color changes from red to dark brown-red. After warming up to ca. -20°C within 1 - 2 hours, 3-phenyl-propanal (450 µl, 3.40 mmol) is added and the reaction mixture allowed to reach room temperature overnight. Hydrolysis with saturated aqueous KF-solution and extractive workup with ether furnishes the crude product, which is purified by flash-chromatography (light petroleum ether / diethyl ether 5:1) to give analytically pure 4-methyl-1-phenyl-5-hexen-3-ol. Yield: 450 mg (2.36 mmol; 70%); diastereomeric ratio 95:5 (determined by capillary gas-chromatography prior to any purification).

(ℓ)-3-Methyl-2-phenyl-4-penten-2-ol [77)]

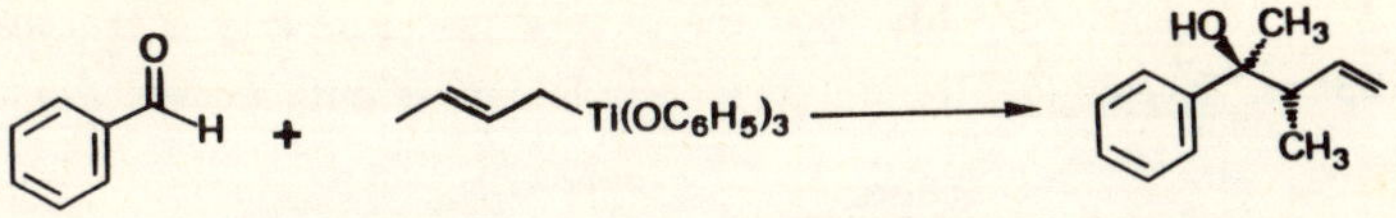

Under argon atmosphere 16 ml (4.24 mmol) chlorotriphenoxy titanium (0.265 M in THF, see section II) is cooled to -80^oC. The red solution is treated with 10 ml (4.10 mmol) 2-butenyl-1-magnesiumhalogenide (0.41 M in THF). The color changes immediately to dark brown-red. After warming up slowly to -30^oC within 1 - 2 hours, acetophenone (400 µl, 3.42 mmol) is added at -100^oC. The solution is allowed to reach room temperature overnight. After hydrolysis with saturated neutral aqueous KF-solution, the reaction mixture is extracted three times with ether. The organic layer is washed thrice with 2 N NaOH and twice with saturated aqueous NaCl-solution, dried (Na_2SO_4) and concentrated to give the crude tertiary alcohol which is purified by flash-crhomatography (light petroleum ether / diethylether 4:1) to yield 469 mg (2.66 mmol, 78%) analytically pure 3-methyl-2-phenyl-4-penten-2-ol; diastereomeric ratio 85:15.

(ℓ)-3-Hydroxy-2-methyl-3-phenylbutanoic acid [77)]

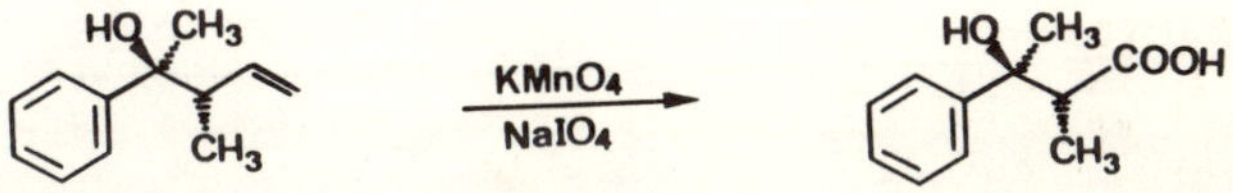

To a solution of 931 mg (5.28 mmol) of 3-methyl-2-phenyl-4-penten-2-ol in 75 ml dioxane and 225 ml H_2O, is added 350 mg $NaHCO_3$, 8.95 g (418 mmol) $NaIO_4$ and 112 mg (0.71 mmol) $KMnO_4$. The purple solution is stirred for 1 hour. Because the $KMnO_4$ is consumed after ca. 1 hour as indicated by a change of the color from purple to red, another 100 mg $KMnO_4$ is added. Stirring is continued for 3 hours. The reaction mixture is worked up with ether. The acid could be separated by extraction with saturated K_2CO_3 solution, careful acidification of the aqueous layer and reextraction with ether. Yield: 616 mg (3.17 mmol, 60%).

3) Heterosubstituted Allylic Organotitanium Reagents

A case in which the regioselectivity of heterosubstituted allylic nucleophiles[125,126] depends strongly upon the metal, is the doubly deprotonated allyl mercaptan, Scheme 19. The dilithio derivative is only moderately

Scheme 19

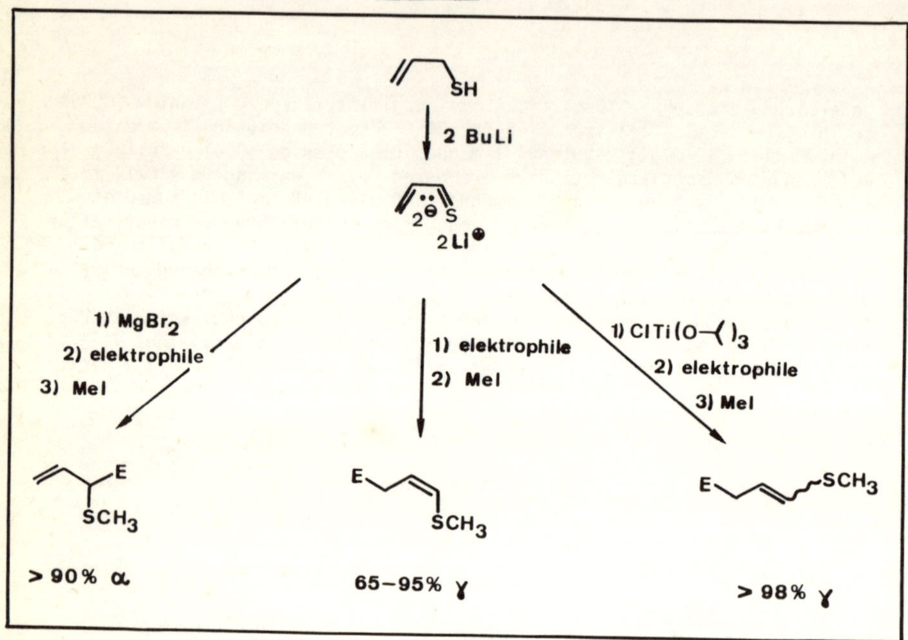

OH R3
R1 R2 CHSCH3
62 a: $R^3 = H$
b: $R^3 = CH_3$

OH R2 R1 SR3
63 a: $R^3 = CH_3$
b: $R^3 = C_6H_5$

R2 R1 O OR
64

R SLi Ti(OCHMe2)3
54 f,g (see table 7)
assumed structure

(Me2CHO)3Ti SR
54 h,i (see table 7)
assumed structure

γ-selective [127], addition of an equiv. $MgBr_2$ causes good α-selectivity[128], and with one equiv. of triisopropoxy titanium chloride a complete (for practical purposes) γ-selectivity of reactions with aldehydes and ketones results [112], see Scheme 19 and *formula* 62a. The 2-methyl-substituted allylmercaptan reacts in the γ-position with a somewhat smaller preference (→ 62b). The α-products 63 are produced in almost as high selectivity as the γ-products 62a, and an equiv. of metalating reagent is saved, if instead of the allylmercaptan the allyl methyl or the allyl phenyl thioethers are lithiated and the resulting allyl lithium reagent transmetallated with $ClTi(OCHMe_2)_3$. The adducts 62 and 63 from aldehydes and ketones assembled in Table 8 demonstrate the generality of the method. Subsequent

---

Table 8. γ-Adducts 62 and α-adducts 63 from aldehydes or ketones and reagents obtained by adding 14 to CH_2=CH-CHSLi_2, CH=C(CH_3)CHSLi or CH_2=CH-CHLiSR (see 54f, 54g, and 54h, 54i [112]). The γ-adducts are E/Z mixtures. Ratios of isomers were determined by ^{1}H-NMR. spectroscopy.

Prod. type	R^1	R^2	yield %	α / γ	% ds of 63
62a	CH_3	H	57	<5/95	
	C_2H_5	H	68	<5/95	
	H_2C=CH	H	39	<5/95	
	C_6H_5CH=CH	H	57	<5/95	
	C_6H_5	H	64	<5/95	
	p-I-C_6H_4	H	57	<5/95	
	p-CN-C_6H_4	H	50	<5/95	
	CH_3	CH_3	60		
	-(CH_2)$_4$-		85		
	H_2C=CH	CH_3	15		
	(CH_3)$_2$C=CH	CH_3	71		
	H_3CO(H)C=CH	CH_3	43		
	-CH=CH-(CH_2)$_3$-		74		

(Table 8 continued)

Prod. type	R^1	R^2	yield %	α / γ	% ds of 63
62a	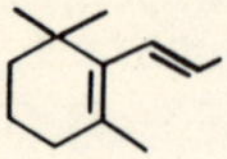	CH_3	74		
	$C_6H_5CH{=}CH-$	CH_3	77		
	C_6H_5	CH_3	77		
	C_6H_5	C_6H_5	54		
62b	$C_6H_5-CH{=}CH$	H	42	2/98	
	C_6H_5	H	63	2/98	
	p-CN-C_6H_4	H	28	10/90	
	CH_3	CH_3	43	20/80	
	$-(CH_2)_4-$		51	10/90	
	$-CH{=}CH-(CH_2)_3-$		40	10/90	
	C_6H_5	CH_3	55	2/98	
63a	C_2H_5	H	89	98/2	3 : 1
	$C_6H_5CH{=}CH$	H	85	98/2	3 : 1
	C_6H_5	H	90	90/10	1 : 1
	p-NO_2-C_6H_4	H	70	98/2	2 : 1
	C_6H_5	CH_3	75	90/10	9 : 1
	C_6H_5	C_6H_5	98	90/10	
63b	C_2H_5	H	88	98/2	7 : 1
	$C_6H_5CH{=}CH-$	H	88	98/2	
	C_6H_6	H	82	98/2	3 : 1
	CH_3	CH_3	85	98/2	
	$-CH{=}CH-(CH_2)_3-$		75	98/2	6 : 1
	C_6H_5	C_6H_5	86	73/2	

synthetic conversions of the α- and γ-products [127-130], for instance alcoholysis of 62a to acetals 64 [112,127], become more interesting by the titanium-mediated improvement of selectivity of their formation. The structures responsible for γ- and α-selectivity of the two sulfur-substituted allylic titanium reagents are tentatively assigned as 54f, 54g, 54h and 54i, respectively (see *formulas* underneath 62, 63 and 64), assuming that the metal is σ-bonded in an allylic position with respect to the site of reaction. This same assumption is also made by other authors who studied RO- and Me_3Si-substituted analogues, see the preparation of 65 [114,121] and 66 [19].

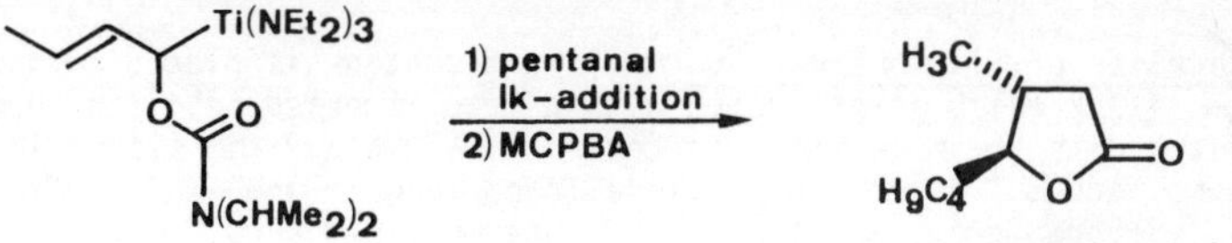

54l → 1) pentanal, lk-addition; 2) MCPBA → 65

87% overall yield

97% ds

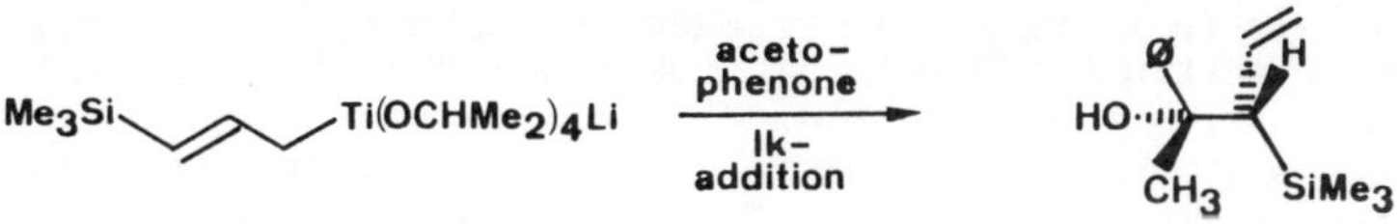

55d → acetophenone, lk-addition → 66

1-Phenyl-4-thiophenyl-1.5-pentadien-3-ol [113]

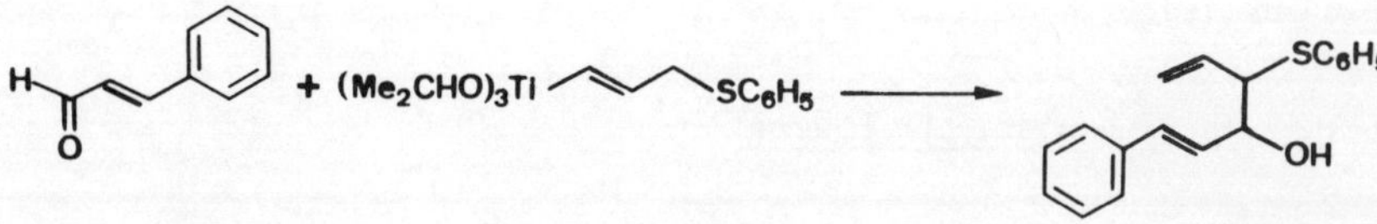

A solution of 0.75 ml (5.11 mmol) of allyl phenyl sulfide in 30 ml dry ether is combined with 0.80 ml (5.34 mmol) TMEDA and treated at -80°C with 3.3 ml (5.21 mmol) of butyllithium (1.58 M in hexane). After stirring for ca. 2 hours at temperatures between -50°C and -20°C, the yellow reaction mixture is cooled to -70°C, 2.5 ml (5.23 mmol) triisopropoxy chlorotitanium is added and the brown-red solution slowly warmed up to ca. -30°C. At -90°C 0.5 ml (3.97 mmol) of cinnamic aldehyde is injected by syringe and the temperature raised to room temperature overnight. The crude product is chromatographed (petroleum ether / ether 4:1) to furnish 980 mg (3.50 mmol, 88%) of analytically pure α-adduct; diastereomeric ratio >6:1 (by ^{13}C-NMR).

1-(3-Thiomethyl-2-propenyl)-cyclopentanol [112)]

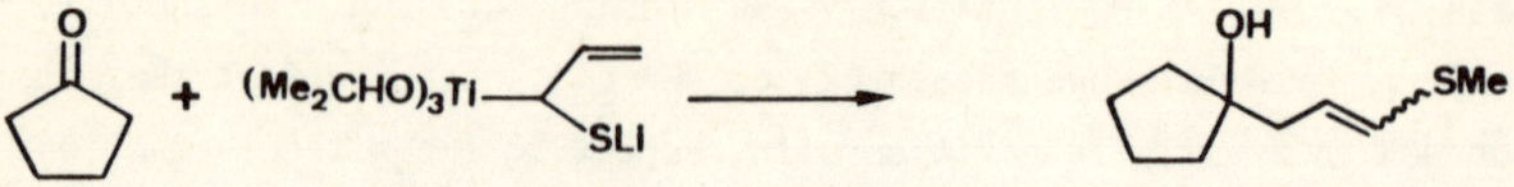

Purification of allyl mercaptan Allyl mercaptan is placed into a one-necked flask which is connected with a trap immersed into liquid nitrogen. The flask with the mercaptan is cooled to $-50^{o}C$ and the system is put under oil pump vacuum. The mercaptan is allowed to warm and condense into the liquid nitrogen trap. The mercaptan thus obtained should be used immediately. On storage and on contact with metal needles of syringes it polymerizes. It is best handled by pressing it through teflon tubes - also for avoiding annoyance by its bad odor.

Metallation of allyl mercaptan and reaction with cyclopentanone: Freshly purified allylmercaptan (1.90 g, 25.6 mmol) is dissolved in 160 ml anhydrous ether and 7.7 ml (51.4 mmol) of TMEDA. The solution is treated with 33 ml (51.2 mmol) of butyllithium (1.55 M in hexane) at $0^{o}C$. After stirring for ca. 5 hours, 12 ml (26 mmol) of triisopropoxy chlorotitanium is added to the yellow reaction mixture at $-80^{o}C$. The color changes immediately to dark brown-red. The solution is allowed to reach ca. $-30^{o}C$, cooled to $-80^{o}C$ and 1.8 ml (20.3 mmol) of cyclopentanone is added. After warming up to room temperature overnight, methyl iodide (2 ml, 32.1 mmol) is added and stirring continued for 3 - 4 hours. Workup consists of hydrolysis with saturated neutral aqueous KF-solution, extraction with ether and washing the combined organic layers with brine. Removal of the solvent affords crude product which is subjected to purification by flash-chromatography (pentane / ether 5:2). Yield: 2.61 g (15.2 mmol, 85%). In the ^{1}H-NMR only γ-adduct can be detected.

2-Methoxy-2-oxaspiro[4.4]nonan [112)]

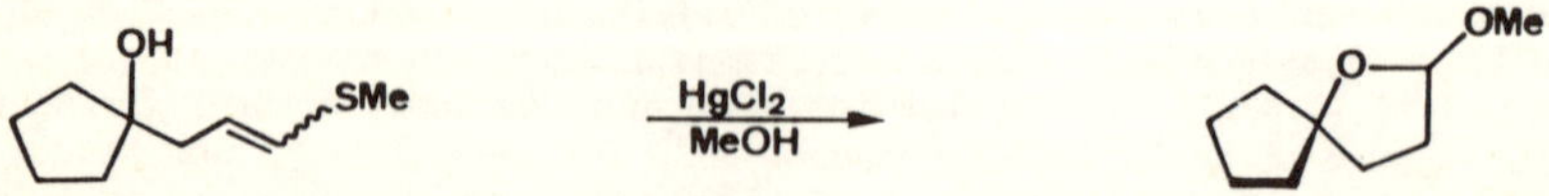

At room temperature, 525 mg (3.05 mmol) of 1-(3-thiomethyl-2-propenyl)cyclopentanol is combined with 14.9 g (10.7 mmol) $HgCl_2$ (20% in anhydrous MeOH).

After stirring for ca. 20 hours, the reaction mixture is poured on light petroleum ether / water (1:2) and the organic layer is extracted twice with petroleum ether. The solvent is removed and the residue purified by flash-chromatography (petroleum ether / ether 19:1). Yield: 0.33 g (70%).

4) Allyl-titanocene Derivatives

There are titanocenes bearing allylic groups with either Ti(III) (π-allyl, η^3, see 56) or Ti(IV) (σ-allyl, η^1, see 57). Their use in organic synthesis has been reviewed just recently [99]. The Ti(III) derivatives are formed from dichloro-titanocene (1a) and two equiv. of allylic *Grignard* reagents, first equation in Scheme 20, or from 1a, a conjugated diene, such as isoprene, and

Scheme 20

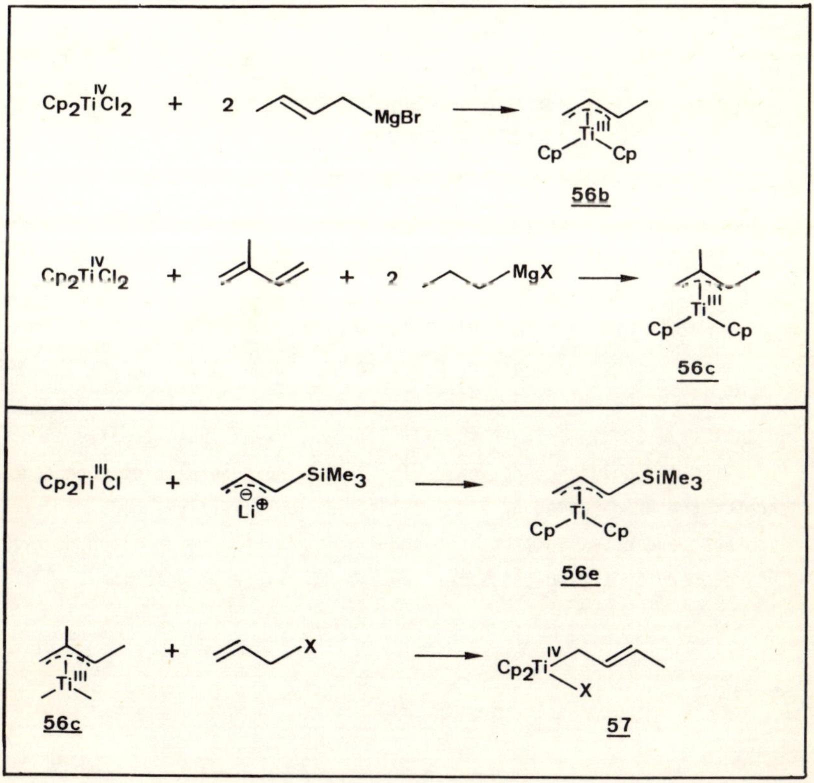

$Cp_2Ti^{IV}Cl_2$ + 2 (2-butenyl)MgBr → 56b (η³-allyl $Ti^{III}Cp_2$)

$Cp_2Ti^{IV}Cl_2$ + isoprene + 2 (propyl)MgX → 56c (η³-allyl $Ti^{III}Cp_2$)

$Cp_2Ti^{III}Cl$ + (allyl anion, $Li^{\oplus}$)-$SiMe_3$ → 56e (η³-allyl-$SiMe_3$ $TiCp_2$)

56c + allyl-X → 57 (Cp_2Ti^{IV}(σ-allyl)X)

propyl *Grignard* reagent, second equation in Scheme 20. In both cases, the organomagnesium compound acts as a reducing reagent. As shown in the third equation of Scheme 20, one can also start from monochloro-titanocene [Ti(III)] and an allylic lithium compound. This appears to be more economical with respect to the organic part of the reagent. Finally, allyl halides can directly be converted to halo-allyl-titanocenes [Ti(IV)] by the "oxidative trans-allylation", see the fourth equation of Scheme 20. The regioselective reactions of these titanocene derivatives with aldehydes occur with excellent diastereoselectivity with relative topicity *lk* [40], see the product 67 obtained with propanal.

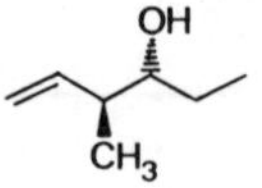

OH
CH_3
67

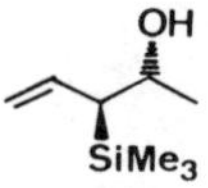

OH
$SiMe_3$
68

93 % ds with 56b
64 % ds with 57a, X = Cl
96 % ds with 57b, X = Br

The formation of the trimethylsilylated homoallylic alcohol 68 follows the same steric course [19], although its relative topicity is specified *ul*, due to change in the priority order. *Peterson* olefination, i.e. elimination of silanol from 68 gives either the *cis*-1.3-pentadiene (base treatment) or the *trans*-isomer (acidic conditions). The following general procedure demonstrates the preparation of allylic titanocenes directly from butadiene or isoprene, and gives instructions about the addition to an aldehyde or ketone. The recovery of titanocene after the reaction is also described.

Homoallylic alcohols from butadienes and aldehydes through titanocene derivatives [117)]

Cp_2TiCl_2 + (R-butadiene) $\xrightarrow{2C_3H_7MgBr}$ Cp_2Ti—(allyl)—R'

1a → 56b or c

56b or c + $R^2R^3C{=}O$ ⟶ $CH_2{=}C(R^1)CH_2C(OH)R^2R^3$

(with recovery of Cp_2TiCl_2)

Propylmagnesiumbromide(16.6 ml, 1 M in ether) is added under argon to a stirred solution of 1a (2.1 g, 8.4 mmol) and butadiene (0.86 g, 15.9 mmol) or isoprene (0.60 g, 8.8 mmol) in ether (40 ml). After 30 min. at 20°C, 0.9 to 1.0 equiv. of an aldehyde or ketone is added by syringe, and the solution is stirred for 1 hour at room temperature. The mixture is quenched with 4 N HCl (10 ml), stirred for 30 min. under argon, and with stirring air is passed through the resulting two layers of ether (pale red color) and water (green) for 15 min. During this operation the green color fades out and 1.7 g of red crystals of 1a precipitate and are collected by filtration. The layers are separated, and the aqueous layer is extracted with ether. The combined ether layers are dried ($MgSO_4$), and the solvent is evaporated under reduced pressure to afford a mixture of 1a and an oil. Pentane is added to the mixture and an additional 100 mg 1a is isolated by filtration. Chromatography of the pentane solution on silica gel affords the product as an oil.

G) Carbometallations of C,C-Multiple Bonds

Carbometallation is the process in which an organometallic compound adds to a C,C double or triple bond. It is thus analogous to the more familiar addition of RM to the polar C,O or C,N double or triple bond, see the first three equations of Scheme 21. With unsymmetrically substituted multiple bonds, the addition is called "*Markownikow* addition" if the metal becomes bonded to the less substituted carbon, cf. fourth equation. Of course, the carbometallation which is applied on the largest scale is polymerization of

Scheme 21

$$\ce{>C=O ->[RM] R-C(<)-O-M}$$

$$\ce{>C=C< ->[RM] R-C(<)-C(<)-M}$$

$$\ce{-C#C- ->[RM] R(-)C=C(-)M -> R(-)C=C(-)E}$$

cis with Cu, Al
trans with Li, Mg, Zn

$$R^1-C\equiv C-H \xrightarrow[\text{«Markownikow»}]{R^2M} (R^1)(R^2)C=C(H)(M) \longrightarrow (R^1)(R^2)C=C(H)(E)$$

$$n\,R-CH=CH_2 \xrightarrow{R^1M} R^1(-\overset{R}{\overset{|}{C}}H-CH_2-)_n-M$$

olefins by organometallic catalysts, including the *Ziegler-Natta* process, see fifth equation of Scheme 21. Attempts to cleanly perform mono-carbometallations are thus sometimes hampered by competing polymerization.

The most useful laboratoy applications appear to be the additions to acetylenes, because they can be used for the stereospecific synthesis of alkenyl derivatives. A recent review article on this topic [131)] and the chapter by *J.F. Normant* in the present volume give all the relevant references and a competent discussion of the various modifications. Although carbometallations can be done with a large number of different metals, the copper derivatives are the most versatile ones and have been studied most extensively [131] for additions to acetylenes. Ironically, one of the most useful components for natural product synthesis, the methyl group, can not be added to these substrates satisfactorily [131]. Once again, titanium and zirconium derivatives come to the rescue: regioselective and *cis*-diastereoselective titanium- and especially zirconium-catalyzed carboaluminations are applicable to "methyl-metallations" (see reference [132] for an extensive survey). The vinyl alanes 69 thus obtained, usually with >95%, often with >98% E-preference, can be converted to many useful products by reaction with

$$R-C\equiv C-H \quad + \quad (CH_3)_3Al \xrightarrow[\underline{2a}]{Cl_2ZrCp_2} \text{(R)(H}_3\text{C)C=CH(AlMe}_2\text{)} \quad \underline{69}$$

$$\xrightarrow[\text{phile}]{\text{Electro-}} \text{(R)(H}_3\text{C)C=CH(E)}$$

$\underline{70}$, E = H, D, I, HgCl, BR_2, COOH, COOEt, CH_2OH, CH_2OMe, CH(OH)Et, CH_2CH_2OH, $CH_2CH_2COCH_3$.

various electrophiles, see the list underneath *formula* 70. A procedure for the *cis*-addition of CH_3 and I to a terminal triple bond follows.

Detailed procedures describing safe techniques of handling alkyl aluminum derivatives may be obtained from the Schering AG (Bergkamen) [133].

(E)-1-Iodo-2.6-dimethyl-1.5-heptadiene [131)]

$$(H_3C)_2C{=}CH{-}CH_2{-}CH_2{-}C{\equiv}CH + (H_3C)_3Al \xrightarrow[\underline{2a}]{Cp_2ZrCl_2}$$

$$(H_3C)_2C{=}CH{-}CH_2{-}CH_2{-}C(CH_3){=}CH{-}Al(CH_3)_2 \xrightarrow{I_2} (H_3C)_2C{=}CH{-}CH_2{-}CH_2{-}C(CH_3){=}CH{-}I$$

Dichloro-bis(η^5-cyclopentadienyl)zirconium (2.92 g, 10 mmol) (2a) is suspended in 1.2-dichloroethane (25 ml) and trimethylalane (1.44 g, 1.92 ml, 20 mmol) (pyrophoric!) [133)] added under nitrogen at room temperature. The dichloro-bis(η -cyclopentadienyl)zirconium dissolves within 10 - 15 min. to produce a yellow solution, to which is added 6-methyl-5-hepten-1-yne (1.08 g, 10 mmol). After stirring for 24 hours at room temperature, iodine (3.04 g, 12 mmol), dissolved in THF (15 ml), is added dropwise at 0°C. After the iodine color has disappeared, the reaction mixture is quenched with water, and ether is added. The organic layer is separated, dried, and distilled to give the title compound; yield: 1.80 g, 72%; b.p. 54 - 55°C/0.55 Torr.

H) Oxidative and Reductive Coupling of Carbonyl Compounds - Redox Umpolung with Titanium Derivatives

With the exception of some allylic Ti-reagents corresponding to d^3-synthons (see 54f, 54g, and 54e, Scheme 19 and Table 8), the conversions of carbonyl

$$R{-}\overset{O}{\overset{\|}{C}}{}^{a1}{-}CH_2^{\,d2}{-}CH_2^{\,a3}{-}CH_3^{\,d4}$$

71

acceptor reactivity (a)

a^1: carbonyl additions

a^3: *Michael* additions

donor reactivity (d)

d^2: enolates as nucleophiles

d^4: dienolates with ω-reactivity

derivatives discussed in the previous sections all used normal reactivity, see 71.

The oxidation potential of Ti(IV) and the reducing properties of lower-valent titanium have been used for coupling of carbonyl compounds. Couplings between atoms of the same reactivity such as the pinacolization [134] and the acyloin condensation [135] or oxidation of enolate derivatives [136-139] furnish products with 1.2n-distances between functional groups, see 72, 73, and 74 in Scheme 22.

Scheme 22

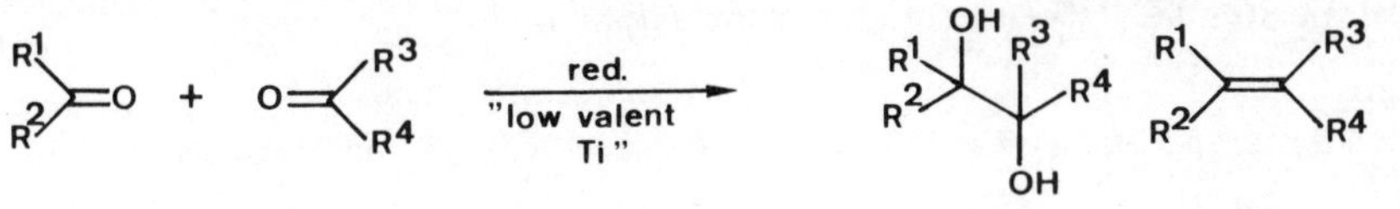

R1(R2)C=O + O=C(R3)R4 —red. "low valent Ti"→ 72 + 73

72 73

preparation of low-valent titanium reagents

$TiCl_3$/LAH [140-145]

$TiCl_3$/K,Li [146-149]

$TiCl_3$/Mg [150]

$CiCl_3$/Zn/Cu [151,152]

$TiCl_4$/Zn [153-155]

$TiCl_4$/Mg/Hg [156]

$CpTiCl_3$/LAH [156]

Al/$AlCl_3$/$TiCl_4$/hexamethylbenzene [148,156]

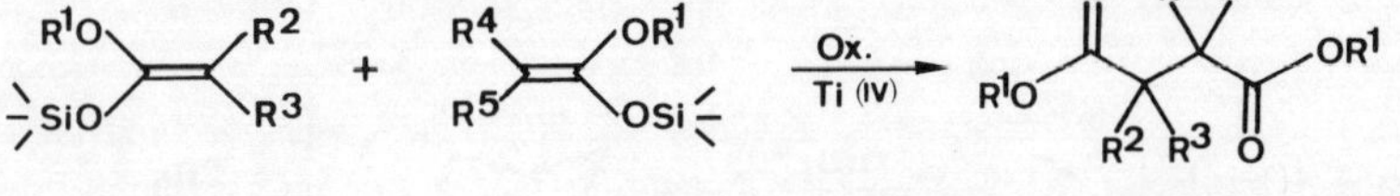

R1O(≡SiO)C=CR2R3 + R4R5C=C(OR1)(OSi≡) —Ox. Ti (IV)→ 74

74

Especially useful is the coupling of ketones and aldehydes directly to olefins 73. Reviews on this topic by two of the discoverers [68,157] have appeared. This coupling is feasible even with formation of highly hindered olefins [154,155]. If two carbonyl compounds of sufficiently different redox potential are used [158], or if one component is cheap and can be used in excess [156], cross-coupling is possible. Intramolecular couplings with formation of one [156,159] or two [149] rings have been described. Some representative and/or especially impressive examples are listed in Table 9, together with references. It is of course possible to do the two steps, i.e. pinacol coupling to 72 and reductive deoxygenation to 73 separately. This means that glycols from sources other than the "Ti(0)"-coupling can be converted to olefins [140] with "low-valent" titanium.

The preparation of these so-called low-valent Ti-reducing reagents is crucial for the success of the reaction. The quality of these reagents is decisive with respect to reaction times, selectivity diol/olefin and yield. Numerous recipes have been recommended by the different authors in the field, see Scheme 22. Probably, zero-valent titanium [160] is the actual reagent with which the reaction begins; depending on the reducing reagents with which this mostly black, pyrophoric (!) material is made, other metals and/or metal salts are present. The driving force is the high affinity of titanium for oxygen, which eventually leads to TiO_2.

There are recent applications of the method in which stereoselectivity was observed, see 75 [152,161] and 76 [162]. Finally, vinylogous extensions of the

2 [aryl ketone, R^1, R^2] → "Ti(O)" ($TiCl_4$/Zn/Py) → 75 — 73:27 to 89:11 pref. of Z-isomer

2 [(R)-3-methylcyclohexanone] → "Ti(O)" ($TiCl_4$/Mg/Hg) → 76 (53%)

Table 9. Coupling of carbonyl compounds with "low-valent titanium" to 1.2-diols or to olefins. Two examples of deoxygenation of glycols to olefins are also given.

starting material	product	yield	conditions	ref.
2 Benzaldehyde	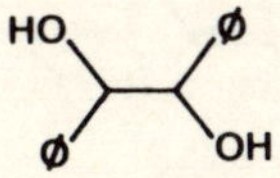 HO, Ø, Ø, OH	98 %	$TiCl_4$/Zn THF, 0°C	153)
$C_6H_5-CO-COOH$ + acetone	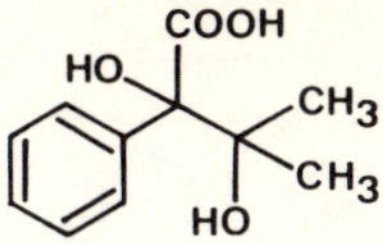 COOH, HO, CH_3, CH_3, HO	85%	2 $TiCl_3$	158)
 O + CH_3CHO	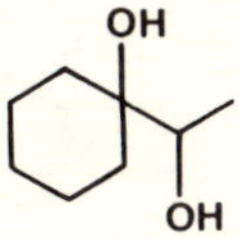 OH, OH	72 %	$CpTiCl_3$/ LAH	156)
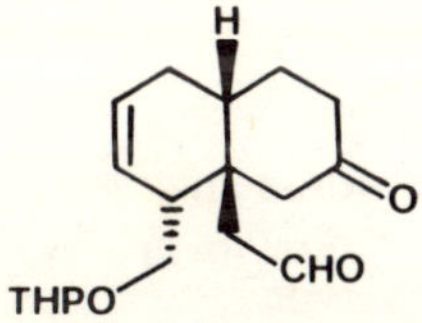 H, O, THPO, CHO	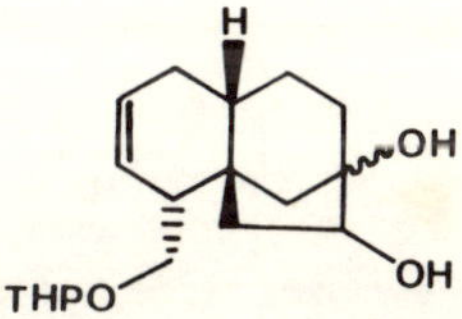 H, OH, THPO, OH	55 %	$CpTiCl_3$/ LAH	156)
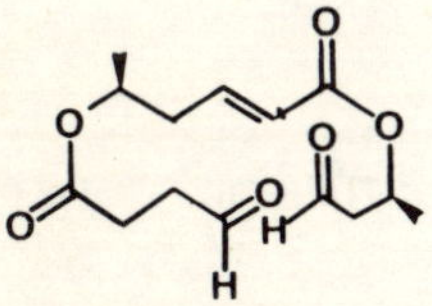 O, O, O, O, O, H, H (→(S,S)-(+)-Grahamimycin A1)	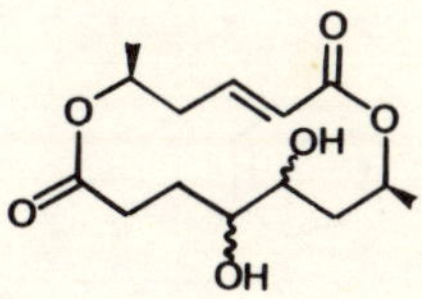 O, O, O, OH, O, OH	35 %	$TiCl_4$ / Zn-Cu THF 5h reflux	159)
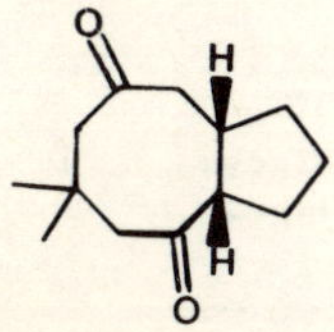 O, H, H, O	 HO, H, HO, H	—	$TiCl_3$/K 5min THF, reflux	149)

(Table 9 continued)

starting material	product	yield	conditions	ref.
2 Benzaldehyde	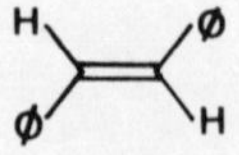 (>99%E)	98%	$TiCl_4$/Zn dioxane, reflux	153)
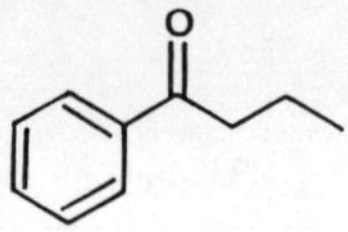	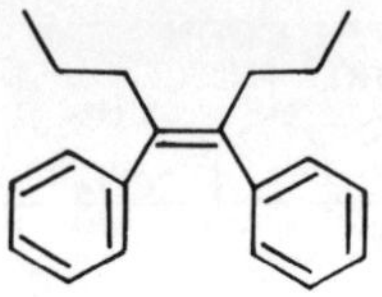	50 %	$TiCl_4$/Zn/ Py THF, reflux	161)
1–Adamantyl–methyl ketone	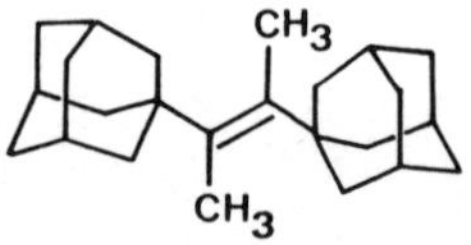 CH_3 CH_3	15%	$TiCl_4$/Zn/Py 3d, THF reflux	154)
retinal	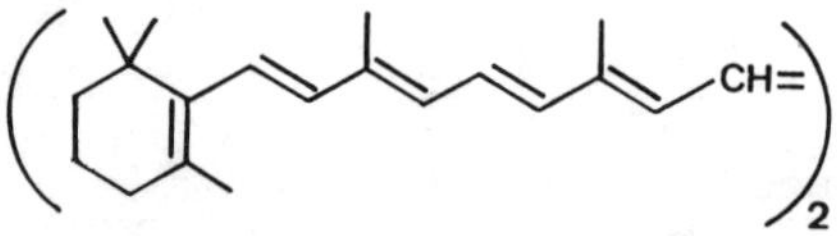 (CH=)$_2$	85%	$TiCl_3$/LAH THF, reflux	140)
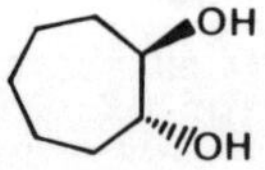 OH OH	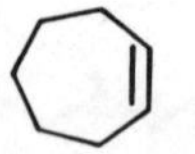	55%	$TiCl_3$/K THF, reflux	146)
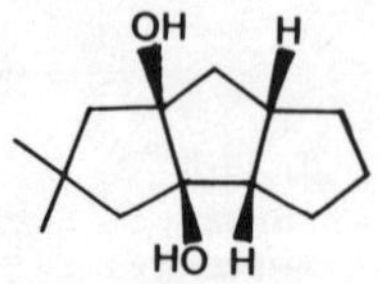 OH H HO H	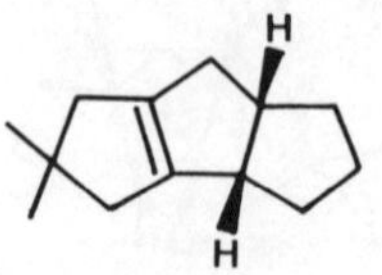 H H	—	$TiCl_3$/K 6 days reflux THF	149)

method are known, see 77 [163] and 78 [164], and similar conditions have been used for a modification of the *Wittig* reaction, see 79, employing dibromomethane as methylene source [165].

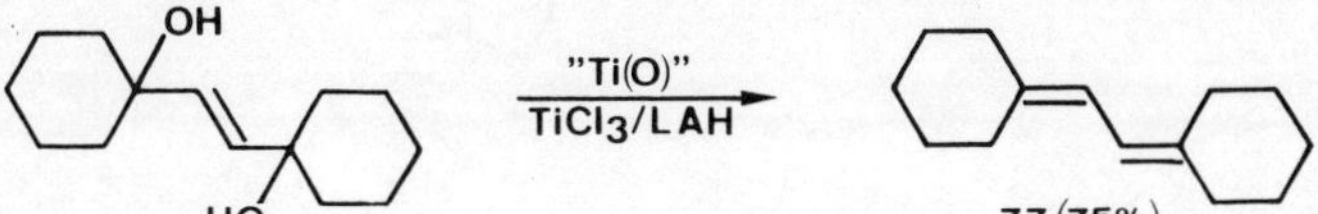

"Ti(O)" $TiCl_3$/LAH

77 (75%)

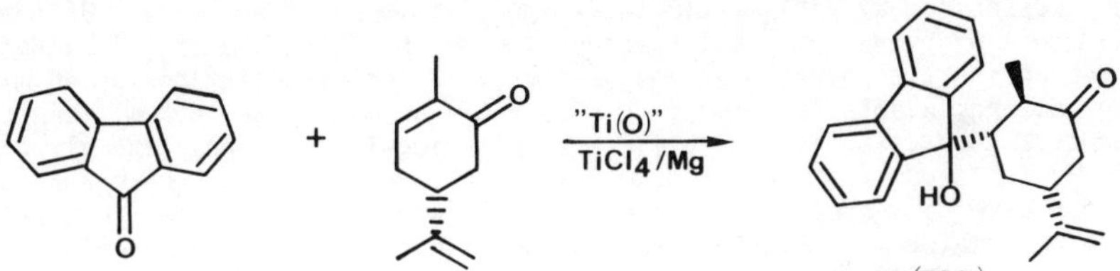

"Ti(O)" $TiCl_4$/Mg

78 (70%)

CH_2Br_2 $TiCl_4$/Zn

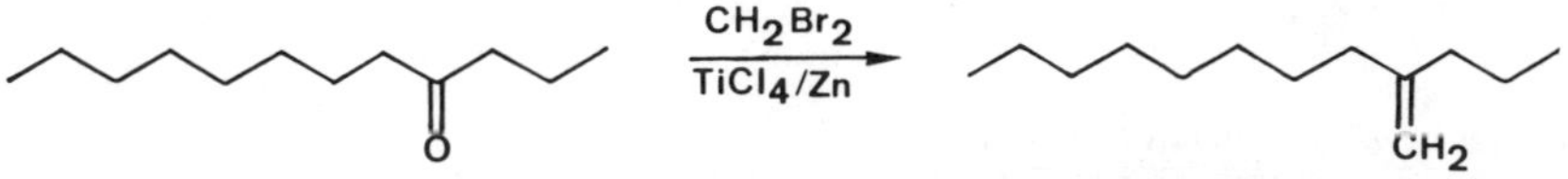

79 (89%)

The oxidation of enolates with coupling by $TiCl_4$ was discovered as an accident during attempted titanium tetrachloride catalyzed reactions with silyl enol ethers of α-branched esters (ketene acetal derivatives). So far, there are only few examples reported in the literature [67,166,167], see for instance the formation of 80 [166] and the procedure below.

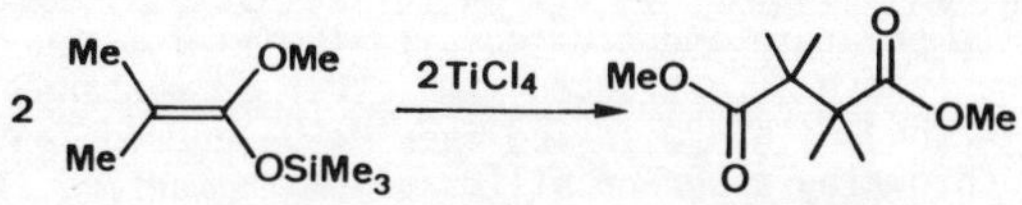

2 Me Me OMe $OSiMe_3$ → 2 $TiCl_4$ → MeO OMe

80 (80%)

Adamantylidene-adamantane [147)]

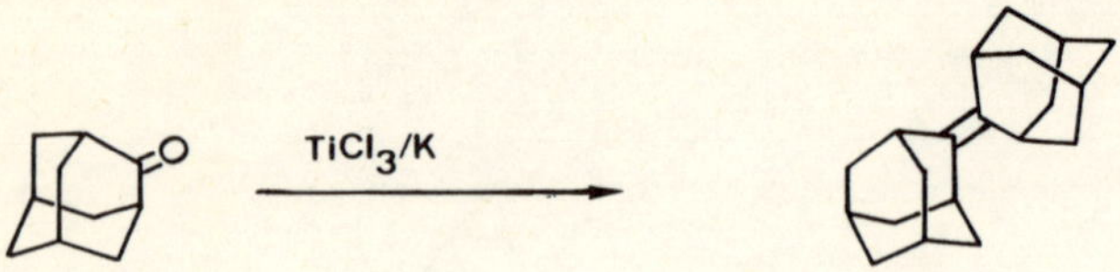

To a suspension of $TiCl_3$ (2.15 g, 14 mmol) in 75 ml of dry THF, 1.92 g (49 mmol) of potassium metal (washed with hexane to remove oil) is added under an inert atmosphere at room temperature. After refluxing for 40 min. and cooling, a solution of 525 mg (3.5 mmol) adamantanone in 5 ml THF is added. The reaction is complete after refluxing for 16 hours. The black slurry is vacuum filtered (medium frit) under an inert atmosphere and the filter cake is washed with hexane. The filtrate is concentrated in a rotatory evaporator to yield the crude product which is purified by column chromatography to yield 91% adamantylidene-adamantane of m.p. 183-185°C. Caution: The black filter cake is pyrophoric when exposed to air and has to be quenched carefully by dropwise addition of methanol under an inert atmosphere.

Tetracyclopropyl-ethylene [154)]

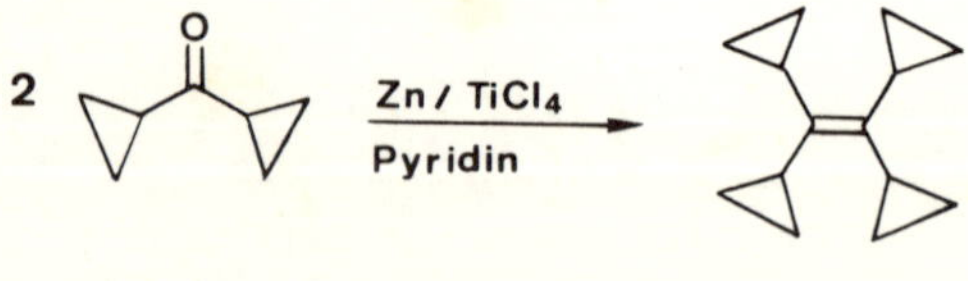

Under an inert atmosphere, neat titaniumtetrachloride (8.2 ml, 75 mmol) is added dropwise to 200 ml anhydrous THF with stirring and cooling in an ice-bath. Zinc powder (10 g, 150 mg-atom) is then added in small portions. After addition of pyridine (5 ml) and dicyclopropyl ketone (7.7 g, 70 mmol), the reaction mixture is heated under reflux for 20 hours. The suspension is cooled to room temperature and treated with 150 ml 10% aqueous K_2CO_3 solution. The aqueous phase is extracted with ether or pentane, the combined organic layers washed with H_2O, dried over $MgSO_4$ and concentrated *in vacuo*. The residue is chromatographed on silica gel with pentane. Yield: 25%, b.p. 70-72°C (0.5 mm Hg).

1-(2-Hydroxy-2-propyl)-cycloheptanol [156)]

A slurry of 70-80 mesh magnesium (1.92 g, 80 mg-atom) and mercuric chloride (0.60 g, 2.21 mmol) in 10 ml THF is stirred under argon for 15 min. at room temperature. The turbid supernatant liquid is removed and the residue is washed with three portions of THF. After addition of 30 ml THF to the dull grey amalgam, the suspension is cooled to -10°C and treated dropwise with titanium tetrachloride (4.4 ml, 40 mmol). A solution of 2.24 g (20 mmol) cycloheptanone and 4.64 g (80 mmol) acetone diluted with 10 ml THF is then added to the yellow-green mixture. The color changes to purple and stirring is continued for 1.5 hours at 0°C. The reaction is quenched with 2 ml saturated aqueous K_2CO_3 solution and stirred for another 15 min. After addition of 100 ml ether and filtration through celite, the filtrate is washed with saturated NaCl solution, dried ($MgSO_4$) and concentrated to afford 5.61 g of a viscous oil. Column chromatography on silica gel (elution with ether/petroleum ether 2:3) gives 2.57 g (75% yield) of 1-(2-hydroxy-2-propyl)-cycloheptanol as colorless crystals; m.p. 51-52°C.

Oxidative dimerization of the O-(trimethylsilyl)ketene-acetal 81 [167)]

81

82

Lithium diisopropyl amide (11.5 ml, 0.913 M in THF) is cooled to -78°C under argon atmosphere and (2S,5S)-5-methyl-2-(*t*-butyl)-1.3-dioxolan-4-one [37) (1.6 g, 10 mmol) is slowly added. After stirring for 30 min., the solution is quenched by the addition of trimethyl chlorosilane. Within ca. 1 hour the temperature is raised to room temperature and stirring is continued for an additional hour. The solvent is removed evaporatively. The residue, which is fairly pure apart from the admixture of lithiumchloride, is directly used without further purification, owing to the high sensitivity of 81 to hydrolysis as well as to thermolysis.

To a solution of 5.7 g (30 mmol) titanium tetrachloride in 50 ml CH_2Cl_2 and 2.7 g (30 mmol) trioxane (which proved to be essential for the dimerisation, although originally meant to achieve hydroxymethylation of the ketene acetal) is added at -78°C the crude trimethylsilyl ketene acetal (ca. 10 mmol, cf. above) dissolved in 10 ml CH_2Cl_2. The resulting deep green reaction mixture is stirred for 3 hours at -78°C and then hydrolyzed by the addition of 100 ml H_2O. Extraction of the purple solution with ether, washing of the combined organic layers with H_2O, drying ($MgSO_4$) and removal of the solvent affords after distillation (120°C, 0.03 mm Hg) 1.2 g (3.8 mmol, 76%) of the dimer in form of colorless crystals; m.p. 123-125°C (from ether / pentane 1:1), $[\alpha]^{25}_D$ = +25.9° (1.95, $CHCl_3$). For the *tert*-butyl and also for the methyl group of compound 82 only one signal is observed in the ^{1}H-NMR, which proves the existence of a C_2-axis.

IV FUNCTIONALIZATIONS AND FUNCTIONAL GROUP INTERCONVERSIONS

The largest part of the present review deals with carbon carbon bond forming reactions which are mediated by titanium or zirconium derivatives. It was demonstrated in the previous chapter III that many classical processes can be made highly selective by turning from derivatives of the main-group metals to those of the early transition metals titanium and zirconium.

This somewhat shorter chapter will now highlight applications of the same elements which commence without elaboration of the carbon skeleton, but which rather change its functionality. It will become evident, that again great improvements in selectivity are possible with derivatives of these two metals. Also, this chapter will demonstrate again the great potential of the simple inexpensive and non-toxic titanates, zirconates and amide analogues in synthesis.

A) Reductions and Oxidations Involving Carbonyl Derivatives

1) *Meerwein-Ponndorf-Verley* Reductions and *Oppenauer* Oxidations with Zirconates

It is surprising, how few applications of these reactions are found in modern synthetic sequences. The complex hydrides of boron [168] and of aluminum [169] with all their modifications, and aluminum hydrides such as DIBAH have become the predominant reagents for carbonyl reduction. For oxidations, the various modifications of chromium(VI) reagents and, more recently, of the *Kornblum* reagent [170] prevail [*Pfitzer-Moffat* [171], *Corey-Kim* [172], *Swern* [173], etc.]

If applied catalytically, the *Meerwein-Ponndorf*-type reductions, for which a secondary alcohol such as isopropanol is used, or the *Oppenauer* oxidations, which can be carried out with acetone or chloral as oxidants, apapear to be more economical, see the accompanying Table 10. Furthermore,

Table 10. Prices per equiv. of hydride or per reduction or oxidation equiv. for some common reducing and oxidizing reagents (from a 1983 catalogue).

Reagent	Mol.Wt.	Price in US $ (largest packing size)	
$LiAlH_4$	38	4.--/$H^{\ominus}$	
$NaBH_4$	38	2.--/$H^{\ominus}$	
DIBAH	142	23.--/$H^{\ominus}$	
$LiBHEt_3$	106	100.--/$H^{\ominus}$	(Superhydride®)
$NaBHEt_3$	122	400.--/$H^{\ominus}$	
$LiBH(iBut)_3$	190	100.--/$H^{\ominus}$	(Li-Selectride®)
$NaCNBH_3$	63	10.--/$H^{\ominus}$	
$LiAlH(O\text{-}t\text{-}C_4H_9)_3$	254	170.--/$H^{\ominus}$	
$NaAlH_2(OCH_2CH_2OCH_3)_2$	202	10.--/$H^{\ominus}$ equiv.	(Vitrid®)
$(CH_3)_2CHOH$	60	--.15/$H^{\ominus}$ equiv.	
$(CH_3)_2CO$	58	--.15/oxid. equiv.	
CCl_3CHO	147	4.50/oxid. equiv.	

for many purposes, these alcohol/carbonyl hydride donors and acceptors are extremely selective in the presence of further functional groups in the substrate. Thus, imino, nitro, cyano, ester, halo, sulfoxide and sulfone groups do not normally interfere with these reactions.

When *Meerwein et al.* first published their experiences with reductions of carbonyl groups by secondary alcohols in the presence of metal alkoxides, they strongly recommended [174)] the use of aluminum alkoxides as catalysts, but they stated that zirconium and tin alkoxides are almost as good. In the course of investigations of organozirconium derivatives as nucleophiles it was found [8,27,175)] that hydride can be transferred from the alkoxide group on the zirconium with a rate comparable to that of the R-group from zirconium, see Scheme 23. This can be avoided by using

Scheme 23

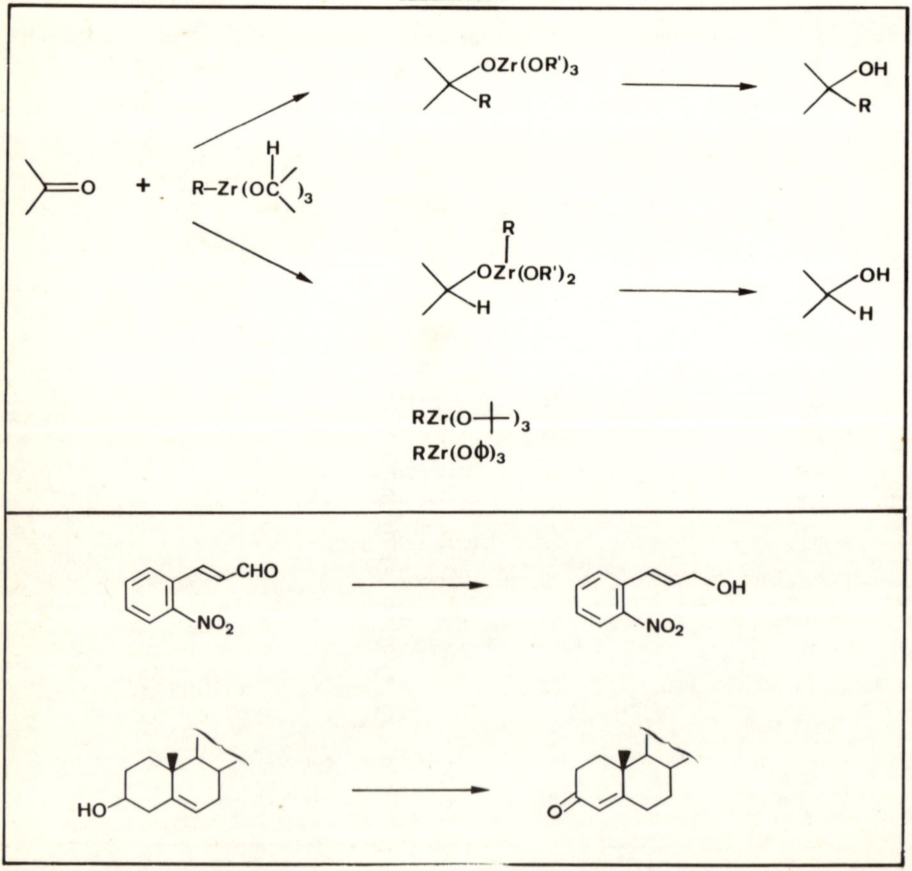

$R{-}Zr(OCH\langle)_3$; $OZr(OR')_3$ / R → OH / R ; R–$OZr(OR')_2$ / H → OH / H

$RZr(O{-}tBu)_3$

$RZr(O\phi)_3$

t-butoxy [23,111] or phenoxy groups on the zirconium [27]. These observations led to a reinvestigation [175] of zirconates as catalysts in reduction/oxidations through hydride transfer from and to carbon. Some results of reductions of aldehydes and ketones are given in Table 11, for oxidations see Table 12, Scheme 23, and the procedures below. It appears that the zirconate-catalyzed reactions may show high functional group selectivity; in contrast to aluminum alkoxides, which are often employed in stoichiometric amounts, the zirconates catalyze rapid conversions at lower concentrations.

Table 11. Reactivities in the zirconate-catalyzed *Meerwein-Ponndorf-Verley* reduction as a function of the p-substituents of benzaldehyde.

p-R-C_6H_4-CHO → (Zr(O-iPr)$_4$, 10 mol %; Isopropanol; RT) → p-R-C_6H_4-CH_2OH

R	time (h)	benzyl alcohol (%)
NO_2	8	~100
C≡N	24	> 90
Cl	24	> 90
H	24	~ 60
OCH_3	24	~ 10
$N(CH_3)_2$	24	<1

Table 12. Reactivities in the zirconate-catalyzed *Oppenauer* oxidation as a function of the oxidizing reagent (top part) and the substrate (lower part).

1-phenylpropan-1-ol → propiophenone: $Zr(O\text{-}tBu)_4$, oxidizing reagent, 6h, rt.

reagent	% conversion
acetone	12
cyclohexanone	64
Cl_3C–CHO	92

1.2 eq. Chloral
5 mol-% $Zr(O\text{-}tBu)_4$
rt.

1-phenylethanol (CH_3) → acetophenone (CH_3): 5min.

menthol (HO) → menthone (O): 24 h

4-Nitrobenzyl alcohol [175)]

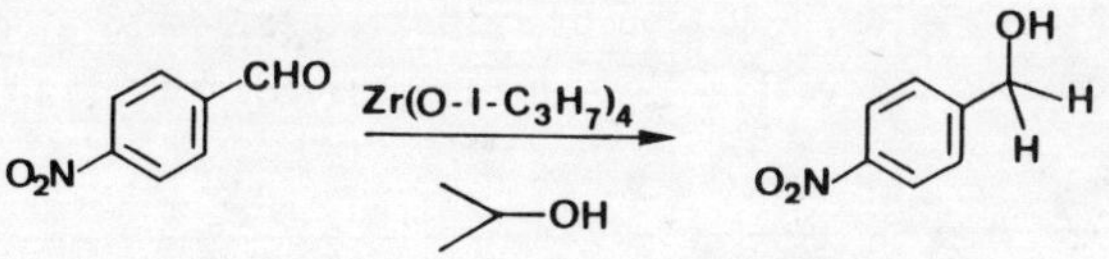

In a dry flask 390 mg (1 mmol) of $Zr(O\text{-}i\text{-}C_3H_7)_4$ dissolved in 70 ml of anhydrous 2-propanol is combined with 1.5 g (9.93 mmol) of 4-nitro-benzaldehyde. After stirring for 24 hours, gas chromatographic analysis showed 94% conversion. The reaction mixture is then hydrolyzed with 50 ml 2 M aqueous HCl. Extraction with ether yields on evaporation of the solvent 1.25 g (8.16 mmol, 82%) of 4-nitro-benzyl alcohol.

Zirconate-catalyzed *Oppenauer* oxidation [175,177)]

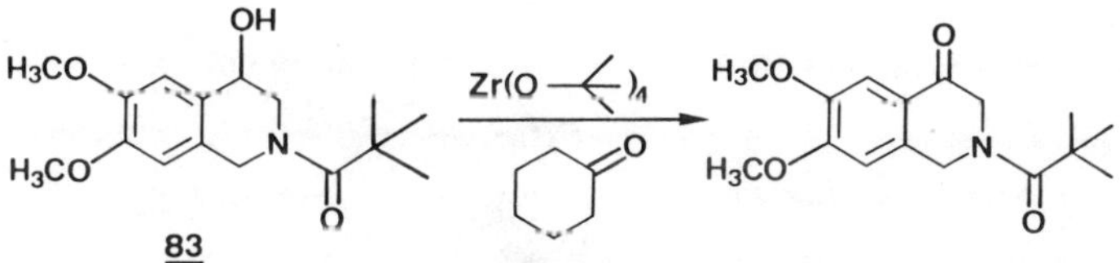

Tetrahydroisoquinoline derivative 83 (1.00 g, 3.41 mmol) is dissolved in 10 ml of cyclohexanone and a catalytic amount of $Zr(O\text{-}t\text{-}C_4H_9)_4$ (ca. 0.5 ml, 1.25 mmol) is added. After stirring at 80°C for 2 hours, the reaction is hydrolyzed with diluted HCl and extracted with ether. Cyclohexanone is removed under oil pump vacuum and the residue is purified by flash-chromatography (CH_2Cl_2/EtOAc 1:1). 450 mg (1.53 mmol, 45%) of the oxidation product is obtained, which can be crystallized from CH_2Cl_2/EtOAc 1:1 to furnish slightly yellow crystals. m.p. 129 - 130°C.

2) Titanium-mediated Reductions

There are some remarcably selective reductions of carbonyl derivatives and analogues by low-valent titanium which are worth mentioning here. Thus, addition of carboxylic acids, acid chlorides, carboxamides, oximes, nitro compounds and sulfoxides to a mixture obtained from $TiCl_4$ and

$NaBH_4$ causes smooth reduction to primary alcohols, amines, and thioethers, respectively [178], see the upper part of Table 13. Also, the Ti(III) derivative from dichloro-titanocene and isopropyl *Grignard* reagent catalyzes the reduction of carboxylic acids to aldehydes by the same magnesium derivative [179], bottom part of Table 13. A catalytic

Table 13. Reductions of carboxylic acids, amides, oximes, nitro groups (see procedure) and of sulfoxides (reverse addition) by a reagent from 1 equiv. $TiCl_4$ and 2 equiv. $NaBH_4$ in dimethoxyethane [178], entries 1 - 7. Reduction of carboxylic acids to aldehydes by $(CH_3)_2CHMgBr$ and cat. amounts of Cp_2TiCl_2 [179], entries 8 - 10.

Entry	Starting material	Product	Yield [%]
1	mandelic acid	C_6H_5-CH(OH)-CH_2OH	60
2	α-phenoxy-propanoic acid	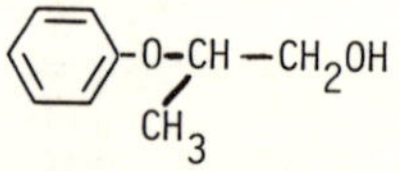 C_6H_5-O-CH(CH_3)-CH_2OH	95
3	N-benzyl-pyrrolidone	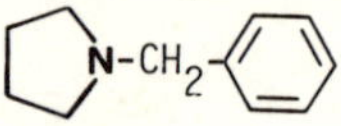 N-benzylpyrrolidine (pyrrolidine N-CH_2-C_6H_5)	93
4	o-iodo-diethylbenzamide	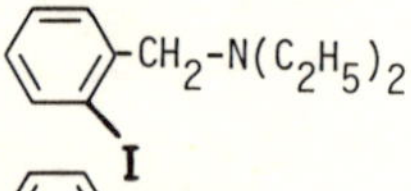 o-I-C_6H_4-CH_2-N$(C_2H_5)_2$	95
5	oxime of o-nitrobenzaldehyde	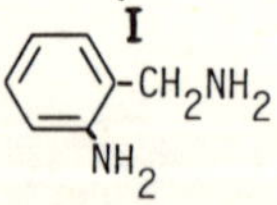 o-NH_2-C_6H_4-CH_2NH_2	82
6	methyl-(2-furylmethyl)sulfoxide	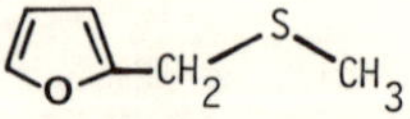 2-furyl-CH_2-S-CH_3	91
7	methyl-(2-pyridyl)sulfoxide	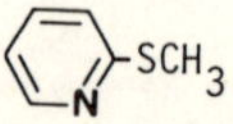 2-pyridyl-SCH_3	89
8	n-C_6H_{13}COOH	n-C_6H_{13}CHO	73
9	$C_6H_5CH_2COOH$	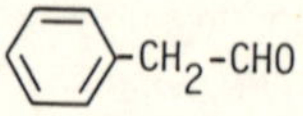 C_6H_5-CH_2-CHO	48
10	benzoic acid	benzaldehyde	55

cycle was proposed [179], in which Cp_2TiH is the actual reducing reagent, and the aldehyde is thought to be protected as $RCH(OMgX)_2$ from further reduction.

General Procedures for Reductions with $TiCl_4/NaBH_4$ [178]; (cf. entries 1 - 7 of Table 13).

Stoichiometry: 1 RCOOH / 1.1 $TiCl_4$ / 3.3 $NaBH_4$; 1 RCOCl / 1.1 $TiCl_4$ / 2.2 $NaBH_4$; 1 $RCONR_2$ / 1.5 $TiCl_4$ / 3 $NaBH_4$; 1 R_2CNOH / 2.1 $TiCl_4$ / 4.2 $NaBH_4$; 1 RSOR' / 1.1 $TiCl_4$ / 2.2 $NaBH_4$.

For the reduction of 10 mmol substrate, a mixture from $TiCl_4$ and $NaBH_4$ in 40 ml dimethoxyethane (glyme) is prepared and stirred in an ice-bath. A solution of the substrate in 10 ml anhydrous glyme is added, and the reaction is quenched with cooling after 14 hours at room temperature by addition of 100 ml H_2O. Extraction with benzene (2x60 ml), in the case of amine products after basification with 28% aqueous ammonia, is followed by drying over Na_2SO_4, evaporation of the solvent and distillation or recrystallization.

General procedure for Reduction of Acids to Aldehydes [179]

$$\mathrm{RCOOH + 2\,Me_2CHMgBr \xrightarrow{Cp_2TiCl_2} RCHO}$$

To stirred 1 M $Me_2CHMgBr$ in ether (63 mmol) is first added Cp_2TiCl_2 (75 mg, 0.3 mmol), and 5 min. later a carboxylic acid (30 mmol), under an argon atmosphere and with ice cooling. After 4 hours at room temperature, 20 ml of 4 N HCl is added, the layers are separated, the aqueous layer extracted (30 ml ether), and the combined organic layers dried ($MgSO_4$) and distilled.

3) *Nef*-Reactions with $TiCl_3$

With the advent of numerous new carbon carbon bond-forming reactions of nitroalkanes [180-182] the importance of the *Nef*-reaction, which converts nitroalkanes to aldehydes or ketones, has grown. As with many other classical reactions, there are literally dozens of procedures for this conversion. One is the treatment of nitroalkanes with titanium trichloride [157,183]. In some cases, this method works when all the others fail. Representative procedures are given along with a list of examples in Table 14.

Table 14. Conversion of nitroalkanes to ketones and aldehydes by treatment of the nitrocompounds with $TiCl_3$ at pH <1 (A) or in a buffered solution at pH ca. 6 (B) [183)]

starting material	product	method/% yield
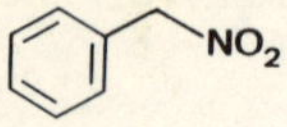	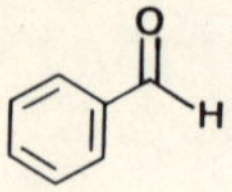	A 80
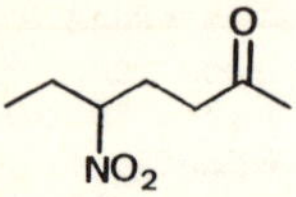	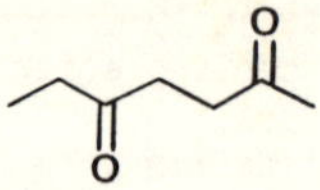	A 85
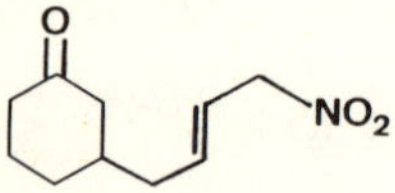	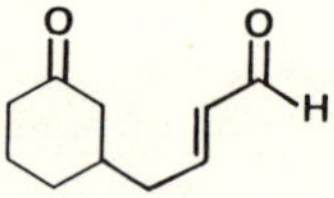	A 60 [184)]
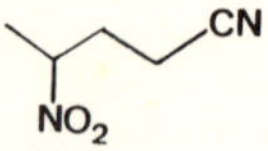	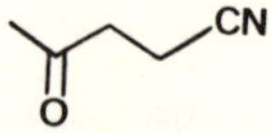	A 55 B 90
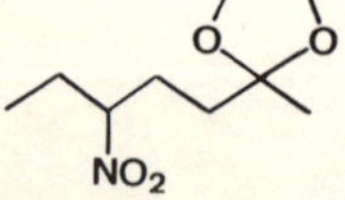	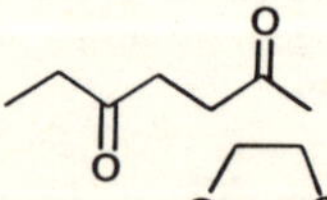	A 40
	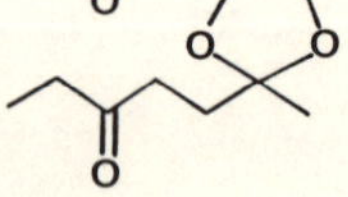	B 70
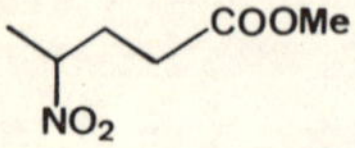	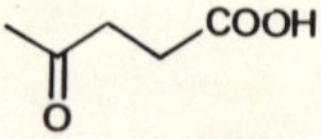	A 40
	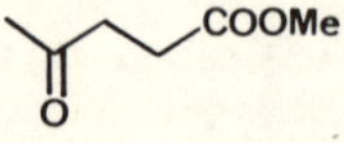	B 90

(Table 14 continued)

starting material	product	method/% yield
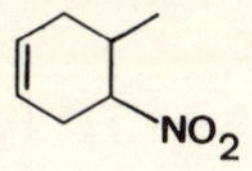 [structure: 4-nitro-5-methylcyclohexene, NO_2]	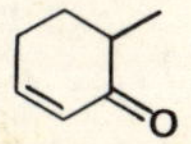 [structure: 6-methylcyclohex-2-en-1-one, O]	A 35
	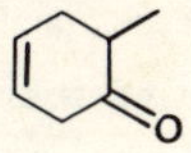 [structure: 6-methylcyclohex-3-en-1-one, O]	B 60

Heptan-2,5-dione [183)] (Method A)

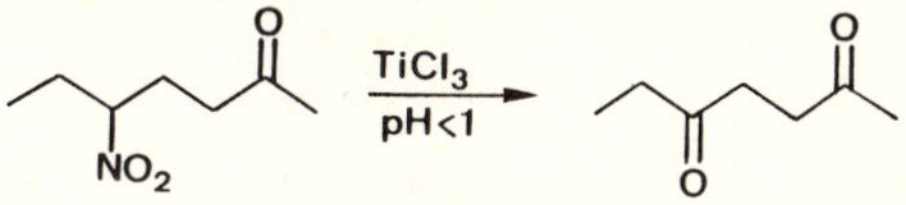

NO_2 — $\xrightarrow[pH<1]{TiCl_3}$ — O, O

A 0.2 M solution of 5-nitroheptan-2-one is combined with four mol-equiv. of $TiCl_3$ (20% aqueous solution) and stirred under nitrogen at room temperature for 24 hours. The reaction mixture is then poured into ether. After separation of the two phases, the aqueous layer is extracted several times with ether. The combined organic extracts are washed with 5% $NaHCO_3$ and with brine, dried (Na_2SO_4) and concentrated. Distillation of the crude product affords heptan-2.5-dione in 85% yield.

2-Methyl-4-cyclohexenone [183)] (Method B)

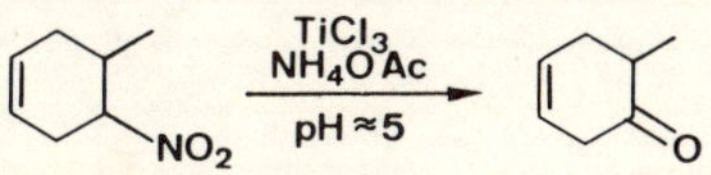

NO_2 — $\xrightarrow[pH \approx 5]{TiCl_3,\ NH_4OAc}$ — O

A 0.5 M solution of 4-nitro-5-methylcyclohexene in methanol is treated with 1 mol-equiv. of $NaOCH_3$ to form the corresponding sodium nitronate. An aqueous solution of $TiCl_3$ and NH_4OAc 1:6 is prepared (ca. 0.4 M with respect to $TiCl_3$). At room temperature 4 mol-equiv. of this buffered $TiCl_3$ solution is added in one portion to the nitronate under a nitrgen atmosphere. After 45 min. the reaction is worked up (cf. procedure above) and 6-methylcyclohex-3-en-1-one is isolated in 66% yield.

B) Reactions of Carbonyl Compounds Involving Dialkylamido-titanium Derivatives

The early work on dialkylamido-derivatives of titanium has been described in an excellent review article [185]. The use of reagents of type $RTi(NR_2)_3$ 22 and $RTi(NR_2)_4M$ 27 as nucleophiles for carbonyl additions has been mentioned occasionally in chapter III (see amination/alkylation of carbonyl derivatives III C2; aldol additions and related processes III F). A procedure describing the preparation of the tetrakis(dimethyl- and diethylamino)-titanium (12, 13) is given in section II. The titanium(IV)dialkylamides, $Ti(NR_2)_4$, react rapidly with various carbonyl derivatives. Depending upon the particular structure of the carbonyl compound, different amino-derivatives can be formed with removal of the oxygen: aminals 84 with non-enolizable aldehydes such as benzaldehyde and pivalaldehyde, but also with acetal-

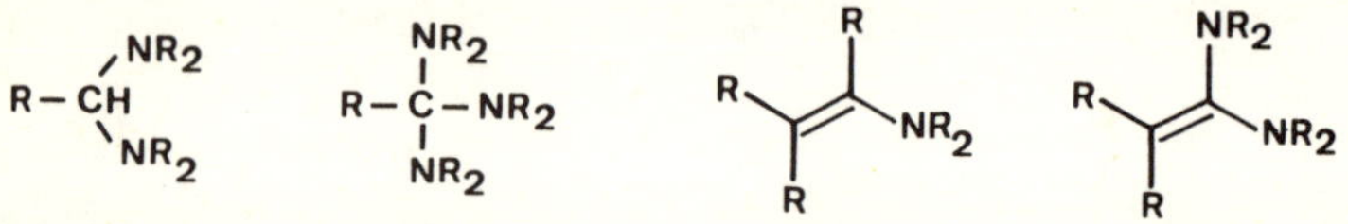

$R{-}CH(NR_2)_2$ **84** $\qquad$ $R{-}C(NR_2)_3$ **85** $\qquad$ $R_2C{=}C(R)NR_2$ **86** $\qquad$ $R_2C{=}C(NR_2)_2$ **87**

dehyde; *ortho*-amides 85 with non-enolizabel carboxylic acid derivatives; enamines 86 with most enolizable aldehydes and ketones; and ketene aminals 87 with enolizable carboxylic esters and amides [186]. This formation of enamines is one of the most efficient methods of preparing this class of substances, and the modification using $TiCl_4$ and secondary amines directly is generally referred to as the *Weingarten* method [187]. Its stoichiometry requires three moles of secondary amine per mole of enolizable carbonyl compound 88. It is especially effective with hindered ketones. Its disad-

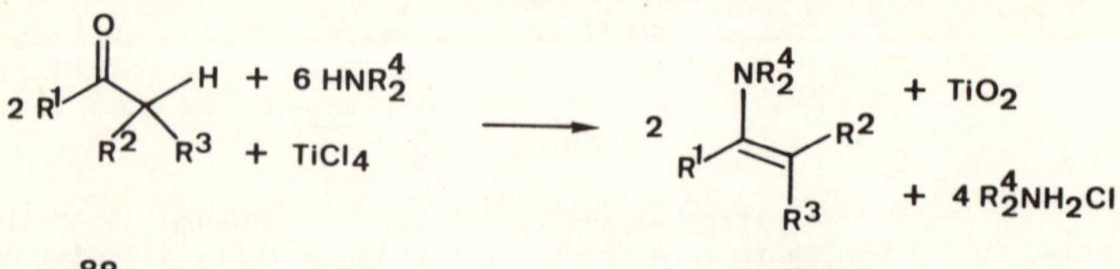

$$2\ R^1C(O){-}CH(R^2)(R^3) + 6\ HNR^4_2 + TiCl_4 \longrightarrow 2\ R^1(NR^4_2)C{=}C(R^2)(R^3) + TiO_2 + 4\ R^4_2NH_2Cl$$

88 $\qquad\qquad$ **86**

vantage is obviously the necessity of using excess amine, if the amine is valuable. A typical procedure follows.

General Procedure for the Synthesis of Enamines with $TiCl_4$ [187)]
(see 88 → 86)

A two-liter four-necked flask is fitted with a mechanical overhead stirrer, a reflux condenser, an N_2 gas line, a thermometer, and a dropping funnel. Under an atmosphere of dry nitrogen, the flask is charged in this order with 500 ml solvent (benzene, pentane, or ether), 0.1 mole of ketone, and a solution of excess amine (4 - 5 equiv.) in 100 ml of the same solvent. To the resulting solution is added over a period of 20 - 60 min. 0.055 mole $TiCl_4$, dissolved in an additional 100 ml of solvent (benzene or pentane), while the temperature is kept between 0° and 10^oC. The resulting mixture is stirred for several hours at 20^oC. The more hindered ketones (pinacolone, 2.5-dimethylcyclohexanone) require longer reaction times. The progress of the reaction is followed by NMR. analysis of aliquots. After the reaction is complete, the reaction mixture is filtered, the solvent is evaporated from the filtrate and the residue is distilled *in vacuo*. All these workup operations must be carried out quickly, under exclusion of moisture.

The rate at which dialkylamido-titanium derivatives transfer an amino group to a carbonyl carbon appears to be faster than that of alkyl transfer from titanium. Thus, the methylating amination (see section III C2) could only be explained with the assumption that the amino group is transferred faster than the methyl group [28)]. In Scheme 24, a surprising reversed selectivity of additions to aldehydes and ketones is demonstrated, which becomes reasonable only if the above assumption is made. Addition of the ate-complex from tetraisopropoxy-titanium and allyl or crotyl *Grignard* reagent to a 1:1-mixture of an aldehyde and a ketone leads to the isolation of the adduct to the aldehyde, with the expected (see section III A1) selectivity. This is true both with aliphatic (left part of Scheme 24) and aromatic carbonyl derivatives (right top part of Scheme 24). On the other hand, as demonstrated in the bottom half of Scheme 24, the ate-complexes from tetrakis(dimethylamino)titanium are perfectly ketone selective, giving rise, after aqueous workup, to the ketone adduct and unreacted aldehyde. These findings [19,115)] are explicable if the primary product from the alkoxy-ate-complex is 89, the result of allylic group transfer, while that with the amino-derivative is 90, i.e. the product of amino-group transfer. The adduct 90 is now a protected form of the aldehyde, and it acts as an allylic group transfer reagent towards the ketone!

Scheme 24. Aldehyde-selective additions of allylic groups from $[(Me_2CHO)_4Ti\text{-}CH_2\text{-}CH{=}CH\text{-}R]$ate-complexes and ketone-selective additions of allylic groups from $[(Me_2N)_4Ti\text{-}CH_2\text{-}CH{=}CHR]$ate-complexes [19,115].

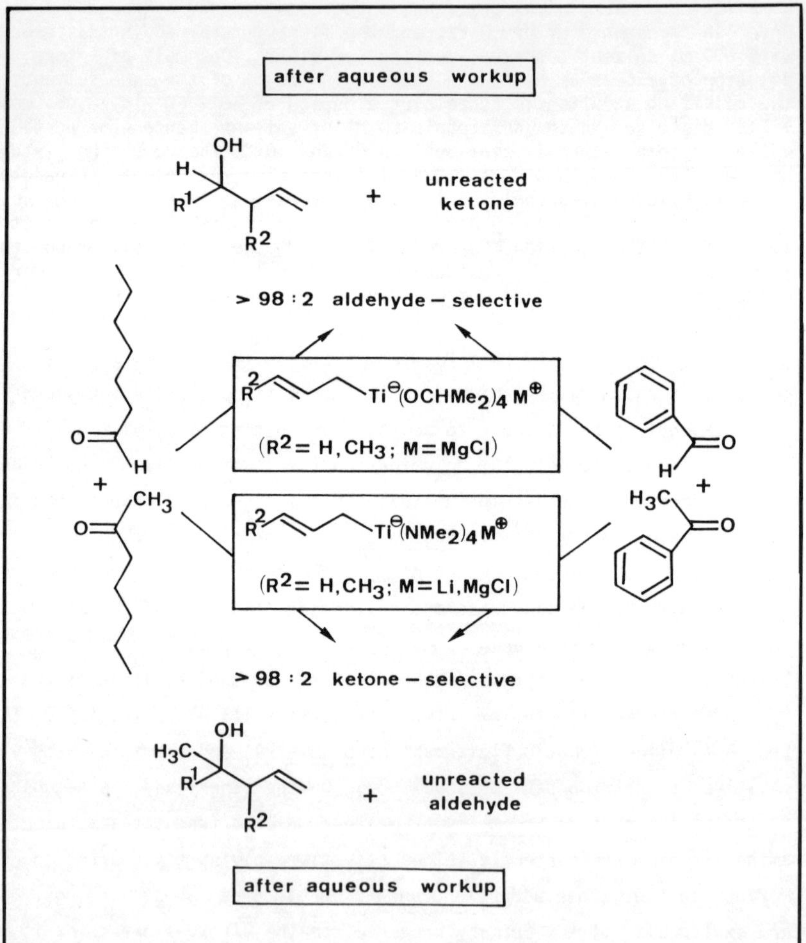

after aqueous workup

OH
H
R1
R2
+ unreacted ketone

> 98 : 2 aldehyde – selective

R^2–CH=CH–CH$_2$–Ti$^{\ominus}$(OCHMe$_2$)$_4$ M$^{\oplus}$

(R^2 = H, CH$_3$; M = MgCl)

O=, H, + CH$_3$, O=

H, =O, + H$_3$C, =O

R^2–CH=CH–CH$_2$–Ti$^{\ominus}$(NMe$_2$)$_4$ M$^{\oplus}$

(R^2 = H, CH$_3$; M = Li, MgCl)

> 98 : 2 ketone – selective

OH
H$_3$C
R1
R2
+ unreacted aldehyde

after aqueous workup

89

90

Scheme 25

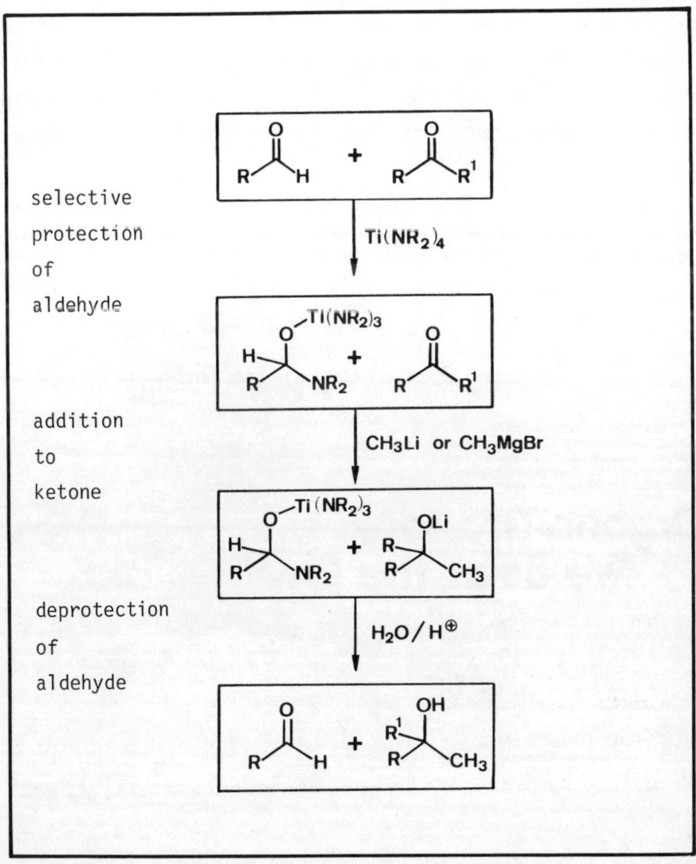

No ate-complex needs to be generated in order to achieve ketone-selective additions of this type [188]. If tetrakis(dialkylamino)titanium is added at $-78^{o}C$ to an aldehyde/ketone mixture prior to methyllithium, again ketone-selective addition of the nucleophile is observed, see Scheme 25. Obviously, $Ti(NR_2)_4$ adds selectively to the aldehyde, leaving behind unchanged ketone, which is trapped by the "second" nucleophile MeLi.

Thus, the expected fast, selective, and irreversible (at low enough temperature) transfer of the amino group to the aldehyde is responsible for the reversal of the normal aldehyde vs. ketone selectivity, which seems surprising at first sight.

C) $Ti(OR)_4$-Catalyzed Transesterifications

The chemical industry offers (see Figures 3 - 5) and uses titanates as cheap bulk chemicals for various applications. Among other transformations, esterifications are catalyzed by titanates. A literature search, inspection of

Fig. 3

Fig. 4

"Call 800-441-9475 to get all the organic titanates you need, fast."

Ollie Yeaton, Sales Manager

Seven versatile DuPont TYZOR® organic titanates can improve your product properties.

To get all the DuPont organic titanates you need, when and where you need them, just call. We're the largest domestic manufacturer of titanates, fully integrated forward from our own titanium ore mines. So you can bet on reliable supply and delivery, whether you need tank trucks, drums or 40-lb. pails.

When TYZOR organic titanates are used to crosslink resins via active hydrogens, the cured resins display improved hardness, electrical properties and solvent resistance. Common applications for crosslinking with TYZOR include printing inks, paints, coatings and gels.

TYZOR is also being used more and more as an esterification catalyst for carboxylic esters.

TYZOR organic titanates have a broad range of reactivity. Water-stable chelates are also available.

40% increase in TYZOR capacity.

In March, 1982, DuPont's already sizeable production capacity for TYZOR will expand by 40%. But right now, for all the organic titanates you need, call 800-441-9475. Or write: Ollie Yeaton, DuPont Company, Room G39404J1, Wilmington, DE 19898, to find out more about how you can use TYZOR organic titanates.

Fig. 5

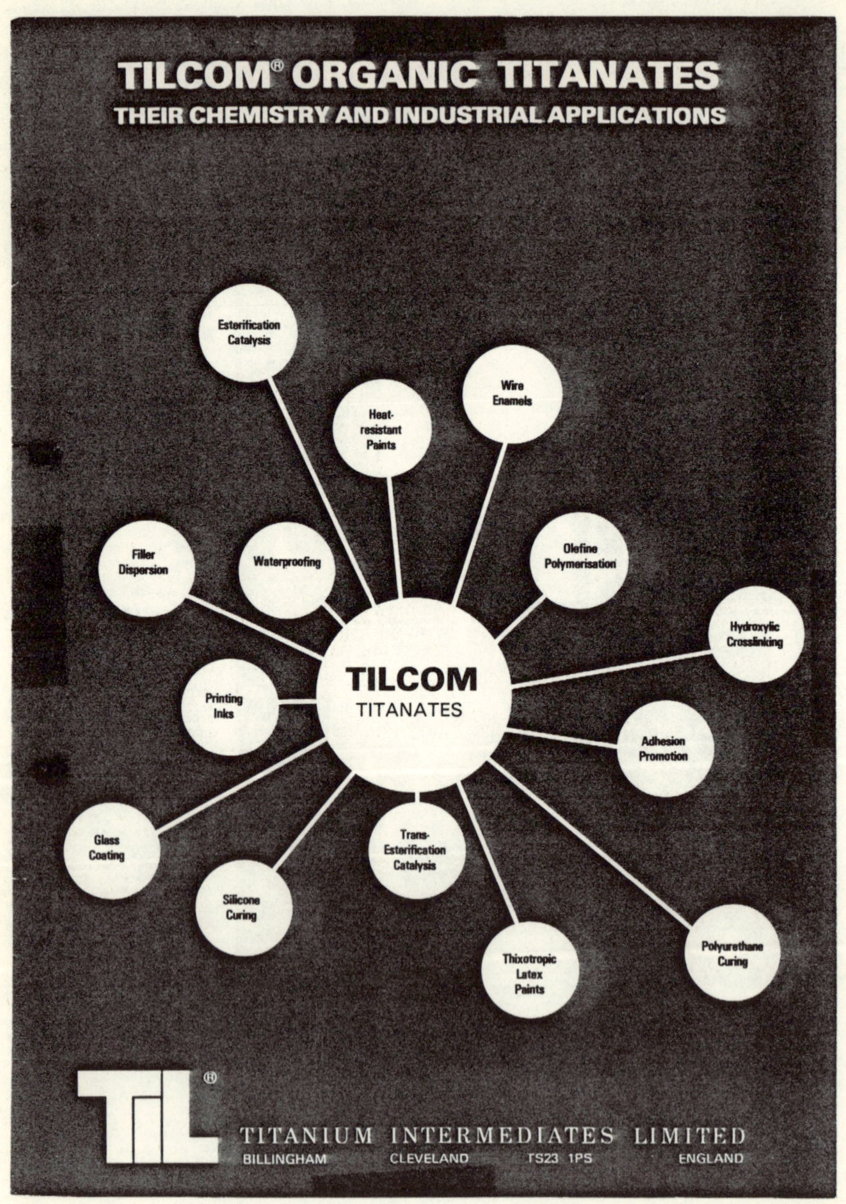
TILCOM® ORGANIC TITANATES
THEIR CHEMISTRY AND INDUSTRIAL APPLICATIONS
Esterification Catalysis
Heat-resistant Paints
Wire Enamels
Filler Dispersion
Waterproofing
Olefine Polymerisation
Hydroxylic Crosslinking
TILCOM
TITANATES
Printing Inks
Adhesion Promotion
Glass Coating
Trans-Esterification Catalysis
Silicone Curing
Thixotropic Latex Paints
Polyurethane Curing
TiL®
TITANIUM INTERMEDIATES LIMITED
BILLINGHAM CLEVELAND TS23 1PS ENGLAND

the eight volumes of *Fieser's* Reagents, and discussions with many colleagues revealed that titanate-mediated esterifications, deacylations, and transesterifications are well known to industrial chemists, but essentially unknown in research laboratories, especially at universities. Accordingly, almost all applications are described in the patent literature and involve rather simple, monofunctional substrates, see the discussion and the references in a recent article [189].

As demonstrated in Scheme 26, the equilibria between esters (RCOOR') and alcohols (R''OH) can be established with titanate catalysis. If an alcohol is used in large excess, as a solvent, a substrate ester is transesterified, a substrate acyl-protected alcohol is deprotected (top arrow in Scheme 26). If, however, an ester is used as a solvent, a substrate ester is also transesterified, but a substrate alcohol is esterified (arrow on the bottom of Scheme 26). These equilibrations can be exploited for selective transesterifications or deprotections of polyfunctional substrates, even in the presence of groups which are very sensitive to acid or base, such as C,C and C,N triple bonds, acetals, β-hydroxy- and β-acyloxy-esters, β-lactams, *t*-butyl-dimethylsilyloxy groups, BOC- [190] and other urethane protecting groups, etc. The titanates may be considered as "neutral *Lewis* acids" with a preference for carbonyl oxygen ($C=\overline{\underline{O}}$) over acetal, ether or alcohol oxygen ($R-\overline{\underline{O}}-R$).

The titanate catalyst used in most cases is tetraethoxytitanium; the small amounts of ethanol introduced into the equilibria do not influence the results if other than ethyl esters are to be made. For preparing methyl esters in methanol as the solvent, simple titanates can not be used, because the very insoluble tetramethoxytitanate precipitates. This can be avoided by using a catalyst obtained from $Ti(OEt)_4$ and one-half equiv. of glycol [191]. Another possibility of preparing methyl esters is to transfer the CH_3O-group from another ester, for instance methyl propionate which is employed as the solvent (bottom part of Scheme 26). Strangely, under these conditions no $Ti(OCH_3)_4$ precipitates [191].

Four characteristic examples are described below in full experimental detail; a list of recent applications is found in Table 15.

Scheme 26

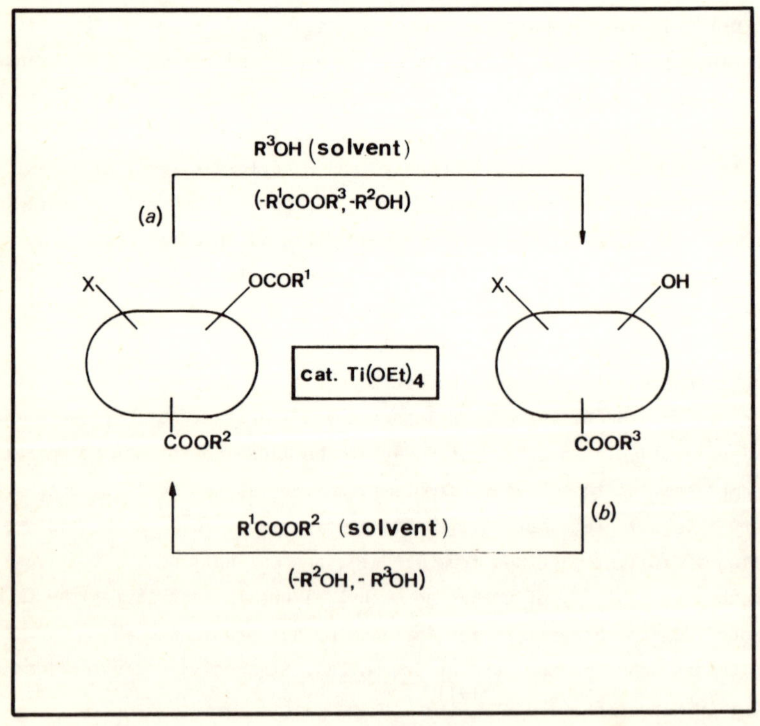

Table 15. Examples of esterification, transesterification, deprotection, and polyester depolymerization to give acid and/or base-sensitive products. The solvent for transesterifications is the alcohol of which the ester is formed; acetylations of alcohols are performed in ethyl acetate. The catalyst is indicated by A [$Ti(OEt)_4$], B [$Ti(O$-i-$C_3H_7)_4$] or C [$(EtO)_3Ti$-OCH_2CH_2O-$Ti(OEt)_3$]. D refers to the alternative method of forming methyl esters in methyl propionate with $Ti(OEt)_4$.

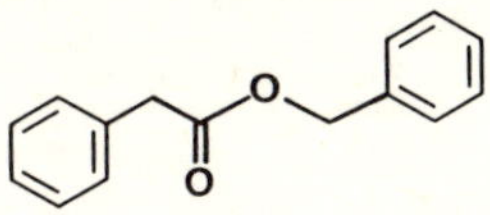

A 85% from ethyl ester [189)]

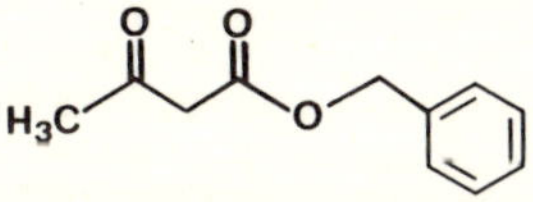

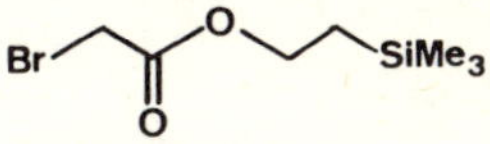

A 71% from ethyl ester [189)]

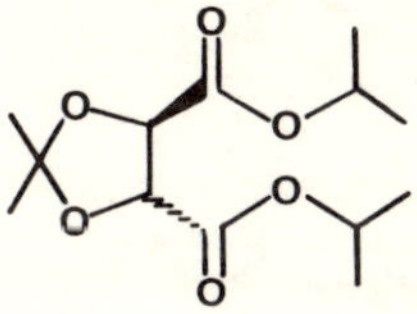

A 56% from t-butyl ester [189)]

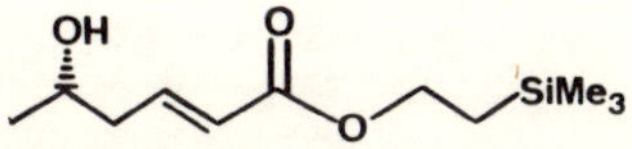

B 91% from methyl ester [189)]

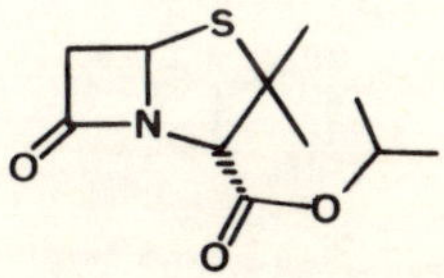

A 74% from methyl ester [189)]

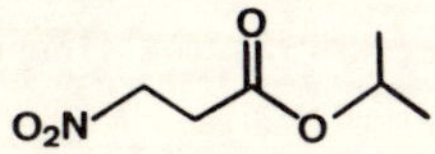

B 91% from methyl ester [189)]

A 50% from methyl ester [189)]

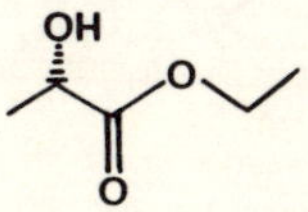

A 60% from propionate [189)]

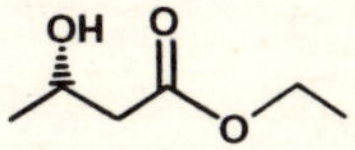

A 90% from 3.5-dinitrobenzoate [189)]

(Table 15 continued)

BOC–Leu–Leu–OMe —B→ BOC-Leu-Leu–OBzl 190)

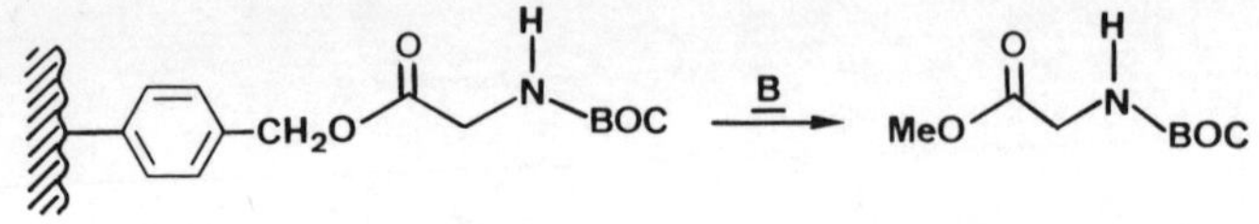

—B→ MeO–C(O)–CH₂–N(H)–BOC 190)

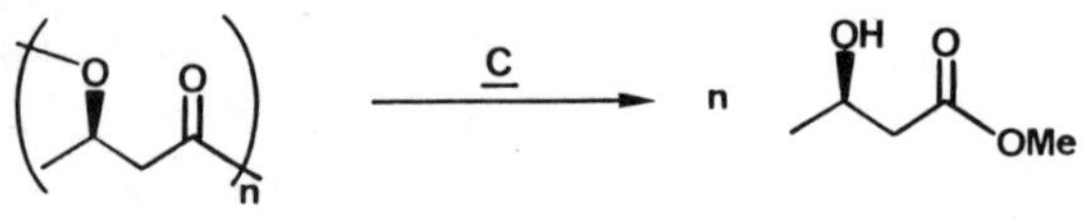

—C→ n 191)

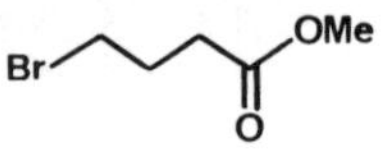

D 83% from ethyl ester 191)

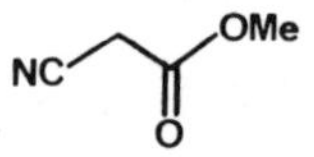

C 88% from ethyl ester 191)

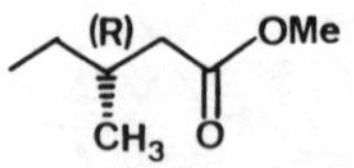

(R)

C 77% from ethyl ester 191)

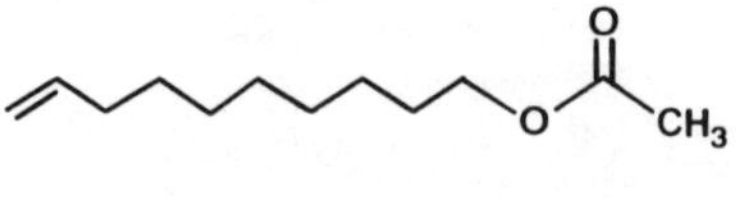

A 94% from alcohol 191)

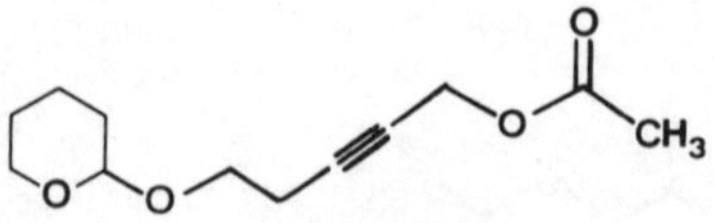

A 70% from alcohol 191)

Isopropyl-2-(*t*-butyl-dimethyl-silyloxy)-butyrate [189)]

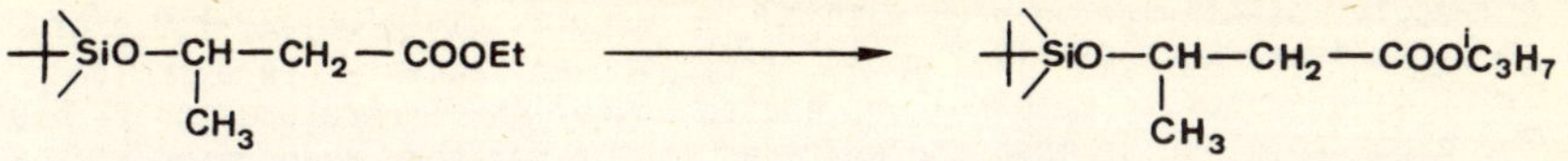

Ethyl 3-(*t*-butyldimethylsiloxy)-butanoate (2 g, 8.1 mmol) is dissolved in 2-propanol (30 ml), $Ti(O\text{-}i\text{-}C_3H_7)_4$ (1.00 g, 3.5 mmol) is added and the mixture is heated to reflux temperature for 6 hours. It is then cooled to ca. 45°C, quenched with 1 M aqueous HCl (temporary turbidity is observed), and extracted with pentane. The organic extract is washed with saturated aqueous $NaHCO_3$, dried ($MgSO_4$) and evaporated at 60°C under reduced pressure to remove the solvent and the residual 2-propanol. 1.76 g (6.76 mmol, 83%) of the transesterified product is obtained, which is pure according to ^{1}H-NMR.

Diethyl malate [189)]

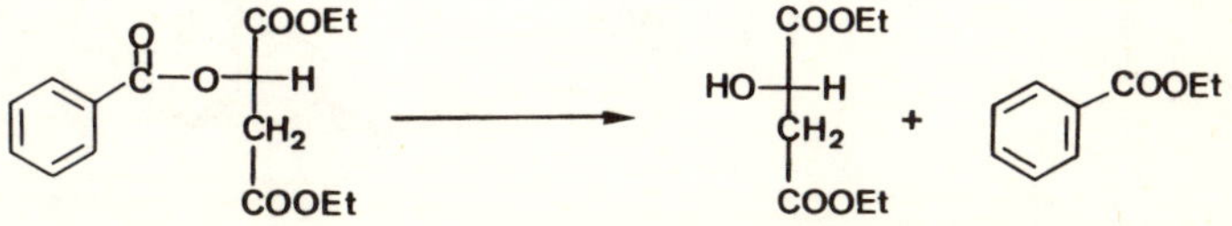

Optically pure (S)-(-)-O-benzoylmalate (2.00 g, 6.79 mmol) is dissolved in absolute ethanol (50 ml) and combined with $Ti(OEt)_4$ (1.00 g, 4.40 mmol). The mixture is heated under reflux for 6.5 hours. It is then cooled to ca. 40°C, hydrolyzed with 1 M aqueous HCl (temporary turbidity is observed) and extracted with ether. After removal of solvent and ethanol, the residue is distilled to yield 0.98 g (6.53 mmol, 96%) ethyl benzoate (b.p. 55°C/ 0.01 Torr) and 0.81 g (4.26 mmol, 64%) of diethyl (S)-(-)-malate (b.p. 90°C/ 0.01 Torr). $[\alpha]^{25}_D = -9.3°$ (neat).

Methyl phenylacetate [190)]

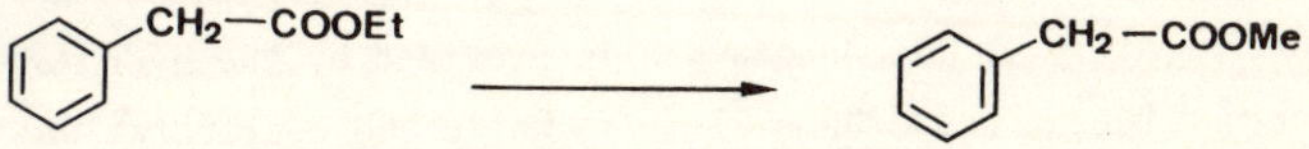

Method i): A solution of 2.0 g (12.2 mmol) of ethyl phenylacetate and ca. 0.5 g (ca. 2 mmol) of $Ti(OEt)_4$ in 50 ml of methyl propionate is refluxed for 110 hours. After cooling to ca. 40°C, the reaction mixture is treated with 30 ml 1 M aqueous HCl and extracted with ether. The combined organic layers are washed with saturated aqueous $NaHCO_3$, dried ($MgSO_4$) and concentrated on the rotatory evaporator. Kugelrohr distillation affords 1.70 g (11.3 mmol, 93%) of methyl phenylacetate, b.p. 155°C/9 Torr.

Method ii): A modified titanium catalyst (gly-Ti) is prepared by mixing one mol-equiv. of ethylene glycol with two mol-equiv. of $Ti(OEt)_4$. 1.0 g of this catalyst and 2.0g (12.2 mmol) of ethyl phenylacetate are dissolved in 50 ml of absolute methanol and heated at reflux temperature for 72 hours. The clear solution is then concentrated at the rotatory evaporator at reduced pressure to ca. 5 ml. After addition of 5 ml 1 M HCl solution, the reaction mixture is extracted four times with ether. The combined organic extract is washed with 1 N HCl, saturated $NaHCO_3$ and twice with brine, dried over $MgSO_4$, filtered and evaporated. The residue is purified by Kugelrohr distillation to yield 1.66 g (11.1 mmol, 91%) of the methyl ester.

3 β-Acetoxy-cholest-5-ene [190)]

HO — $Ti(OEt)_4$ / EtOAc → AcO

Cholest-5-en-3-ol (2.0 g, 5.17 mmol) and a catalytic amount of $Ti(OEt)_4$ (ca. 0.5 g), dissolved in 50 ml of ethyl acetate, is heated to reflux for 24 hours. After cooling to ca. 40°C and hydrolysis with dilute 1 M aqueous HCl, the reaction mixture is worked up with ether. After recrystallization from ethanol, 1.81 g (4.22 mmol, 82%) of 3 β-acetoxy-cholest-5-ene is obtained, m.p. 110-114°C, $[\alpha]^{25}_D = -41°$ (c = 2.23, $CHCl_3$).

D) Additions to C,C Multiple Bonds and Eliminations - the Hydrozirconation

One of the earliest synthetic applications of zirconium derivatives as reagents was the hydrozirconation, which was reviewed [192)] in 1976. In many respects, the hydrozirconation duplicates the hydroboration [193)], in others it is complementary to it. An example is provided by the hydrometallation of isomeric octenes, see Scheme 27 [193)]. While the hydroboration is reversible only at elevated temperatures, addition/elimination of the zirconocene hydride Cp_2ZrClH (2c, preparation see section II) to double bonds is so facile, that all three octenes furnish the same 1-octyl-zirconium derivative at room temperature. As is also evident from the examples given in Scheme 27, the hydrozirconation is strictly directed by steric factors [193)]: its thermodynamically-controlled additions lead to primary organo-zirconocenes, tetrasubstituted and cyclic trisubstituted olefinic double bonds do not react, and the addition to acetylenes produces that vinylzirconium compound in which the zirconium is bonded to the carbon with the smaller substituent, with hydrogen in the *cis*-position [193].

The hydrozirconation products are subjected to further transformations with replacement of the zirconium from the organic group. Two such transformations have been discussed in previous sections, the carbonylation (III D)

Scheme 27

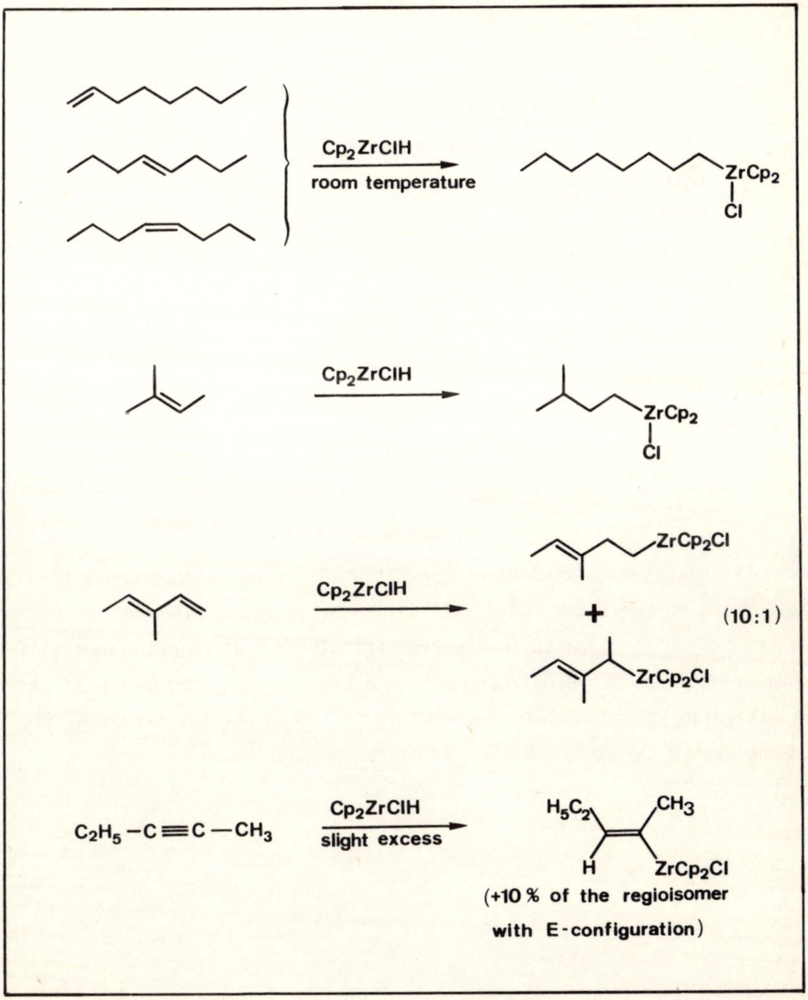

Cp_2ZrClH, room temperature

$ZrCp_2$ / Cl

Cp_2ZrClH

$ZrCp_2$ / Cl

$ZrCp_2Cl$

Cp_2ZrClH

+ (10:1)

$ZrCp_2Cl$

$C_2H_5-C\equiv C-CH_3$ $\xrightarrow[\text{slight excess}]{Cp_2ZrClH}$ H_5C_2, CH_3, H, $ZrCp_2Cl$

(+10 % of the regioisomer with E-configuration)

Scheme 28

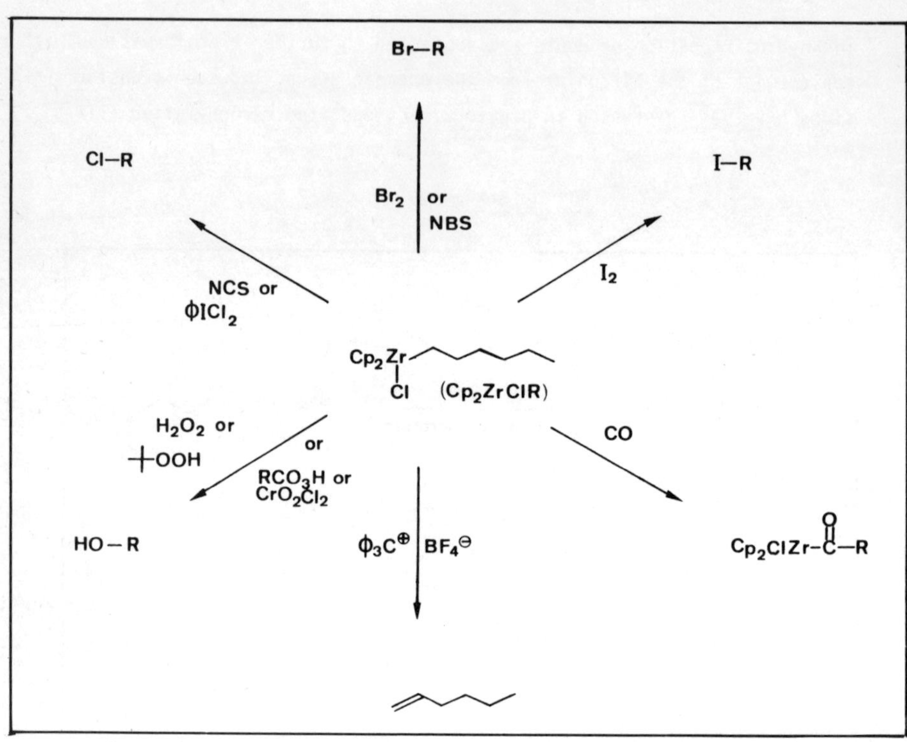

and the nickel-catalyzed *Michael* additions of vinyl-zirconocenes to α.β-unsaturated carbonyl compounds (III B). A carbozirconation step is also most likely to be involved in the Cp_2ZrCl_2-catalyzed carboaluminations (III G). Some reactions of hydrozirconation intermediates with formation of hetero-substituted carbon skeletons are shown in Scheme 28, and a typical procedure demonstrates the reversibility of hydrozirconations.

1-Triacontanol from 15-triacontene [194)]

$$C_{14}H_{29}-CH=CH-C_{14}H_{29} \xrightarrow[\text{2) } t\text{-}C_4H_9OOH]{\text{1) } Cp_2ZrClH} H_{59}C_{29}-CH_2OH$$

The hydrozirconation reagent 2c is prepared *in situ* in the following way: A solution of 2.92 g (10 mmol) of Cp_2ZrCl_2 (2a) in 30 ml THF is combined within 10 min. at 23°C under inert atmosphere with an equiv. amount of Vitriole® (70% toluene solution, see section II).

After 2 hours, 2.1 g (5 mmol) of neat 15-triacontene is added by syringe through a serum capped side arm of the reaction flask. A suspension results which is kept under argon at 40°C for 4 days. After cooling to room temperature, the mixture is heated with 2.75 ml of 3.7 M t-C_4H_9OOH in 1.2-dichloro-ethane for 1 hour. For workup, 3 ml of H_2O are added, the mixture is filtered, and the filtrate is concentrated under reduced pressure to ca. 50 ml. Dilution with acetone (300 ml) gives 2 g of crude 1-triacontanol, which is recrystallized from 1.2-dichloroethane to give 1.5 g (68%) of 97% pure triacontanol, m.p. 89-90°C.

E) Stereoselective Epoxidations with Titanium and Vanadium Catalysts

Epoxides (91) are useful reagents in general organic synthesis, and versatile building blocks for natural product synthesis in particular. They are

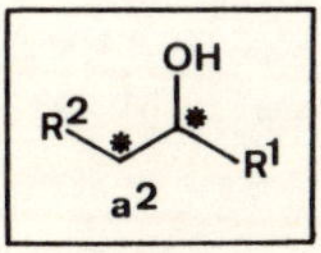

OH
R2 R1
a2
93

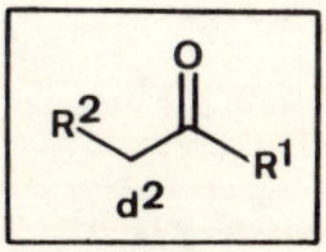

O
R2 R1
d2
94

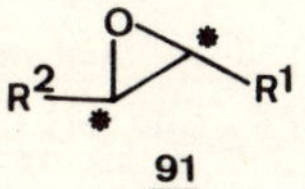

R2 O R1
91

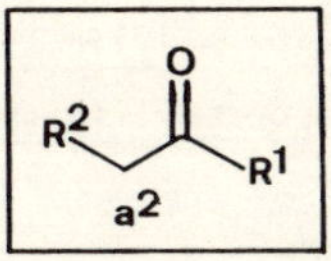

O
R2 R1
a2
95

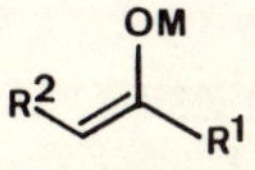

OM
R2 R1
92

a^2-reagents [62], i.e. electrophiles in the β-position of an oxygen functionality. The 1.2-bifunctionality pattern makes epoxides the counterparts of enol derivatives 92, which are d^2-reagents, cf. the synthon boxes 93 and 94. Since the alcohols resulting from nucleophilic opening of epoxides can be oxidized to carbonyl derivatives, epoxides also correspond to aldehyde, ketone, acid, or ester d^2-synthons 95, depending on the nature of the R^1-group and of the particular transformation. But epoxides do not only provide an umpolung of carbonyl d^2-reactivity, they also can supply products with a given configuration: where enol derivatives 92 have two sp^2-carbon atoms, epoxides 91 have two asymmetric centers which may have a certain relative configuration *and* a certain sense of chirality. Enol derivatives may only react diastereoselectively under special circumstances, see section on aldol additions above, while epoxides are building blocks with centers of chirality [126]. Thus, if carbonyl compounds have been said to be "virtually the backbone of organic synthesis" [195], the epoxides correspond to at least "one of the main muscles".

"Stereochemical Aspects of the Synthesis of Epoxides" [196] and "Stereoselectivity in Reactions of Epoxides" [197] have been reviewed ten years ago. In continuation of his earlier studies on transition metal catalyzed oxidations, and especially on vanadium(V) and molybdenum(VI) mediated epoxidations by *t*-butyl hydroperoxide, which were reviewed with "practical considerations" in Aldrichimica Acta [198,199], *K.B. Sharpless* has now discovered a titanate/tartrate-catalyzed enantioselective epoxidation [200]. This has received the deserved attention by synthetic organic chemists and will be reviewed here briefly, together with some recent advances in diastereoselective epoxidations of homoallylic alcohols with vanadium(V)/*t*-butyl hydroperoxide.

1) Asymmetric Epoxidations of Allylic Alcohols with *t*-Butyl Hydroperoxide (TBHP) / Diethyl or Diisopropyl Tartrate (DET, DIPT) / Tetrakis-isopropoxy-titanium

Our previous review in volume no. 2 of this series, *Modern Synthetic Methods 1980* [126], ended with a list of epoxides, see Scheme 29. These are all available in both enantiomeric and, where applicable, diastereomeric forms from simple precursors such as lactic acid, malic acid, and tartaric acid. With the new methodology [37,201] outlined in Scheme 30,

Scheme 29 from ref. [126)]

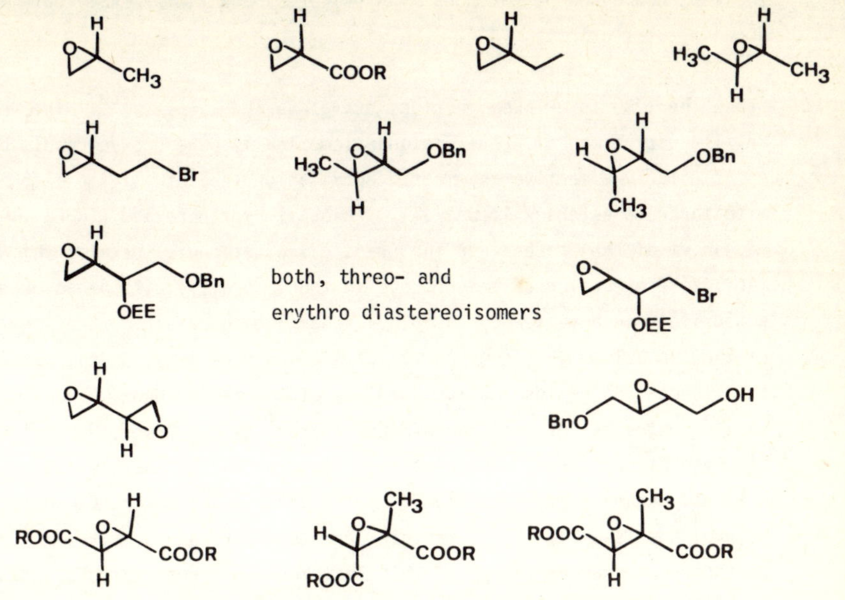

both, threo- and erythro diastereoisomers

Scheme 30

R^1	R^2	% Ausb.	% ds		
CH_3	C_2H_5	82	97	■	
	$(CH_2)_2CH_3$	72	97	■	■
	$(CH_2)_3CH_3$	68	97	■	
	$(CH_2)_6CH_3$	82	95	■	■
	$CH_2-CH=CH_2$	77	98		
	$CH_2C_6H_5$	81	96		
$CH_2C_6H_5$	C_2H_5	40	>95		
	$(CH_2)_2CH_3$	42	>95		
	$CH_2-CH=CH_2$	40	>95		
C_6H_5	$(CH_2)_2CH_3$	84	95		

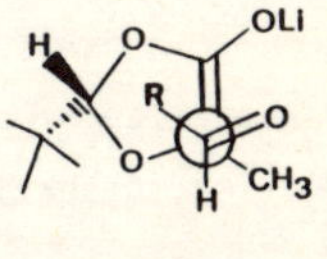

$R=C_2H_5$

the list could now be extended by a number of enantiomerically pure 1.1-disubstituted oxiranes [167].

Since the 1980 Interlaken meeting, *Sharpless* and *Katsuki* [200] discovered an alternative way of preparing enantiomerically pure epoxyalcohols by a spectacularly effective asymmetric epoxidation (AE) of allylic alcohols with tartrate esters / $Ti(OAlkyl)_4$ / *t*-butyl hydroperoxide (TBHP). The mechanism of the process and the practical experimental details which facilitate execution of the process in the laboratory will be covered extensively in a review by *Sharpless* scheduled to appear during August of 1983 in Volume 16 (Issue no. 3) of Aldrichimica Acta. Another review on "Stereo- and Regioselective Openings of Chiral 2.3-Epoxy Alcohols. Versatile Routes to Optically Pure Natural Products and Drugs. Unusual Kinetic Resolutions", the manuscript of a lecture held at the Eleventh International Carbohydrate Symposium, August 1982, in Vancouver, Canada, will be published shortly [202]. We thank Professor *K. Barry Sharpless* for providing a manuscript copy of this last-mentioned article to us prior to publication; we are also grateful for copies of the dissertations of *B.E. Rossiter* [203] and *S.S. Woodard* [204].

The steric course of the asymmetric epoxidation is described in Scheme 31. With "normal" CIP-priority sequence [105], the relative topicity [40] of transfer of the epoxide oxygen under the influence of the tartrate ester is specified ***lk***, i.e. the Si-face of the sp^2-center bonded to CH_2OH reacts if the tartrate has (S.S)-, the Re-face if it has (R.R)-configuration. Without quoting examples - there must be more than 100 by now, including ca. 40 applications in natural product syntheses [202] - the reasons for rapid adoption of the method are obvious [202]: 1) *simplicity* (all ingredients are inexpensive and commercially available), 2) *reliability* (it succeeds with most allylic alcohols), 3) *high enantiomeric purity* (generally >90% ee, usually >95% ee (i.e. ratio of enantiomers or enantioselectivity 97.5/2.5), 4) *absolute configuration of products predictable* (see Scheme 31), 5) *relatively insensitive to sense of chirality of centers in the substrate* (see discussion in next paragraph), 6) *versatility of 2.3-epoxy alcohols as intermediates* (see above, introduction to section E).

Thus, all positions R^1, R^2, and R^3 in the achiral allylic alcohol shown

Scheme 31

ASYMMETRIC EPOXIDATION WITH
RELATIVE TOPICITY *lk*

preferred with (S.S)-(-)-tartrate

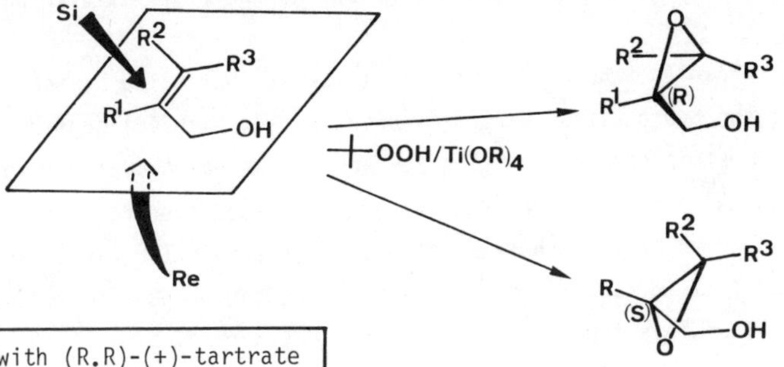

Si

R^2

R^3

R^1

OH

Re

$\vdash$OOH/Ti(OR)$_4$

preferred with (R.R)-(+)-tartrate

(i) Topicities and configurations for priority sequence in allylic alcohol: C-O > C=C > R^1; in product, epoxide: O > epoxy-C > CH_2OH; in dozens of published examples, we found only two exceptions to the priority sequence given for the allylic alcohol: $R^1 = R^2 = C_6H_5$, $R^3 = H$ and $R^1 = c\text{-}C_6H_{11}$, $R^2 = R^3 = H$ [200].

(ii) Reactivity increases with degree of substitution of the double bond; E-propenyl reacts for instance 70-100 times faster than vinyl [202,204].

(iii) Rate of reaction is larger with E- (R^3 = H) than with Z-allylic alcohol (R^2 = H).

(iv) Increasing size of R^3 decreases selectivity, thus with $R^3 = CH_2R$ the selectivity is still high - albeit in a slow reaction, while with $R^3 = C_6H_5$ or $CH(CH_3)_2$, the selectivity drops [202].

in Scheme 31 may be varied, except that with $R^3 \neq H$ the reaction is rather slow. Since (R.R)-(+)- and (S.S)-(-)-tartaric acids [126] are equally readily available in the amounts necessary for this asymmetric synthesis, both enantiomeric epoxides can be obtained. Tartrate ester employed in stoichiometirc amounts can be recovered in the workup procedure (see below).

The reactions of allylic alcohols with centers of chirality are still subject to stereochemical control with relative topicity *lk*. Two cases have been studied more thoroughly: allylic alcohols with a center of chirality at the carbinol carbon and at the other allylic position (C-4).

As shown in Scheme 32, the *lk*-attack is controlled by *ul*-1.2-induction, if the hydroxylated center is substituted. With (S)-configuration on the asymmetric carbon atom, the Re-face of the neighbouring trigonal center reacts faster, if the (R.R)-(+)-tartrate-modified reagent is employed. Depending on the substrate structure, the relative rates range from ca. 15:1 to ca. 150:1 [205] which allows the preparation of allylic alcohols of this type with >99% ee by kinetic resolution [205]. Routinely, the reaction is run to a conversion of ca. 55%, and the unchanged enantiomer is recovered. On the other hand, of two enantiomeric allylic alcohols of the type shown in Scheme 32, one can selectively be epoxidized by (S.S)-tartrate-derived reagent (R in Scheme 32), the other one by (R.R)-tartrate-derived reagent (S in Scheme 32), producing enantiomeric epoxyalcohols of the same relative configuration (Re-attack on the S-enantiomer by R.R-reagent and Si-attack on the R-enantiomer by S.S.-reagent are preferred). The *ul*-1.2-induction observed here is also the preferred relative topicity in V($\underline{\bar{V}}$)-catalyzed epoxyalcohol syntheses, see next section.

In contrast to the situation in Scheme 32, a center of chirality on the other side of the double bond, see Scheme 33, does not have enough influence to override the intrinsic *lk*-preference of the reagent. The other enantiomer of the alcohol can also be attacked from the diastereotopic Re-face by the (R.R)-reagent or from the diastereotopic Si-face by the (S.S)-reagent. Thus, if both enantiomeric alcohols are available, all four possible stereoisomers with *trans*-configuration on the oxirane ring can be prepared, see bottom of Scheme 33 (cf. ref. [218]).

Scheme 32

Topicity and chirality specified with CHR(OH) > C= > H and O > C= > R^1

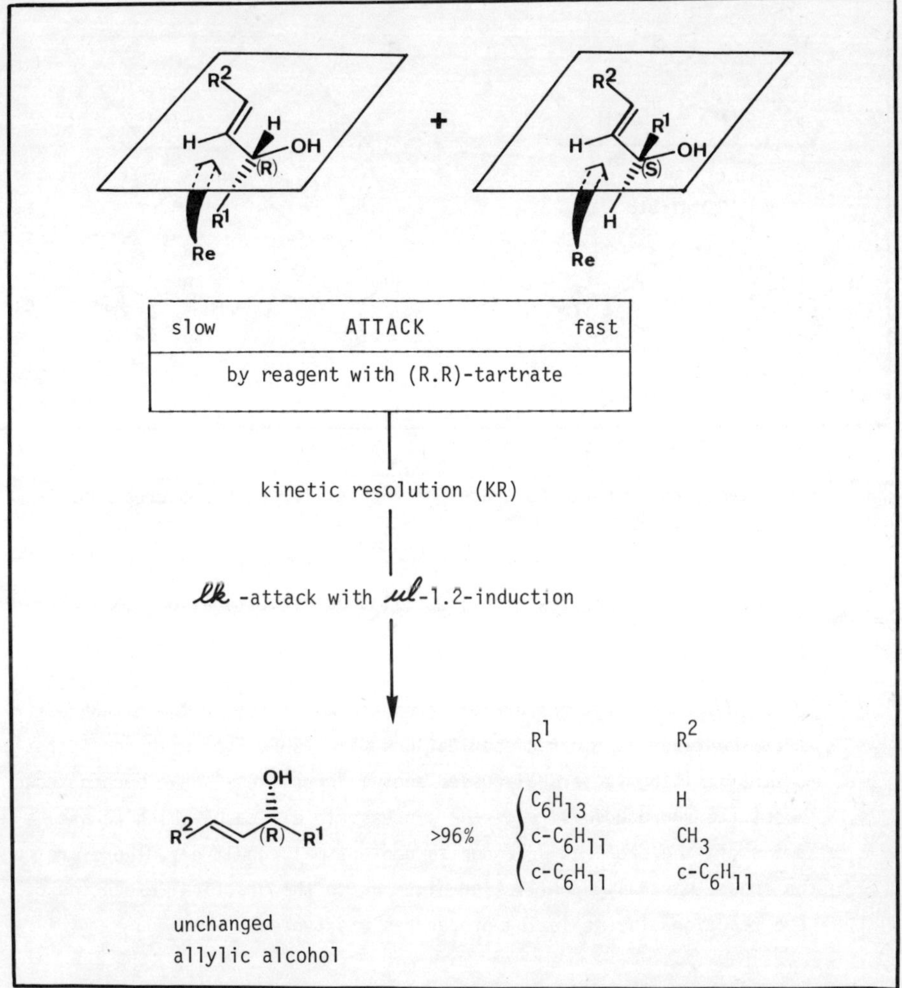

The asymmetric epoxidations of allylic alcohols described seems to be a case in which a non-enzymatic conversion reaches levels of selectivity which are usually considered typical of enzymic transformations. Furthermore, the substrate selectivity, i.e. the sensitivity to changes of substrate structure, is very low in this process. It is hard to imagine that

Scheme 33

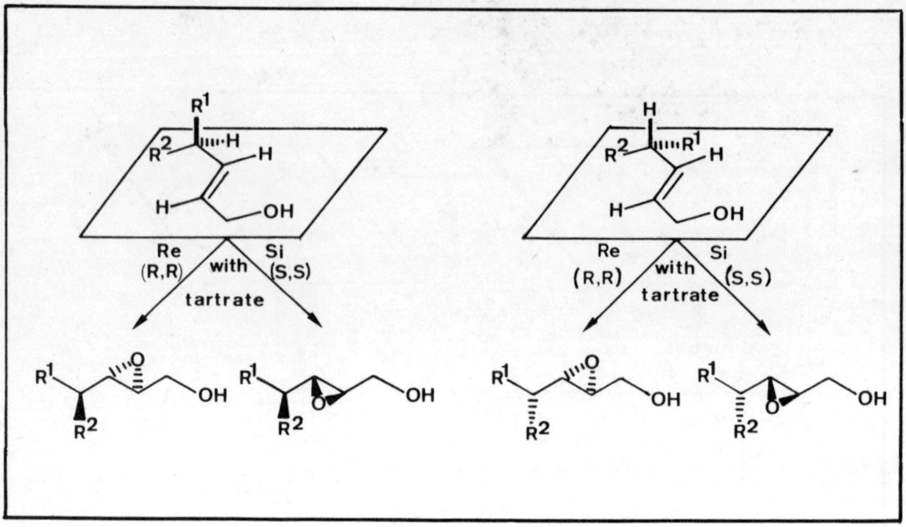

an enzyme can convert substrates of such different structures equally selectively!

The mechanism of the reaction is still under investigation.

Practical considerations are of foremost importance in the titanate/tartrate-mediated asymmetric epoxidations with TBHP. Even with consideration of the limiting factors discussed above, "complaints" have become known about the reproducibility of the process. In all cases which we are aware of, the problems were due to non-optimal conditions. There are numerous, partially conflicting hints as to the practical execution of the reaction. The following procedures are typical.

Solutions of *t*-Butyl-hydroperoxide (TBHP) [198,203]

General Remarks: Absolutely anhydrous TBHP is explosive and should not be prepared without experience with peroxides and without extreme precautions [206,207]. Although handled on industrial scale, absolute TBHP is not commercially available and is excluded from conventional shipment. The 90% aqueous TBHP is less dangerous, but is also not recommended here. The best grade to start with is 70% aqueous TBHP, which should contain as little as possible hydrogen peroxide, di-*t*-butyl peroxide, and *t*-butanol. For the asymmetric epoxidations, the preparation of anhydrous solutions of TBHP is necessary. The preferred solvent is 1.2-dichloroethane. Solutions of anhydrous TBHP are safe to handle and can be stored in a refrigerator, but must not be concentrated.

Preparation of a Dichloroethane Solution: In a 2-L separatory funnel, 500 ml (3.6 mol) 70% aqueous TBHP and 850 ml 1.2-dichloroethane are combined, and the upper aqueous layer of ca. 125 ml is removed. The 1225 ml organic layer containing 3.5 mol TBHP is placed into a 2-L flask, SiC boiling chips are added, and the flask is equipped with a water separator for heavier than water solvents. The flask is heated (steam or oil bath!) and the azeotrope containing ca. 20% water distilled off. After water ceases collecting in the trap, the solution is allowed to cool, the water separator is replaced by a short path distillation head, and heat is reapplied. Ca. 225 ml solvent are removed by distillation, leaving ca. 1 L of 3.5 M TBHP. Solutions in CH_2Cl_2 and other solvents can be prepared similarly. Store in a refrigerator.

Determination of TBHP Concentration: By iodometric titration [207]: the peroxide content is determined with an aliquot of the solution. - By NMR spectroscopy: a TBHP solution in $C_2H_4Cl_2$ is measured in $CDCl_3$ (TMS as internal standard). The *t*-C_4H_9 signal (ca. 1.25 δ) integration A and the dichloroethane signal (ca. 3.70 δ) integration B give the molarity of the solution as $M = A\,[0.10\,A + 0.18\,B]^{-1}$.

(2S,3S)-Epoxy-geraniol [200]

TBHP; $Ti(OCHMe_2)_4$ (8); (R,R)-(+)-DET

A dry 500 ml flask is charged (-23°C cooling bath; argon atmosphere; magnetic stirring) with 200 ml dry CH_2Cl_2, 5.94 ml (5.68 g, 20 mmol) 8 and 3.7 ml (4.5 g, 22 mmol) diethyl (+)-tartrate (10% excess). After stirring for 5 min., 3.47 ml (3.08 g, 20 mmol) geraniol and ca. 11 ml of ca. 3.5 M TBHP (40 mmol, 2 equiv.) as anhydrous solution in $C_2H_4Cl_2$, CH_2Cl_2 or CCl_4 are added. The homogeneous solution is sealed and kept in a freezer (ca. -20°C) overnight (ca. 18 hours). The reaction can be monitored by TLC. The flask is returned to the CCl_4 / dry ice bath (-23°C) and its stirred contents quenched with 50 ml 10% aqueous tartaric acid. The aqueous layer solidifies, the bath is removed after 30 min., after stirring at room temperature for one hour the layers are separated. The organic layer is washed with water, dried (Na_2SO_4) and concentrated.

The residue contains diethyltartrate or other tartrate esters which may have been used, and TBHP. The peroxide content is so small in the present case that it does not have to be removed [198]. For hydrolysis of the ester, the crude product is dissolved in 150 ml ether and combined (stirring, ice bath) with 60 ml of 1 N NaOH. The two-phase system is stirred for no longer than 30 min., the organic phase is dried (Na_2SO_4), concentrated, and the residue (ca. 4.3 g) chromatographed on silica gel to give 2.6 g (77%) epoxide of $[\alpha]_D$ -6.36° (1.5, $CHCl_3$), corresponding to ca. 95% ee.

For water soluble epoxyalcohols, a modified procedure is recommended (see below). With more reactive allylic alcohols (without substituent *cis* to CH_2OH) catalytic amounts (0.1 equiv.) of both the titanate and the tartrate suffice. Diisopropyl tartrate is often preferable, but is more difficult to hydrolyze without destroying epoxide product.

Workup for water-soluble product epoxyalcohols [208]

At the end of the reaction period (see previous procedure), 4 equiv. of CH_3SCH_3 are added at -20°C, and after 40 min. the cold reaction mixture is slowly added to twice its volume of a vigorously stirred, saturated (ca. 5%) aqueous NaF solution at room temperature. After stirring for 5 - 20 hours, NaCl is added to saturate the aqueous phase. Filtration of the gel-like mixture through Celite®, separation of the phases, three extractions of the aqueous phase with CH_2Cl_2, drying of the combined organic layers over Na_2SO_4 and concentrating yields the epoxyalcohol mixed with the tartrate ester which are separated by chromatography on silica gel or, depending on the sensitivity of the epoxide and the relative boiling points, by fractional distillation.

Kinetic Resolution of Allylic Alcohols (Scheme 32) [205]

For instance:

OH (R) + OH (S) —TBHP, (R,R)-DIPT, <u>8</u>→ OH (R)

A solution of 1 equiv. <u>8</u>, 1.2 - 1.5 equiv. (R.R)-(+)-DIPT, 1 equiv. d,l-allylic alcohol in dry CH_2Cl_2 (10 ml/mmol of alcohol) is stirred at -20°C under inert atmosphere. Anhydrous TBHP (4 - 6 M in CH_2Cl_2, 0.6 equiv.) is added, and the homogeneous mixture is stored in a freezer (-20°C) for 15 hours to 12 days. The reaction must be monitored and should not be stopped before 55% conversion. The cold reaction mixture is poured into twice its volume of acetone containing 0.3 ml H_2O/mmol of the titanate <u>8</u>. The mixture is allowed to warm to room temperature with stirring. Filtration, concentration, tartrate hydrolysis as described above and chromatography yields the product.

For the cyclohexenyl methyl carbinol 15 hours reaction time furnishes (R)-alcohol of >96% ee.

(+)-(2S,3R)-2.3-Epoxy-1-tridecanol [203)]

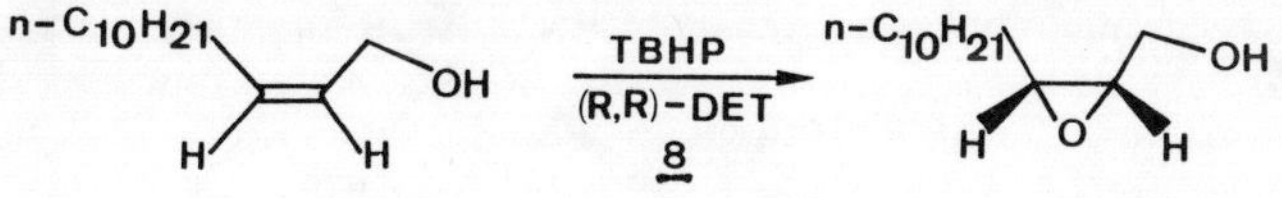

Under inert argon atmosphere, L-(+)-diethyl tartrate (11.3 g, 55 mmol) and *cis*-2-tridecen-1-ol (10.5 g, 50 mmol) are dissolved in 250 ml CH_2Cl_2 and cooled to -78°C. After adding Ti(O-*i*-C_3H_7)$_4$ (15 ml, 50 mmol) by syringe and stirring for several min., 26 ml (83 mmol) of anhydrous TBHP (3.2 M in CH_2Cl_2) is slowly added. The reaction mixture is warmed to -35°C and stirred for 3 days. It is quenched by combining with 150 ml of 10% tartaric acid and shaking vigorously. This is repeated followed by washing the organic layer with H_2O and brine. On concentration *in vacuo* a wet white solid is obtained, which is dissolved in 200 ml ether, combined with 100 ml of 1 M aqueous LiOH and stirred for 30 min. at 0°C. The ether solution is washed with brine, dried and concentrated to give a white mass. This is chromatographed on a Waters prep. 500 (petroleum ether / ether 1:1) to give 8.3 g (38.7 mmol, 77%) of fluffy white crystals.

2) Epoxidations of Allylic and Homoallylic Alcohols by TBHP/Vanadium(V)

Diastereoselective epoxidations of olefins containing alcohol functions with various distances between the two functional groups have been found to be more or less diastereoselective, both with peracids and with transition metal (V^{5+}, Mo^{6+} etc.) catalyzed *t*-butyl hydroperoxide as reagents, see a recent review on metal-catalyzed oxidations of olefins and acetylenes [198)]. The preferred relative topicity *ul* of the 1.2-induction in titanate-induced epoxidations of α-branched *trans*-allylic alcohols was discussed in the previous section, see Scheme 32. A somewhat smaller *ul*-preference had also been found in VO(acac)$_2$-catalyzed TBHP epoxidations of the same allylic alcohol [198,209,210)], while with a trimethylsilyl substituent in the 2-positon, the selectivities are much larger [211)], see Table 16. *cis*-Substituents on the double bond of open-chain systems somehow direct the vanadium-catalyzed oxidation to the other diastereotopic face of the olefinic double bond, compare Table 16 with Table 17. Finally, with cyclic allylic alcohols, the result is a *cis*-

Table 16. Epoxidations of allylic alcohols without substituents in the *cis*-position of the carbinol center by TBHP/V^{5+} reagent.

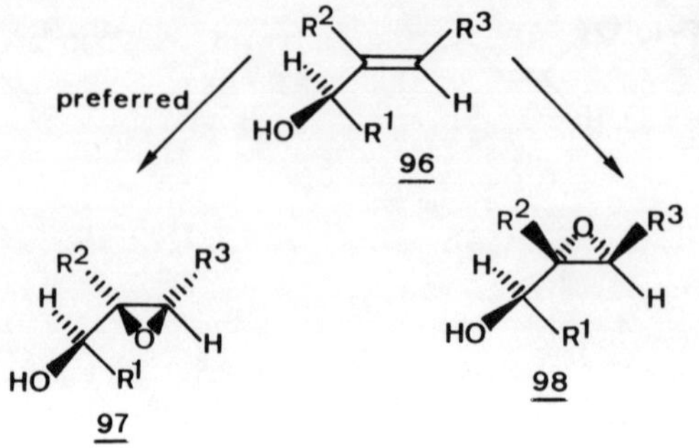

R^1	R^2	R^3	preferred rel.topicity	97/98	ref.
CH_3	H	H	*ul*	80:20	198,209)
CH_3	CH_3	H	*ul*	95:5	198,209)
CH_3	H	CH_3	*ul*	71:29	198,209)
				60:40	210)
n-C_3H_7	H	C_2H_5	*ul*	63:37	210)
i-C_3H_7	H	H	*ul*	85:15	210)
t-C_4H_9	H	CH_3	*lk*	81:19	210)
CH_3	Me_3Si	H	*ul*	>99:1	211)
$CHMe_2$	Me_3Si	H	*ul*	>99:1	211)
CH_3	Me_3Si	CH_3	*ul*	>99:1	211)
$CHMe_2$	Me_3Si	CH_3	*ul*	>99:1	211)

epoxidation, see Scheme 34. Except with some cyclic substrates and with the bulky trimethylsilyl substitution, the selectivities are not as high as those which were encountered with the titanate/tartrate-mediated epoxidations, dealt with in the preceding section.

Table 17. Epoxidations of acyclic allylic alcohols with substituents in the *cis*-position of the carbinol center by TBHP/V^{5+} reagent.

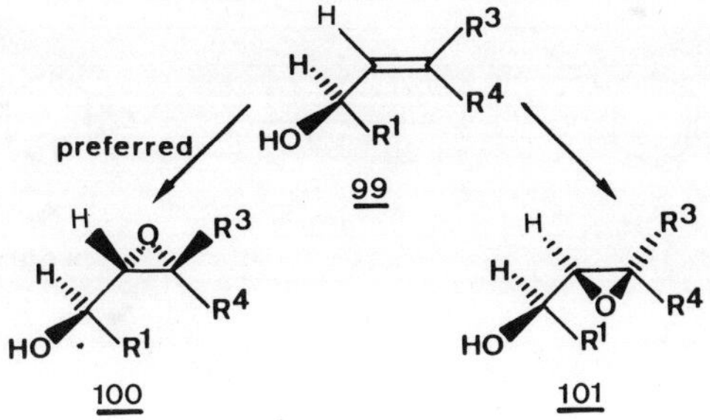

R^1	R^3	R^4	preferred rel.topicity	100/101	ref.
CH_3	H	CH_3	*lk*	71:29	198,209)
CH_3	CH_3	CH_3	*lk*	86:14	209)
CH_3	H	$SiMe_3$	*ul*	96:4	211)
CH_3	CH_3	$SiMe_3$	*ul*	>99:1	211)
i-C_3H_7	H	$SiMe_3$	*ul*	>99:1	211)
C_6H_5	H	$SiMe_3$	*lk*	>99:1	211)

This is not so when we turn to the homoallylic alcohols [215]. Their epoxidation by the V^{5+}/TBHP-reagent is highly selective in most cases, see Table 18. Only bulky substitution at the carbinol center and *trans*-configuration of the double bond [215] causes the diastereoselectivity to fall below 80%; usually it is above 95%. The R^2-group (see Table 18) on the carbon atom between the carbinol center and the proximate olefin sp^2-center appears to be responsible for the stereochemical course of the reaction.

Scheme 34. $VO(acac)_2$-catalyzed *syn*-epoxidations of cyclic allylic alcohols by TBHP.

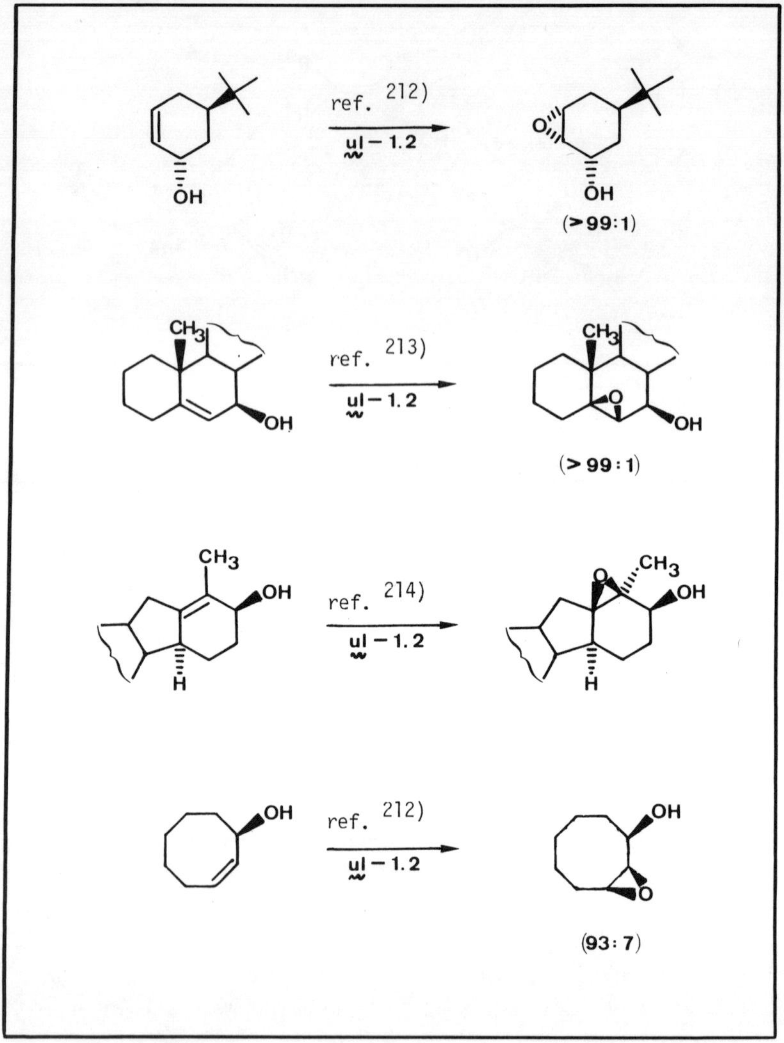

ref. 212) ul – 1.2 (>99:1)

ref. 213) ul – 1.2 (>99:1)

ref. 214) ul – 1.2

ref. 212) ul – 1.2 (93:7)

More convincing than for the simple allylic alcohols, see models 102 [198,209] and 103 [210], is the mechanistic picture 104 proposed [215] for this homoallylic case. A tetrahedral coordination of the vanadium is assumed, with a pseudo-axial and a pseudo-equatorial ligand on the metal. The better fit with Z- than with E-configuration on the double bond is thought to arise from repulsive interaction with the O-*t*-butyl group, see arrow ① in 104. The group located in *cis*-position on the remote olefinic

Table 18. Epoxidations of homoallylic [215] and of bis-homoallylic [216] alcohols by TBHP/V^{5+} reagent leading to products with up to four consecutive asymmetric centers in high diastereoselectivity.

lk-1.2 induction

R^1	CH_3	C_6H_5	$CHMe_2$	CH_3
R^2	CH_3	CH_3	CH_3	$CHMe_2$
% ds	82	83	75	83

	R^1	H	CH_3	C_6H_{13}	CH_3	$CHMe_2$
	R^2	CH_3	H	H	CH_3	CH_3
	R^3	CH_3	C_2H_5	$(CH_2)_7COOR$	C_6H_{13}	CH_3
	% ds	>99.5	92	96	99	>99.5
rel.topicity	1.2	*lk*	--	--	*lk*	*lk*
	1.3	--	*lk*	*lk*	*ul*	*ul*

(Table 18 continued)

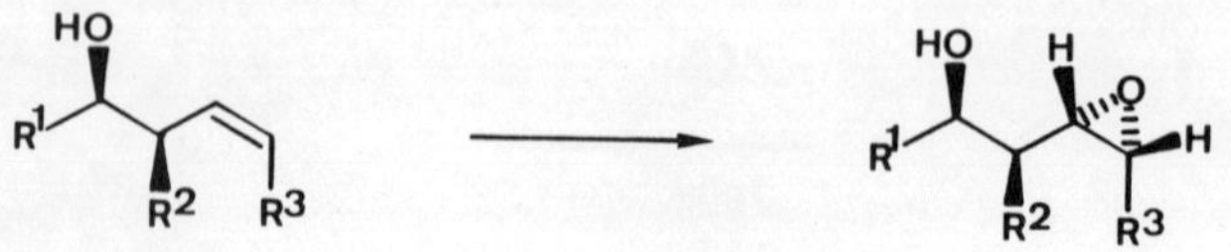

R^1	CH_3	CH_3	$CHMe_2$	C_5H_{11}	CH_3
R^2	CH_3	CH_3	CH_3	CH_3	C_5H_{11}
R^3	CH_3	C_6H_{13}	CH_3	CH_3	CH_3
% ds	99	98	67	94	>99.5
rel.topicity 1.2	*lk*	*lk*	*lk*	*lk*	*lk*
rel.topicity 1.3	*lk*	*lk*	*lk*	*lk*	*lk*

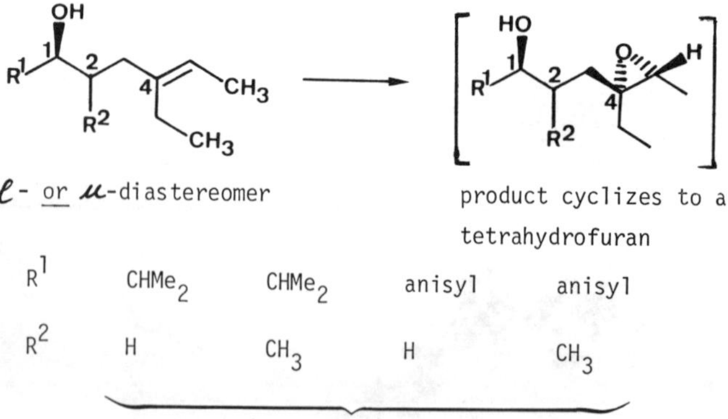

l- or *u*-diastereomer

product cyclizes to a tetrahydrofuran

R^1	$CHMe_2$	$CHMe_2$	anisyl	anisyl
R^2	H	CH_3	H	CH_3

>88% Re attack of oxygen on C(4)

independant of configuration at C(2)

carbon forces a substituent in the allylic position into a quasi-equatorial position, arrow ② in 104. This in turn fixes the position of a substituent on the carbinol center as being either pseudo-axial or -equatorial, depending upon the relative configuration along the C(1), C(2) bond of the homoallylic system. If the substituent is larger than CH_2R and is forced into an axial position, the 1.3-repulsion indicated by arrow ③ in 104 becomes responsible for lower selectivity.

Similar considerations may be used [215)] to discuss the still appreciable diastereoselectivity observed with bis-homoallylic alcohols, with which

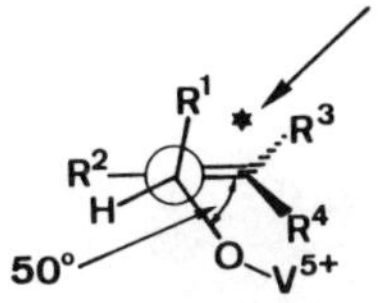

102 a

addition from below

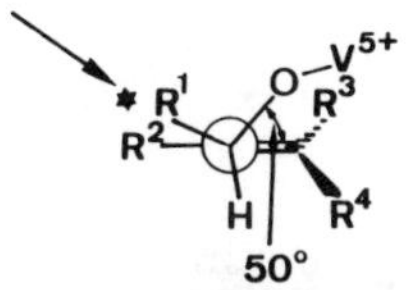

102 b

addition from above

103 a

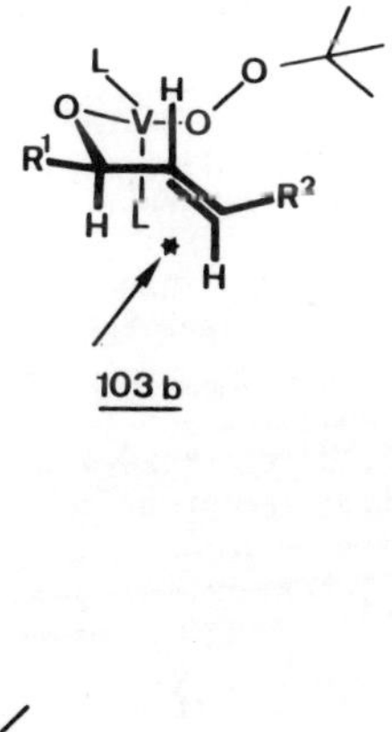

103 b

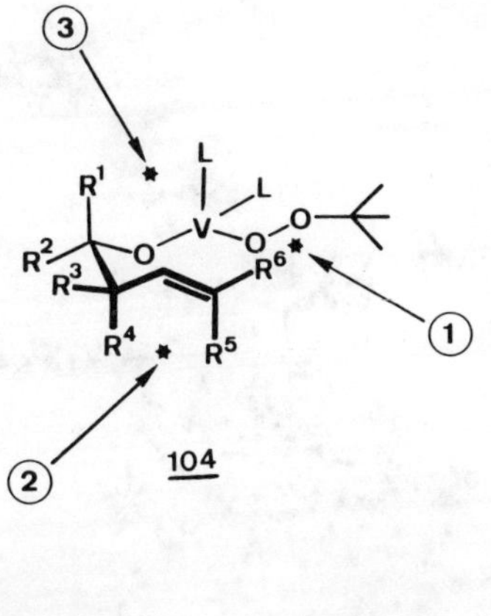

104

only the configuration at the carbinol center seems to matter for the stereochemical course of reaction [216], see bottom of Table 12.

These results are the more important, because there are hardly any other methods which allow remote stereocontrol of this type and quality.

O-Silylated allylic and homoallylic alcohols have just recently [217] been shown to be also epoxidized reasonably selectively by TBHP/V^{5+}.

General procedure for V^{5+}-catalyzed epoxidation of homoallylic alcohols [203,215]

$$R^1\text{-CH(OH)-CH}(R^2)\text{-CH=CHR}^3 \xrightarrow[O=V(O_2C_5H_7)_2]{(CH_3)_3C\text{-OOH}} R^1\text{-CH(OH)-CH}(R^2)\text{-}\overset{O}{\widehat{CH-CH}}\text{-}R^3$$

To a ca. 0.1 M solution of the homoallylic alcohol in anhydrous CH_2Cl_2 is added ca. 1-2% of vanadium (IV) oxide bis(2.4-pentadionate) and 1.5 mol-equiv. of anhydrous 1 M *tert.*-butyl hydroperoxide (TBHP) at ice-bath temperature. The solution is stirred for 16 hours at room temperature. The remaining TBHP is best hydrolyzed by the addition of 1 mol-equiv. of freshly prepared 10% Na_2SO_3 at 0°C. Stirring is continued for an additional 2 - 3 hours with gradual warming to room temperature. The layers are separated and the organic phase is washed twice with water, once with brine, dried ($MgSO_4$) and concentrated. The crude products are purified by distillation or column chromatography.

V ACKNOWLEDGMENTS

We thank *Anneliese Hungerbühler, Martin Schiess, Peter Schnurrenberger, Albert K. Beck,* and *Silvia Sigrist* for preparing the manuscript, Professor *Walter L. Meyer* (on leave from the University of Arkansas) for his help with the correction of the typed version, and the *Dynamit Nobel AG* (D-Troisdorf) for generous supplies of titanates and zirconates used in our own investigations. We were fortunate to attract able and enthusiastic collaborators to join us part of the way through our continuing search for applications of transition metal derivatives in organic synthesis; their names are given in the list of references. Financial support and supply of starting materials *(Dr. H. Braunschweiger, F. Seemann)* by the *Sandoz AG* (Basel) is gratefully acknowledged.

LIST OF REFERENCES

1) See e.g. A. Segnitz in Houben-Weyl-Müller, "Methoden der organischen Chemie", 13/7, p. 326 (1975).

2) R.H. Grubbs, Prog. Inorg. Chem. 24, 1 (1978); N. Calderon, J.P. Lawrence & E.A. Ofstead, Adv. Organomet. Chem. 17, 449 (1979).

3) J. Chatt, L.M. da Câmara Pina & R.L. Richards, "New Trends in the Chemistry of Nitrogen Fixation", Academic Press, London 1980.

4) J.J. Eisch & R.B. King, "Organometallic synthesis", Vol. 1, Academic Press, London 1965, and ref. cited therein.

5) B. Kautzner, P.C. Wailes & H. Weigold, J. Chem. Soc. Chem. Commun. 1969, 1105; P.C. Wailes & H. Weigold, J. Organomet. Chem. 24, 405 (1970); D.B. Carr & J. Schwartz, J. Am. Chem. Soc. 101, 3521 (1979).

6) F.N. Tebbe, G.W. Parshall & G.S. Reddy, J. Am. Chem. Soc. 100, 3611 (1978).

7) S.H. Pine, R. Zahler, D.A. Evans & R.H. Grubbs, J. Am. Chem. Soc. 102, 3270 (1980).

8) B. Weidmann, Ch.D. Maycock & D. Seebach, Helv. Chim. Acta 64, 1552 (1981).

9) D.C. Bradley & I.M. Thomas, J. Chem. Soc. 1960, 3857.

10) M. Schiess, unpublished results, ETH Zürich 1982/83.

11) C.Dijkgraaf & J.P.G. Rousseau, Spectrochim. Acta A 24, 1213 (1968).

12) B. Weidmann, Dissertation ETH Zürich No. 7203 (1982).

13) M.T. Reetz, R. Steinbach & K. Kesseler, Angew. Chem. Suppl. 1982, 1899.

14) L. Widler & D. Seebach, Helv. Chim. Acta 65, 1085 (1982).

15) H. Bürger & H.J. Neese, Z. Anorg. Allg. Chem. 370, 275 (1969).

16) K. Clauss, Liebigs Ann. Chem. 711, 19 (1968).

17) H. Holloway, Chem. and Ind. 1962, 214.

18) B. Weidmann & D. Seebach, Angew. Chem. 95, 12 (1983); ibid. Int. Ed. Engl. 22, 31 (1983).

19) M.T. Reetz, Top. Curr. Chem. 106, 1 (1982).

20) B. Weidmann & D. Seebach, Helv. Chim. Acta 63, 2451 (1980).

21) M.T. Reetz, R. Steinbach, J. Westermann & R. Peter, Angew. Chem. 92, 1044 (1980); ibid. Int. Ed. Engl. 19, 1011 (1980).

22) B. Weidmann, L. Widler, A.G. Olivero, Ch.D. Maycock & D. Seebach, Helv. Chim. Acta 64, 357 (1981).

23) Ch.D. Maycock, unpublished results, ETH Zürich 1981.

24) B. Weidmann, unpublished results, ETH Zürich 1981.

25) Th. Kauffmann, E. Antfang, B. Ennen & N. Klas, Tetrahedron Lett. 23, 2301 (1982).

26) M.T. Reetz, R. Steinbach, J. Westermann, R. Urz, B. Wenderoth & R. Peter, Angew. Chem. 94, 133 (1982); ibid. Int. Ed. Engl. 21, 135 (1982); ibid. Suppl. 1982, 257.

27) Ch. Papaspyrou, unpublished results, ETH Zürich 1982.

28) M. Schiess & D. Seebach, Helv. Chim. Acta 65, 2598 (1982).

29) M.T. Reetz, R. Steinbach, B. Wenderoth & J. Westermann, Chem. and Ind. 1981, 541.

30) M.T. Reetz, J. Westermann & R. Steinbach, Angew. Chem. 92, 931 and 933 (1980); ibid. Int. Ed. 19, 900 and 901 (1980).

31) M.T. Reetz, J. Westermann & R. Steinbach, J. Chem. Soc. Chem. Commun. 1981, 237.

32) M.T. Reetz, R. Steinbach & B. Wenderoth, Synth. Commun. 11, 261 (1981).

33) M.T. Reetz, B. Wenderoth, R. Peter, R. Steinbach & J. Westermann, J. Chem. Soc. Chem. Commun. 1980, 1202.

34) T. Kauffmann, R. König, C. Pahde & A. Tannert, Tetrahedron Lett. 22, 5031 (1981).

35) J.G. Dillard, Inorg. Chem. 8, 2148 (1969); J.F. Hanlan & J.D. McCowan, Can. J. Chem. 50, 747 (1972).

36) T.S. Patterson, J. Chem. Soc. 69, 1002 (1896).

37) D. Seebach & R. Naef, Helv. Chim. Acta 64, 2704 (1981).

38) J.J. Sims, L.H. Selman & M. Cadogan, Org. Synth. 51, 109 (1971).

39) D. Seebach, T. Weller, G. Protschuk, A.K. Beck & M.S. Hoekstra, Helv. Chim. Acta 64, 716 (1981).

40) D. Seebach & V. Prelog, Angew. Chem. 94, 696 (1982); ibid. Int. Ed. Engl. 21, 654 (1982).

41) D. Seebach, H. Dörr, B. Astani & V. Ehrig, Angew. Chem. 81, 1002 (1969); ibid. Int. Ed. Engl. 8, 982 (1969).

42) D. Seebach, G. Crass, E.M. Wilka, D. Hilvert & E. Brunner, Helv. Chim. Acta 62, 2695 (1979).

43) T. Mukaiyama, K. Soai, T. Sato, H. Shimizu & K. Suzuki, J. Am. Chem. Soc. 101, 1455 (1979).

44) J.P. Mazaleyrat & D.J. Cram, J. Am. Chem. Soc. 103, 4585 (1981).

45) A.K. Beck, unpublished results, ETH Zürich 1982/83.

46) M. Carmack, C.J. Kelly, J. Org. Chem. 33, 2171 (1968).

47) C. Blandy & D. Gervais, Inorg. Chim. Acta 52, 79 (1981).

48) G.M. Whitesides, C.P. Casey & J.K. Krieger, J. Am. Chem. Soc. 93, 1379 (1971).

49) J. Schwartz, M.J. Loots & H. Kosugi, J. Am. Chem. Soc. 102, 1333 (1980); F.M. Dayrit & J. Schwartz, J. Am. Chem. Soc. 103, 4466 (1981), and ref. cited therein.

50) D.W. Hart, T.F. Blackburn & J. Schwartz, J. Am. Chem. Soc. 97, 679 (1975).

51) M.J. Loots & J. Schwartz, J. Am. Chem. Soc. 99, 8045 (1977).

52) F.W. Hartner & J. Schwartz, J. Am. Chem. Soc. 103, 4979 (1981).

53) E. Thorn-Csányi, Nachr. Chem. Tech. Lab. 29, 700 (1981).

54) S.H. Pine, R. Zahler, D.A. Evans, & R.H. Grubbs, J. Am. Chem. Soc. 102, 3270 (1980).

55) J.W. Sam Stevenson & T.A. Breyson, Tetrahedron Lett. 23, 3143 (1982).

56) S. Murai & N. Sonoda, Angew. Chem. 91, 896 (1979); ibid. Int. Ed. Engl. 18, 837 (1979); P. Pino, F. Piacenti & M Bianchi in "Organic Synthesis via Metal Carbonyls", Vol. 2, p. 43, Wiley-Interscience, 1977.

57) I. Ugi, Angew. Chem. 94, 826 (1982); ibid. Int. Ed. Engl. 21, 810 (1982).

58) J.P. Collman, Acc. Chem. Res. 8, 342 (1975).

59) I.I. Lapkin & T.N. Povarnitsyna, Zh. Obshch. Khim. 38, 99 (1968).

60) L.S. Trzupek, T.L. Newirth, E.G. Kelly, N.E. Sbarbati & G.M. Whitesides, J. Am. Chem. Soc. 95, 8118 (1973).

61) Reviews: M.P. Periasamy & H.M. Walborsky, Org. Prep. Proc. Int. 11, 293 (1979), and ref. cited therein.

62) D. Seebach, Angew. Chem. 91, 259 (1979); ibid. Int. Ed. Engl. 18, 239 (1979).

63) C.A. Bertelo & J. Schwartz, J. Am. Chem. Soc. 97, 228 (1975).

64) M. Passerini, Gazz. Chim. Ital. 51, 126 and 181 (1921). See also [57] and ref. cited therein.

65) T. Mukaiyama, K. Watanabe & M. Shiono, Chem. Lett. 1974, 1457.

66) B. Crociani, M. Nicolini & R.L. Richards, J. Organomet. Chem. 101, C 1 (1975).

67) M.T. Reetz, Angew. Chem. 94, 97 (1982); ibid. Int. Ed. Engl. 21, 96 (1982).

68) T. Mukaiyama, "Titanium Tetrachloride in Organic Synthesis", Angew. Chem. 89, 858 (1977); ibid. Int. Ed. Engl. 16, 817 (1977).

69) I. Fleming, "The Alkylation of Enolates and Enole Derivatives: The Use of Silyl Enole Ethers", Chimia 34, 265 (1980).

70) J.K. Rasmussen, "O-Silylated Enolates - Versatile Intermediates for Organic Synthesis", Synthesis 1977, 91.

71) E.W. Colvin, "Silicone in Organic Synthesis", Butterworths, London 1981.

72) T.H. Chan & I. Fleming, "Electrophilic Substitution of Organosilicone Compounds - Applications ot Organic Synthesis", Synthesis 1979, 761.

73) T. Mukaiyama & A. Ishida, Chem. Lett. 1975, 319.

74) M.T. Reetz, in "30 Jahre Fonds der Chemischen Industrie 1950-1980", p. 29, Verband der Chemischen Industrie e.V. 1980.

75) M.T. Reetz & R. Peter, Tetrahedron Lett. 22, 4691 (1981).

76) L. Widler, projected Ph.D. thesis, ETH Zürich 1983.

77) L. Widler & D. Seebach, Helv. Chim. Acta 65, 1972 (1982).

78) M.T. Reetz & W.F. Maier, Angew. Chem. 90, 50 (1978); ibid. Int. Ed. Engl. 17, 48 (1978).

79) T. Mukaiyama & M. Hayashi, Chem. Lett. 1974, 15.

80) T. Mukaiyama, K. Banno & K. Narasaka, J. Am. Chem. Soc. 96, 7503 (1974).

81) K. Narasaka, K. Soai, Y. Aikawa & T. Mukaiyama, Bull. Chem. Soc. Jpn. 49, 779 (1976).

82) I. Paterson & I. Fleming, Tetrahedron Lett. 1979, 995.

83) M.T. Reetz, S. Hüttenhain, P. Walz & U. Löwe, Tetrahedron Lett. 1979, 4971.

84) M. Miyashita, T. Yanami & A. Yoshikoshi, J. Am. Chem. Soc. 98, 4679 (1976).

85) A. Ishida & T. Mukaiyama, Bull. Chem. Soc. Jpn. 51, 2077 (1978).

86) I. Fleming & T.V. Lee, Tetrahedron Lett. 22, 705 (1981).

87) C.H. Heathcock, "The Aldol Addition Reaction" in "Asymmetric Synthesis", Vol. 2, chapter 2, Ed. J.D. Morrison, Academic Press Inc., New York 1983.

88) C.H. Heathcock, in "Comprehensive Carbanion Chemistry", Vol. II, Ed. T. Durst & E. Buncel, Elsevier, in preparation.

89) C.H. Heathcock, "The Aldol Condensation as a Tool for Stereoselective Organic Synthesis", Pure Appl. Chem. (1983), in press.

90) C.H. Heathcock, "Acyclic Stereoselection via the Aldol Condensation" in "Asymmetric Reactions and Processes in Chemistry", p. 55, Ed. E.L. Eliel & S. Otsuka, ACS-Symposium Series No. 185, 1982.

91) D.A. Evans, J.V. Nelson & T.R. Taber, "Stereoselective Aldol Condensations" in "Topics in Stereochemistry", Vol. 13, p. 1, Wiley-Interscience, New York 1982.

92) C.H. Heathcock, "Acyclic Stereocontrol Through the Aldol Condensation", Science 214, 395 (1981).

93) S. Masamune in "Organic Synthesis Totday and Tomorrow", p. 197, Ed. B.M. Trost & R. Hutchinson, Pergamon Press, New York 1981.

94) D.A. Evans, J.M. Takacs, L.R. McGee, M.D. Ennis, D.J. Mathre & J. Bartroli, "Chiral Enolate Design", Pure & Appl. Chem. 53, 1109 (1981).

95) G. Solladié, "Asymmetric Synthesis Using Nucleophilic Reagents Containing a Chiral Sulfoxide Group", Synthesis 1981, 185.

96) Y. Yamamoto & K. Maruyama, "The Metal Effect on the Stereoselection in Acyclic Systems - The Reaction of Allylic Organometallic Compounds and

of Metal Enolates", J. Synth. Org. Chem. Jpn. 40, 332 (1982).

97) M. Shibasaki & S. Ikegami, "Chelation-Controlled Stereo and Regioselective Reactions: Application to Natural Product Synthesis", J. Synth. Org. Chem. Jpn. 40, 602 (1982).

98) R.W. Hoffmann, "Diastereogenic Addition of Crotylmetal Compounds to Aldehydes", Angew. Chem. 94, 569 (1982); ibid. Int. Ed. Engl. 21, 555 (1982).

99) F. Sato, "Applications of Bis(η-cyclopentadienyl)titanium Compounds in Organic Synthesis", J. Synth. Org. Chem. Jpn. 40, 744 (1982).

100) T. Hiyama, "Chromium Reagents for Selective C-C Bond Formation", J. Synth. Org. Chem. Jpn. 39, 81 (1981).

101) D. Seebach, A.K. Beck, T. Mukhopadhyay & E. Thomas, "Diastereoselective Synthesis of Nitroaldol Derivatives", Helv. Chim. Acta 65, 1101 (1982).

102) D. Seebach & J. Goliński, "Diastereoselective *Michael* Additions", Helv. Chim. Acta 64, 1413 (1981).

103) S. Blarer, W.B. Schweizer & D. Seebach, "Diastereo- and Enantioselective *Michael* Additions", Helv. Chim. Acta 65, 1637 (1982).

104) Y. Kishi *et al.* , J. Am. Chem. Soc. 104, 7362, 7364, 7367 and 7369 (1982).

105) V. Prelog & G. Helmchen, Angew. Chem. 94, 614 (1982); ibid. Int. Ed. Engl. 21, 567 (1982).

106) M. Schlosser & B. Schaub, J. Am. Chem. Soc. 104, 5821 (1982).

107) Y. Yamamoto & K. Maruyama, Tetrahedron Lett. 21, 4607 (1980).

108) D.A. Evans & L.R. McGee, Tetrahedron Lett. 21, 3975 (1980).

109) D.A. Evans & L.R. McGee, J. Am. Chem. Soc. 103, 2876 (1981).

110) M.T. Reetz, R. Steinbach & K. Kesseler, Angew. Chem. 94, 872 (1982); ibid. Int. Ed. Engl. 21, 864 (1982); ibid. Suppl. 1982, 1899.

111) R. Imwinkelried, Master's thesis, ETH Zürich 1982/83.

112) L. Widler, Th. Weber & D. Seebach, Chem. Ber., 116 (1983) in preparation.

113) Th. Weber, Master's thesis, ETH Zürich 1982.

114) R. Hanko & D. Hoppe, Angew. Chem. 94, 378 (1982); ibid. Int. Ed. Engl. 21, 372 (1982); ibid. Suppl. 1982, 961.

115) M.T. Reetz & B. Wenderoth, Tetrahedron Lett. 23, 5259 (1982).

116) E. Klei, J.H. Teuben, H.J. de Liefde Meijer, E.J. Kwak & A.P. Bruins, J. Organomet. Chem. 224, 327 (1982).

117) F. Sato, S. Iijima & M. Sato, Tetrahedron Lett. 22, 243 (1981).

118) F. Sato, K. Iida, S. Iijima, H. Moriya & M. Sato, J. Chem. Soc. Chem. Commun. 1981, 1140.

119) E.J. Corey & D. Enders, Chem. Ber. 111, 1337 and 1362 (1978); D. Enders, H. Eichenauer & R. Pieter, Chem. Ber. 112, 3703 (1979); K.G. Davenport, H. Eichenauer, D. Enders, M. Newcomb & D.E. Bergbreiter, J. Am. Chem. Soc. 101, 5654 (1979).

120) D. Hoppe & A. Brönneke, Tetrahedron Lett. 24 (1983), in press.

121) D. Hoppe, Nachr. Chem. Tech. Lab. 30, 1030 (1982)

122) R.D. Rieke & S. Bales, J. Am. Chem. Soc. 96, 1775 (1974).

123) M. Ishiguro, N. Ikeda & H. Yamamoto, J. Org. Chem. 47, 2225 (1982).

124) D.E. Bergbreiter & E. Pendergrass, J. Org. Chem. 46, 219 (1981).

125) Review: D. Seebach & K.H. Geiss, J. Organomet. Chem. Library 1, 1 (1976).

126) D. Seebach & E. Hungerbühler, in "Modern Synthetic Methods 1980", Ed.: R. Scheffold, Salle + Sauerländer, Aarau 1980.

127) K.H. Geiss, B. Seuring, R. Pieter & D. Seebach, Angew. Chem. 86, (484 (1974); ibid. Int. Ed. Engl. 13, 479 (1974); K.H. Geiss, D. Seebach & B. Seuring, Chem. Ber. 110, 1833 (1977).

128) D. Seebach, K.H. Geiss & M. Pohmakotr, Angew. Chem. 88, 449 (1976); ibid. Int. Ed. Engl. 15, 437 (1976); M. Pohmakotr, K.H. Geiss & D. Seebach, Chem. Ber. 112, 1420 (1979).

129) K.H. Geiss, Dissertation, Giessen 1975.

130) P. Pohmakotr, Dissertation, Giessen 1978.

131) J.F. Normant & A. Alexakis, Synthesis 1981, 841.

132) E. Negishi, Pure & Appl. Chem. 53, 2333 (1981).

133) "Aluminiumalkyle, Hinweise zur sicheren Handhabung", Schering, Industrie-Chemikalien, D-4619 Bergkamen.

134) H.O. House, "Modern Synthetic Reactions", 2nd Edition, Benjamin, Menlo Park 1972.

135) J.J. Bloomfield, D.C. Owsley & J.M. Nelke, Org. React. 23, 259 (1976).

136) Y. Ito, T. Konoike & T. Saegusa, J. Am. Chem. Soc. 97, 649 (1975).

137) Y. Kobayashi, T. Taguchi & E. Tokuno, Tetrahedron Lett. 1977, 3741; Y. Ito, T. Konoike & T. Saegusa, J. Am. Chem. Soc. 97, 2912 (1975).

138) Y. Ito, T. Konoike, T. Harada & T. Saegusa, J. Am. Chem. Soc. 99, 1487 (1977).

139) R.M. Dessau & E.I. Heiba, J. Org. Chem. 39, 3457 (1974).

140) J.E. McMurry & M.P. Fleming, J. Am. Chem. Soc. 96, 4708 (1974).

141) H. Wynberg, K. Lammertsma & L.A. Hulshof, Tetrahedron Lett. 1975, 3749.

142) R.F. Langler & T.T. Tidwell, Tetrahedron Lett. 1975, 777.

143) D.S. Bomse & T.H. Morton, Tetrahedron Lett. 1975, 781.

144) A.L. Baumstark, E.J.H. Bechara & M.J. Semigran, Tetrahedron Lett. 1976, 3265.

145) P.D. Mollere, K.N. Houk, D.S. Bomse & T.H. Morton, J. Am. Chem. Soc. 98, 4732 (1976).

146) J.E. McMurry & M.P. Fleming, J. Org. Chem. 41, 896 (1976).

147) J.E. McMurry, M.P. Fleming, K.L. Kees & L.R. Krepski, J. Org. Chem. 43, 3255 (1978).

148) S.C. Welch & M. Walters, J. Org. Chem. 43, 4797 (1978).

149) J.E. Pauw & A.C. Weedon, Tetrahedron Lett. 23, 5485 (1982).

150) S. Tyrlik & I. Wolochowicz, Bull. Soc. Chim. Fr. 1973, 2147.

151) J.A. Marshall, R.E. Bierenbaum & K.H. Chung, Tetrahedron Lett. 1979, 2081.

152) L. Castedo, J.M. Saá, R. Suau & G. Tojo, J. Org. Chem. 46, 4292 (1981).

153) T. Mukaiyama, T. Sato & J. Hanna, Chem. Lett. 1973, 1041.

154) D. Lenoir, Synthesis 1977, 553.

155) D. Lenoir & P. Lemmen, Chem. Ber. 113, 3112 (1980).

156) E.J. Corey, R.L. Danheiser & S. Chandrasekaran, J. Org. Chem. 41, 260 (1976).

157) J.E. McMurry, Acc. Chem. Res. 7, 281 (1974).

158) A. Clerici & O. Porta, J. Org. Chem. 47, 2852 (1982).

159) D. Ghiringhelli, Tetrahedron Lett. 24, 287 (1983).

160) R.D. Rieke, Acc. Chem. Res. 10, 301 (1977).

161) J. Leimner & P. Weyerstahl, Chem. Ber. 115, 3697 (1982).

162) B.P. Mundy, R. Srinivasa, Y. Kim & T. Dolph, J. Org. Chem. 47, 1657 (1982).

163) H.M. Walborsky & H.H. Wüst, J. Am. Chem. Soc. 104, 5807 (1982).

164) J.M. Pons & M. Santelli, Tetrahedron Lett. 23, 4937 (1982).

165) K. Takai, Y. Hotta, K. Oshima & H. Nozaki, Tetrahedron Lett. 1978, 2417.

166) S. Inaba & I. Ojima, Tetrahedron Lett. 1977, 2009.

167) G. Calderari, Master's thesis, ETH Zürich 1982.

168) H.C. Brown & S. Krishnamurthy, Tetrahedron 35, 567 (1979). See also [169].

169) A. Hajós, "Complex Hydrides", Elsevier, Amsterdam 1979.

170) N. Kornblum, J.W. Powers, G.J. Anderson, W.J. Jones, H.O. Larson, O. Levand & W.M. Weaver, J. Am. Chem. Soc. 79, 6562 (1957).

171) K.E. Pfitzner & J.G. Moffatt, J. Am. Chem. Soc. 87, 5661 (1965). For a review see: J.G. Moffatt, "Sulfoxide - Carbodiimid and Related Oxidations" in "Oxidationen", Vol. 2, p. 1, Ed. R.L. Augustine & D.J. Tredor, Marcel Dekker, New York 1971.

172) E.J. Corey & C.U. Kim, J. Am. Chem. Soc. 94, 7586 (1972); E.J. Corey & C.U. Kim, Tetrahedron Lett. 1973, 919; E.J. Corey & J.W. Suggs, Tetrahedron Lett. 2647 (1975).

173) A.J. Mancuso, S.L. Huang & D. Swern, J. Org. Chem. 43, 2480 (1978).

174) H. Meerwein, B.v. Bock, Br. Kirschnick, W. Lenz & A. Migge, J. prakt. Chem. N.F. 147, 211 (1937).

175) A. Planta, Master's thesis, ETH Zürich 1982.

176) J.F. Eastham & R. Teranishi, Org. Synth. Coll. Vol. IV, 192 (1963).

177) M. Bös, unpublished results, ETH Zürich 1982.

178) S. Kano, Y. Tanaka, E. Sugino & S. Hibino, Synthesis 1980, 695.

179) F. Sato, T. Jinbo & M. Sato, Synthesis, 1981, 871.

180) Review: D. Seebach, E.W. Colvin, F. Lehr & T. Weller, Chimia 33, 1 (1979).

181) P. Knochel & D. Seebach, Nouv. J. Chim. 5, 75 (1981); P. Knochel & D. Seebach, Tetrahedron Lett. 22, 3223 (1981); ibid. 23, 3897 (1982); P. Knochel & D. Seebach, Synthesis 1982, 1017.

182) D. Seebach, A.K. Beck, F. Lehr, T. Weller & E.W. Colvin, Angew. Chem. 93, 422 (1981); ibid. Int. Ed. Engl. 20, 397 (1981); F. Lehr, J. Gonnermann & D. Seebach, Helv. Chim. Acta 62, 2258 (1979). See also [101].

183) J.E. McMurry & J. Melton, J. Org. Chem. 38, 4367 (1973); see also [157].

184) R. Henning, Dissertation, Giessen 1978; D. Seebach, R. Henning & F. Lehr, Angew. Chem. 90, 479 (1978); ibid. Int. Ed. Engl. 17, 458 (1978).

185) H. Bürger & H.J. Neese, Chimia 24, 209 (1970).

186) H. Weingarten & W.A. White, J. Am. Chem. Soc. 88, 850 (1966). H. Weingarten & W.A. White, J. Org. Chem. 31, 2874 (1966).

187) W.A. White & H. Weingarten, J. Org. Chem. 32, 213 (1967).

188) M.T. Reetz, Lecture presented at the University of Zurich, Dec. 17, 1982.

189) D. Seebach, E. Hungerbühler, R. Naef, P. Schnurrenberger, B. Weidmann & M. Züger, Synthesis 1982, 138.

190) H. Rehwinkel & W. Steglich, Synthesis 1982, 826.

191) P. Schnurrenberger, M.F. Züger & D. Seebach, Helv. Chim. Acta 65, 1197 (1982). See also D. Seebach & M.F. Züger, Helv. Chim. Acta 65, 495 (1982).

192) J. Schwartz & J. Labinger, Angew. Chem. 88, 402 (1976); ibid. Int. Ed. Engl. 15, 333 (1976).

193) H.C. Brown, "Organic Syntheses via Boranes", John Wiley & Sons, New York 1975.

194) T. Gibson, Tetrahedron Lett. 23, 157 (1982).

195) J.D. Roberts & M.C. Caserio, "Basic Principles of Organic Chemistry",

p. 426, Benjamin, New York 1965.

196) G. Berti, in "Topics in Stereochemistry", Vol. 7, p. 93, Ed.: N.L. Allinger & E.L. Eliel, John Wiley & Sons, New York 1973.

197) D.N. Kirk, Chem. and Ind. 1973, 109.

198) K.B. Sharpless & T.R. Verhoeven, Aldrichimica Acta 12, 63 (1979).

199) See also the review on "The Catalytic Epoxidation of Olefins with Organic Hydroperoxides", J. Sobczak & J.J. Ziolkowski, J. Mol. Catalysis 13, 11 (1981).

200) T. Katsuki & K.B. Sharpless, J. Am. Chem. Soc. 102, 5974 (1980).

201) D. Seebach, M. Bös & R. Naef, J. Am. Chem. Soc. 105 (1983), in press.

202) Eleventh International Carbohydrate Symposium, 1982, Vancouver: K.B. Sharpless, C.H. Behrens, T. Katsuki, A.W.M. Lee, V.S. Martin, M. Takatani, S.M. Viti, F.J. Walker & S.S. Woodard, Pure & Appl. Chem., (1983), in press.

203) B.E. Rossiter, "Selective Epoxidation of Olefins, Allylic Alcohols, and Homoallylic Alcohols with *tert*-Butyl Hydroperoxide and Molybdenum, Vanadium, and Titanium Catalysts. Application to Asymmetric Synthesis of (+)-Disparture and (-)-γ-Amino-β-(R)-Hydroxybutyric Acid", Dissertation, Stanford University 1981.

204) S.S. Woodard, "A Mechanistic Inverstigation of Asymmetric Epoxidation Catalyzed by Chiral Titanium-Tartrate Complexes, Dissertation, Stanford University 1981.

205) V.S. Martin, S.S. Woodard, T. Katsuki, Y. Yamada, M. Ideda & K.B. Sharpless, J. Am. Chem. Soc. 103, 6237 (1981).

206) W. Weigert, "Wasserstoffperoxid und seine Anwendung", Hüthig-Verlag, Heidelberg 1978.

207) R. Criegee, in Houben-Weyl, "Methoden der Organischen Chemie", Vol. VIII, p. 1 (1952).

208) B.E. Rossiter, T. Katsuki & K.B. Sharpless, J. Am. Chem. Soc. 103, 464 (1981).

209) B.E. Rossiter, T.R. Verhoeven & K.B. Sharpless, Tetrahedron Lett. 1979, 4733.

210) E.D. Mihelich, Tetrahedron Lett. 1979, 4729.

211) A.S. Narula, Tetrahedron Lett. 23, 5579 (1982).

212) T. Itoh, K. Jitsukawa, K. Kaneda & S. Teranishi, J. Am. Chem. Soc. 101, 159 (1979).

213) D. Baldwin & J.R. Hanson, J. Chem. Soc. Perkin Trans. 1, 1941 (1975).

214) A. Murai, N. Iwasa & T. Masamune, Chem. Lett. 1977, 235.

215) E.D. Mihelich, K. Daniels & D.J. Eickhoff, J. Am. Chem. Soc. 103, 7690 (1981).

216) T. Fukuyama, B. Vranesic, D.P. Negri & Y. Kishi, Tetrahedron Lett. 1978, 2741.

217) T. Hiyama & M. Obayashi, Tetrahedron Lett. 24, 395 (1983).

218) T. Katsuki, A.W.M. Lee, P. Ma, V.S. Martin, S. Masamune, K.B. Sharpless, D. Tuddenham, F.J. Walker & S.M. Viti, J. Org. Chem. 47, 1373 and 1378 (1982); A.W.M. Lee, V.S. Martin, S. Masamune, K.B. Sharpless & F.J. Walker, J. Am. Chem. Soc. 104, 3515 (1982).

VITAMIN B_{12} AND RELATED Co-COMPLEXES AS CATALYSTS IN ORGANIC SYNTHESIS

Professor R. Scheffold, Dr. G. Rytz and Dr. L. Walder,
Institut für Organische Chemie, Universität Bern, Freiestr. 3, CH-3012 Bern.

CONTENTS

1. INTRODUCTION

The uniqueness and complexity in structure and reactivity of vitamin B_{12} and related corrinoids have presented major challenges in all areas of the natural and life sciences during the past half century.

It is the aim of this chapter to emphasize the field of applications of vitamin B_{12} and related synthetic cobalt complexes as catalysts in organic synthesis. Among the numerous publications on the chemistry of organocobalt complexes [for an excellent review see Pratt and Craig [1)] as well as Dodd and Johnson [2)]] the selection is restricted to reports concerning non-enzymatic catalysis of potential value in synthesis. Some recent and unpublished results from our laboratory are included.

Up to now only few applications of vitamin B_{12} as catalyst in organic synthesis and no large-scale industrial process are known. Nevertheless we believe that this compound will be of importance to the chemist with a practical interest in organic synthesis for the following reasons:

- vitamin B_{12} is a natural, nontoxic, chiral and enantiomerically pure catalyst
- coenzyme B_{12} in conjunction with a specific enzyme is known to promote reactions of no precedent in organic chemistry
- further work on the catalytic action of vitamin B_{12} and related compounds might contribute to the understanding of the (partially still unclear) mechanisms of coenzyme B_{12}-catalyzed reactions in biochemical transformations
- following Eschenmoser's most fascinating hypothesis on the prebiotic synthesis of corrinoids [3)], such macrocyclic cobalt complexes might be considered to belong to the group of "first hour" catalysts taking part in the development of organic chemistry on the earth.

2. COBALT COMPLEXES AND THEIR SYNTHESIS

2.1. Vitamin B_{12}

A complete and comprehensive review of all major chemical, biochemical and medical aspects on vitamin B_{12} was published 1982 by Dolphin [4]. Other recent monographs appeared by Friedrich [5], Zagalak and Friedrich [6], by Pratt [7] on the inorganic chemistry of vitamin B_{12} and by Babior [8] on the biochemistry and pathophysiology. The stereochemical course of coenzyme B_{12}-dependent rearrangements was recently reviewed by Rétey and Robinson [9].

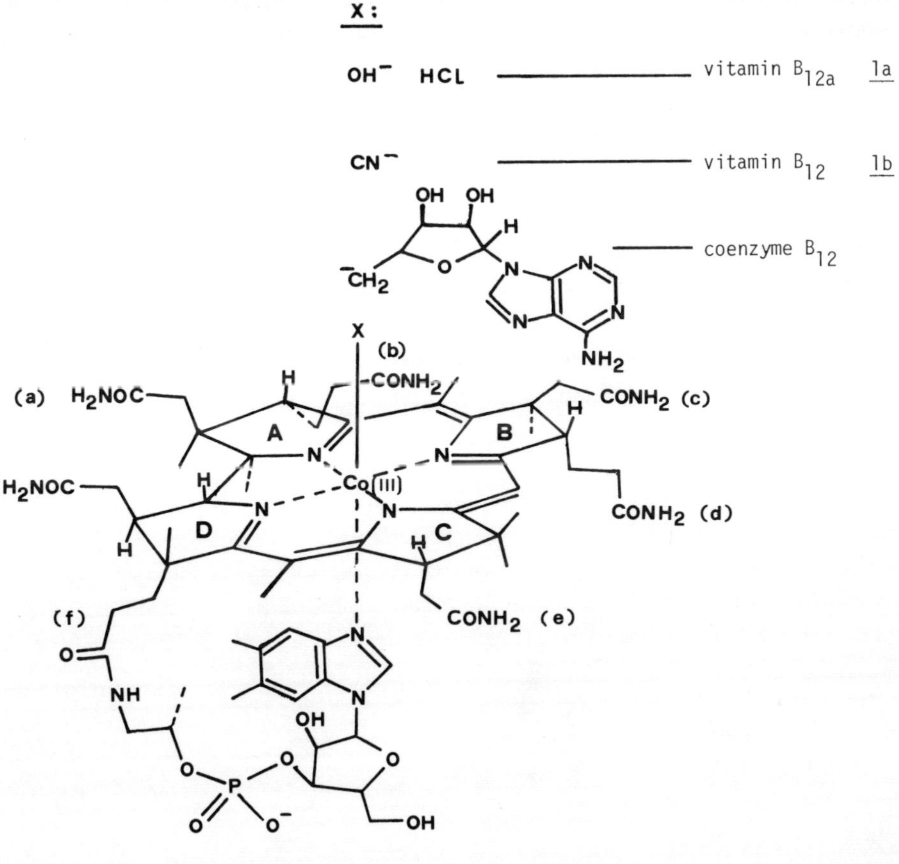

X:

OH⁻ HCL ———— vitamin B_{12a} 1a

CN⁻ ———— vitamin B_{12} 1b

(5'-deoxyadenosyl) ———— coenzyme B_{12}

1a vitamin B_{12a} = hydroxycobalamin -hydrochloride $C_{62}H_{90}ClCoN_{13}O_{15}P$ mw. 1382,9

1b vitamin B_{12} = cyanocobalamin $C_{63}H_{88}CoN_{14}O_{14}P$ mw. 1355,4

The common structural feature of compounds belonging to the vitamin B_{12} family is the cobaltcorrin nucleous. Derivatives differing in axial substituents and side chain functional groups have been reviewed by Mervyn and Smith [9].

Corrinoids are synthesized by microorganisms. They are present in small amounts in every animal tissue (e.g. 1 mg/kg in beef liver [11]),in biomasses such as municipal sludges [12,13] in very high dilution even in soil and water [4]. The industrial production of vitamin B_{12} (cf. recent reviews [14,15])is based on fermentation processes from carbohydrates by different strains of *Propionibacteria* or *Pseudomonas denitrificans*. It is isolated from the broth (20 to 60 mg/liter) via adsorption on different supports or by extraction purified by chromatography and then either precipitated or crystallized. Derivatives of cyanocobalamin like coenzyme B_{12} are manufactured by condensing a 5'-ester of adenosine with reduced vitamin B_{12} [16]. The world marked for cyanocobalamin is estimated for 1980 [14] to be 12000 kg/year, a major part of which is directed to the synthesis of other cobalamins. The pharmaceutical compositions take 6500 kg, 5500 kg are used in animal feed. The bulk selling price of cyanocobalamin fluctuates between $ 6 and $ 1.80 per gram, hydroxycobalamine hydrochloride being slightly more expensive. The biggest producer of vitamin B_{12} is RHONE POULENC in France.

As catalyst in organic reactions hydroxycobalamine hydrochloride (vitamin B_{12a}) is used as such, cyanocobalamin (vitamin B_{12}) is "activated" as described by Fischli [17].

2.2. [HDP]cobalt complexes

Easily accessible synthetic models of vitamin B_{12} are the compounds derived from the [hexadecamethyl-diazo-hexahydroporphinato]cobalt kation ([HDP]cobalt cation) 2b [18,19]. The ligand of this cation is on the oxidation level of a corphin [20]. Addition of nucleophiles e.g. water or alcohols lead to 1-hydroxy-[HDP]cobalt(III) complexes (Z=OH) or 1-alkoxy[HDP]cobalt(III) complexes (Z=OR)[21]. The axial ligands X and Y might be exchanged as usual in Co(III)-chemistry [7]. On reduction 1-hydro[HDP]cobalt(I) (X=Y = no ligand, Z=H) is obtained [21,22]. [HDP]cobalt complexes are soluble in organic solvents, easy to separate and crystallize. The ligand system is hard to solvolyze, the central metal atom cannot be removed or exchanged [21].

The superposition of projections of coenzyme B_{12} (open circles) and of a [HDP]cobalt derivative (X = 4-methoxycyclohexyl, Y = trifluoromethylsulfonyl, Z = hydrogen, filled circles) [23] reveals similarities and differences in the structures of the two compounds. The close resemblance within the conjugated perimeter extending from N(21) to N(24) is obvious.

X=Y=Br
dibromo[HDP]cobalt(III)kation 2b

X=Y=Br , Z=OH
1-hydroxy-dibromo[HDP]cobalt(III) 2a

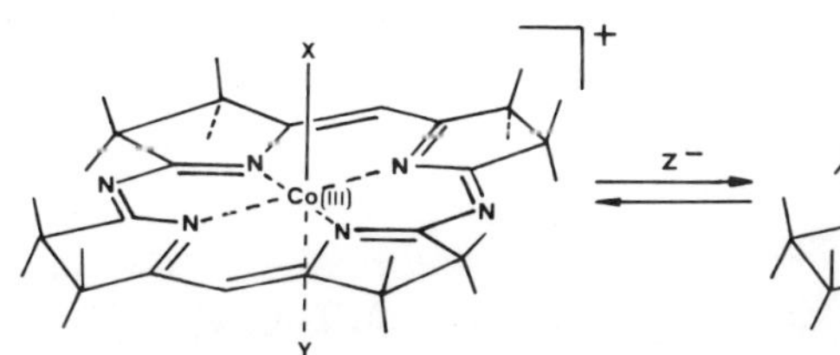

Scheme 1
superposition of projections of
coenzyme B_{12} (open circles)
and
[HDP]cobalt(III) (filled circles)
X = 4-methoxycyclohexyl
Y = CF_3SO_3
Z = H

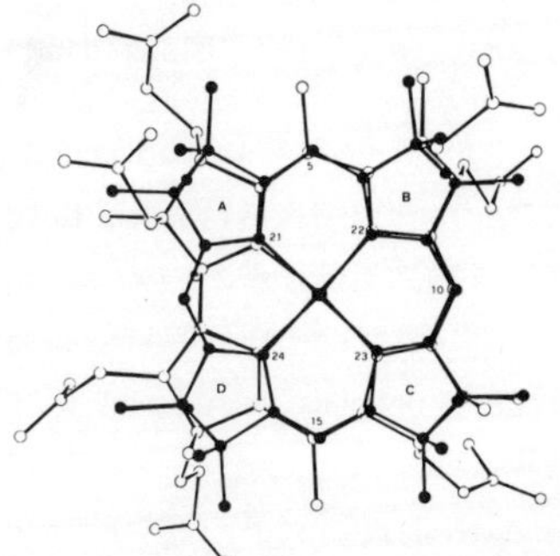

The compound 2a (X=Y=Br) is prepared by a fully convergent synthesis starting from eight equivalents of acetone via four equivalents of tetramethylsuccinic dinitril [24], two equivalents of the bicyclic "bisamidine"[25] and final cyclocondensation in presence of the central metal ion [18] in five steps and almost 50 % "over all" yield. Likewise a number of other [HDP]complexes e.g. the nickel- [26], palladium- [27], copper- [28] complexes were synthesized. The structure [29] and reactivity [30] of the [HDP]nickel complexes as well as their application as reagents in asymmetric synthesis [31] are described.

Scheme 2 Synthesis of 1-hydroxy-dibromo[HDP]cobalt(III) 2a.

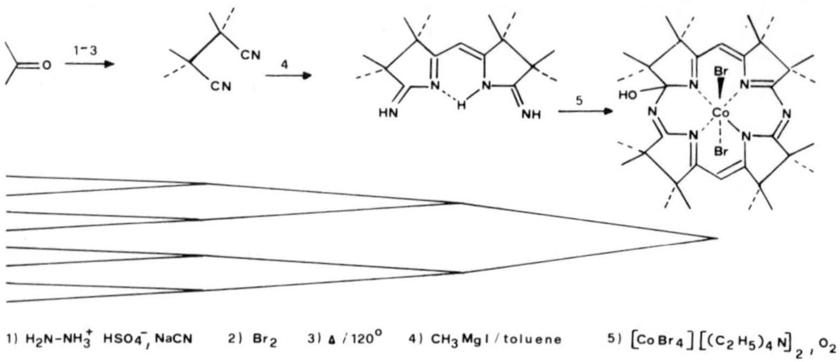

1) $H_2N\text{-}NH_3^+\ HSO_4^-$, NaCN 2) Br_2 3) Δ / 120° 4) CH_3MgI / toluene 5) $[CoBr_4][(C_2H_5)_4N]_2$, O_2

[HDP]Co-complexes containing peripheral CD_3-groups are prepared from d_6-acetone.

Dibromo[1-hydroxy-2,2,3,3,7,7,8,8,12,12,13,13,17,17,18,18-hexadecamethyl-10,20-diazo-octahydroporphinato]cobalt(III) (2a).

a) Preparation of the "bisamidine" [25]

In a 200 ml three-necked round-bottomed flask, equipped with a mechanical stirrer and a distillation bridge, 10.0 g (73.5 mmol) of tetramethylsuccinicdinitrile [24] were dissolved under a N_2 atmosphere in 80 ml of abs. toluene by heating moderately with an oil bath. To the rapidly stirred solution a Grignard reagent, prepared in the usual way from 5.0 g (205 mmol) of magnesium turnings and 25.0 g (177 mmol) of methyliodide, was added and the oil bath temperature was raised up to 120°. When the boiling temperature

reached 100-110°, the bridge was replaced by a reflux condenser. The yellow mixture was heated to reflux for 2 h, then poured into a rigorously stirred two phase system of 200 g of ice, 50 g of NH_4Cl and 200 ml of CH_2Cl_2. The organic phase was separated and washed with 200 ml of cold 1 M NaOH. The aqueous phases were extracted with CH_2Cl_2 (2x100 ml). The combined organic phases were dried (Na_2SO_4) and solvent was evaporated.
Yield: 10.0 g of crude product. Crystallisation from CH_2Cl_2/heptane afforded 7.5 g (71 %) of "bisamidine", mp 190-192°. For analysis a sample was recrystallized three times from CH_3OH/acetic ester and sublimed at 150°/0.005 torr: mp 199-201°. [For large-scale synthesis of "bisamidine" and spectroscopic data see [25)]].

b) Preparation of 2a [18)] $C_{34}H_{51}Br_2CoN_6O$ mw. 778,6

A solution of the "bisamidine" (0.577 g, 2.00 mmol) in 5 ml abs. t-butanol was added dropwise to a stirred suspension of bis-tetraethylammonium-tetrabromocobaltate [32)] (0.639 g, 1.00 mmol) in 5 ml abs. t-butanol under nitrogen at 60°. The mixture was refluxed for 48 h; then air was bubbled through during 0.5 h, whereby solvent was mainly removed. The black residue was suspended at room temperature in 20 ml of water, filtered through celite on a G3 glass frit and washed with 80 ml of water. The product was eluted with dichloromethane (4 times 20 ml); the solution was filtered through cotton wool, the solvent evaporated and the crude product was dried at 80° (8 h, 0.01 torr).
Yield: 0.733 g of 2a (94 %) as a black powder, which was used for further experiments. For analysis a sample was recrystallized three times from isopropanol/dichloromethane and once from hexane. [For spectral data see [18)]].

c) Other derivatives of 2

1-hydroxy-dicyano[HDP]cobalt(III) 2b $C_{36}H_{51}CoN_8O$ mw. 670,8
2b is easily prepared from 2a by exchanging the axial ligands with NaCN.[18)]

dibromo[HDP]cobalt(III)-trifluoracetate 2c $C_{36}H_{50}Br_2CoF_3N_6O_2$ mw. 874,6
2c is obtained from 2a on dissolution in CF_3COOH/CH_2Cl_2 and evaporation of the solvents [18)].

1.3. Cobalt phthalocyanine (CoPc)

There have been published several monographs and reviews: on synthesis, production and applications as pigments and dyestuffs by Moser and Thomas [33], on chemistry, electronic structure and physical properties by Lever [34] and recently by Berezin [35] on chemistry, thermodynamic-, kinetic- and electro-optical properties. The monograph by Kropf and Steinbach [36] as well as the review by Didorov and Maslow [37] concerne the application of phthalocaynines as catalysts.

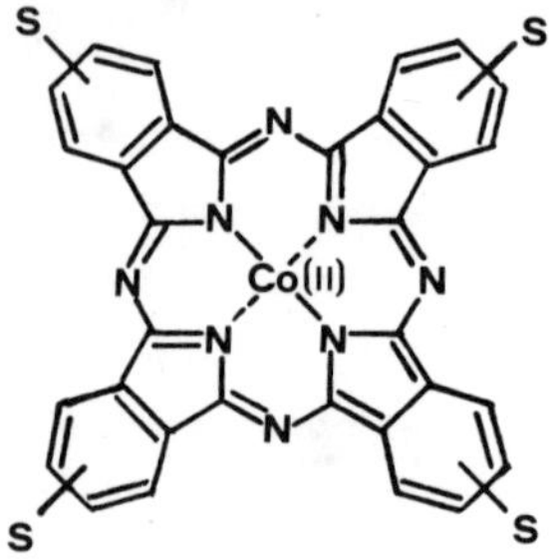

CoPc <u>3a</u> (S = H)
$C_{32}H_{16}CoN_8$ mw. 571,4

$CoPc(SO_3H)_4$ <u>3b</u> (S = SO_3H in m-position)
$C_{32}H_{16}CoN_8O_{12}S_4$ mw. 891,7

$CoPc(SO_3H)_{2,6}$ <u>3c</u>
$C_{32}H_{16}CoN_8(SO_3)_{2,6}$ mw. 654,8

In Co(II)Pc the d^7 central metal atom is square planar coordinated by four nitrogen atoms. The seventh electron of the paramagnetic metal ion is presumably in the d_{z^2} orbital of Co and this seems to prevent the close association of axial ligands [38]. Co(II)Pc is insoluble in water and most organic solvents. It is soluble in concentrated sulfuric acid, quinoline and chloronaphtalene. In pyridine it is slightly soluble and one pyridine is loosely coordinated [39]. Derivatives of CoPc bearing substituents at the benzene nuclei, normally at the 4-positions, are well-known. Of special interest as catalysts are the water-soluble sulfonates $Co(II)Cp(SO_3H)_x$ [40] and the polymer-attached Co(II)Pc-derivatives [41].

Reduction of Co(II)Pc affords the $[Co(I)\text{-}Pc]^-$ anion. The Li- and Na-salts are soluble in alcohols, acetonitrile, DMSO, DMF, acetone, pyridine. They can be crystallized from THF and dioxane [42]. In an impressive series of publications Eckert and Ugi discussed the reactivity of salts of the $[Co(I)Pc]^-$ anion as supernucleophile [42-44], as reagent for the reductive cleavage of 2-halo alkyl esters [45-47] and β-halogenated urethanes [48-50] and as catalyst for the reduction of functional groups [51-52].

Co(II)Pc is easily prepared by reaction of phthalonitrile with metallic cobalt [53] or by condensation of phthalic anhydride, urea and cobalt chloride [50,54].

Preparation of cobalt(II)phthalocyanine 3a [54]

In a 1,5 L sulfonation flask equipped with a short reflux condenser (without water cooling) and an efficient mechanical stirrer phthalic anhydride (236,8 g), urea (384 G), anhydrous cobalt dichloride (54 g) and ammonium molybdate (1,2 g) are placed together with nitrobenzene (700 ml). The mixture is heated with stirring for 1 h at 130°, then 1 h at 150°, then 1 h at 170° and finally for 16 h at 170-175°. The reaction mixture is then filtered through a large G3 frit and the precipitate is washed with ethanol (1 L) and then with acetone (1 L). The crude product is first treated with 1N NaOH/H_2O (2 L) for 2 h at 90-95°, filtered off and then treated with 1N HCl/H_2O (2 L) for 2 h at 90-95° and filtered off. After drying 3a (205 g, 89 %) is obtained as a dark violet powder.

Partially sulfonated Co(II)phthalocyanine Co(II)[Pc(SO_3H)$_{2,6}$] 3c [54]

Cobalt(II)phthalocyanine 3a is placed in a 1,5 L sulfonation flask equipped with a short reflux condenser and a mechanical stirrer. Fuming sulfuric acid 26 % anhydride (800 g) is added with external cooling and the solution is stirred for 1 h at room temperature. The reaction mixture is then heated to 60° within 20 min and kept at 60-61° for 12 h. After cooling the solution is poured slowly into a stirred mixture of ice (2 kg) and brine (1 L) and the suspension filtered through a large G3 glass frit. The precipitate is washed first with brine (3 L), then with 3 % HCl/H_2O (4 L) and dried. 3b (145 g) is obtained as water soluble dark blue powder. The sulfur content corresponds to 2,6 SO_3H groups per Co(II)Pc.

The synthesis of highly pure 4,4',4",4"'-cobalt(II)tetrasulfophthalocyanine 3b is described by Weber and Busch [40].

2.4. Cobaloximes $Co(DH)_2$

Cobaloximes have been introduced as vitamin B_{12} model compounds [55]. They are likely the most thoroughly examined synthetic cobalt complexes; for reviews see Schrauzer [56], Dodd and Johnson [2] as well as Brown [57]. Cobaloximes 4 are easily prepared from the corresponding oxims. A number of interesting intramolecularly bridged cobaloxims have been synthesized by Rétey [58]. They have been studied as models for the active site of the coenzyme B_{12} requiring methylmalonyl CoA mutase. The most often used cobaloxims with R = CH_3 are generally sparingly soluble in water and most organic solvents. They are hydrolized only slowly in acidic and basic media. Cobaloximes can be reduced to the corresponding Co(II) and Co(I) species, the latter behave as supernucleophiles [59] and can easily be alkylated at the central metal atom.

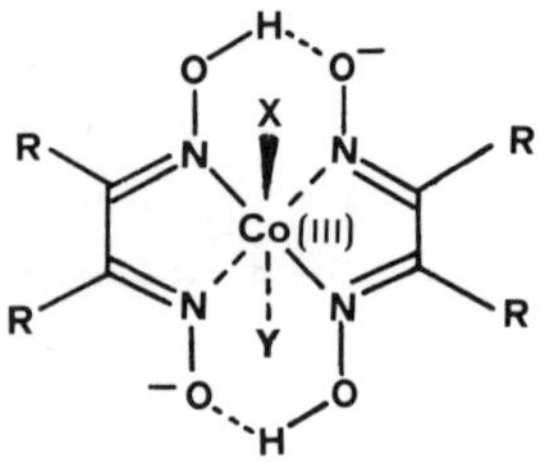

$Co(DH)_2Cl$,py 4a (R = CH_3, X = Cl , Y = py)

$C_{13}H_{19}ClCoN_5O_4$ mw. 403,7

Chloro(pyridino)bis(dimethylglyoximato)cobalt(III) [60] Cl, Py-$Co(DH)_2$ 4a

To a hot solution of 5.0 g (0.021 mol) of $CoCl_2 \cdot 6H_2O$ and 5,5 g (0.047 mol) of dimethylglyoxime in 200 ml of 95 % ethanol, 3.44 g (0.043 mol) of pyridine is added. After cooling to 20°, air is blown through the solution for 30 min and the mixture is allowed to stand for 1h . The brown crystals are collected, washed with water, ethanol and diethylether (3 times 50 ml each) and dried at room temp. in vacuo. Yield: 5.9 g (70 %, based on $CoCl_2 \cdot 6H_2O$).

2.5. Cyano cobalt complexes

The cyanocobalt complexes are included in this article since they are easily prepared, structurally simple and since their reactivity exhibits analogies to the inorganic chemistry of vitamin B_{12}. Furthermore, it has to be considered that these complexes might be formed at prebiotic conditions. The knowledge of their reactivity is therefore not only prerequisite to develop efficient catalytic systems, but may also contribute to an understanding of early catalytic processes. An excellent review covering the reactions catalyzed by pentacyanocobaltate(II) is given 1967 by Kwiatek [61]. For the chemistry of transition metal cyanides cf. reviews by Chandwick and Sharpe [62], Ford-Smith [63] and the comprehensive monograph by Sharpe [64].

Scheme 3 Synthesis and Selected Interconversions of Cyano Complexes of Co(III), Co(II) and Co(I)

$$Co(II)_{solv.} + 5\,CN^- \rightleftharpoons \underset{\underline{5a}}{[Co(II)(CN)_5]^{3-}} \quad 1)$$

$$\underset{\underline{5a}}{[Co(II)(CN)_5]^{3-}} + e^- \rightleftharpoons \underset{\underline{5b}}{[Co(I)(CN)_5]^{4-}} \quad 2)$$

$$\underset{\underline{5b}}{[Co(I)(CN)_5]^{4-}} + H_2O \rightleftharpoons \underset{\underline{5c}}{[Co(CN)_5H]^{3-}} + OH^- \quad 3)$$

$$\underset{\underline{5b}}{[Co(I)(CN)_5]^{4-}} \rightleftharpoons \underset{\underline{5d}}{[Co(I)(CN)_4]^{3-}} + CN^- \quad 4)$$

$$\underset{\underline{5b}}{[Co(I)(CN)_5]^{4-}} + 0{,}5H_2 \rightleftharpoons \underset{\underline{5c}}{[Co(CN_5)H]^{3-}} \quad 5)$$

$$2\underset{\underline{5a}}{[Co(II)(CN)_5]^{3-}} + H_2O \longrightarrow \underset{\underline{5c}}{[Co(CN)_5H]^{3-}} + \underset{\underline{5e}}{[Co(III)(CN)_5OH]^{3-}} \quad 6)$$

Solutions of pentacyanocobaltate(II) 5a are simply prepared by addition of cyanides to Co(II) salts in a molar ratio of 5:1 [61)] cf. Eq. 1. The bistetraethylammonium salt of 5a was obtained as a crystalline compound [65)]. Solutions of 5a in water [66)] or acetonitrile [65)] have a magnetic moment of 1,72 or 1,77 BM/Co resp., corresponding to one unpaired electron per ion. 5a is reduced by different reducing agents e.g. by hydrated electrons [67)] or by cathodic reduction at -1,2 V (S.C.E.) in highly alkaline solutions [68)] to the dimagnetic pentacyanocobaltate(I) 5b cf. Eq. 2. In aqueous solution 5b is protonated to give 5c cf. Eq. 2. The pK_a of 5c is 17 to 19 [68)]. In alkaline aqueous solution 5b dissociates rapidly to the diamagnetic tetracyanocobaltate(I) 5d [69)] cf. Eq. 4. Alternative ways to get 5c are the hydrogenation of 5a [70)] cf. Eq. 5 or the slow aging process of 5a in aqueous solutions [71)] cf. Eq. 6.

The reactivity of cyanocobalt complexes with organic substrates depends strongly on their constitution and oxydation state.

$[Co(II)(CN)_5]^{3-}$ (5a) reacts with organic halides in a stepwise mechanism via radical formation to pentacyanoorganocobaltate(III) [72)] cf. Eq. 7.

$$\underset{5a}{2[Co(II)(CN)_5]^{3-}} + RX \longrightarrow [Co(III)(CN)_5R]^{3-} + [Co(III)(CN)_5X]^{3-} \quad 7)$$

$[Co(CN)_5H]^{3-}$ (5c) obtained according Eqs. 1 and 5 is a typical hydrogenation catalyst used for the hydrogenolysis of organic halides [61)] and the hydrogenation of activated olefins [61,73)]. In these reactions free radicals are considered as intermediates. Organocobalt complexes of the type $[Co(II)(CN)_5R]^{3-}$ are not reduced at these reaction conditions and their formation would therefore desactivate the catalyst. Indeed this catalyst is working only for a few catalytic cycles. This drawback has recently been overcome by the use of phase-transfer conditions [73)].

$[Co(I)(CN)_5]^{4-}$ (5b) and $[Co(I)(CN)_4]^{3-}$ (5d) are strong nucleophiles and reducing reagents involved e.g. in carbonylation reactions [69)] and the selective cyanation of vinyl halides catalyzed by tetracyanocobaltate (5d) [74)].

2.6. Other Cobalt Complexes

Considering the large number of known synthetic macrocyclic cobalt complexes[75] only a small selection of representative compounds can be presented here.

a) X,Y[1,3-bis(diacetylmonoxime-imino)propane]cobalt(III)

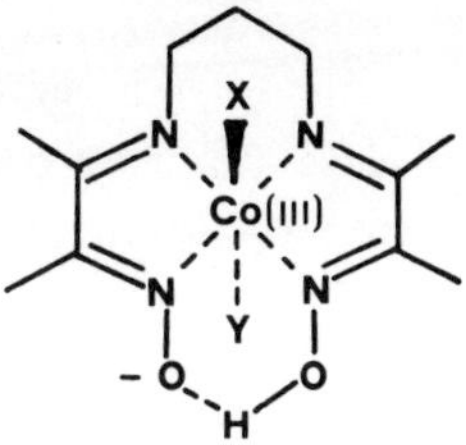

X,Y-Co(III)[(DO)(DOH)pn] 6
"Costa"-complex

B_{12}-models of the Co(III)[(DO)(DOH)]-type were synthesized by Costa [76] by treating 1-diacetylmonoxime-imino-3-diacetylmonoximato-imino-propane with an aqueous solution of Co(II)halide and subsequent oxidation by air. The grey-green product is precipitated, filtered and washed. The alkylated products are obtained by the reaction of the Co(I) compound with alkyl halides or from the Co(III) compound by treatment with Grignard reagent [for bis-organo-cobalt compounds see [77,78]].

b) [bis(salicylaldehyde)ethylenediimine]cobalt(II)

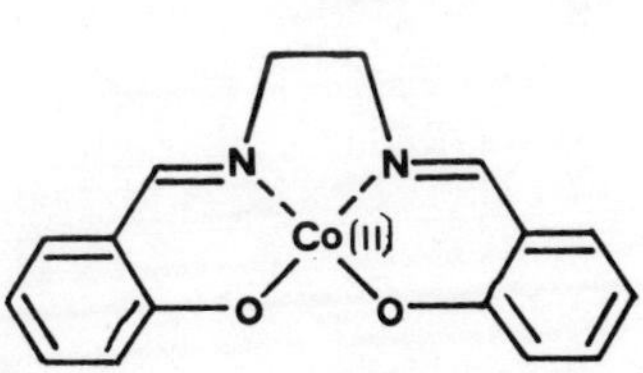

Co(II)salen 7
(other name: salcomine)

Cobalt chelates of bis(salicylaldehyde)ethylene-diimide have been extensively studied [79]. The compound 7 is prepared by heating the free ligand salen H_2 with cobalt(II)acetate in DMF under N_2[80]. X,Y-Co(III)salen (X=Y=NH_3, py, PPh_3 etc.) is synthesized from the same starting materials in hot methanol in presence of air and the appropriate base [81].

c) [di(salicylal)-3,3'-diimino-di-n-propylamine]cobalt(II)

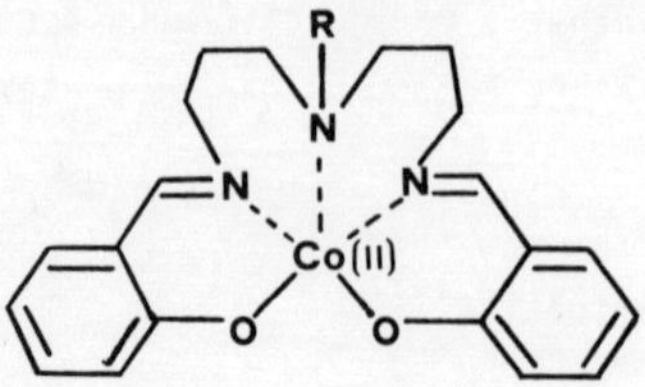

R = H, Co(II)salpr 8

The pentadentate Schiff base prepared from 3,3'-diamino-di-n-propylamine and salicylaldehyde combines with Co(II)acetate to give yellow-brown crystalline Co(II)salpr 8 [82], which has been found to catalyze autoxidation reactions [83].

d) Cobalt(II)α,β,γ,δ-tetraphenylporphyrin

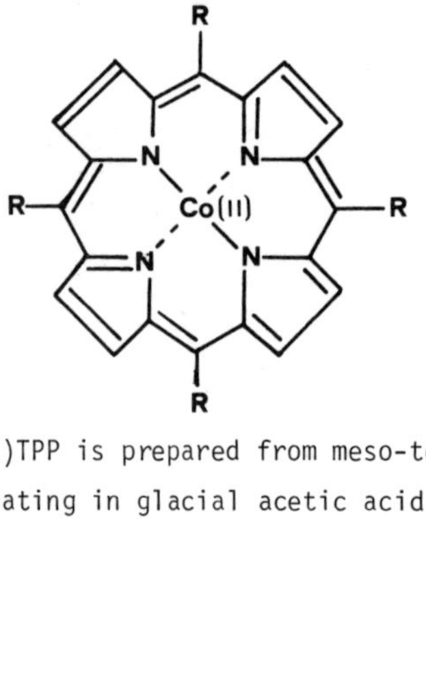

R = phenyl Co(II)TPP 9

Co(II)TPP is prepared from meso-tetraphenyl-porphyrin [84], and Co(II)acetate by heating in glacial acetic acid [85]. For a review see [86].

3. PROPERTIES AND REACTIONS OF COBALT COMPLEXES

3.1. Oxidation, Reduction

There are two important reasons, why transition metals contribute so much to catalyst systems:

- variability of coordination number
- variability of oxidation state

In case of macrocyclic tetradentate Co-complexes the number of additional axial ligands depends crucially on the oxidation state of the central metal atom. In vitamin B_{12} and related Co-complexes generally Co(III) is ligated by two axial ligands (X,Y), Co(II) by one (Y) and Co(I) by none [87-89] according to scheme 3. This trend of decreasing coordination number has been qualitatively described by the X-Co-Y three center 4 → 5 → 6 electron bonding model, which combines axial ligand σ orbitals with metal orbitals of substantial d_{z^2} character[90]. The unsaturated macrocycle represents by itself an electroactive subunit of cobalt chelates, which may undergo reduction.

Reduction potentials $E^{o'}_{III/II}$, $E^{o'}_{II/I}$ and $E^{o'}_{I/L}$ of cobalt chelates were determined by electrochemical methods. They are depending on the axial ligands, the solvent and the square planar macrocycle. In schemes 3 and 4, $E^{o'}$ are represented by white zones, their width being due to the variation of $E^{o'}$ caused by different axial ligands and solvents. The white zones separate stability zones of the different oxidation states (dark areas) as a function of electrode potential. Potential regions of unknown electrochemistry or solvent/electrolyte discharge (white areas) are separated from stability zones by S-shaped lines.

Scheme 3 Reduction potentials and related stability-zones.

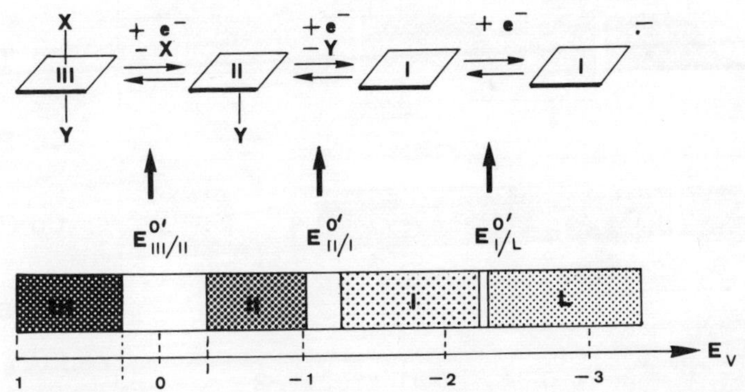

Scheme 4a Ranges of thermodynamic stability of Co-complexes at variable electrode potential

a) Non-alkylated Co-complexes

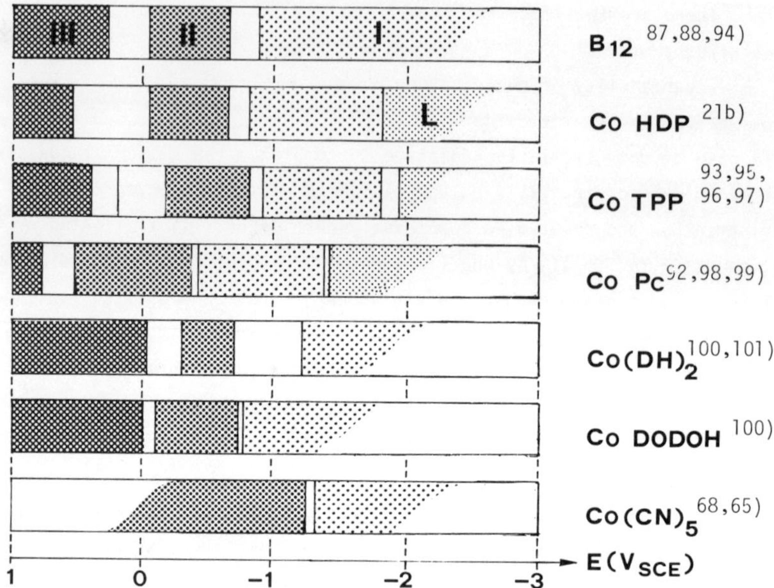

Scheme 4b

b) Alkylated Co-complexes (containing one (Co-C)-bond)

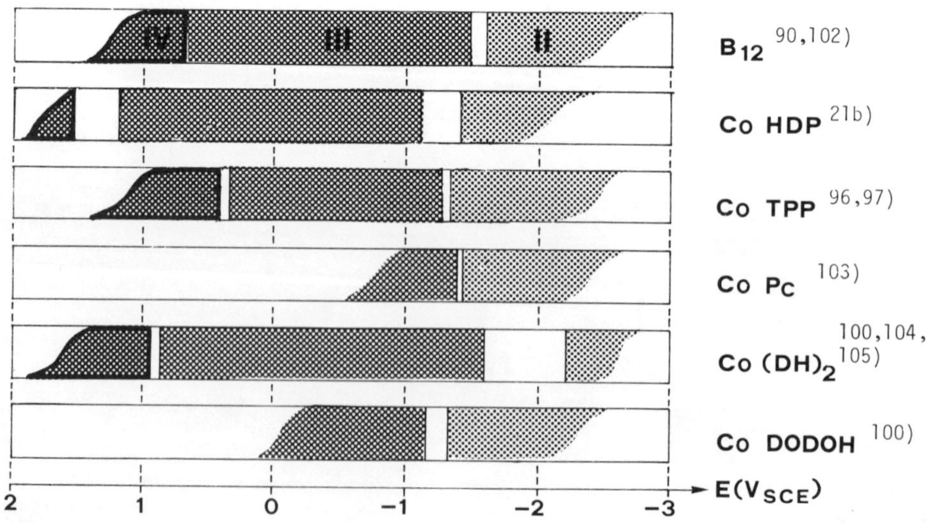

As metal reduction III → II and II → I is coupled with axial ligand expulsion, $E^{o'}_{III/II}$ and $E^{o'}_{II/I}$ depend on the complex formation constant of axial ligands with Co in its different oxidation state. A strongly coordinating axial ligand requires a more negative potential for Co metal reduction than a weekly coordinating one. Approximate formal reduction potentials (white zones) for Co chelates with different macrocyclic ligand systems as well as for the pentacyano cobaltate anion are shown in scheme 4. Reported $E^{o'}$ values for different axial bases and various solvents/electrolytes (all adapted to the S.C.E. reference system) lie within the white zones.

The first macrocyclic ring reduction is indicated. Its $E^{o'}_{I/L}$ is strongly depending on the degree of unsaturation of the macrocycle and may involve one or more electrons.

If one axial ligand is an alkyl residue, reversible coordination is excluded. Usually irreversible chemical follow-up reaction are observed upon reduction (or oxidation) of these organocobalt compounds. In most cases the one electron reduced (or oxidized) Co(II) [or Co(IV)] complex has been shown (by fast electrochemical techniques) to exist prior to irreversible expulsion of the alkyl residue. Surprisingly uniform $E^{o'}_{III/II}$ values for the first reduction step are observed (scheme 4b). Their significant shift to a more negative potential (with respect to the non-alkylated complexes) reflect the unique ligand strength exhibited by the alkylanion moiety.

3.2. Formation of the (Co-C) bond.

After 1961 when Lenhert and Hodgkin [106] have shown that coenzyme B_{12} exhibits a (Co-C) bond, hundreds of organocobalt complexes have been synthesized and studied literature in this field has been reviewed[1,2,57] and only some selected aspects will be discussed here.

Formation of the (Co-C) bond may take place by one of the following three reaction types:

$$[Co(I)] \quad + \quad R^{+} \longrightarrow [Co(III)\text{-}R] \qquad 8)$$

$$[Co(II) \quad + \quad R^{\cdot} \longrightarrow [Co(III)\text{-}R] \qquad 9)$$

$$[Co(III)] \quad + \quad R^{-} \longrightarrow [Co(III)\text{-}R] \qquad 10)$$

(R^+, $R^\cdot$ and R^- stand for electrophilic, radical-like and nucleophilic carbon species).

a) Alkylations of Co(I) complexes (Eq. 8).

This reaction type, which consists in a concomitant increase in the oxidation state and coordination number is well known in transition metal chemistry as oxidative addition [107]. Vitamin B_{12s} (cob(I)alamin) <u>1(I)</u> is a very powerful nucleophile [108] and reacts rapidly with alkyl halides to form alkyl-B_{12} derivatives. From kinetic studies Schrauzer et. al. [59] concluded that the rate law corresponds to an S_N2 type substitution at carbon. Second order rate laws were also determined for alkylations of many B_{12} model compounds. Table 1 shows Pearson's nucleophilic reactivity constants n_{CH_3J} [109] (n_{CH_3J} is defined as log k_N/log K_{CH_3OH}; k_N being the 2nd-order rate constant for the methylation of N by CH_3J in CH_3OH at room remp.) of some Co(I) complexes compared to "normal" nucleophiles.

Table 1 Pearsons's nucleophilic reactivity constants of selected nucleophiles and Co(I) complexes.

Nucleophile N		n_{CH_3J}	Ref.
CH_3OH		0	109)
Cl^-		4,37	109)
$S_2O_3^{2-}$		8,95	109)
Co(I)Pc	3(I)	10,8	42)
Co(I)DODOHpn	6(I)	12,3-13,2 [a)]	59)
$Co(I)(DH)_2$	4(I)	13,3-14,3 [a)]	59)
B_{12s}	1(I)	14,4	59)
Co(I)salen	7(I)	14,6	110)

a) Value depending on axial base Y added.

The absolute reactivity of Co(I) nucleophiles with RX is up to 10^5-10^7 times greater than those e.g. of J^- or $S_2O_3^{2-}$. The reactivity pattern of RX follows generally the sequence X : J>Br>Cl and R: prim. >sec.[59].

The stereochemical course of the substitution of X by Co(I) at saturated carbon is not uniform. Inversion of configuration has been determined in the reaction of cobaloxime(I) 4(I) with cyclohexane epoxides, bromides and tosylates[111] as well as with open-chain alkyl derivatives[112-114]. Predominant (83%), but no net inversion has been found in the reaction of B_{12s} 1(I) with (1R)- and (1S)-[1-$^2H_1$]-ethyl tosylate by Rétey et. al.[115]. Alkylations of Co(I)HDP 2(I) with the conformationally rigid 3α and 3β-iodocholestane lead exclusively to the thermodynamically more stable isomer with Co in 3β-position[116]. Even in case of the alkylation of 2(I) with either cis- or trans-4-methoxycyclohexylbromide the thermodynamically favoured trans-product was preferentially formed in a ratio of 10:1[23] (cf. scheme 5). The configuration of the main product was confirmed by X-ray analysis[23].

Scheme 5

Reductive alkylation of Co[HDP] 2a [via Co(I)] by cis- and trans-4-methoxycyclohexylbromide

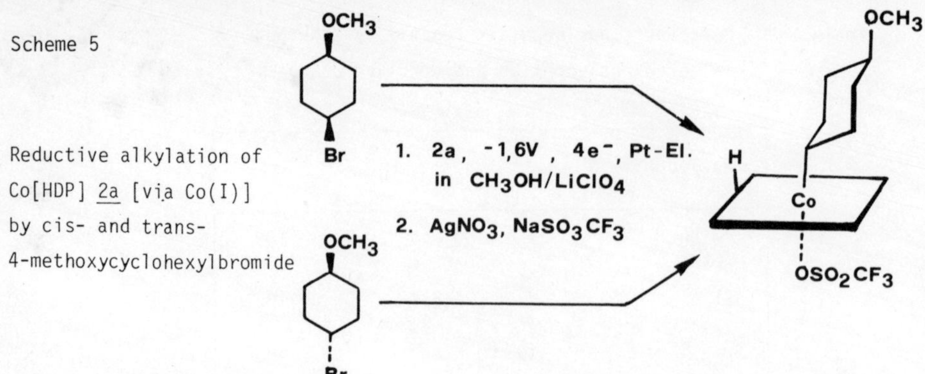

Substitutions of leaving groups at saturated carbon in cyclopropyl derivatives [117] and bridge-head alkyl halides [118] by Co(I) complexes proceed with retention of configuration. Retention about a double bond has been determined in the alkylation of Co(I)$(DH)_2$ 4(I) by vinyl halides [119].

Several mechanisms ranging from S_N2 to one-electron transfer mechanisms are considered to explain these kinetic and stereochemical facts [120].

The (Co(III)-C) bond can also be formed through addition reactions. Comments on the reaction of olefins with Co(I) complexes are based upon the fact that the latter might be protonated. The pK_a of hydrido (tri-n-butylphosphine)-cobaloxime 4b (4 with X=H, Y=$P(C_4H_9)_3$) has been published to be 10.58 [121]. Acrylonitrile affords with 4b in neutral media the organocobaloxim with Co in 2-position of the propionitrile, whereas in an alkaline medium the 3-substituted propionitrile is formed exclusively [122]. The stereochemical course of such additions has been investigated by Gaudemer et al. [123, 124]. Co(I)$(DH)_2$ 4(I) is also known to react with alkines [122]. Syn-addition of 4(I) to phenylacetylene has been observed [125].

b) Alkylations of Co(II) complexes (Eq. 9) [126].

Alkylations of Co(II) complexes like B_{12r} 1(II)[127] and several B_{12}-model compounds [128] by R-X show second-order kinetics with reaction rate v v = k[Co(II)][RX]. The reactivity of RX increases with increasing stability of the radical R· (e.g. along the sequence $CH_2ClCOOR < CHCl_2COOR < CCl_3COOR$)

and along the sequence R-Cl<R-Br<R-J [127]. Disproportionation of B_{12r} and subsesquent alkylation of B_{12s} has to be excluded [129]. Halpern et. al. [127] proved that these alkylations of Co(II) complexes proceed by the following radical mechanism (Eqs. 11-13).

$$[Co(II)] \quad + \quad R\text{-}X \longrightarrow [Co(III)\text{-}X] \quad + \quad R\cdot \qquad \text{rate determining} \qquad 11)$$

$$[Co(II)] \quad + \quad R\cdot \longrightarrow [Co(III)\text{-}R] \qquad \text{fast} \qquad 12)$$

$$2[Co(II)] \quad + \quad R\text{-}X \longrightarrow [Co(III)\text{-}X] \quad + \quad [Co(III)\text{-}R] \qquad 13)$$

c) Alkylations of Co(III) complexes (Eq. 10).

Reaction of Co(III) complexes with nucleophilic carbon species is another method for the synthesis of organo-cobalt compounds. Formation of the [Co(III)-C]-bond takes place with organolithium- and organomagnesium reagents[60,130,131] or with carbanions, stabilized by charge-delocalisation [132] (e.g. malonates). The natural β-methyl-cobalamin 1 (X = CH_3) was obtained from B_{12a} 1 (X = OH) with a new methylating reagent $(CH_3SiF_6)(NH_4)_3$ [133]. Electron-rich olefins like vinylethers were shown to react with Co(III) complexes like B_{12a} 1a or $Co(III)(DH)_2$ 4 in alcohols and in presence of base (Eq. 14)[134a]. Evidence for an intermediate olefin-π-complex of Co(III) has been provided [134b].

$$[Co(III)] \quad + \quad CH_2{=}CH\text{-}OR \xrightarrow[N(Et)_3]{ROH} [Co(III)]\text{-}CH_2\text{-}CH(OR)_2 \qquad 14)$$

3.3. Cleavage of the (Co-C) bond.

Cobalt-carbon bond cleavage of alkyl-Co complexes [Co-R] may be induced by:

- Reduction or oxidation of [Co-R].
- Electrophilic, nucleophilic or radical attack at the group R.
- Modification within the group R.
- Charge-transfer interactions of the macrocycle with additional reagents.
- Axial ligand exchange.
- Light or heat.
- Steric interaction between macrocycle and alkyl group.

An actual cleavage is often caused by combined parameters.

In scheme 6 the formal routes of (Co-C) bond cleavage have been compiled [135]. Depending on the oxidation state, the group R may react as electrophile in [Co(IV)-R] (path a), as radical in [Co(III)-R] (path d) or as nucleophile in [Co(II)-R] (path g). Although the reactivity of organocobalt complexes is mainly controlled by its oxidation state, other parameters may favour decay routes b, c, e, f.

Scheme 6 Stability and formal cleavage pathways of the (Co-C) bond at different oxidation states.

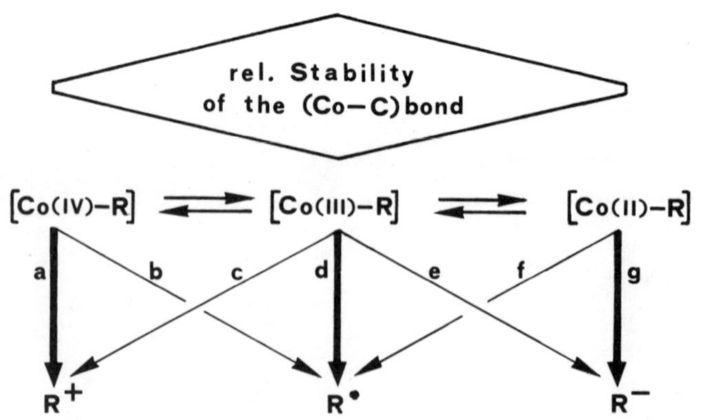

Organocobalt complexes on the oxidation level Co(III) are usually stable enough to be isolated. The (Co-C) bond of these compounds may be cleaved as a consequence of an attack by a reagent in a bimolecular reaction. On the other hand the corresponding organocobalt complexes on the oxidation level Co(II)or(IV)are very unstable, so that 1st order decay followed by trapping of the free alkyl species by the reagent is favoured. Loss of stereochemical information at the carbon originally bond to Co may result from such an alternative pathway.

Redox reactions between oxidizing or reducing reagents and [Co(III)-R] may compete with bimolecular substitution reactions.

Charge transfer complexes of the macrocycle of [Co(III)-R] with electrophiles such as tetracyanoethylene or transition metals e.g. $Pt(II)Cl_4^{2-}$ have been shown to enhance the rate of organocobalt complex decomposition [136, 137]. These results may be interpreted as a partial oxidation or reduction, both effecting a labilisation of the (Co-C) bond, cf. [138].

Cleavage of the (Co-C) bond in biological transformations [139, 140] or under non-biological conditions [141, 142] have been reviewed by several authors.

Cleavage of the (Co(III)-R) bond:

path d: Co(III)-R dissociation energies have been determined for various organocobalt(III) chelates and have been found to span the range of 20-30 kcal/mol [143]. The apparent stability of organocobalt(III) complexes can be rationalized by reversible fission and bond formation, a process often invoked for vitamin B_{12} coenzyme - catalized reactions [144]. Radical scavengers speed up the observed cleavage rate, e.g. unstable Sn(III) or Au(II) species as well as Cr(II) or cobaloxime(II) effect transalkylation [145-148]. A common 2nd order homolytic displacement (path d) has been proposed. However, a reaction sequence involving reduction [Co(III)-R] → [Co(II)-R], followed by cleavage has been considered as an interpretation of the Cr(II)-induced cleavage [147] (path f or g). Inversion of configuration has been reported for the cobaloxime(II) induced cleavage [148]. Homolytic cleavage of the (Co-C) bond is observed in photolysis or thermolysis and literature has been compiled in different review articles [141]. Even in case of photolysis the cleavage is reversible and radical scavengers have been found to improve quantum yields [149]. β-elimination processes - not included in scheme 6 - have been invoked to explain olefin formation resulting from thermolysis of cobaloximes [123]. However, intermediate radical formation followed by hydrogen abstraction has been suggested by Halpern [126].

path e: Metal cations such as Pb(IV), Tl(III), Hg(II) and Pd(II) have been reported to displace the alkyl group of alkylcobalamins in a bimolecular electrophilic substitution. The above mentioned reagents exhibit a standard reduction potential >0.8 V [142]. Therefore, an alternative pathway involving oxidation [Co(III)-R] → [Co(IV)-R] followed by cleavage (path a or b) can not be excluded. As a matter of fact, the halogenation of alkylcobalt(III) complexes - originally interpreted as electrophilic substitution (path e) [151] - has recently been shown to proceed via an oxidation followed by nucleophilic attack [152] or oxidation followed by radical attack [153] according [Co(III)-R] → [Co(IV)-R] followed by path a or b. Partial to net inversion of configuration has been observed for these reactions. Organocobalt(III) complexes are usually stable in presence of acid [141] (H^+ is unable to oxidize [Co(III)-R]) unless the alkyl group is able to undergo β-elimination e.g. β-hydroxyethylcobaloximes undergo fast acid induced decomposition [154]. Strong axial ligands may exchange the axial alkyl residue (decomposition of Co(III)-R on path e), if intramolecular stabilization of the negative charge is possible, e.g. fragmentation of triethylaminoethylcobalamine in presence of cyanide [155].

path c: Direct nucleophilic attack at the α-carbon of n-alkylcobalt chelates is of minor importance. Dodd et al. have reported transalkylation between cobaloxime(I) and alkylcobaloxime(III) with inversion of configuration (path c) [148]. One electron reduction of the alkylcobalt(III) by cobalt(I) followed by radical cleavage (path f) has been ruled out by the authors. Towards other nucleophiles as RS^- simple n-alkylcobalamins have been shown not to be susceptible [150]. On the other hand acylcobalamins exposed to amines are cleaved according to path c [156]. Electron-withdrawing substituents in the β-position of ethylcobaloximes cause sensitivity towards bases, e.g. β-cyanoethylcobaloxime undergoes base induced β-elimination leading to Co(I) and acrylonitrile [122].

Cleavage of the (Co(II)-R) bond:

path f and g: [Co(III)-R] complexes are reducible in the potential range of -1.2 to -1.6 V (vs SCE) (chapter 3.1.). One-electron reduced complexes (corresponding to Co(II) complexes) have been shown to exist. Compared to their parent [Co(III)-R] compounds they exhibit a reduced stability, which is depending dramatically on the nature of the macrocycle, the alkyl moiety and the axial base. The 1st order rate constant for decomposition of [Co(II)-R] in case of methyl-cobalamin has been estimated to be 1200 s^{-1} at -30° C [91]. At room temperature the half-life time of methyl Co(II)salen is ∿0.005 to 0.07 s [135] and the half-life time of one electron reduced methyl Co(II)TPP has been estimated to exceed 50 s [96]. Radical or anionic decay mechanisms (path f or g) have been proposed to explain coulometric results and recovered reduction products [135, 157] as well as visible spectra recorded after pulse radiolysis [158]. The fact that under reducing conditions eventually formed free radicals may further be reduced to the anion complicates the distinction between path f and g. The stereochemical course of the reductive hydrogenolysis of alkylcobalamin [159] and alkyl [Co(III)HDP] [116, 23] under protic conditions occures with retention of configuration.

Cleavage of the (Co(IV)-R) bond:

The one-electron oxidized complexes [Co(IV)-R] can be obtained by electrochemical [135, 160, 161, 162] or chemical oxidation [142, 152, 161, 163]. They exhibit a limited lifetime, which is depending on the structure of the macrocycle and the alkyl moiety, e.g. oxidized methylcobalamin decays much faster than alkylcobaloximes [142].

path a: Nucleophilic attack on [Co(IV)-R] with inversion of configuration at the originally Co-bonded carbon has been reported for alkylcobaloximes and alkyl Co(IV)HDP [152, 21b]. For alkyl cobaloximes 2nd order rate constants for the nucleophilic substitution reaction have been measured. Intramolecular nucleophilic displacement involving the macrocycle has been observed [163] or postulated on the basis of kinetic results [162].

path b: In case of methylcobalamin [142] and MeCo(IV)salen [162] homolytic cleavage has been proposed. The possibility of oxidation of the free radical followed by nucleophilic attack [141] or direct homolytic displacement of Co(III) in [Co(IV)-R] have been postulated [153].

3.4. Mechanistic hypothesis [23]

Under reducing conditions vitamin B_{12} and related Co complexes exhibit uncommon reactivity involving two particular steps:

- Formation of the (Co-C) bond by oxidative addition.
- Reductive cleavage of the (Co-C) bond.

a) Nucleophilicity of square-planar Co d^8-complexes and formation of the (Co-C) bond.

Extended Hückel calculations (with charge iteration to self consistency at Co, SCCC model [164]) have been performed for Co(I)HDP 2(I) and Co(I)corrin 1(I) models with molecular x-and y-axis defined by pairs of opposite nitrogen atoms. The calculations show three high lying occupied MO's , which are close in energy: one is a π-MO with mainly d_{xz} and d_{yz} contributions from Co, the second is a σ-MO with d_{xy}-character, the third is a σ-MO with d_{z^2}-character. This σ-MO with d_{z^2}-character is responsible for the nucleophilic reactivity in the z-direction [165].

Corresponding calculations for an isoelectronic Ni(II) HDP model show that the highest MO that might be involved in nucleophilic reactions in the z-direction, is a π-MO mainly ligand in character exhibiting big coefficients at the four ligating N-atoms [30]. The corresponding orbital with essentially (Ni)d_{z^2}-character is considerably lower in energy ($\sim$2,5 ev).

The observed nucleophilic reactivities correspond to frontier orbital control: Co(I) complexes react with electrophiles at the central metal atom, whereas Ni(II)HDP complexes react with electrophiles at a ligand N-atom [31c], cf. also[166]

b) Reductive cleavage of the (Co-C) bond [167].

One-electron reduced cobalt alkyl chelates have been shown to exhibit a measurable lifetime, which is depending on the structure of their macrocycle (cf. 3.3.).

SCCC-calculations have been performed for one-electron reduced models of CH_3,NH_3[Co-Corrin] and CH_3,NH_3[Co HDP]. The results show only small quantitative differences for the two models.

Scheme 7 Correlation diagram of frontier orbitals of $[CH_3,NH_3CoHDP]^{\bar{\cdot}}$ *vs.* (Co-C) distance.

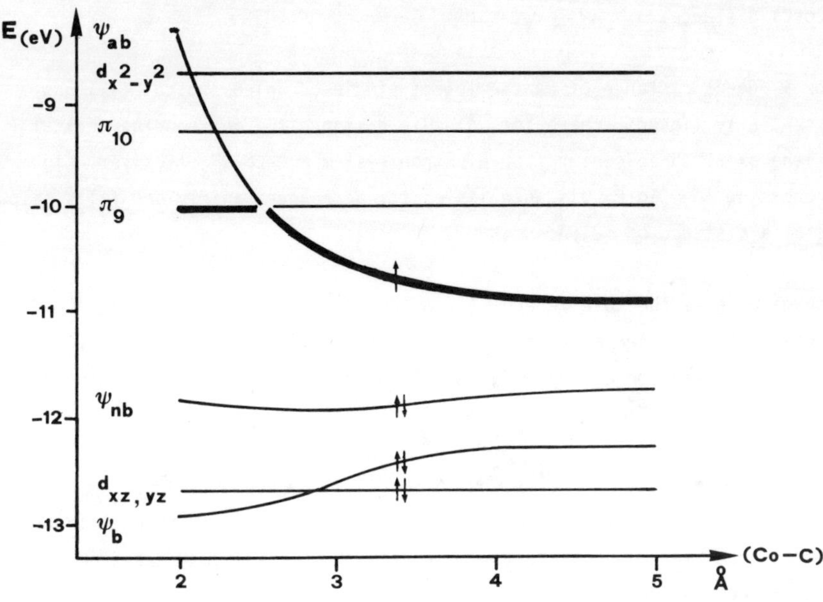

Scheme 8 Correlation diagram of the (SCCC) total energies of two electronic configurations of $[CH_3NH_3CoHDP]^{\bar{\cdot}}$ *vs.* (Co-C) distance.

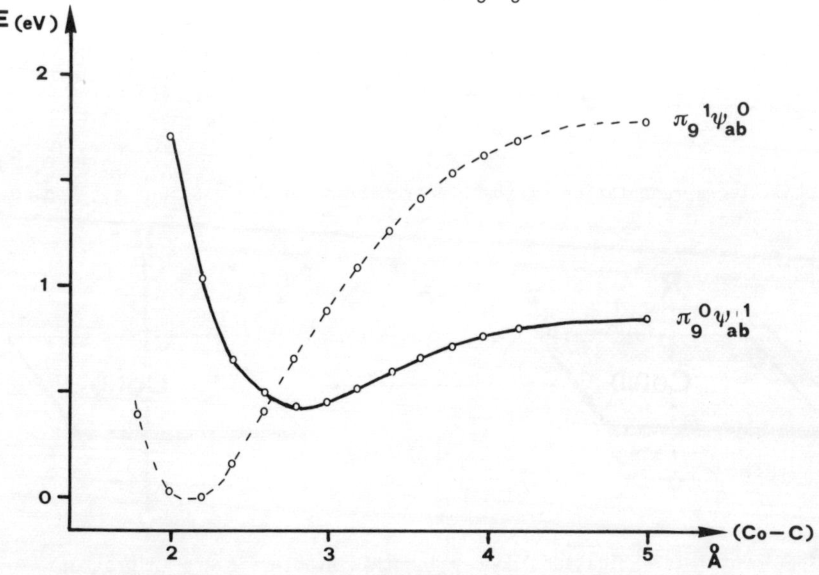

Scheme 7 shows the energy of the highest occupied and the lowest empty orbital of the "one-electron intermediate" $[CH_3,NH_3,Co\ HDP]^-$ as a function of the (Co-C) distance.

For a (Co-C) distance of 2Å the lowest single-occupied orbital (π_9) has essentially ligand π-character. At this distance the antibonding orbital (ψ_{ab}) of the axial three-center four-electron system H_3C-Co-NH_3 lies very high: On increasing the (Co-C) distance its energy decreases rapidly and falls below π_9 at a distance of ∿2,5Å.

Scheme 8 shows the total energy (from SCCC model) as a function of (Co-C) distance for the two electronic configurations $\pi_9^1\ \psi_{ab}^0$ and $\pi_9^0\psi_{ab}^1$. At a (Co-C) distance longer than ∿2,5Å the configuration $\pi_9^0\psi_{ab}^1$ is more stable than $\pi_9^1\psi_{ab}^0$.

Based on these results the following mechanism is proposed:
During formation of the "one-electron intermediate" an electron is transferred from the reducing agent (or cathode) into the ligang π-system. On thermal excitation, i.e. on stretching the (Co-C) bond, the odd electron is transferred from the ligand π-system into the antibonding (R-Co-Y)-MO (ψ_{ab}) [168]. This eventually leeds to the (Co-C) bond rupture.

Population of an empty ligang π-MO with an electron, subsesquent transfer into ψ_{ab} and breaking of the (Co-C) bond may not only be achieved by reduction, but also by photochemical excitation of alkyl-B_{12} and alkyl-Co HDP complexes.

Schematic representation of the electron-transfer on (C-C)bond stretching.

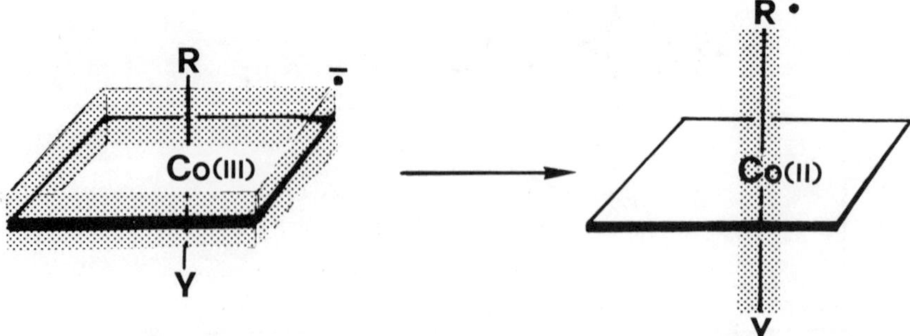

Experimental investigations concerning this hypothesis are in progress.

3.5. Catalytic cycles.

Till now fourteen biochemical transformations catalyzed by Co corrinoid-dependent enzymes are known [169]. Some of them consist of stereospecific rearrangements of vicinal groups. Though their reaction mechanisms are still not understood in detail, the catalytic activity of the Co corrinoids is connected with the reversible formation and cleavage of (Co-C) bonds [170].

In 3.2. and 3.3. possible (Co-C) bond formation and cleavage reactions under non-biological conditions have been presented. Generally any combination of these two reaction types could be used to transform a substrate (S) into a product (P) by a catalytic cycle (scheme 9).

Scheme 9

Generalized catylytic cycle.

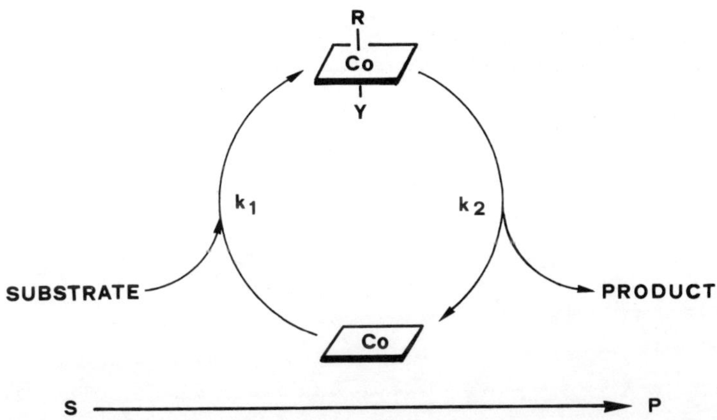

However, experimental conditions limit the choice of combinations, e.g. it should be difficult to run a homogeneous catalysis with reductive formation of the (Co-C) bond and its oxidative cleavage. In an actual catalysis the overall transformation $S \rightarrow P$ should also proceed with a reasonable rate. For several reactions k_1 and k_2 are available from literature. The overall rate can be calculated from a steady-state treatment for the intermediate [Co-R] concentration assuming irreversible formation and cleavage steps and fast decay of the "activated" R-fragment to final products.

Most synthetic reactions catalyzed by B_{12} and related compounds are performed under reducing conditions, maintained by cathodes (Hg [96, 116, 171-73)], Pt [96, 174)], C [175)], or chemical reducing agents such as BH_4^- [52, 124, 176-78)], H_2 [55, 178)] and Zn [173, 180)]. Scheme 10 shows the potential range, where such a catalysis is possible. It is limited on the positiv side by the potential needed either for the reduction of $[B_{12}\text{-}R]$ in case of the dark-reaction [$\sim$ -1,5 V (SCE)] or for the reduction of $B_{12r} \rightarrow B_{12s}$ in case of the photochemically induced (Co-C) bond cleavage. On the negative side the potential is limited by the undesired direct reduction of the organic electrophile or the solvent.

Scheme 10 Potential range for B_{12}-catalyzed reduction of organic electrophiles undergoing oxidative addition with Co(I).

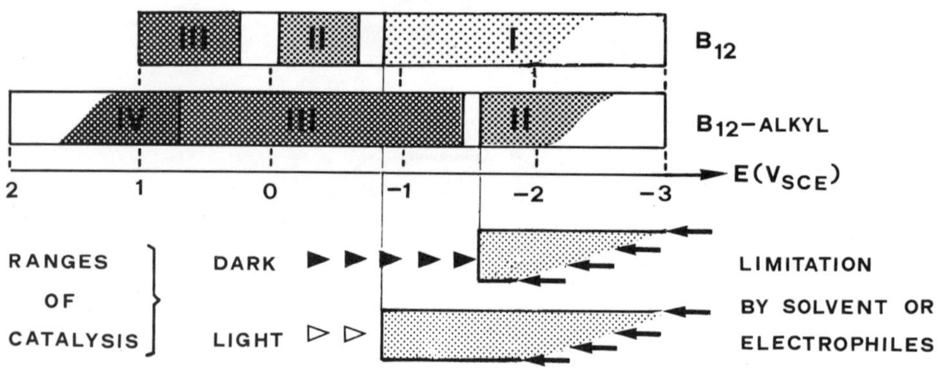

Catalytic reduction of an alkyl halide RX in presence of B_{12} and in absence of light is discussed starting from reduced B_{12} (cf. scheme 10). The +I oxidation state is maintained by a potential, which is more negative than $\sim$ -0.9 V (SCE). Alkylation of Co(I) by RX, present in excess, leads to [Co(III)-R]. This species can only be reduced at a potential more negative than $\sim$ -1.5 V (SCE). If such a negative potential is chosen, the (Co-C) bond is cleaved and, assuming a two-electron reduction during one turnover [171)], the initial Co(I) complex is regenerated as well as a potential anionic species R^-, which undergoes further reactions (cf. scheme 11). It depends on the reaction conditions (structure of R, presence of electrophiles etc.) whether elimination, proton capture or addition takes place.

Scheme 11 Catalytic reductive activation of R-X in the dark.

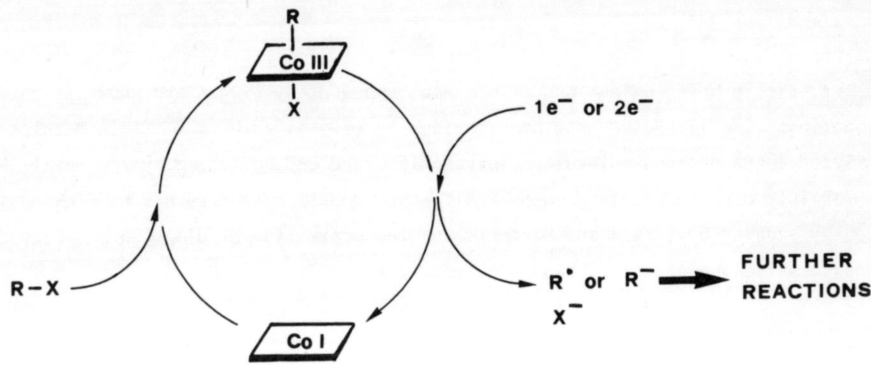

If the reduction potential falls between ∿ -0.9 and ∿ -1.4 V, catalysis in the dark is stopped since all the catalyst exists in the alkylated form, which is not reducible under these conditions.

In case of exposure to visible light Co(I) is alkylated as discussed above, however, the (Co-C) bond is now cleaved homolytically by irradiation (cf. scheme 10). This leads to Co(II), which is easily reducible at ∿ -0.9 V regenerating Co(I). The R˙ moiety may undergo further reaction or reduction (cf. scheme 12).

Scheme 12 Catalytic reductive and photochemical activation of R-X.

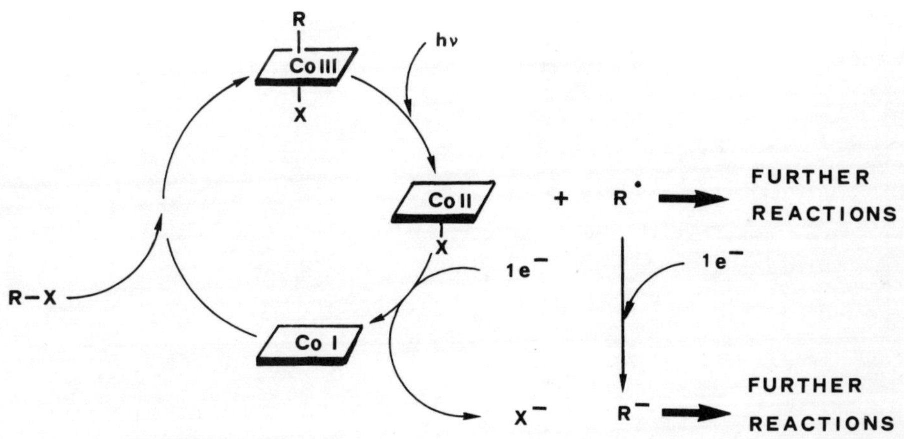

4. APPLICATION IN ORGANIC SYNTHESIS.

4.1. Oxidations.

Liquid-phase oxidations of organic compounds by oxygen are known as autoxidations. The literature has been reviewd by several authors [181]. In metalcatalyzed autoxidations molecular dioxygen is bound and activated by the metal. Reversible binding of dioxygen by Schiff-type cobalt complexes has been extensively studied by Calvin and Martell [79]. The activation of dioxygen proceeds by successive one-electron transfers from metal chelates to O_2.

Activation of molecular oxygen by cobalt complexes [181-183].

$$O_2 \underset{E^0 = -0{,}75\ V(SCE)}{\overset{e^-}{\rightleftharpoons}} O_2^{\bar{\bullet}} \underset{E^0 = -2{,}0\ V(SCE)}{\overset{e^-}{\rightleftharpoons}} O_2^{2-} \overset{2e^-}{\rightleftharpoons} 2\,O^{2-}$$

dioxygen; superoxide ion (moderate reducing agent) (very weak oxydant) (pK_a of $HO_2^{\bullet}$ in H_2O = 4,88); peroxide ion $\overset{2H^+}{\rightleftharpoons}$ H_2O_2; $2\,O^{2-}$ $\overset{4H^+}{\rightleftharpoons}$ $2\,H_2O$

$$O_2 \overset{Co(II)}{\rightleftharpoons} Co(III){-}O{-}O^{\bullet} \overset{Co(II)}{\rightleftharpoons} Co(III){-}O{-}O{-}Co(III)$$

superoxocobalt(III) μ-peroxocobalt(III)

$$O_2 \overset{Co(I)}{\rightleftharpoons} Co(II){-}O{-}O^{\bullet} \rightleftharpoons Co(III)\!<\!\!\begin{array}{c}O\\|\\O\end{array}$$

superoxocobalt(II) peroxocobalt(III)

All three types of dioxygen-adducts of cobalt chelates superoxo-[184], μ-peroxo[185] and peroxo-cobalt chelates[186] are known. The autoxidation of dipolar protic substrates like e.g. phenols, acetoins, indols etc. are believed to proceed via deprotonation (by superoxocobalt(III)), formation of intermediates with a (Co-C) bond, insertion of O_2 into the (Co-C) bond and fragmentation of the organoperoxocobalt(III) complex with cleavage of the (O-O) bond. For a critical discussion of proposed mechanisms see Sheldon and Kochi[181a]. The most frequently used catalysts are Co(II)salen derivatives[187], Co(II)Pc and more recently polymer-supported derivatives of Co(II)salen[188] and Co(II)Pc[189].

Phenols (unsubstituted in p-position) are autoxidized in presence of Co(II)salen derivatives[189-191] or Co(II)Pc[192] in DMF[189-191], $CHCl_3$[190], CH_3OH[189] or aqueous organic media[193] to the corresponding p-benzoquinones (BQ) or p-diphenoquinones (DPQ) in high yield but variable proportions depending on the reaction conditions (Eq. 15).

15)

Nighinaga et. al.[194] reported the catalytic autoxidation of 4-alkyl-2,6-di-tert. butylphenols with Co(II)chelates in methanol. The corresponding p-quinols are obtained in quantitative yield (Eq. 16).

16)

Bayston and Winfield[195] have shown that the oxidation of CO by O_2 in aqueous solution is catalyzed by vitamin B_{12a} (Eq. 17). Ligands, which displace the aquo group, (e.g. CN^-) inhibit the reaction. The overall rate of the reaction reaches three maxima at pH 5,7, 5,1 and <1,1, the first due to protonation of hydroxocobalamin and the third due to replacement of the benzimidazole ligand by an aquo group.

In presence of catalytic amounts of vitamin B_{12a} formaldehyde is easily autoxidized in aqueous solutions at pH 5,8 - 7,8 to formic acid [196] (Eq. 18). Formic acid is further oxidized at pH 6 to CO_2 [197] (Eq. 19). Vitamin B_{12a} is also a catalyst for the autoxidation of ascorbic acid to dehydroascorbic acid at pH 5 in water [198] (Eq. 20). In these reactions vitamin B_{12a} is a much more active catalyst than cobaloxim [196].

Autoxidations catalyzed by vitamin B_{12a} (1a)

$$CO \xrightarrow{O_2,\ H_2O\ \ pH\ 5{,}1-5{,}7} CO_2 \qquad 17)$$

$$CH_2O \xrightarrow{O_2,\ H_2O,\ pH\ 5{,}8-7{,}2} HCOOH \qquad 18)$$

$$HCOOH \xrightarrow{O_2,\ H_2O,\ pH\ 6} CO_2 + H_2O \qquad 19)$$

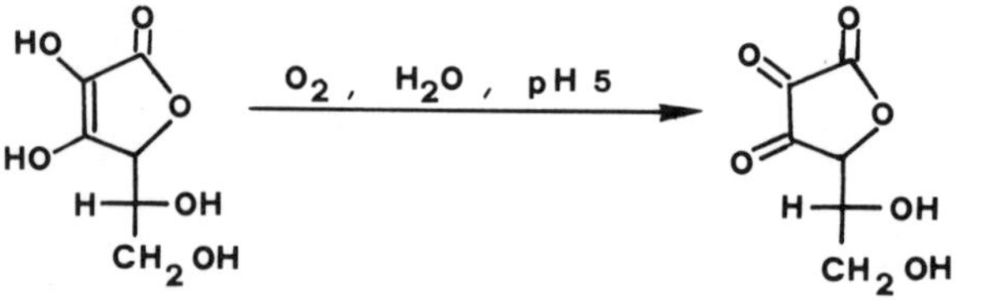

20)

In close analogy to the first step of the enzymatic conversion of L-trypthophan to nicotinic acid, 3-substituted indoles like e.g. tryptophan are oxidatively cleaved by air in CH_2Cl_2 and in presence of Co(II)salen [199] or Co(II)Pc [200] to yield formylkynurein (Eq. 21). Oxidation of indoline to indole in 83% yield is achieved by bubbling O_2 through a solution of indoline in methanol containing catalytic amounts of Co(II)salen [201] (Eq. 22).

Autoxidations of indole derivatives

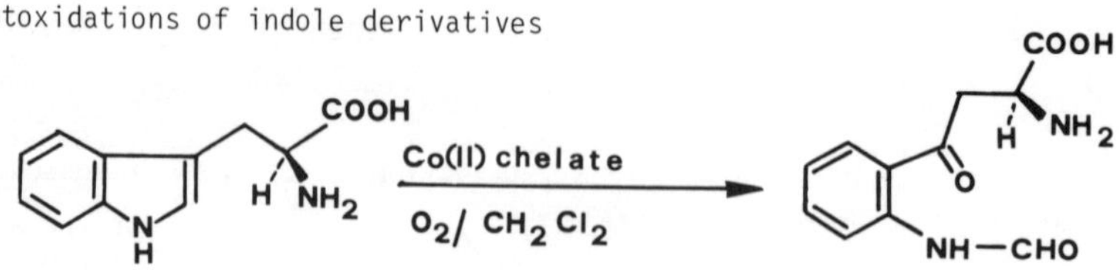

21)

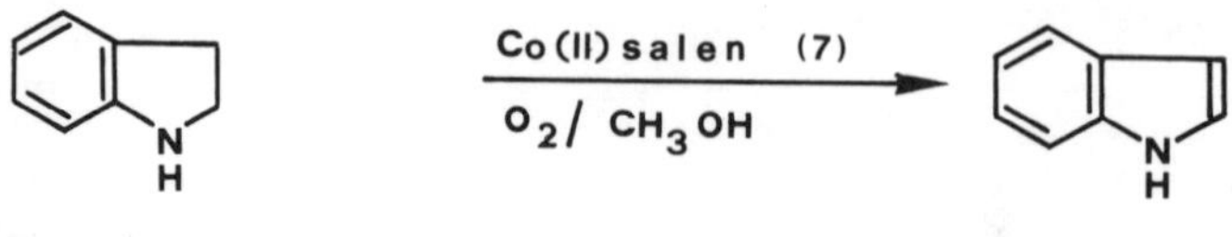

22)

Thiols are easily autoxidized in presence of catalytic amounts of Co complexes. Thus, 2-mercapto-4-methylthiazole, dissolved in 1,08 N NaOH containing a little CoPc-disulfonate and shaken under 58 psi O_2 for 2 hrs. is oxidized to the thiazole sulfinate salt, which after hydrolysis with HCl at 40° gives the 4-methylthiazole in 81% yield [202] (Eq. 23). For the oxidation of 2-mercaptobenzothiazole cf. [203].

$$\text{4-methyl-2-mercaptothiazole} \xrightarrow[\text{1,08 N NaOH}]{\text{CoPc(S)}_2\,/\,O_2} \text{thiazole-2-}SO_2^- \xrightarrow{H^+/H_2O} \text{4-methylthiazole} \qquad 23)$$

Numerous publications and patents concerne the catalytic oxidation of mercaptanes in petroleum distillates or gases (sweetening process). Metal phthalocyanines, e.g. partially sulfonated CoPc combined with bases on different supports e.g. charcoal [204], in aqueous caustic adsorbents [205] as well as polymer attached Co-phthalocyanines [206] are used as catalysts.

Co-salts and Co-phthalocyanines [36] play an important role as catalysts in the autoxidation of alkylbenzenes [207] and alkanes [208]. Recently it has been reported by Drago et. al. that terminal olefins are autoxidized in presence of catalytic amounts of Co(II)bis(salicylidene-γ-iminopropyl)methylamine 8 (R-CH_3) to the corresponding ketones and alcoholes [209] (Eq. 24). This is a rare example of an autoxidation of linear olefins by a first-row transition-metal, Pd(II)-Cu(II)-catalysts were already known to oxidize linear olefins by the well known Wacker process [210]. Reaction conditions: 40 mmol alkene in 45 mL solvent , ethanol or 1:1 ethanol-benzene) 0.73 mmol catalyst, 30 psi O_2, 50 to 70°; the presence of some $RhCl(PPh_3)_3$ cause synergistic enhancement of the production of ketone.

$$R\text{-}CH{=}CH_2 \xrightarrow[O_2\,/\,\text{EtOH}]{\text{Co-catalyst } 8(R{=}CH_3)} R\text{-}CO\text{-}CH_3 + R\text{-}CH(OH)\text{-}CH_3 \qquad 24)$$

Styrene is readily oxidized by sodium hypochlorite in presence of catalytic amounts of metal tetraphenylporphyrins e.g. Mn(TPP)OAc or Co(TPP)Br under mild phase-transfer conditions [211] (Eq. 25). Reaction conditions: To 0,025 mmol of catalyst and 0.05 mmol of benzyldimethyltetradecylammonium chloride were added 10 mL of CH_2Cl_2 and 1 mmol of olefin; finally 5 mL of 0,35 M aqueous sodium hypochlorite solution (pH 12) were added and the mixture stirred at room temperature.

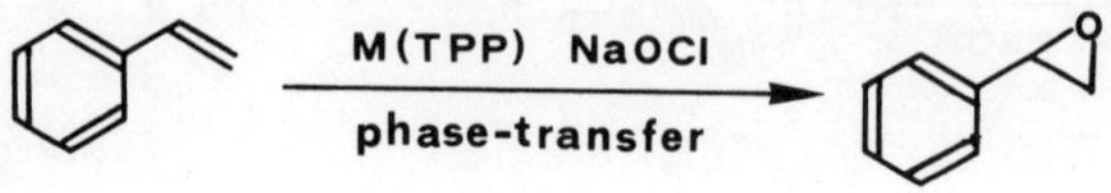

M(TPP) NaOCl
phase-transfer

25)

In relation to the cytochrome P-450 oxygenase, which catalyzes the hydrooxylation of a variety of organic substrates, it has been found by Mansuy et. al. [212] that metal tetraphenylporphyrins catalyze the oxydation of alkanes by alkylhydroperoxides. Thus, cyclohexane is converted by cumylhydroperoxide in presence of some Fe(TPP)Cl or Co(II)TPP 9 into cyclohexanol and cyclohexanone (Eq. 26). The reactivity of the Co-catalyst decreases slowly during the reaction, the Fe-catalyst remains active.

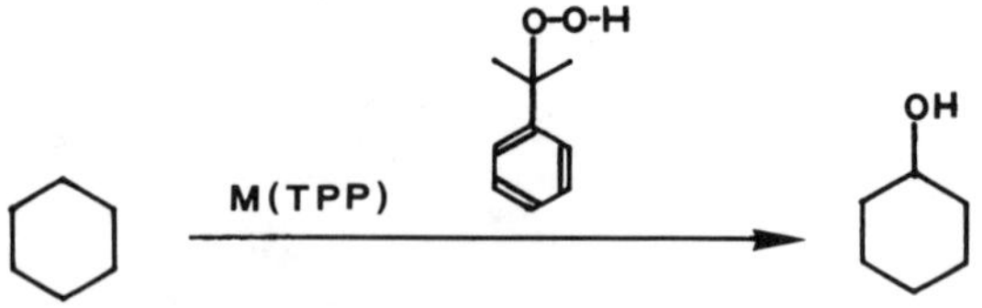

O-O-H

M(TPP)

OH

26)

4.2. Hydrogenation of olefins.

Numerous methods for the hydrogenation of olefins are already known. However, selective and specific reactions under mild conditions are still of potential interest.

Under acidic reaction conditions hydridocobalamin [108] reacts with olefins containing isolated double bonds [56]. Evidence for the formation of the corresponding alkylcobalamins according to Markownikow's rule has been presented, e.g. propylene is reported to afford sec. propylcobalamin [213]. Non-activated olefins have been hydrogenated in presence of catalytic amounts of vitamin B_{12} 1 in aqueous acetic acid and an excess of zinc [159, 180, 214, 215] cf. Eq. 27 [214] and 28 [159].

citronellol $\xrightarrow[\text{CH}_3\text{COOH / H}_2\text{O, Zn, 20°, 17 h}]{\text{10 mol \% B}_{12}}$ (86 %) 27)

α-pinene, β-pinene $\xrightarrow[\text{CH}_3\text{COOH / H}_2\text{O, Zn / 20°, days}]{\text{10 mol \% B}_{12}}$ 69 % + 2 %; Retention 97 % 28)

Olefins containing isolated double bonds were also shown to be hydrogenated by $NaBH_4$ in presence of CoPc 3 as catalyst [52] (Eq. 29).

$$C_8H_{17}\text{-HC=CH-}C_7H_{14}\text{-COOEt} \xrightarrow[\text{CoPc (10 mol \%)}\ 40\ h/20^o]{NaBH_4} C_{17}H_{35}COOEt \quad 100\% \qquad 29)$$

Hydrogenation of activated olefins (conjugated dienes, styrenes, α,β-unsaturated aldehydes, ketones, acids and acid derivatives), catalyzed by $[Co(CN)_5]^{3-}$ 5, has been reviewed [61]. More recently phase-transfer reaction conditions have been found to accelerate the reaction and to prevent catalyst desactivation [73]: Homogeneous hydrogenation of conjugated dienes, catalyzed by $K_3[Co(CN)_5H]$ (∿ 10 mol %), leads to monoenes, generally with E configuration in an overall 1,4-addition of hydrogen (Table 2). However, stereospecificity is not very high.

Table 2 Hydrogenation of conjugated dienes catalyzed by $K_3[Co(CN)H]$ **5c** under phase-transfer conditions [73].

Diene	Reaction-time (h)	Products	Yield (%)		
	6	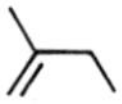	(<1)	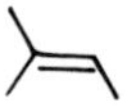	(99)
	6	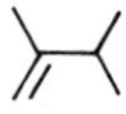	(20)	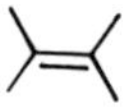	(80)
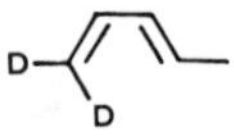	3	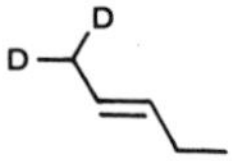	(53)	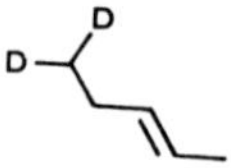	(47)
		no reaction			
	12		(100)		

Under the same conditions exclusive hydrogenation of the (C-C) double bond in α,β-unsaturated ketones may take place. Aldehydes are less reactive; substituents in β-position diminish the reactivity (Table 3).

Isolated (C-C) double bonds and carbonyl functional groups are not attacked by $K_3[Co(CN)_5H]$. Asymmetric hydrogenation of various activated olefins by the same catalyst in presence of an optically active α-amino acid was shwon to be not very effective [216].

Table 3 Hydrogenation of α,β-unsaturated ketones catalyzed by $K_3[Co(CN)_5H]$ 5c under phase-transfer conditions [73].

Substrate	Reaction time (h)	Product	Yield (%)
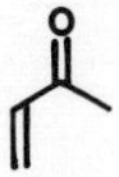	24		(60)
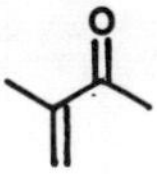	8	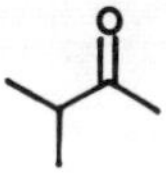	(85)
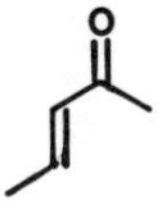 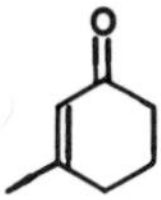		**no reaction**	
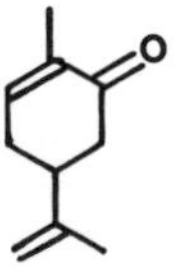	20	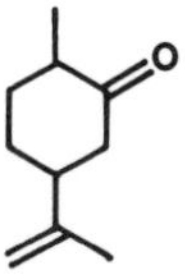	(93)
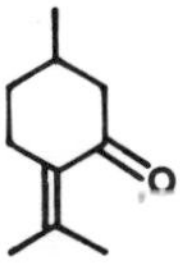	720	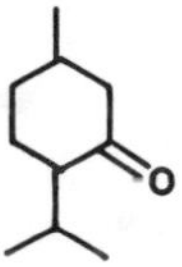	(71)

Conjugated olefins are also hydrogenated by $Co(II)(DH)_2X$ 4 in presence of H_2 [56]. In an almost neutral medium the first step of the catalysis is assumed to be an α-addition of the intermediate Co hydride species to the activated double bond (cf. 3.2.) favoured by electron withdrawing substituents [124, 179]. In alkaline media, the Co-catalyzed hydrogenation is reported to proceed via β-attack of Co(I). As a second step in both cases a reductive cleavage of the (Co-C) bond with proton capture (e.g. from the solvent) is formulated. Examples are listed in table 4.

Asymmetric induction during catalytic hydrogenation has been observed in presence of optically active bases X (e.g. quinine), $Co(II)(DX)_2X$ 4 and H_2. Under these conditions α-phenylacrylophenone affords 1-methyl-2-oxo-stilbene. An enantiomeric excess of 49% has been reported [217].

Table 4 Hydrogenation of activated olefins catalyzed by $Co(II)(DH)_2$ py 4 / H_2.

Substrate	Catalyst mol %	Reaction-time (h)	Product Yield (%)	Ref.
$H_2C{=}CHCN$	100	—	—	179)
$H_2C{=}C(CH_3)CN$	30	4,3	$H_3C\text{-}CH(CH_3)CN$ (87)	179)
$H_2C{=}C(CH_3)CO_2CH_3$	10	2	$H_3C\text{-}CH(CH_3)CO_2CH_3$ (98,5)	179)
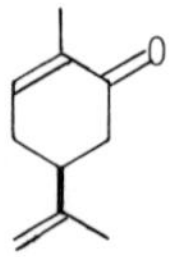 [structure]	25	2	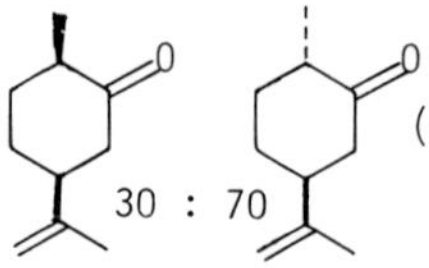 [structures] 30 : 70 (60)	124)
$H_3CO_2C(H)C{=}C(CH_3)CO_2CH_3$	25	3	$H_3CO_2C\text{-}CH_2\text{-}CH(CH_3)CO_2CH_3$ (90)	124)

Fischli et al. have reported on cob(I)alamin catalyzed hydrogenations of allylic alcohols [124, 218], α,β-unsaturated carbonyl derivatives, nitriles [17, 180], esters [218, 219, 220], acids [218], amides [218, 220] and ketones [218]. Results of enantioselective hydrogenations are compiled in table 5.

The observed stereochemistry of the reduction is explained by β-attack of Co(I) to the activated olefin. In some cases reduction of the carbonyl function has also been observed [17, 218, 221].

Table 5 Enantioselective hydrogenation of α,β-unsaturated carbonyl derivatives catalyzed by cob(I) alamin (10 mol %).

R	yield (%)	e.e. (%)	time (h)	yield (%)	e.e. (%)	time (h)
$COOC_2H_5$	86	21.6(S)	70	81	0	93
COOH	80	20.1(S)	90	80	0	88
$CONH_2$	83	23.1(S)	44	85	0	89
CON(morpholino)	66	23.8(S)	136	41	0	336
$CON(CH_3)_2$	25	15.6(S)	300	30	0	336
$COCH_3$	63	32.7(S)	20	69	5.9(R)	20
CH_2OH	94	0	24	82	0	46

Hydrogenation of the (C-C) double bond of α,β-unsaturated carbonyl derivatives (nitriles, esters, amides) is achieved in protic media by $NaBH_4$ in presence of CoPc 3 (Table 6) [52]. It depends on the reaction conditions, whether functional groups are reduced too (cf. 4.4.).

Table 6 Hydrogenation of activated olefins with $NaBH_4$ in EtOH/20° in presence of CoPc 3 (10 mol %) [52].

Substrate	Reaction time (h)	Product	Yield (%)
C_6H_5-CH=CH-CN	1	C_6H_5-CH_2-CH_2-CN	(82)
C_6H_5-CH=CH-CO_2Et	0.5	C_6H_5-CH_2-CH_2-CO_2Et	(98)
$(CH_3)_2$=CH-CO-NH_2	165	$(CH_3)_2$CH-CH_2-CO-NH_2	(80)

Reduction of α,β-unsaturated ketones with Zn in protic neutral media catalyzed by a sulfonated CoPc [54)] has been investigated in our laboratory [222)]. Specific transformation of carvone to dihydrocarbone (Eq. 30) has been shown to take place within 22 hrs with a yield of 91% (cf. experimental procedure). In this reaction vitamin B_{12a} <u>1a</u> is less efficient and affords only 15% dihydrocarvone under the same conditions.

Zn / NH_4Cl, $EtOH/H_2O$ 20°, 4 mol % $CoPcS_{2,6}$ <u>3c</u>; 91% (30)

Experimental procedure for the reduction of Carvone [222)].

Activated zinc (313 mg, 4.8 mmol) [17)] and ammoniumchloride (174 mg, 3.3 mmol) were put under N_2 in a 50 mL two-necked round-bottomed flask, equipped with a magnetic stirrer and a reflux condenser. Then a mixture of ethanol/water (16 mL, 1:1) was added. 46 mg (0,07mmol) of partially sulfonated Co-phthalocyanine <u>3c</u> [54)] were added to the stirred suspension. The colour changed from blue to yellowish-brown within 3 or 4 min. (-)-carvone (244 mg, 1.6 mmol) were injected. After stirring for 22 hrs at room temperature 91% (GC) dihydrocarvone have been detected. The solution was filtered through cotton wool, eluted with petroleum ether (3x20 mL) and dried over sodium sulfate. The solvent was evaporated and the product distilled at 5 Torr/82^{0} C. Isolated yield: 173 mg (71%) of dihydrocarvone (trans:cis = 2.5:1). Analogous reactions performed in absence of catalyst or NH_4Cl showed no formation of dihydrocarvone.

Hydrogenation of acetylene [223, 224)] and nitrogen [223)] by a Co porphyrin [223)] and a CoPc [224)] respectively has been reported.

Since it has been found that $[Co(I)Pc]^-$ doesn't add to (C-C) double bonds [44)], catalytic hydrogenations of olefins do probably not involve intermediates with (Co-C) bonds [222b)]. It is therefore still questionable, whether alkyl Co complexes are necessarily intermediates in the main route of (C-C) double bond reduction although corresponding organocobalt species have been isolated under special (i.e. non-catalytic) conditions. Hydrogen evolution from acidic solutions under reducing conditions is catalyzed by Co complexes. Activated hydrogen species may be responsible for the catalytic hydrogenation of olefins, cf. also [225)].

4.3. Hydrogenolysis of organic halides.

Methods for the hydrogenolysis of organic halides and related compounds have been reviewed [226]. Besides chemical reducing agents, complex hydrides and electrochemical reduction, several catalytic methods are known. Catalysts commonly emploied are of the noble metal type. Less frequently used are homogeneous or heterogeneous cobalt salt-catalysts, e.g. $CoH(BH_4)$ [227] or $[Co(CN)_5H]^{-3}$ [61], respectively. The latter one is known to reduce saturated organic halides [61] or non-allylic halides [228] to the corresponding alkanes or olefins. Organocobalt complexes have been ruled out as intermediates in the catalytic cycle (it has been stated that the formation of organopentacyanocobaltate does stop catalysis) [61]. On the other hand are organocobalt species in case of macrocyclic cobalt complexes absolutely necessary intermediates according to the catalytic cycle in scheme 11. A first application has been described by Williams [171]. Vitamin B_{12} was found to catalyze the electrochemical reduction of alkyliodides at mercury electrodes. The significant amount of alkylmercuryiodide and dialkylmercury compounds apart from alkanes as products and coulometric results have been interpreted by a radical cleavage mechanism of the intermediate (Co-C) bond. More recently higher yields in hydrogenolized products with CoPc, vitamin B_{12} and CoHDP as catalysts have been reported. Selected examples are displaied in table 7.

Table 7 Co-catalyzed hydrogenolysis of organic halides in protic solvents (replaced halide is underlined).

Substrate	Catalyst	Reducing agent	Yield (%)	Ref.
n-$C_{12}H_{25}$-Br	CoPc <u>3</u>	$NaBH_4$	(97)	52)
2,4-$(CH_3)_2C_6H_3$-Cl	CoPc <u>3</u>	$NaBH_4$	(63)	52)
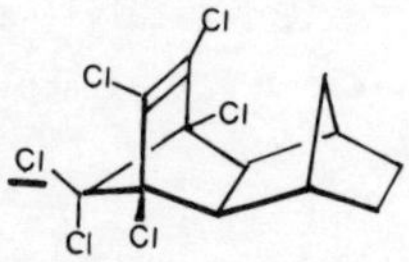	B_{12} <u>1b</u>	$NaBH_4$	(quant.)	227)
Bromocyclododecane	B_{12} <u>1b</u>	Zn	(82)	214)
3-Iodocholestane	CoHDP <u>2</u>	Zn or electrochem.	(quant.)	116a)
4-Methoxycyclohexylbromide	B_{12a} <u>1a</u>	Zn or electrochem.	(93-97)	23)

The stereochemical course of hydrogenolysis can be deduced from the results observed in (Co-C) bond formation step (spanning the range of net inversion to euqilibration in the organocobalt intermediate) and its reductive cleavage (retention of configuration) (cf. chapter 3.2. and 3.3.).

The overall and step-by-step stereochemical course has been studied in case of the epimeric 4-methoxycyclohexylbromides in D_2O. The results shown in scheme 13 have been interpreted by the formation of the thermodynamically more stable organocobalt complex and its reductively induced cleavage and protonation in D_2O with retention of configuration [23].

Scheme 13 Stereochemical course of the hydrogenolysis of cis- and trans-4-methoxycyclohexylbromide in D_2O.

OCH3
Br
OCH3
Br
1. 2a, -1,6V, 4e⁻, Pt-El. in CH3OH/LiClO4
2. AgNO3, NaSO3CF3
OCH3
H
Co
OSO2CF3
-1,9V, 2e⁻, Pt-El.
THF/D2O/LiClO4
OCH3
D
+
OCH3
D
4,5 : 1

In the cobalt-catalized electrochemical reaction the rate of hydrogenolysis can be followed by the current-time dependence. The catalyst is undergoing electron transfer reactions at the surface of the electrode and chemical reaction in solution. This situation has been calculated on the base of the "reaction layer model" [96, 229]. Such studies will help to optimize cell design, catalyst concentration and other experimental parameters for most efficient future application of this method.

Experimental procedure for the hydrogenolysis of alkylhalides catalyzed by CoPc 3a (also valuable for the reduction of functional groups, cf. 4.4., table 8) [52].

$NaBH_4$ (2.7 g, 70 mmol) and CoPc (0.5 g, 0.9 mmol) were dissolved in 50 mL EtOH under a N_2 atmosphere. To the deeply colored solution (or suspension) 10 mmol substrate was added and stirred at 20-25° (for 21 hrs in case of n-$C_{12}H_{25}Br$, 2 hrs in case of 2,4-$(CH_3)_2$ C_6H_3-CH_2Cl). The reaction mixture was neutralized with 5 N HCl at 0° until gas evolution stopped (5-10 min., pH 6-7). Then the suspension was centrifuged. The precipitate was washed 3 times with MeOH. The catalyst was recovered after 3 times washing with H_2O followed by drying at 20 to 200°. In case of aliphatic amines the residue was treated with 1N NaOH/ether. Volatile amines were isolated from the ether phase as hydrochlorides. Other products were obtained by evaporation of the organic phase, which was dried with Na_2SO_4. The products were purified, in case of neutral substances by dissolving them in hexane/ether followed by filtration, in case of unvolatile amines by dissolving them in H_2O/HCl followed by filtration.

4.4. Reduction of functional groups.

Reagents exhibiting specific reactivity may be used for the transformation of one distinct functional group in presence of other functionalities with only slightly different chemical properties. There exist many examples for applications of Co complexes in this field and only some of them occuring under reducing reaction conditions will be discussed here.

Aliphatic and aromatic nitro compounds are reduced by BH_4^- or H_2 in presence of catalytic amounts of a macrocyclic Co complex (cf. table 8). Mechanistic studies have been published concerning the borohydride reduction of nitrobenzene catalyzed by vitamin B_{12a} 1a [230]. It has been suggested that the B_{12}-catalyzed steps consist of additions of B_{12s} to unsaturated systems containing a multiply-bonded nitrogen. The remaining steps, e.g. the reduction of phenylhydroxylamine leading to aniline, are achieved directly by borohydride. In accordance with these statements it has later been published that $[Co(I)Pc]^-$ also adds to nitro groups [44].

Synthetic applications have been performed by Eckert [51] using CoPc 3 in non-catalytic quantities however. Under these mild conditions the following functional groups are not attacked: Arenes, arylhalides, activated and non-activated (C-C) double bonds, cyanides, isocyanides, aldehydes, ketones, esters and amides. Thus an aromatic nitroaldehyde is reduced by Li[CoPc] and the resulting non-isolated aminoaldehyde undergoes further condensation with a ketone in a Friedländer-type reaction (Eq. 31) [51]. Based on these results first steps in the synthesis of the antitumor-active alcaloid camptothecin have been done [51].

$$\text{2-nitrobenzaldehyde} + \text{acetone} \xrightarrow[\text{CH}_3\text{OH} \quad 63\ \text{h}]{6\ [\text{Co Pc}]^-} \text{2-methylquinoline}\ (77\ \%) \qquad 31)$$

Experimental procedure for the reductive annelation Eq. 31.

Under N_2 [CoPc]Li·4.5 THF [50] (12-13 g, 13-14 mmol) is dissolved in 70 mL N_2-saturated methanol. To the green solution an excess of acetone (∿50 mmol) and 2-nitrobenzaldehyde (302 mg, 2.0 mmol) are added. The mixture is stirred at room temperature for 63 hrs. Water (5 mL) is added and during 5 min. CO_2 and air are bubbled through at once. The blue precipitate is centrifuged (5 min., 3000 r.p.m.) and washed with ethanol and twice with ether. The combined solutions are concentrated and extracted (hexane/HCl 0.5 N, 50 mL). The organic layer is extracted twice with water. The combined aqueous phases are washed with hexane, neutralized with NaOH (5N) and extracted with ether. The ether extract is dried (KOH) and concentrated to get directly 2-methylquinoline (77%).

There has also been reported on the reduction of nitro [61, 232] and nitroso [52, 61] compounds, catalyzed by $[Co(CN)_5]^{3-}$ 5a [61] and CoTPP 9 [232] respectively.

The reductive methylation of amines in presence of formaldehyde and hydrogen has been found to be catalyzed by vitamin B_{12} 1b or $Co(DH)_2 \cdot 2H_2O$ 4 (Eq. 32) [233]. In the same manner thiols are transformed to methylthioethers [233].

$$\text{PhNHR} + CH_2O \rightleftharpoons \text{PhN(R)}CH_2OH \xrightarrow[H_2,\ CH_3OH,\ 2h]{\text{cat. } B_{12a}} \text{PhN(R)}CH_3 \qquad 32)$$

R = H, CH_3 ; 10 %, ($R = CH_3$)

Further examples of functional group reductions in presence of catalytic amounts of a Co complex are compiled in tables 8 and 9.

Table 8 Co-catalyzed reductions of aliphatic and aromatic nitro compounds.

Substrate	Catalyst (mol %) [a]	React. time (h)	Red. agent	Solvent	Product (yield, %)	Ref.
$C_6H_5NO_2$	vitamin B_{12} <u>1a</u> (0.14)	48	KBH_4	$MeOH:H_2O$ 9:1	Aniline (70) Azobenzene (30)	230)
$n\text{-}C_3H_7NO_2$	CoPc <u>3a</u> (10)	189	$NaBH_4$	EtOH	$n\text{-}C_3H_7NH_2$ (76)	52)[b]
$p\text{-}NO_2\text{-}C_6H_4\text{-}CH_2\text{-}CN$	CoPc <u>3a</u> (10)	6	$NaBH_4$	EtOH	$p\text{-}NH_2\text{-}C_6H_4\text{-}CH_2\text{-}CN$ (85)	52)[b]
$p\text{-}NO_2\text{-}C_6H_4\text{-}CH{=}CH(CO_2Et)$	CoPc <u>3a</u> (10)	120	$NaBH_4$	EtOH	$p\text{-}NH_2\text{-}C_6H_4\text{-}CH_2\text{-}CH_2(CO_2Et)$ (64)	52)[b]
$p\text{-}NO_2\text{-}C_6H_4\text{-}CH{=}CH(CO_2Me)$	$Co(DH)_2$ <u>4</u> (50)	1	H_2	MeOH	$p\text{-}NH_2\text{-}C_6H_4\text{-}CH_2\text{-}CH_2(Co_2Me)$ (78)	231)

a) based on 1 mol (=100%) of substrate.

b) cf.4.3. for experimental procedure.

Table 9 Selected examples of Co-catalyzed functional group reductions.

Substrate	Catalyst (mol %) [a]	React. time (h)	Red. agent	Solvent	Product (yield,%)	Ref.
c-C_6H_{10} =NOH	CoPc 3a (10)	48	$NaBH_4$	EtOH	c-C_6H_{11}-NH_2 (78)	52)
C_6H_5-CH_2-C(COOH)=NOH	$[Co(CN)_5]^{3-}$ 5 (100)	8	H_2	H_2O	C_6H_5-CH_2-CH(COOH)-NH_2 (82)	234)
C_6H_5-CH=N-i-C_3H_7	CoPc 3a (10)	72	$NaBH_4$	EtOH	C_6H_5-CH_2NH-i-C_3H_7 (92)	52)
p-NO_2-C_6H_4-CH_2-CN	CoPc 3a (10)	156	$NaBH_4$	EtOH	p-NH_2-C_6H_4-CH_2-CH_2-NH_2 (77)	52)
C_6H_4-CH_2-CH_2-CN	vit.B_{12} 1b (10)	6	Zn	CH_3COOH/H_2O	C_6H_5-CH_2-CH_2-CHO (88)	17)
1-Naphtonitrile	vit.B_{12} 1b (10)	5.5	Zn	CH_3COOH/H_2O	(1-naphtyl)methylamine (75)	221)
Cyclododecanecarbo-nitrile + piperidine	vit.B_{12} 1b (10)	116	Zn	CH_3COOH	1-(cyclododecylmethyl)-piperidine (85)	235)
C_6H_5-CH_2-CH_2-CHO	vit.B_{12} 1b (10)	137	Zn	CH_3COOH/H_2O	C_6H_5-CH_2-CH_2-CH_2OH (30)	17)
10,11-Epoxyundecyl-acetate	heptamethyl-cob(I)yrinate (10)	52	Zn	CH_3COOH	undecylacetate (51.5) 10-undecenylacetate (22.3)	215)
Benzil	$Co(DH)_2$.2 qui-nine 4 (10)	(?)	H_2	C_6H_6	S(+)benzoin (98), 61.5% e.e.	236)
Azobenzene	$Co(DH)_2$ 4 (20)	0.5	H_2	alcohol	hydrazobenzene (99)	231)

a) based on 1 mol (=100%) of substrate.

4.5. Reductive elimination.

Alkyl-Co(III) complexes containing a (potential) leaving group in β-position with respect to the central atom are usually stable. If the leaving group is activated, they may undergo β-elimination. A well-known example is the cleavage of the (Co-C) bond in adenosylcobalamin 1c in acidic media [237] (Eq. 33).

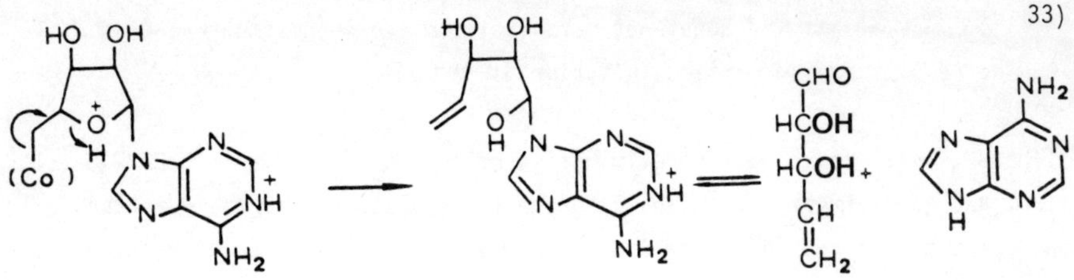

33)

Alkyl-Co(III) complexes bearing a leaving groupe in β-position immediately undergo β-elimination on reduction with formation of olefins and Co(II)-species, which may be further reduced to a Co(I) species (Eq. 34).

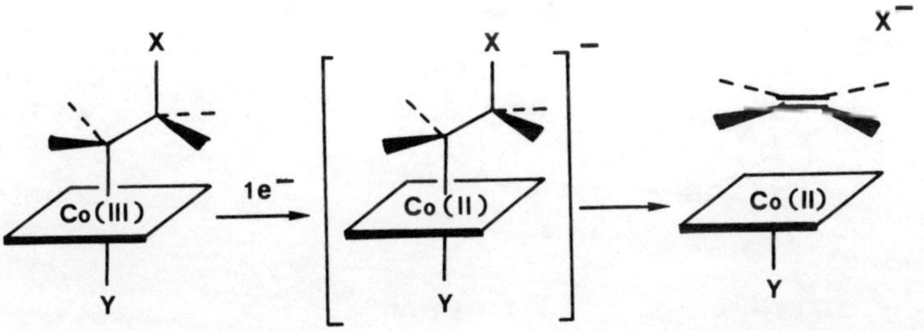

34)

Vicinal leaving groups are reductively eliminated at very mild conditions if the formation and cleavage of the (Co-C) bond are combined in a catalytic cycle (Eq.35)

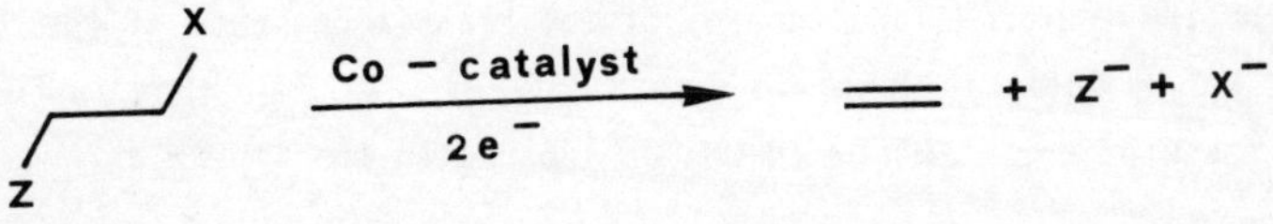

$$Z\text{-}CH_2\text{-}CH_2\text{-}X \xrightarrow[2e^-]{Co-catalyst} CH_2{=}CH_2 + Z^- + X^- \qquad 35)$$

4.5.1. Stereochemistry and cis- trans-isomerisation of olefins.

In order to establish the stereochemistry of this catalytic reductive elimination Philippe [238)] studied vitamin B_{12}-catalyzed reduction of *erythro*- and *threo*-3-bromo-2-butanol . The starting materials were prepared according to known procedures [239)]. The reductive elimination was carried out in a variety of solvents at room temperature at pH 6,5 using activated Zn-dust as reducing agent and vitamin B_{12a} 1a as catalyst.The 2-butenes were isolated and identified by GC-analysis. The results are summarized in tables 10 and 11.

The B_{12}-catalyzed reductive elimination of the *threo*-isomer in $EtOH/H_2O$ 1:1 was fast and gave almost pure trans-2-butene (entry 1), whereas the reaction in absence of catalyst was very slow and led to a 1:1 mixture of both 2-butenes (entry 4). The *erythro*-isomer on the other hand gave almost pure trans-2-butene in the slow uncatalyzed reduction (entry 5) and a 1:1 mixture of both 2-butenes in the fast catalyzed reaction (entry 5).

Scheme 14 Newman-projections of *erythro*- and *threo*-3-bromo-2-butanole (butane conformation antiperiplanar)

CH3 HO H H Br CH3

erythro

CH3 HO H Br H CH3

threo

In the thermodynamically favoured conformation of the *erythro*-isomer, Br and OH are antiperiplanar and therefore ideally oriented for an anti-elimination induced by Zn. In the *threo*-isomer the antiperiplanar arrangement of Br and OH for the anti-elimination by Zn would afford a less stable conformation (for a stereochemical study of Zn-promoted elimination of halohydrins cf. House et. al. [240]). On the other hand the conformational energy of the transition state of the S_N2 displacement of Br^- by Co(I) is lower in case of the *threo*-isomer than in the *erythro*-isomer (assumption: (C-H) or (C-OH) coplanar with the linear arrangement of Br-C-Co).

Table 10 Reductive elimination of *threo*-3-bromo-2-butanol catalyzed by vitamin B_{12a} <u>la</u>.

Entry	Reducing agent	Catalyst mol % a)	Solvent	T °C	Ratio of 2-butenes cis:trans
1	Zn/NH_4Cl	1,4	$EtOH/H_2O$	20	1 : 99
2	Zn/NH_4Cl	1,4	DMSO	20	3 : 97
3	Zn/NH_4Cl	1,4	DMF	20	18 : 72
4	Zn/NH_4Cl	-	$EtOH/H_2O$	60	54 : 46

Table 11 Reductive elimination of *erythro*-3-bromo-2-butanol catalyzed by vitamin $B_{12}a$ <u>la</u>.

Entry	Reducing agent	Catalyst mol % a)	Solvent	T °C	Ratio of 2-butenes cis:trans
5	Zn/NH_4Cl	1,4	$EtOH/H_2O$	20	45 : 55
6	Zn/NH_4Cl	1,4	DMSO	20	98 : 2
7	Zn/NH_4Cl	1,4	DMF	20	96 : 4
8	Zn/NH_4Cl	-	$EtOH/H_2O$	60	0 : 100 b)

a) based on 1 mol (= 100%) of substrate

b) contamined by 6% of non-olefinic compounds

It is interesting to note that the vitamin B_{12}-catalyzed reactions in DMSO yield almost pure trans-2-butene from the *threo*-isomer (entry 2) and almost pure cis-2-butene from the *erythro*-isomer (entry 6). This observation is probably due to the ability of the solvent to promote S_N2-reactions. We therefore favour a mechanism for the B_{12}-catalyzed reaction (in DMSO), which consists of a (Co-C) bond formation with inversion at the carbon atom followed by a reduction of the complex with subsequent anti-elimination (scheme 15). Until now the configuration of the intermediate alkyl Co(III) complexes of this reactions have not been determined and the discussed mechanism remains hypothetic.

Scheme 15 Proposed mechanism for the vitamin B_{12}-catalyzed elimination of bromohydrins (shown for the *erythro*-3-bromo-2-butanol).

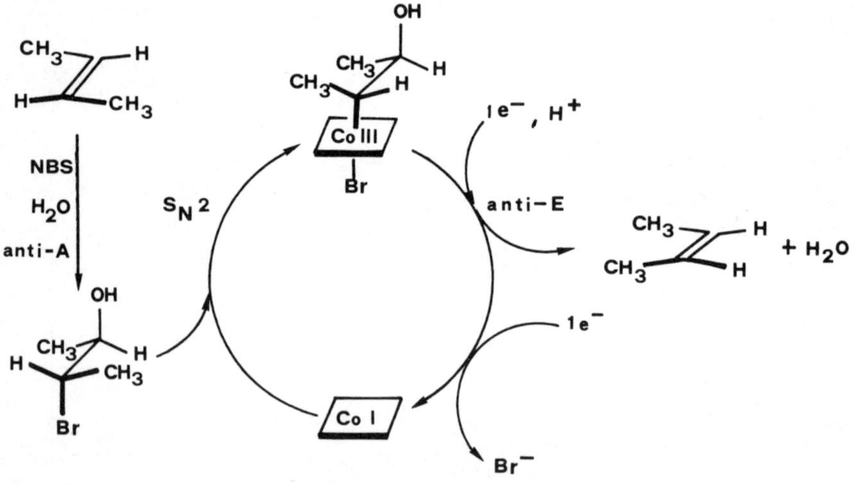

The sequence of the two reactions consisting of:

a) anti-addition of HOBr (or generally XY) to an olefin followed by the

b) "over-all" syn-elimination of X and Y (according scheme 15)

represents a two-step procedure for the cis-trans isomerisation of olefins (methods for cis-trans isomerisations of olefins have recently been reviewed by Sonnet [241]). This two-step procedure has been applied for the interconversion of cis- into trans-stilbene and trans- into cis-stilbene as well as for the interconversion of cis-cyclooctene into the strained trans-cyclooctene [238].

4.5.2. Reductive cleavage of protecting groups.

Since the introduction of the 2,2,2-trichloroethylester as protecting group in the cephalosporin synthesis by Woodward et al.[242] a wide variety of methods based essentially on reductive β-elimination have been developed. The β-haloethyl group and analogous groups have been used to protect alcohols [243], amines [244], carboxylic acids [45, 245], phosphoric acids [246] and carbonyl compounds [247]. Several modifications based on reductive cleavage (Zn/CH_3OH, Zn/CH_3COOH, Zn/phosphate buffer solution [248] or electrochemical reduction [245, 249]) reaction with strong nuleophiles (HSe^- and Se^{2-} [250]), LiCoPc [45]) and Co-catalyzed reductive cleavage (with vitamin B_{12a} <u>1a</u> [173], CoHDP <u>2b</u> [173] and CoPc [251]) have been reported for the deprotection reaction.

The stability of the β-haloethyl group towards bases and nucleophiles increases along the series: $-CH_2CCl_3 < -CH_2CH_2I < -CH_2CH_2Br < -CH_2CH_2Cl$. The β-chloroethyl group would therefore be the most interesting protecting group, provided that it may be removed by a mild procedure.

In table 12 some representative examples of protections and deprotections of compounds containing different functional groups are listed. Alcohols or amines are protected as β-haloethyl-carbonates or carbaminates respectively. Acids are protected as β-haloethylesters and carbonyl compounds as ketals containing the halide in β-position with respect to the ketal oxygen.

The catalyzed reductive elimination according to scheme 15 consists of decomposition of the β-haloethylcarbonates- and carbaminates with liberation of the corresponding alcohol or amine respectively; as by-products ethylene, CO_2 and the halide anion are formed. The decomposition of a β-haloethylester leads to the corresponding carboxylate, ethylene and halide anion. The β-haloethylketal undergoes reductive elimination with formation of an allylhemiacetal, which on hydrolysis, leads to the carbonyl compound; as by-products allylalcohol and halide anion are formed.

As shown in table 12 and the following experimental procedures, the protection is easy to realize and the vitamin B_{12}-catalyzed reductive deprotection proceeds under extremely mild conditions (pH 6 to 7): other functional groups e.g. esters, β-lactames and ketales remain unattacked. As electron source zinc or electrochemical methods were emploied. It has been shown by Beer[252] that mercury as cathode material can be replaced by graphite or glassy carbon. First results showed that immobilized Co-complexes on a graphite surface allow deprotection of β-bromoethylesters with high yield implying several hundreds of turnovers per cobalt[253].

Experimental procedures for the protection and deprotection of functional groups (cf. table 12).

a) Protection[238]

Entry 2: To a stirred solution of vanillin (4.56 g, 30 mol) and 4.29 g (30 mmol) of β-chloroethyl chloroformiate in 20 mL of dry THF (N_2 atmosphere) were added 5 mL of pyridin at 0^{0}. After 2 hrs at RT, the acylation was complete (TLC). The white precipitate was filtered off and the reaction mixture partitioned between Et_2O and H_2O. The organic layer was successively washed with dilute HCl (1N) and brine. The solvent was removed under vacuum and the residue (yellow oil) was sufficiently pure (TLC, Silicagel, Et_2O, Rf = 0.55) for ketalisation, Yield: 6.90 g (89%). Following the procedure of Sterzycki[253], the ketal was prepared with ethylene glycol. The crude product was crystallized from Et_2O/pentane (1:5) to afford 72% of the ketal-carbonate. m.p. = 82-83^{0}.

Entry 3: To 1.34 g (10 mmol) of 2-phenyl propanal in a 100 ml flask were added 2,32 g (15 mmol) of 3-bromo-1,2-propanediol, 100 mg of p-TsOH and 50 mL of benzene. The reaction mixture was refluxed for 20 hrs with removal of water using a Dean-Stark trap. The product was isolated after washing with water and removal of the solvent: 2.70 g (100%) crude bromo ketal, which was shown to be 92% pure by GC analysis. A small sample was purified by flash chromatography.

Entry 4: A mixture of 1.54 g (10 mmol) of 4-tert.-cyclohexanone and 21 mg (0.1 mmol) of Et_4NBr in 20 mL α-epichlorhydrin was refluxed for 6 hrs. The excess of α-epichlorhydrin was then evaporated under reduced pressure. The residue was diluted with CH_2Cl_2 (100 ml) and washed with a sat. $NaHCO_3$ solution. The organic layer was separated, dried over Na_2SO_4 and the solvent was removed. Fractional distillation of the yellow oily crude product gave 2.23 g (91%) of the chloroethylethylene ketal. b.p. = 67^{0}/0.01 Torr.

Table 12: Vitamin B_{12a}-catalyzed reductive removal of the β-haloethyl group and related protecting groups.
Methods: b-1) Zn-reduction, b-2) electrochemical reduction.

Entry:	Functional Group	Protected as	Method and Reaction cond.	Yield (% isol.)	Ref.
	Alcohols, Phenols				
1	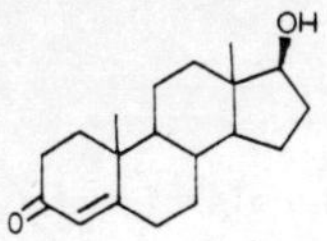	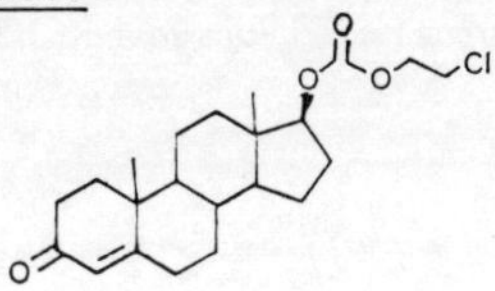	b-1) RT, 2h	(92)	238)
2	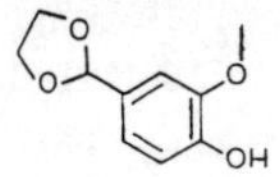	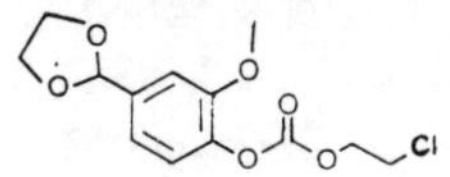	b-1) RT, 6h	(86)	238)
	Carbonyl and 1,3-dicarbonyl compounds				
3	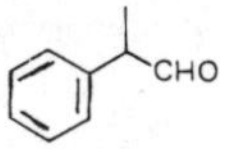	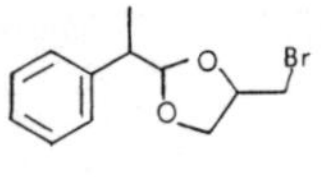	b-1) RT, 26h	(90)	238)
4	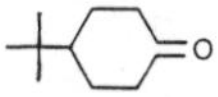	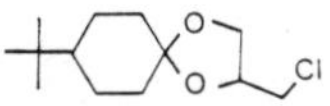	b-1) RT, 20h	(92)	238)
5	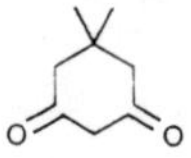	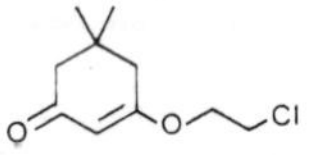	b-1) RT, 18h	(96)	238)
	Amines				
6	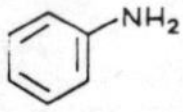	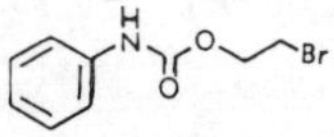	b-1) RT, 15h	(96)	173)
7	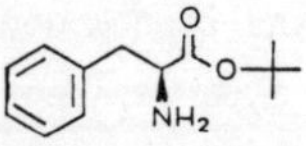	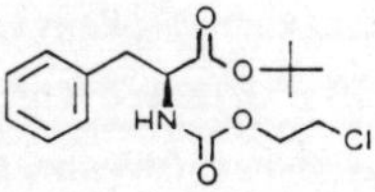	b-1) RT, 18h	(77)	238)
	Carboxylic Acids				
8	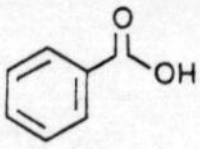	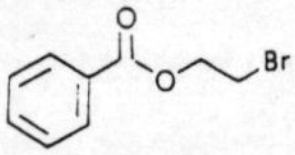	b-1) RT, 30h	(98)	173)
9	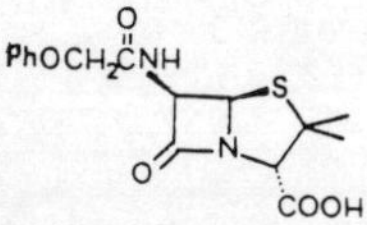	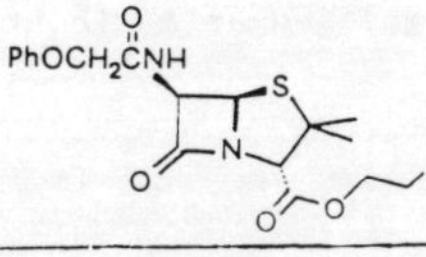	b-2) 0^{0}, 17h	(95)	173)

Entry 5: A 100 mL flask was charged with 2.10 g (15 mmol) of dimedon, dissolved in 20 mL of 2-chlorethanol and 20 mg of p-TsOH. The mixture was refluxed (150^0) for 4 hrs. Excess of the solvent was removed in vacuo. Ether (40 mL) was added and the mixture washed with sat. $NaHCO_3$ in brine. Removal of the solvent after usual work-up gave 3.05 g (100%) of a sufficiently pure product.

Entry 6: A solution of 0.72 g (5 mmol) of β-chloroethyl chloroformiate in 5 mL Et_2O was added dropwise to a stirred solution of 1.10 g (5 mmol) L-phenylalanine t-butylester and 0.6 g (5,5 mmol) Na_2CO_3 in 10 mL H_2O at RT. The mixture was allowed to react overnight. The layers were separated and the aqueous phase was extracted with Et_2O (3x20 mL). The ether extract was washed with brine (1x20 mL), dried (Na_2SO_4), filtered and evaporated in vacuo and afforded 1.63 g (100%) of the oily carbaminate.

b) Deprotection

b-1) Typical procedure for the vitamin B_{12}-catalyzed deprotection with Zn-reduction (Zn dust was activated by successive washings with HCl (1N), water, methanol and dried in vacuo.)

Entry 7: A solution of vitamin B_{12a} <u>1a</u> (20 mg, 0.014 mmol, 1 mol % based on 1 mol (=100%) of substrate) in MeOH was purged with argon for 1/2 h. Subsequent to deoxygenation, 1.30 g (20 mmol) of Zn dust, 1.07 g of NH_4Cl were added and the mixture was allowed to react for 15 min, at RT, under argon. A solution of 2-chlorethoxycarbonyl-phenylalanine-tert.-butyl-ester (430 mg, 1.30 mmol) in 5 mL MeOH was injected through a rubber septum. The reaction was complete in about 2 hrs (green coloration by vitamin $B_{12\ s}$). The suspension was filtered, acidified with a solution of citric acid (10%) and extracted with ether. Akaline work-up (addition of saturated $NaHCO_3$) gave 223 mg (78%) of the unmasked amino-acid t-butylester characterized by IR, NMR and MS. (In a blank experiment without vitamin B_{12a} 93% of unchanged starting material was isolated after 16 hrs at RT).

b-2) Typical procedure for the vitamin B_{12}-catalyzed deprotection by electrochemical reduction.

Apparatus: An electrolysis cell with a glass frit as cell divider, a mercury pool cathode and a platinum anode were used. The electrolyte solutions were degassed prior to use and pre-electrolyzed at -1.50 V (SCE) in order to remove electroactive impurities till the current dropped to a steady value of ca. 0.4 mA. A constant potential of -1.90 V (SCE) was applied to the mercury pool during electrolysis.

Entry 9: Penicillin V-β-bromoethyl ester (0.75 g, 1.73 mmol) and B_{12a} 1a (95 mg, 0.07 mmol) were dissolved in 25 mL of electrolyte (0.1 N $LiClO_4$, 0.13 N NH_4ClO_4 in 4 : 1 ethanol/water) and electrolyzed at a stirred Hg electrode in the cathode compartment of an electrolysis cell at - 1.7 V (SCE) and 1° C under argon. After 17 h, the reaction solution was diluted with water (50 mL) and worked up with CH_2Cl_2: 0.57 g (95%) penicillin V (identified by TLC and by IR and NMR spectra).

4.5.3. Olefin synthesis.

A well-known procedure for olefin synthesis is the method developed by Boord Eq. 36 (for a review see Schlosser [255]).

$$CH_2{=}CH{-}OR \xrightarrow{X_2} XCH_2{-}CH(X){-}OR \xrightarrow{R'MgX} XCH_2{-}CH(R'){-}OR \xrightarrow{2e^-} CH_2{=}CH{-}R' \qquad 36)$$

According to Boord the α-halide of α,β-dihaloethers (easily obtained by halogen addition to vinyl ethers) is first substituted by an alkyl group; the halide and the alkoxygroup in the β-halo ether are then reductively eliminated. Crombie [256] et al. as well as Ansell et al. [257] extended the discribed methodology for the synthesis of 3- and 4-alkenols starting from 2,3-dihydrofurane or 2,3-dihydropyrane respectively as starting materials (for a review see Arora and Ugi [258]). The reducing agents usually applied in the reductive elimination-step were Zn in boiling alcohol or Na in THF or ether. Under these conditions mixtures of cis- and trans-olefins might be formed [259].

It has been found that the vitamin B_{12}-catalyzed elimination (using Zn or the cathode as electron-source) provides a very mild and stereospecific version for the last step of the olefin synthesis.

The described reaction sequence has been applied in a convergent and stereospecific synthesis of propylure (scheme 16), a sex attractant of the pink bollworm moth *Pectinophera Gossypiella* (for earlier synthesis see [260]).

Scheme 16 Synthesis of propylure [238].

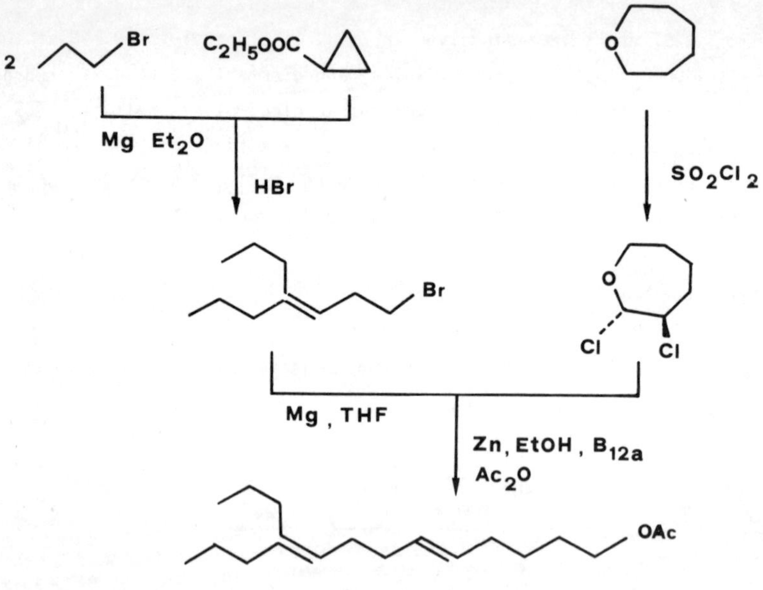

2 Br C_2H_5OOC O

Mg Et_2O

HBr

SO_2Cl_2

Br O Cl Cl

Mg, THF

Zn, EtOH, B_{12a}

Ac_2O

OAc

The two prestructs were prepared straightforwardly: the homoallyl bromide [261] was obtained from cyclopropane-carboxylic acid ethylester by treatment with two equivalents of n-propylmagnesiumbromide followed by Julia's [262] homoallylic rearrangement with 48% HBr in a one-pot procedure; trans-1,2-dichloro oxepane was prepared by chlorination of oxepane with sulfuryl chloride [238]. The combination of the two prestructs was achieved by reaction of the Grignard-reagent (prepared from the bromide and activated magnesium [263]) with the dichloro oxepane. Reductive elimination with Zn in EtOH in presence of 1 mol % of vitamin B_{12a} <u>1a</u> gave the alcohol, which was finally acetylated. Likewise trans-5-decenyl acetate, the sex-attractant of the peach borer moth *Anarsia Lineatella* [264] was synthesized from butyl bromide and trans-2,3-dichloro oxepane in three steps with 67% yield [238].

4.6. Reductive (C-C) bond formation.

1,4-additions of organic halides to activated olefins by means of organometallic intermediates are well-known (cf. e.g. [265]). Generally these reactions involve two or even more steps (Eq. 37):

a) formation of the reagent (often via highly reactive intermediates as e.g. organolithium or organomagnesium derivatives)

b) addition of the reagent to the activated olefin.

Methods for direct reductive (C-C) bond formation [267a] e.g. coupling of α,β-unsaturated carbonyl derivatives with other electrophiles (allyl bromides) [267b] have been published.

$$\text{R–Y} \xrightarrow{\ a\ } \text{R–M} + \text{=}\!\!\diagup^{\text{EWG}} \xrightarrow{\ b\ } \text{R–CH}_2\text{–CH}_2\text{–EWG} \qquad 37)$$

EWG = electron-withdrawing group

$$\text{R–Y} + \text{=}\!\!\diagup^{\text{EWG}} \xrightarrow[-\text{Y}^-]{2e^- + \text{Co–catalyst} + \text{H}^+} \text{R–CH}_2\text{–CH}_2\text{–EWG} \qquad 38)$$

The catalytic version (Eq. 38) performed in presence of macrocyclic Co-complexes under reducing conditions, has been found to be a new, elegant and mild method for (C-C) bond formation [172].

In both cases an "Umpolung" of one electrophile takes place prior to the coupling reaction (Eq. 37 and Eq. 38).

In absence of the catalyst either RY or the activated olefin [266] undergo individual direct reduction at a sufficiently negative electrode potential (cf. scheme 10). The purpose fulfilled by the catalyst is to trigger the specific (C-C) bond formation starting from RY. This occurs via the reduced Co-complex, oxidative addition of RY leading to [Co-R] and reductive cleavage of the (Co-C) bond (cf. scheme 11). The resulting activated R group then combines with the olefin.

4.6.1. Reductive coupling of alkyl halides and activated olefins.

Electrochemical reduction of bromo-cyclohexenones bearing RY as well as an activated olefin within the same molecule leads to intramolecular coupling catalyzed by vitamin B_{12a} 1a or Co HDP 2b (table 13)[172].

Table 13 Electroreduction of bromocyclohexenones catalyzed by B_{12a} 1a or Co HDP 2b (yields, indicating isolated products, are equal in both cases). Reduction potential: Between -1.5 and -1.6 V (vs. SCE).

	Substrate	Products Yield	
endo trig	2-($(CH_2)_n$–Br)-cyclohex-2-enone	bicyclic ketone with $(CH_2)_n$ ring	2-($(CH_2)_n$–H)-cyclohex-2-enone
	5-ring (n = 3)	—	90 %
	6-ring (n = 4)	95 %	—
	7-ring (n = 5)	70 %	10 %
exo trig	3-($(CH_2)_n$–Br)-cyclohex-2-enone	spiroketone with $(CH_2)_n$ ring	3-($(CH_2)_n$–H)-cyclohex-2-enone
	4-ring (n = 3)	—	90 %
	5-ring (n = 4)	95 %	—
	6-ring (n = 5)	45 %	40 %

The results, which reveal two competing reactions (1,4-addition leading to bicyclic- or spiroketones and proton-capture leading to products exhibiting a hydrogenolyzed side-chain), are in agreement with the general rules for *Exo-Trig* and *Endo-Trig* ring closure [268]. Cyclisation predominates if the reactive centre at the end of the side chain can easily adopt a spatial position favourable for a 1,4-attack leading to six- and seven-membered rings by endocyclic closure or to five- and six-membered rings by exocyclic closure.

In order to clear up, which one of the two electrophilic groups present within the same molecule is attacked by Co(I), intermediates were isolated. At a potential of -1.0 V (vs. SCE) the reaction has been stopped on the level of the alkyl Co intermediate (cf. scheme 10). The constitution of these crystalline complexes has been determined by analysis and spectral data in case of vitamin B_{12} [269] as well as of Co HDP [172]. Their structure reveals that the alkyl halide moiety has been attacked exclusively. Upon further reduction of the isolated alkyl Co complexes in presence of the (intramolecular) Michael olefin and a proton source, products as listed in table 13 have been detected.

Intermolecular coupling reactions have been studied in case of 3-halocholestanes and different Michael acceptors [270]. Results are compiled in table 14.

It is noteworthy that under these mild reaction conditions an unprotected carbonyl group in the alkyl halide is not attacked (cf. entry 5 to 7, table 14). The α:β ratio of the coupling products is obviously related to the reactivity of the Michael acceptor. Highly reactive acceptors lead predominantly to less stable 3α-isomers, whereas less reactive acceptors give mainly 3β-isomers. Iodocholestanes react more rapidly than bromocholestanes (e.g. Zn/NH_4Cl as reducing agent in THF/CH_3OH at room temperature is sufficient for iodocholestanes, whereas a reaction temperature of 60° is necessary for the corresponding bromides).

For reactions involving allyl halides, only the chlorides are suitable, the bromides and iodides are too reactive [271].

General procedure for coupling reactions by controlled potential electrolysis catalyzed by macrocyclic Co complexes [172].

Solutions of 1 mmol of a bromocyclohexenone (table 13) and 0,05 mmol of vitamin B_{12a} <u>1a</u> (or 0,2 mmol of Co HDP <u>2b</u>) in 25 mL of electrolyte (0.1 N $LiClO_4$, 0.05 N NH_4Br in DMF) were electrolyzed at -1.6 V (*vs.* SCE) [272] in a divided (H-type) cell at a stirred Hg-pool cathode under argon at 20° in the dark. Products were isolated by extraction with pentane.

Table 14 Intermolecular Coupling of 3-Halo-steroids and Activated Olefins under Standard Conditions a) 270)

Entry	Halo-steroid	Activated Olefin	Applied reduction potential V (*vs* SCE)	Product	Total yield % b)	Ratio of Isomers c) 3α : 3β
1	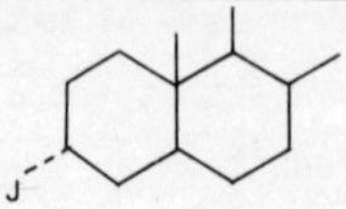 J	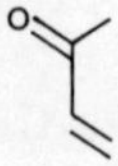	-1,4	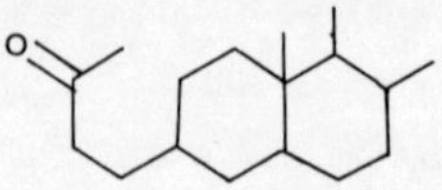	98 (65)	75 : 25
2		 CN	-1,4	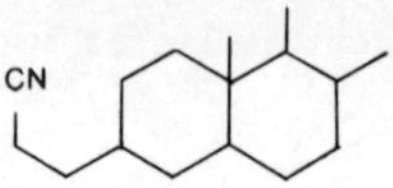 CN	95 (30)	70 : 30
3		 CN	-1,6	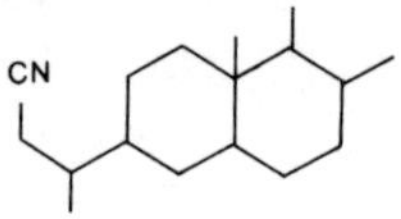 CN	73 (12)	35 : 65 d)
4		 COOEt	-1,7	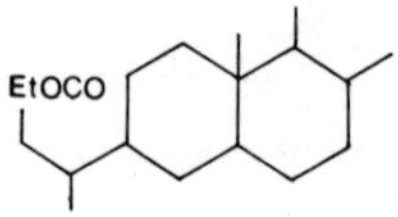 EtOCO	72 (12)	5 : 95 d)
5	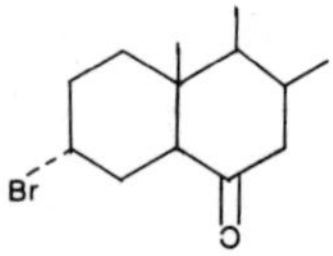 Br	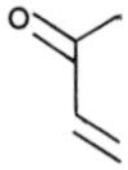	-1,4	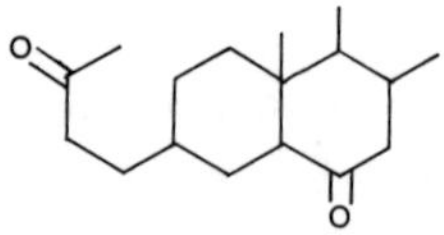	76	80 : 20
6		 CN	-1,4	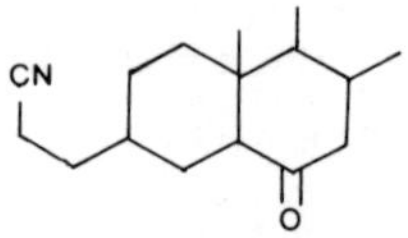 CN	86	72 : 28
7	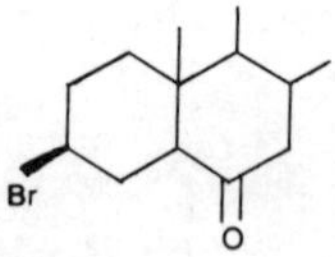 Br	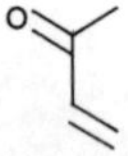	-1,4	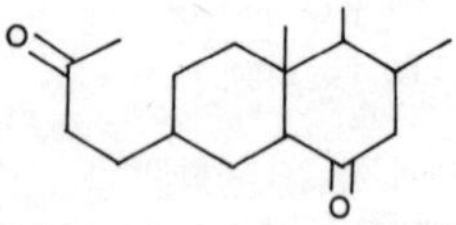	51	80 : 20

a) Standard condition for electrochemical reduction : molar ratio RY:olefin=1:5, 4 mol % vitamin B_{12a} 1a in electrolyte (0.1 N $LiClO_4$, 0,05 N NH_4Br in DMF), electrolysis at the indicated potential.
Standard condition for Zn-reduction: reducing agent = activated Zn-powder.

b) The yield of isolated material is indicated (values in brackets concerne isolated yield in the Zn-version).As by-product cholestane is formed.

c) Determined by NMR-spectrometry.

d) Four diastereomers are formed due to the additional centre of chirality.

4.6.2. Reductive coupling of acyl derivatives and activated olefines.

Reagents used for nucleophilic acylation exhibit inverted reactivity at the intrinsicly electrophilic carbonyl carbon. There exist several useful methods to achieve "Umpolung" of a carbonyl group [273]. Generally two types of nucleophilic acylating reagents can be distinguished:

- reagents with masked carbonyl functionality,
- acylmetallic compounds.

Metalloaldimines, anions of cyanohydrins or 1,3-dithianes and metalated enol derivatives, as well as thiazoliumsalt derivatives [274] belong to the first class of reagents. The acylmetallic compounds exhibiting nucleophilic reactivity are less known. Acyliron(0)carbonyls react with electrophiles, e.g. acids to aldehydes [275], primary alkylhalides or olefines to ketones [276] and with enones to 1,4-diketones [277]. Similar reactions have been reported for carbonylnickelate complexes [278]. These methods involving acyltransitionmetals are usually not performed in a catalytic manner (apart from related hydroformylation reactions catalized by Co carbonyls or Rh carbonyls [279]).

Macrocyclic acylCo(III) complexes can be synthesized from Co(I) and carboxylic acid dervatives e.g. the corresponding anhydrides or chlorides. The acetylCo(III)salen [280], benzoylCo(III)phthalocyanine [40] and acetylcobalamin(III) [281] have been characterized. In contrast to the acylironcarbonyl compound, electrophilic reactivity of the carbonyl carbon persists in case of acylcobalt(III) complexes [156, 282]. However, inverted reactivity is achieved upon reduction or photolysis of these complexes [283].

Recently the application of macrocyclic cobalt complexes as catalysts for nucleophilic acylation has been studied in our laboratory. The catalytic cycles according to scheme 11 and 12 in chapter 3.5. have been adapted to this case (RX is replaced by the anhydride $O(COR)_2$ and [Co-R] by [Co-COR]. The acyl-Co(III) complex is cleaved either reductively (scheme 11) or photochemically (scheme 12). In presence of electrophiles E the corresponding coupling products $R\text{-}C\genfrac{}{}{0pt}{}{\nearrow O}{\searrow E}$ derived form $R\text{-}\dot{C}{=}O$ or $R\text{-}C^{\ominus}{=}O$ were obtained with reasonable yields [283], (organic radicals are known to abstract easily $H^{\bullet}$ form DMF which is used as solvent [284]).

A closer look to the kinetics of the light promoted vitamin B_{12}-catalized nucleophilic acylation of acrylonitrile at -1,0 V (SCE) reveals some interesting details [285]: In scheme 17 the cathodic current, which is a direct measure for the overall rate of catalysis, is shown as a function of light irradiation of the cathodic compartment. Obviously, only in presence of light catalysis is occuring. (Reduction of Co(III)acyl derivatives affords potentials more negative than - 1,5 V (SCE)).

Scheme 17 Light induced catalytic coupling of acetanhydride and acrylonitrile at -1.0 V (SCE)

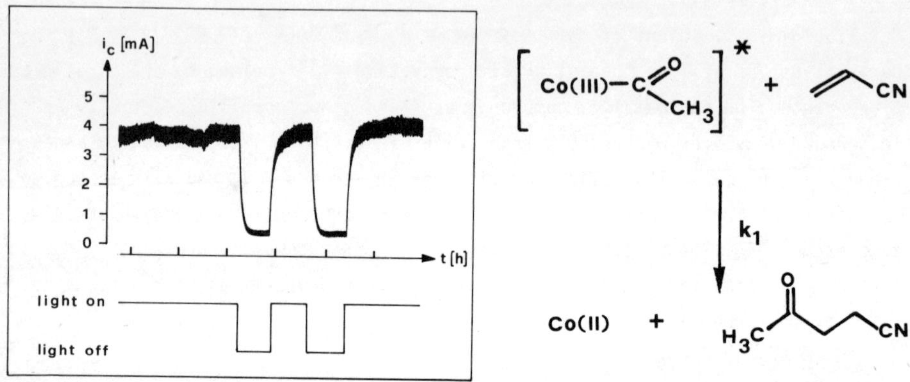

The exponential increase and decrease of the current response upon switching the light of a 100 W lamp implies that the current is due to the reduction of electroactive species generated in solution. The plateau of the catalytic current (4 m A in scheme 17) was found to depend on the acrylonitril concentration. The observed linear dependence of reciprocal current and reciprocal acrylonitril concentration is shown in scheme 18.

Scheme 18 Dependence of reciprocal current on reciprocal conc.of acrylonitrile

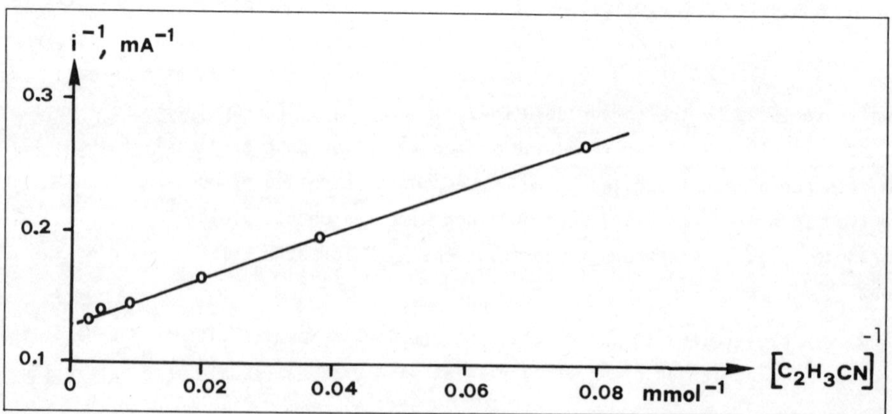

This current-substrate relationship is in agreement with a bimolecular reaction involving photoexcited acylocobalamin and acrylonitrile (scheme 17), as shown by the following analysis:

The overall rate of catalysis as a function of the rate of the individual steps within the catalytic cycle has been calculated from a steady-state-treatment for the different Co species formed during one catalytic turnover. In the implified Eq. 39 kinetics of formation of the acylcobalamin , the efficiency of its photo-excitation and the rate of electrochemical regeneration of Co(I) are hidden in term A. The assumed bimolecular reaction (cf. scheme 17) implies the experimentally observed $1/i \sim 1/_{[Acr]}$ relationship.

$$\frac{1}{i} = A + \frac{1}{F \cdot n \cdot V \cdot k_1 \cdot [Acr] \cdot [B_{12}]} \qquad 39)$$

i	cathodic current	k_1	2nd order rate constant (reaction in scheme 17)
F	Faraday	[Acr]	acrylonitrile concentration
V	cell volume	$[B_{12}]$	total catalyst concentration

In table 15 and 16 first results of the preparative application of this B_{12}-catalized acylation of Michael acceptors have been compiled. In contrast to the thiazolium salt-catalized nucleophilic acylation (Stetter-reaction), which is restricted to α,β-unsaturated ketones, esters, nitriles, sulfones and Mannich bases, in the vitamin B_{12}-catalized version α,β-unsaturated aldehydes (table 15) are as selectively acylated as other Michael acceptors. Best yields have been observed for acid anhydrides as acylating reagent. Acid chlorides are too electroactive (cf. 3.5.) and activated esters as cyano methylesters or dicarboethoxymethylesters do not undergo oxidative addition at a reasonable rate. Unsubstituted and β-monosubstituted Michael acceptors are both easily acylated. No induction of chirality has been observed in the latter case. β,β-disubstituted acceptors studied so far are not acylated. 1,2-dioxocompounds and α-hydroxyketones resulting from coupling reactions of two acyl-moieties or from nucleophilic attack of the acylcobalamin at the carbonyl carbon respectively are not observed.

In the absence of light the (Co-C) bond cleavage may be induced reductively (either electrochemically at more negative potentials or with metallic Zn). Generally the same products as in the light promoted reaction are obtained, but yields are slightly lower than in case of the photochemical version.

Table 15: Reductive photochemical acylation of α,β-unsaturated nitriles and ketones [283)]

Substrates		B_{12a} 1a (mol%)	Product	isol.Yield (%)
$(CH_3CO)_2O$	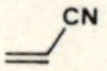 CN	4	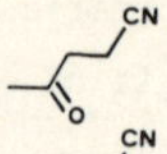 CN, O	60
CH_3COCl	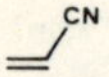 CN	4	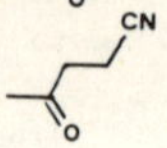 CN, O	6
$(CH_3CO)_2O$	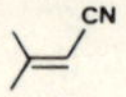 CN	4	no product formed	-
"	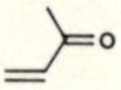 O	4	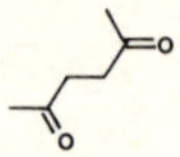 O, O	63
"	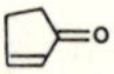 O	4	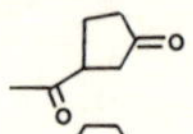 O, O	42
"	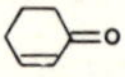 O	4	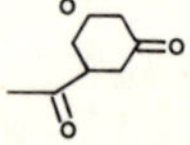 O, O	40

Table 16: Reductive photochemical acylation of α,β-unsaturated aldehydes [283)]

Substrates		B_{12a} 1a (mol%)	Product	isol. Yield (%)
$(CH_3CO)_2O$	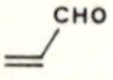 CHO	2	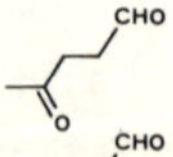 CHO, O	47
"	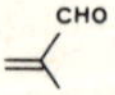 CHO	2	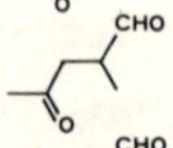 CHO, O	34
"	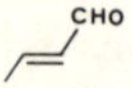 CHO	2.5	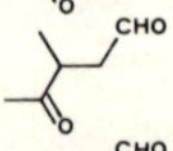 CHO, O	50
"	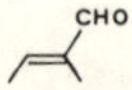 CHO	5	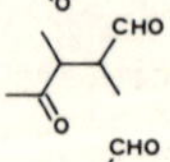 CHO, O	30
$(n\text{-}C_7H_{15}CO)_2O$	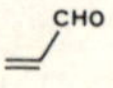 CHO	14	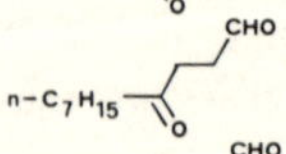 CHO, n-C_7H_{15}, O	71
$(n\text{-}C_7H_{15}CO)_2O$	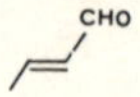 CHO	14	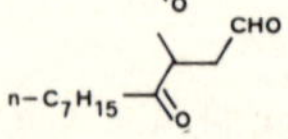 CHO, n-C_7H_{15}, O	80

Simple protonation of the potential ayclanion with formation of the corresponding aldehyde is possible. n-Octylacid anhydride was electrochemically reduced to octanal (20%) in presence of acetic acid and 4 mol % of vitamin $B_{12\,a}$ 1a.

Experimental procedures for nucleophilic acylation:

1) General procedure, electrochemical reduction under light irradiation.

$LiClO_4$ (6 mmol) and hydroxocobalamin hydrochloride (0,2 mmol) dissolved in 20 mL deareated DMF was electrolized under argon at -1.0 V (vs. SCE) in a conventional divided cell at a stirred mercury-pool electrode (12 cm^2) and at room temperature. After approximately 1 h the solution turned dark green. The deareated acid anhydride (5 mmol) and Michael acceptor (7-10 mmol) were added (the solution turned yellowish after addition of the anhydride). Upon irradiation (100 W unfocused incandescent bulb at 30 cm distance), the current grew immediately to a value of approximately 10 mA. After approximately 24 h current declined rapidly. At this point, the reaction mixture was diluted with 20 mL H_2O, then extracted with ether or pentane (5 times 50 mL) followed by evaporation of the solvent and resolution of the reaction mixture by column chromatography (silicagel, ether).

2) General procedure, Zn reduction in the dark.
(Example: Acetylation of acrylonitrile)

A suspension of activated Zn-dust [17)] (77 mmol) in 100 mL DMF was stirred under argon for 1 h. Then 0.5 g hydroxocobalamin hydrochloride (0.36 mmol) was added. When the reaction mixture turned deep green, 5,0 g acetic anhydride (0.5 M, 50 mmol) and 40 g acrylonitrile (0.75 M, 75 mmol) were added. The colour of the solution turned immediately yellow. After 24 hours stirring at room temperature the reaction mixture was poured on ice/water and filtered. The solution was extracted with CH_2Cl_2 (5 times 50 mL). After evaporation the residue was treated with 5 g $NaHSO_3$ (saturated solution) and extracted with CH_2Cl_2. The water layer was treated with saturated Na_2CO_3 and extracted with CH_2Cl_2. After evaporation levulinic nitrile (1.48 g, 30%) was obtained. (In case of long-chain anhydrides, e.g. C_7C_8, the resolution was advantageously performed using Girard-T reagent).

6. EPILOGUE.

In the preface to his classical "opus magnum" on vitamin B_{12} [4], *Professor David Dolpin* states: *"Solutions to B_{12}-related problems present some of the principal scientific achievements of the last half century; each decade has recorded a milestone towards an understanding of the nature and function of these systems"*. Making use of B_{12} in organic synthesis, we didn't hesitate to profane this vitamin and feel now inclined to quote *Hölderlin's Hyperion: "Es ist nichts Heiliges, was nicht ... durch allberechnende Barbaren ... zum ärmlichen Behelf herabgewürdigt ist"*.

Two things remain to be expressed: The excuse for having omitted much relevant work and the hope that further work on B_{12}-catalyzed reactions (specially its electrochemical versions with carbon electrodes) will contribute to the development of clean technologies using natural, non-toxic catalysts.

ACKNOWLEDGMENTS.

First of all, *R.S.* would like to thank the undergraduate and graduate students and the postdoctoral fellows for their enthusiastic collaboration. Their names are given in the list of references. We thank *Dr. A. Pugin, CIBA-GEIGY AG, Basel,* for the procedure of the Co-phthalocyanine-synthesis and supplies of starting materials. Financial support by the *Schweizerischer Nationalfonds zur Förderung der wissenschaftlichen Forschung* as well as *CIBA-GEIGY AG, Basel,* are gratefully acknowledged. Last but not least we are greatly obliged to *Mrs. Rosmarie Rohner* and *Miss Sabine Scheffold* for their invaluable help in preparing the manuscript and the drawings.

6. REFERENCES

1) J.M. Pratt and P.J. Craig, Adv. in Organomet. Chem. 11, 331-446 (1973).

2) D. Dodd and M.D. Johnson, J. Organomet. Chem. 52, 1-232 (1973).

3) a) A. Eschenmoser, Chem. Soc. Reviews 5, 377 (1976);
b) ibid. Jahrbuch der Akad. der Wissenschaften in Göttingen 1977, pp 29-43;
c) ibid. Nova Acta Leopoldina, Neue Folge Nr. 247, Bd. 55, J.A. Barth, Leipzig 1982, pp 5-47;
d) ibid. J. Chem. & Eng. News Sept. 14 (1981) p. 36.

4) D. Dolphin, "B_{12}" Vol. 1: "Chemistry", Vol. 2: "Biochemistry and Medicine", J. Wiley & Sons, New York, 1982.

5) W. Friedrich, "Fermente, Hormone, Vitamine" Vol. III/2 "Vitamine B_{12} und verwandte Corrinoide", G. Thieme, Stuttgart 1975.

6) B. Zagalak and W. Friedrich Eds., "Vitamin B_{12}" (Proc. 3rd Europ. Symp. on Vitamin B_{12} and Intrinsic Factor, Zürich), W. de Gruyter, Berlin, New York, 1979.

7) J.M. Pratt, "Inorganic Chemistry of Vitamin B_{12}", Academic Press, London, New York, 1972.

8) B.M. Babior, "Cobalamin, Biochemistry and Pathophysiology", J. Wiley & Sons, New York 1975.

9) J. Rétey and J.A. Robinson "Stereospecificity in Organic Chemistry and Enzymology", Verlag Chemie, Weinheim, 1982, pp 185-207.

10) L. Mervyn and E.L. Smith, Prog. Ind. Microbiol. 5, 151-201 (1964).

11) L. Mervyn, Comprehens. Biochem. 21, 153 (1971).

12) W. Friedrich and K. Bernhauer, Angew. Chem. 71, 284 (1959).

13) W. Friedrich in ref. 5) pp 163-168.

14) G. Barrère, B. Geneste and A. Sabatier, Pour la Science Nr. 49, 56-65 (1981).

15) J. Florent and L. Ninet in H.J. Peppler and D. Perlmann (Eds.) "Microbiol. Technology", Academic Press, London 2nd ed., Vol. 1, pp 497-519.

16) a) Hoffmann-La Roche, Belg. Pat 631, 589 (1963).
b) Glaxo Group, British Pat. 963, 373 (1964).

17) A. Fischli, Helv. 61, 2560 (1978).

18) G. Rytz and R. Scheffold, Helv. 63, 733 (1980).

19) For the nomenclature of [HDP]complexes compare: R. Scheffold, J. Löliger, H.-U. Blaser and P. Geisser, Helv. 58, 49 (1975).

20) R. Schwesinger, R. Waditschatka, J. Rigby, R. Nordmann, W.B. Schweizer, E. Zass and A. Eschenmoser, Helv. 65, 600 (1982) and lit. cited therein.

21) a) G. Rytz, Thesis, University of Bern, 1979.
b) L. Walder, Thesis, University of Bern, 1979.

22) P.Engel, G.Rytz, L.Walder, U.Vögeli and R.Scheffold in ref. 6 , p. 171.

23) Ch. Bosshard, Thesis, University of Bern, 1982.

24) C.G. Overberger, M.T. O'Shaugnessy and H. Shalit, J. Am. Chem. Soc. 71, 2661 (1949); C.G. Overberger and M.B. Berenbaum, Org. Syntheses 32, 48 (1952); C.G. Overberger, Pao-Tung Huang and M.B. Berenbaum, ibid. 32, 16 (1952); 32, 50 (1952).

25) a) R. Scheffold, J. Löliger, H.-U. Blaser and P. Geisser, Helv. 58, 49 (1975);
b) U.P. Schlunegger, S. Leutwyler, F. Friedli and R. Scheffold, Helv. 58, 65 (1975).

26) J. Löliger and R. Scheffold, Helv. 60, 2644 (1977).

27) P. Geisser, P. Schmid and R. Scheffold, Chimia 34, 279 (1980).

28) R. Scheffold, K. Meier, G. Rytz and L. Walder, 1st Europ. Conf. on Organomet. Chem., Jablonna, Poland, 1976, Conf. paper pp 132.

29) H.C. Mez, J. Löliger, U. Vogel, K. Meier and R. Scheffold, Helv. 64, 1098 (1981).

30) K. Meier, R. Scheffold and G. Calzaferri, Helv. 64, 2361 (1981).

31) a) U. Vogel, K. Meier, J. Löliger and R. Scheffold, Helv. 64, 1113 (1981).
b) K. Meier, R. Scheffold and P. Engel, Helv. 64, 1496 (1981).
c) K. Meier and R. Scheffold, Helv. 64, 1505 (1981).

32) N.S. Gill and F.B. Taylor, Inorg. Synth. 9, 140 (1967).

33) F.H. Moser and A.L. Thomas "Pththalocyanine Compounds". Am. Chem. Soc. Monograph Series, Reinhold Publ. Corp. New York 1963.

34) A.B.P. Lever, Adv. Inorg. Chem. Radiochem. 7, 27-144 (1965).

35) B.B. Berezin, "Coordination Compounds of Porphyrins and Phthalocyanins", J. Wiley & Sons, New York, 1981.

36) H. Kropf and F. Steinbach, "Katalyse an Phthalocyninen", Symposium 10. Mai 1972, Hamburg, G. Thime Verlag Stuttgart 1973.

37) A.N. Sidorov and V.G. Maslow, Uspekhi khimii, 44, 577 (1975).

38) J.F. Gibson, D.J.F. Ingram and D. Schonland, Discuss. Faraday Soc. 26, 72 (1958).

39) J.S. Griffith, Discuss. Faraday Soc. 26, 81 (1958).

40) For the preparation and magnetic properties of cobalt(II)-4,4',4"'-tetrasulfophthalocyanine see J.H. Weber and D.H. Busch, Inorg. Chem. 4, 469 (1965).

41) T.A.M. Maas, M. Kuijer and J. Zwart, J.C.S. Chem. Comm. 86 (1976).

42) H. Eckert and J. Ugi, Angew. Chem. 87, 847 (1975); ibid. Int. Ed. Engl. 14, 825 (1975).

43) H. Eckert and J. Ugi, J. Organomet. Chem. 118, C 55, C 59 (1976).

44) H. Eckert and A. Schier, Angew. Chem. 91, 841 (1979); ibid. Int. Ed. Engl. 18, 794 (1979).

45) H. Eckert and J. Ugi, Angew. Chem. 88, 717 (1976); ibid. Int. Ed. Engl. 15, 681 (1976).

46) H. Eckert, Synthesis 332 (1977).

47) H.A. Keller, R.G.K. Schneiderwind, H. Eckert and J. Ugi, Angew. Chem. 93, 581 (1981); ibid. Int. Ed. Engl. 20, 577 (1981).

48) H. Eckert, I. Lagerlund and J. Ugi, Tetrahedron 33, 2243 (1977).

49) H. Eckert, M. Listl and J. Ugi, Angew. Chem. 90, 388 (1978); ibid. Int. Ed. Engl. 17, 361 (1978).

50) H. Eckert und J. Ugi, Liebigs Ann. Chem. 278 (1979).

51) H. Eckert, Angew. Chem. 93, 216 (1981); ibid. Int. Ed. Engl. 20, 208 (1981).

52) H. Eckert and Y. Kiesel, Angew. Chem. 93, 477 (1981); ibid. Int. Ed. Engl. 20, 473 (1981).

53) I.M. Heilbron and R.P. Linstead, US. Patent 2, 124, 419, ICI (1938).

54) A. Puget, Ciba-Geigy AG, Basel, personal communication.

55) G.N. Schrauzer, Acc. Chem. Research 1, 97 (1968).

56) G.N. Schrauzer, Angew. Chem. 88, 465 (1976); ibid. Int. Ed. Engl. 15, 417 (1976).

57) K.L. Brown in ref. 4, Vol. 1 , pp. 245-294.

58) a) J. Rétey, Helv. 54, 2747 (1971);
b) M.W. Bartlett and J. Dunitz, Helv. 54, 2753 (1971);
c) H. Flohr, U.M. Kempe, W. Pannhorst and J. Rétey, Angew. Chem. 88, 443 (1976); ibid. Int. Ed. Engl. 15, 427 (1976);
d) H. Flohr, W. Pannhorst and J. Rétey, Angew. Chem. 88, 613 (1976); ibid. Int. Ed. Engl. 15, 561 (1976);
e) H. Flohr, W. Pannhorst and J. Rétey, Helv. 61, 1561 (1978);
d) J. Rétey, pp 439-460 in ref. 4).

59) G.N. Schrauzer and E. Deutsch, J. Am. Chem. Soc. 91, 3341 (1969).

60) G.N. Schrauzer, Inorg. Synth. 11, 61 (1968).

61) J. Kwiatek, Catalysis Reviews, Vol.1, M.Dekker Inc. New York , 1968 pp.37-72.

62) B.M. Chadwick and A.G. Shapre, Advan. Inorg. Chem. Radiochem. 8, 83 (1966).

63) M.H. Ford-Smith, "The Chemistry of Complex Cyanides", H.M.Stationary Office, London, 1964.

64) A.G. Shapre, "The Chemistry of Cyano Complexes of the Transition Metals", Academic Press, London 1976.

65) D.A. White, A.J. Solodar and M.M. Baizer, Inorg. Chem. 11, 2160 (1972).

66) G.A. Mills, S. Weller and A. Wheeler, J. Phys. Chem. 63, 403 (1959).

67) G.D. Venerable and J. Halpern, J. Am. Chem. Soc. 93, 2176 (1971).

68) J. Hanšlik and A.A. Vlcek, J.C.S. Chem. Comm. 47 (1969); Inorg. Chem. 8, 669 (1969).

69) G. Guastalla, J. Halpern and M. Pribanic, J. Am. Chem. Soc. 94, 1575 (1972).

70) a) M. Jguchi, J. Chem. Soc. Japan 63, 634 (1942); CA 41, 2975d (1947);
b) B. de Vries, Konikel. Ned. Akad. Wetenshap. Proc. Ser. B 63, 443 (1960);
c) N.K. King and M.E. Winfield, J. Am. Chem. Soc. 83, 3366 (1961).

71) W.P. Griffith and G. Wilkinson, J. Chem. Soc. 2757 (1959).

72) J. Halpern and J. Maher, J. Am. Chem. Soc. 87, 5361 (1965).

73) D.L. Reger, M.M. Habib and D.J. Fanth, J. Org. Chem. 45, 3860 (1980); D.L.Reger, M.M.Habib and D.J.Fanth , Tetrahedron Lett. 115 (1979).

74) T. Funabiki, H. Hosomi, S. Yoshida and K. Tarama, J. Am. Chem. Soc. 104, 1560 (1982).

75) Review: G.A. Melson, Ed. "Coordination Chemistry of Macrocyclic Compounds", Plenum Press, New York 1979.

76) G. Costa, G. Mestroni and E. de Savorgnani, Inorg. Chim. Acta 3, 323 (1969).

77) G. Costa, G. Mestroni, T. Licari and E. Mestroni, Inorg. Nucl. Chem. Letters 5, 561 (1969).

78) G. Costa, G. Mestroni and C. Covevar, J.C.S. Chem. Comm. 706 (1971).

79) Review: A.E. Martell and M. Calvin, "Chemistry of the Metal Chelate Compounds" Prentice Hall, Englewood Cliffs, New Yersey 1954 pp 337-352.

80) C. Floriani and F. Calderazzo, J. Chem. Soc. A, 946 (1969).

81) G. Costa, G. Mestroni and L. Stefani, J. Organometal. Chem. 7, 493 (1967).

82) R.H. Bailes and M. Calvin, J. Am. Chem. Soc. 69, 1886 (1947).

83) see chapter 4.1. Oxidations.

84) A.D. Adler, F.R. Longo, J.D. Finarelli, J. Golmacher, J. Assour and L. Korsakoff, J. Org. Chem. 32, 476 (1967).

85) G.D. Dorough, J.R. Miller and F.M. Huennekens, J. Am. Chem. Soc. 73, 4315 (1951).

86) a) J.W. Buchler in "The Porphyrins" Ed. D. Dolphin, Academic Press, New York, Vol. I, 1978, pp 468;
b) R.H. Felton, ibid. Vol. V, 1978, pp 76.

87) D. Lexa, J.M. Savéant and J. Zickler, J. Am. Chem. Soc. 99 , 2786 (1977).

88) D. Lexa and J.M.Savéant, J. Am. Chem. Soc. 98, 2652 (1976).

89) L. Walder, G. Rytz, P. Engel, U. Vögeli and R.Scheffold , Helv. in prep.

90) J.F. Endicott, J. Lilie, J.M. Kuszaj, B.S. Ramaswamy, W.G. Sahmonsees. M.G. Simic, M.D. Glick and D.P.Rillema, J.Am.Chem.Soc. 99, 429 (1977).

91) D. Lexa and J.M. Savéant, J. Am. Chem. Soc. 100 , 3221 (1978).

92) L.D. Rollmann and R.T. Iwamoto, J. Am. Chem. Soc. 90, 1455 (1968).

93) K.M. Kadish, L.A. Bottomley, S. Kelly, D. Schaeper and L.R. Shine, Bioelectrochem. Bioenerg. 8, 213 (1981).

94) D.Faure, D.Lexa and J.M.Savéant, J. Electroanal. Chem. 140, 285, 297 (1982).

95) F.A. Walker, K.M. Kadish, J. Am. Chem. Soc. 98, 3484 (1976).

96) D. Lexa, J.M. Savéant and J.P. Soufflet, J. Electroanal. Chem. 100, 159 (1979).

97) M. Perrée-Fauvet, A. Gaudemer, P. Boucly and J. Devynck, J. Organometal. Chem. 120, 439 (1976).

98) D.W. Clack, N.S. Hush and I.S. Woolsey, Inorg. Chimica Acta 19, 129 (1976).

99) J.A.R. van Veen, J.F. van Baar, C.J. Kroese, J.G.F. Coolegem, N. de Wit and H.A. Colijn, Ber. Bunsenges. Phys. Chem. 85, 693 (1981).

100) C.M. Elliott, E. Hershenhart, R.G. Finke and B.L. Smith, J. Am. Chem. Soc. 103, 5558 (1981). Reduction potentials reported here do not agree with those reported in ref. 90 and 104.

101) G. Costa, G. Mestroni, A. Puxeddu and E. Reisenhofer, J. Chem. Soc. (A), 1970, 2870.

102) H.B. Mark, private communication.

103) R. Scheffold, unpublished results.

104) G. Costa, A. Puxeddu and G. Tauzher, Inorg. Nucl. Chem. Letters 4, 319 (1968).

105) M.D. Le Hoan, Y. Robin, J. Devinck, L. Bied-Charreton and A. Gaudemer, J. Organometal. Chem. 52, 1 (1973).

106) P.G. Lenhert and D.C. Hodgkin, Nature 192, 937 (1961).

107) J.K. Stille, this volume pp. 8-16.

108) At neutral conditions cob(I)alamin is not protonated and exhibits highly nucleophilic reactivity: cf. P.K. Das, H.A.O. Hill, J.M. Pratt and R.J.P. Williams, Biochim. Biophys. Acta 161, 646 (1967). The pK_a of protonated cob(I)alamin has been determined to be 1: cf. ref. 88.

109) R.G. Pearson, H. Sobel and J. Songstad, J. Am. Chem. Soc. 90, 319 (1968).

110) G.N. Schrauzer and J.W. Sibert, J. Am. Chem. Soc. 92, 1022 (1970).

111) F.R. Jensen, V. Madan and D.H. Buchanan, J. Am. Chem. Soc. 92, 1414 (1970).

112) D. Dodd and M.D. Johnson, J.C.S. Chem. Comm. 571 (1971).

113) H.L. Fritz, J.H. Espenson, D.A. Williams and G.A. Molander, J. Am. Chem. Soc. 96, 2378 (1974).

114) P.L. Bock and G. Whitesides, J. Am. Chem. Soc. 96, 2826 (1974).

115) M. Fountoulakis, J. Rétey, W.E. Hull and B. Zagalak in ref. 6 p. 169.

116) a) L. Walder, G. Rytz, K. Meier and R. Scheffold, Helv. 61, 3013 (1978);
b) G. Rytz, L. Walder and R. Scheffold in ref. 6, p. 173.

117) J. Schäffler and J. Rétey, Angew. Chem. 90, 906 (1978); ibid Int. Ed. Engl. 17, 845 (1978) and references cited therein.

118) H. Eckert, D. Lenoir and I. Ugi, J. Organometal. Chem. 141, C 23 (1977).

119) D. Dodd, M.D. Johnson, B.S. Meeks, D.M. Titchmarsh, K.N.V. Duong and A. Gaudemer, J. Chem. Soc. Perkin II 1261 (1976).

120) cf. e.g. the discussions in ref. 59, 117, 118 and M. Okabe and M. Tada, Chem. Lett 831 (1980).

121) G.N. Schrauzer, R.J. Windgassen and J. Kohnle, Chem. Ber. 98, 3324 (1965).

122) G.N. Schrauzer and R.J. Windgassen, J. Am. Chem. Soc. 89, 1999 (1967).

123) K.N.V. Duong, A. Ahond, C. Merienne and A. Gaudemer, J. Organometal. Chem. 55, 375 (1973).

124) M.N. Ricroch and A. Gaudemer, J. Organometal. Chem. 67, 119 (1974).

125) M. Naumberg, K.N.V. Duong and A. Gaudemer, J. Organometal. Chem. 25, 231 (1970).

126) Review: J. Halpern in ref. 4, Vol. 1, pp. 501-541.

127) H.-U. Blaser and J. Halpern, J. Am. Chem. Soc. 102, 1684 (1980).

128) J. Halpern, Ann. N.Y. Acad. Sci. 239, 2 (1974).

129) R.L. Birke, G.A. Brydon and M.F. Boyle, J. Electroanal. Chem. 52, 237 (1974).

130) F. Wagner and K. Bernhauer, Ann. N.Y. Acad. Sci. 112, 580 (1964).

131) P.Abley, E.R.Dockal and J.Halpern, J. Am. Chem. Soc. 95, 3166 (1973)

132) D. Cummins and E.D. McKenzie, Inorg. Nucl. Chem. Lett. 12, 521 (1976)

133) D. Autissier, P. Barthélemy and L.Pénasse, Bull. Soc. Chim. France 192 (II, 1980).

134) a) R.B. Silverman and D. Dolphin, J. Am. Chem. Soc. 96, 7094 (1974); b) ibid. 98, 4626 (1976).

135) For a similar scheme cf. G. Costa, A. Puxeddu and E. Reisenhofer, Bioelectrochem. Bioenerget. 1, 29 (1974).

136) Y.T. Fanchiang, J.C.S. Chem. Commun. 1369 (1982).

137) Y.T. Fanchiang, W.P. Ridley and J.M. Wood, J. Am. Chem. Soc. 101, 1442 (1979).

138) J.K. Kochi, "Organometallic Mechanisms and Catalysis". Academic Press, N.Y. 1978, pp. 466.

139) J.M. Pratt in ref. 4, Vol. 1, p. 325-92.

140) B.T. Golding in ref. 4, Vol. 1, p. 543-82.

141) H.P.C. Hogenkamp in ref. 4, Vol. 1, p. 295-323.

142) J.M. Wood and Y.T. Fanchiang, in ref. 6, p. 539-556.

143) J. Halpern, Acc. Chem. Res. 15, 238 (1982).

144) R.H. Abeles and D. Dolphin, Acc. Chem. Res. 9, 114 (1976).

145) L.J. Dizikes, W.P. Ridley and J.M. Wood, J. Am. Chem. Soc. 100, 1010 (1978).

146) G. Agnes, S. Bendle, H.A.O. Hill, F.R. Williams and R.J.P. Williams, J.C.S. Chem. Commun. 850 (1971).

147) J.H. Espenson and T.D. Sellers, Jr., J. Am. Chem. Soc. 96, 94 (1974).

148) D. Dodd, M.D. Johnson and B.L. Lockman, J. Am. Chem. Soc. 99, 3664 (1977).

149) J.F. Endicott and G.J. Ferraudi, J. Am. Chem. Soc. 99, 243 (1977).

150) G. Agnes, H.A.O. Hill, J.M. Pratt, S.C. Ridsdale, F.S. Kennedy and R.J.P. Williams, Biochem. Biophys. Acta 252, 207 (1971).

151) F.R. Jensen, V. Madan and D.H. Buchanan, J. Am. Chem. Soc. 93, 5283 (1971).

152) R.H. Magnuson, J. Halpern, I.Y. Levitin and M.E. Vol'pin, J.C.S. Chem. Commun. 44, (1978).

153) J.P. Kitchin and D.A. Widdowson, J.C.S. Perkin I 1384 (1979).

154) G.N. Schrauzer and R.J. Windgassen, J. Am. Chem. Soc. 89, 143 (1967).

155) H.P.C. Hogenkamp, J.E. Rush and C.A. Swenson, J. Biol. Chem. 240, 3641 (1965).

156) K. Bernhauer and E. Irion, Biochem. Z. 339, 530 (1964).

157) D. Lexa and J.M. Savéant, preprint (1982).

158) R. Blackburn, M. Kyaw, G.O. Philips and A.J. Swallow, J.C.S. Faraday I, 71, 2277 (1975).

159) A. Fischli and P.M. Müller, Helv. 63, 1619 (1980).

160) A.V. Benedetti, E.R. Dockal, H.L. Chum and T. Rabockai, J. Electroanal. Chem. 133, 45 (1982).

161) J. Halpern, M.S. Chan, J. Hanson, T.S. Roche and J.A. Topich, J. Am. Chem. Soc. 97, 1606 (1975).

162) M.E. Vol'pin, I.Y. Levitin, A.L. Sigan, J. Halpern and G.M. Tom, Inorg. Chim. Acta 41, 271 (1980).

163) F. Gaudemer and A. Gaudemer, Tetrahedron Lett. 21, 1445 (1980).

164) R. Hoffmann, J. Chem. Phys. 39, 1397 (1963).

165) For MO-calculations of Co(I) Pc complexes cf. R. Taube, Z. Chem. 6, 8 (1966); the electronic structure of Co(II) complexes with Schiff bases is discussed by C. Daub, C.W. Schläpfer and A. von Zelewsky, "Structure and Bonding", 36, 129-171 (1979).

166) a) The porphinoid factor F 430 from *Methanobacterium thermoautotrophicum* has recently been determined by Eschenmoser et. al. to be the Ni(II) complex of a (tetrahydro) derivative of the corphin system (containing a conjugated 9-center, 10-π-electron unit within the macrocycle) cf. A. Pfaltz, B. Jaun, A. Fässler, A. Eschenmoser, R. Jaenchen, H.H. Gilles, G. Diekert and R.K. Thauer, Helv. 65, 828 (1982).

b) Some synthetic square-planar Ni complexes have been found by Pletcher et. al. to exhibit catalytic properties at reducing conditions, cf. C. Gosden, K.P. Healy and D. Pletcher, J. Chem. Soc. Dalton, 972 (1978); J.Y. Becker, J.B. Kerr, D. Pletcher and R. Rosas, J. Electroanal. Chem. 117, 87 (1981); C. Gosden, J.B. Kerr, D. Pletcher and R. Rosas, ibid. 117, 101 (1981) and lit. cited therein.

In our laboratory it has been found that the [HDP] complexes of Ni are several powers of ten less active catalysts than the corresponding Co-complexes.

167) For other theoretical considerations concerning (Co-C) bond cleavage cf.

a) B.T. Golding in ref. 4, pp. 543-582.

b) B.T. Golding and L. Radom, J.C.S. Chem. Comm. 939 (1973); J. Am. Chem. Soc. 98, 6331 (1976).

c) L. Salem, O. Eisenstein, N.T. Anh, H.-B. Bürgi, A. Devaquet, G. Segal and A. Veillard, Nouv. J. Chim. 1, 335 (1977).

d) G.N. Schrauzer, L.P. Lee and J.W. Sibert, J. Am. Chem. Soc. 92, 2997 (1970).

168) According to our calculations it is not excluded that the odd electron may be directly transferred from the reducing agent into the antibonding (R-Co-Y)-MO (ψ_{ab}) providing that Y is either strongly electronegative or the (Co-C) bond is already lengthened in the Co(III) complex (e.g. by steric interactions) or the electron-transfer takes place *via* the ligand Y.

169) G.N. Schrauzer, Angew. Chem. 89, 239 (1977); ibid. Int. Ed. Engl. 16, 233 (1977).

170) J.M. Pratt, ref. 4, Vol. 1, p. 325-92.

171) H.A.O. Hill, J.M. Pratt, M.P. O'Riordan, F.R. Williams and R.J.P. Williams, J. Chem. Soc. A, 1859 (1971).

172) R. Scheffold, M. Dike, S. Dike, T. Herold and L. Walder, J. Am. Chem. Soc. 102, 3642 (1980).

173) R. Scheffold and E. Amble, Angew. Chem. 92, 643 (1980); ibid. Int. Ed. Engl. 19, 629 (1980).

174) H. Kageyama, M. Hidai and Y. Uchida, Bull. Chem. Soc. Jpn. 46, 2901 (1973).

175) J.A. Cotting and H.F. Beer, unpublished results from our laboratory.

176) H. Eckert, G.N. Schrauzer and I. Ugi, Tetrahedron 31, 1399 (1975).

177) D. Zanette and F. Nome, J. Org. Chem. 44, 2308 (1979).

178) A.E. Brearley, H. Gott, H.A.O. Hill, M. O'Riordan, J.M. Pratt and R.J.P. Williams, J. Chem. Soc. (A) 612 (1971).

179) S. Takeuchi, Y. Ohgo and J. Yoshimura, Bull. Chem. Soc. Jpn. 47, 463 (1974).

180) A. Fischli, Helv. 65, 2697 (1982) .

181) a) R.A. Sheldon and J.K. Kochi, "Metal-Catalyzed Oxydations of Organic Compounds", Academic Press, New York, 1981.
b) A.E. Martell, Acc. Chem. Res. 15, 155 (1982).
c) R.D. Jones, D.A. Summerville and F. Basolo, Chem. Rev. 79, 139-179 (1979).
d) T. Matsuura, Tetrahedron 33, 2869-2905 (1977).

182) T.G. Spiro, Ed., "Metal Ion Activation of Dioxygen", Wiley, New York, 1980.

183) D.T. Sawyer, M.J. Gibian, M.M. Morrison and E.T. Seo, J. Am. Chem. Soc. 100, 627 (1978).

184) G.A. Rodley and W.T. Robinson, Nature (London) 235, 438 (1972).

185) M. Calligans, G. Nardis, L. Randaccio and A. Ripamonti, J. Chem. Soc. A, 1069 (1970).

186) B. Bosnich, W.G. Jackson, J.T.D. Lo and J.W. Mc Laren, Inorg. Chem. 13, 2605 (1974).

187) D.L. Tomjy, L.H. Vogt and J.G. Wirth, J. Org. Chem. 35, 2029 (1970).

188) R.S. Drago, J. Gaul, A. Zombek and D.K. Straub, J. Am. Chem. Soc. 102, 1033 (1980).

189) H.M. van Dort and H.J. Geursen, Recl. Trav. Chim. Pays-Bas, 86, 520 (1967).

190) L.H. Vogt, J.G. Wirth and H.L. Finkbeiner, J. Org. Chem. 34, 273 (1969).

191) a) Br. Patent 1,268,653 (1972) to BASF.
b) Ger. Patent 2,460,665 and 2,517,870 (1975) to Shell.

192) V. Kothari and J.J. Tazuma, J. Catal. 41, 180 (1976).

193) T.J. Fullerton and S.P. Ahern, Tetrahedron Lett. 139 (1976).

194) a) A. Nishinaga, K. Nishizawa, H. Tomita and T. Matsuura, J. Am. Chem. Soc. 99, 1287 (1977).
b) A. Nishinaga, H. Tomita and T. Matsuura, Tetrahedron Lett. 21, 1261, 2833, 4849 (1980).

195) J.H. Bayston and M.E. Winfield, J. Catal. 9, 217 (1967).

196) Y. Hohokabe and N. Yamazaki, Bull. Chem. Soc. Japan, 44, 2142 (1971).

197) Y. Hohokabe and N. Yamazaki, Bull. Chem. Soc. Japan, 44, 2145 (1971).

198) J. Utsumi, K. Harada and H. Miura, Jap. Patent 7237,425 (1972).

199) A. Nishinaga, Chem. Lett. 273 (1975).

200) M.N. Dufour-Ricroch and A. Gaudemer, Tetrahedron Lett. 4079 (1976).

201) A. Inada, Y. Nakamura and Y. Morita, Chem. Lett. 1287 (1980).

202) H.S. Kao, W. Sklarz and L.M. Weinstock (Merck & Co.) US. 4,111,948 (1978).

203) V.A. Ignatov, A.N. Lazovenko and V.F. Borodkin, Izv. Vyssh. Ucheb. Zavad. Khim. Khim Teknol. 23, 39 (1980).

204) D.H. Carlson and J.R. Deering (UOP Inc.) US. 4,142,964 (1979) and many earlier patents.

205) A.M. Kuliev, G.A. Agaev, V.G. Kochetkov and V.S. Chernomyrdin, Gazov Prom-st. 27 (1980).

206) J.H. Schutten and J. Zwart, J. Mol. Catal. 5, 109 (1979).

207) Y. Kamiya, J. Catal. 33, 480 (1974); A. Onopchenko, J.G.D. Schulz and R. Seekircher, J. Org. Chem. 37, 1414 (1972).

208) A. Onopchenko and J.G.D. Schulz, J. Org. Chem. 38, 909 (1973); ibid. 45, 3716 (1980).

209) A. Zombeck, D.E. Hamilton and R.S. Drago, J. Am. Chem. Soc. 104, 6782 (1982).

210) J. Tsuji, J. Schimizu and K. Yamamoto, Tetrahedron Lett. 34, 2975 (1976) and lit. cited therein.

211) E. Guilmet and B. Meunier, Tetrahedron Lett. 21, 4449 (1980); for the synthesis of propylenoxide by air oxydation of propylene, catalyzed by $CoPc/Li_2CO_3$ at 160-230^0 cf. V. Ragini, P. Carmiti and U. Rainaldi, Chim. Ind. (Milan) 58, 473 (1976).

212) D. Mansuy, J.-F. Bartoli, J.-C. Chottard and M. Lange, Angew. Chem. 92, 938 (1980); ibid. Int. Ed. Engl. 19, 909 (1980).

213) J.H. Grate and G.N. Schrauzer, J. Am. Chem. Soc. 101, 4601 (1979).

214) A. Fischli and P.M. Müller, Helv. 63 , 529 (1980).

215) A. Fischli, Helv. 65 , 1167 (1982).

216) Y. Ohgo, S. Takeuchi and J. Yoshimura, Bull. Chem. Soc. Jpn. 43, 505 (1970).

217) J.P. Collman and L.S. Hegedus, "Principles and Applications of Organotransition Metal Chemistry", University Science Books,Mill Valley CA, 1980 p. 354.

218) A. Fischli and D. Süss, Helv. 62, 2361 (1979).

219) A. Fischli and D. Süss, Helv. 62, 48 (1979).

220) A. Fischli and J.J. Daly, Helv. 63, 1628 (1980).

221) A. Fischli, Helv. 62, 882 , (1979).

222) a) R. Peyer and Ch. Weymuth, unpublished results.
b) In aprotic solvents no (Co-C) bond formation was observed between electrochemically prepared $[Co(I)Pc]^-$ and carvone.

223) E.B. Fleischer and M. Krishnamurthy, J. Am. Chem. Soc. 94, 1382 (1972).

224) M. Ichikawa, R. Sonoda and S. Meshitsuka, Chem. Lett. 709 (1973).

225) R.G. Banks, P.K. Das, H.A.O. Hill, J.M. Pratt and R.J.P. Williams, Discuss. Faraday Soc. 46, 80 (1968).

226) A.R. Pinder, Synthesis 425 (1980).

227) D. Bieniek, P.N. Mozo, W. Klein and F. Korte, Tetrahedron Lett. 4055 (1970).

228) T. Funabiki, Y. Yamazaki, K. Tarama, J.C.S. Chem. Comm. 63 (1978).

229) L. Walder, G. Rytz, R. Orlinksi, fifth EUCHEM Conf. on Organic and Organomet. Electrochem. Ireland, 1983.

230) A.E. Brearley, H. Gott, H.A.O. Hill, M.O. Riordan, J.M. Pratt and R.J.P. Williams, J. Chem. Soc. (A) 612 (1971).

231) Y. Ohgo, S. Takeuchi and J. Yoshimura, Bull. Chem. Soc. Jpn. 44, 283 (1971).

232) H.W. Whitlock, Jr. and B.K. Bower, Tetrahedron Lett. 4827 (1965).

233) a) G.N. Schrauzer and R.J. Windgassen, Nature 241, 492 (1967).
b) G.N. Schrauzer and R.J. Windgassen, J. Am. Chem. Soc. 89, 3607 (1967).

234) M. Murakami and J.-W. Kang, Bull. Chem. Soc. Jpn. 36, 763 (1963).

235) A. Fischli, Helv. 61, 3028 (1978).

236) Y. Ohgo, S. Takeuchi and J. Yoshimura, Bull. Chem. Soc. Jpn. 44, 583 (1971).

237) H.P.C. Hogenkamp and H.A. Barker, J. Biol. Chem. 236, 3097 (1961).

238) M. Philippe, Ph.D. Thesis, University of Berne, unpublished results from our laboratory.

239) S. Winstein and H.J. Lucas, J. Am. Chem. Soc. 61, 1576 (1939).

240) H.O. House and R.S. Ro, J. Am. Chem. Soc. 80, 182 (1958).

241) P.E. Sonnet, Tetrahedron, 36, 557 (1980).

242) R.B. Woodward, K. Heusler, J. Gosteli, W. Oppolzer, R. Ramage, S. Ranganathan and H. Vorbrücken, J. Am. Chem. Soc. 88, 852 (1966).

243) a) T.B. Windholz and D.B.R. Johnston, Tetrahedron Lett. 2555 (1967);
b) R.M. Jacobson and J.W. Clader, Synth. Commun. 9, 57 (1979).

244) a) J. Grimshaw, J. Chem. Soc. 7136 (1965);
b) H. Yajima, H. Watanabe and M. Okamoto, Chem. Pharm. Bull., 18, 2185 (1981)
c) H. Eckert and I. Ugi, Liebigs Ann. Chem. 278 (1979) and lit. cited therein.

245) B. Marinier, Y.C. Kim and J.M. Navarre, Can. J. Chem. 51, 208 (1973).

246) F. Eckstein, Angew. Chem. 77, 912 (1965); ibid. Int. Ed. Engl. 4, 876 (1965).

247) a) E.J. Corey, E.J. Trybulsky and J.W. Suggs, Tetrahedron Lett. 4577 (1976).
b) B. Bernet and A. Vasella, Helv. 62, 1990, 2400, 2411 (1979).
c) J.L. Isidor and R.M. Carlson, J. Org. Chem. 38, 554 (1973).

248) G. Just and K. Grozinger, Synthesis, 457 (1976).

249) a) F.Kasafirek, Tetrahedron Letters 2021 (1972).
b) J.Engels, Liebigs Ann.Chem. 557 (1980) and lit. cited therein.

250) T.L. Ho, Synth. Commun. 8, 359 (1978).

251) H. Eckert and Y. Kiesel, Synthesis, 947 (1980) and lit. cited therein.

252) H.F. Beer, unpublished results from our laboratory.

253) J.-A. Cotting, unpublished results from our laboratory.

254) R. Sterzycki, Synthesis, 724 (1979).

255) M. Schlosser in Houben-Weyl: Methoden der organischen Chemie, 4th. Edit. Thieme Stuttgart 1972, Vol. V/1b, pp. 212-219.

256) a) L. Crombie and S.H. Harper, J. Chem. Soc. 1707, 1714 (1950);
b) L. Crombie, S.H. Harper, J. Gold and B.J. Stokes, J. Chem. Soc. 136 (1956).

257) M.F. Ansell and M.E. Selleck, J. Chem. Soc. 1238 (1956); M.F. Ansell and B. Gadsby, J. Chem. Soc. 3388 (1958).

258) A.S. Arora and I.K. Ugi in Houben-Weyl: Methoden der organischen Chemie, 4th Edit. Thieme Stuttgart 1972, Vol. V/1b, pp. 760-762.

259) N.A. Porter, A.N. Roe and A.T. McPhail, J. Am. Chem. Soc. 102, 7576 (1980); cf. also lit. 256 b).

260) a) J.F. Normant, A. Alexakis and G. Cahiez, Tetrahedron Lett. 23, 2027 (1978).
b) A.Y. Meyers and E.W. Collington, Tetrahedron 27, 5979 (1971).
c) K. Utimoto, M. Kitai, M. Naruse and H. Nozaki, Tetrahedron Lett. 48, 4233 (1975).
d) R.-C. Cookson and N.W. Hughes, J. Chem. Soc. Perkin Trans. 1, 2738 (1973).
e) M. Stoll and J. Flament, Helv. 52, 1996 (1969).

261) W. Biernacki and A. Gdula, Synthesis 37 (1978).

262) M. Julia, S. Julia and R. Guégan, Bull. Soc. Chim. Fr. 1972 (1960).

263) Y.H. Lai, Synthesis, 585 (1981).

264) W. Roelofs, J. Kochansky, E. Anthon, R. Rice and R. Carde, Environ. Entomol. 4, 580 (1975).

265) Among numerous methods the most famous is the cuprate addition. cf. G.H. Posner, "An Introduction to Synthesis Using Organocopper Reagents", J. Wiley + Sons, New York 1980. For a recent improvement cf. S.H. Bertz, G. Dabbagh and G.M. Villacorta, J. Am. Chem. Soc. 104, 5824 (1982).

266) For uncatalyzed reductive coupling cf. M.M. Baizer, Ed., "Organic Electrochemistry", Marcel Dekker, New York 1973, pp. 399-411, 679-704.

267) a) H.J. Schäfer, Angew. Chem. 93, 978 (1981); ibid. Int. Ed. Engl. 20, 911 (1981).

b) S. Satoh, H. Suginome and M. Tokuda, Tetrahedron Lett. 1895 (1981).

268) J. Baldwin, R.C. Thomas, L.I. Kruse and L. Silberman, J. Org. Chem. 42, 3846 (1977).

269) Th. Herold, unpublished results from our laboratory.

270) Z. Chilmonczyk, unpublished results from our laboratory.

271) S. Shirali, unpublished results from our laboratory.

272) The potential difference between the Ag/Ag^+ electrode and the SCE in DMF was found to be -0.36 V.

273) O.W. Lever, Jr., Tetrahedron 32, 1943 (1976).

274) for a review cf. W. Kreiser, Nachr. Chem. Tech. Lab. 29, 172 (1981).

275) M.P. Cooke, Jr., J. Am. Chem. Soc. 92, 6080 (1970).

276) M.P. Cooke, Jr., R.M. Parlman, J. Am. Chem. Soc. 97, 6863 (1975).

277) M.P. Cooke, Jr., R.M. Parlman, J. Am. Chem. Soc. 99, 5222 (1977).

278) E.J. Corey, L.S. Hegedus, J. Am. Chem. Soc. 91, 4926 (1969).

279) C. Masters, "Homogeneous Transition-metal Catalysis" London, Chapman and Hall Ltd. (1981).

280) G. Costa, G. Mestroni, G. Pellizer, J. Organometal. Chem. 11, 333 (1968).

281) O. Müller, G. Müller, Biochem. Z. 336, 299 (1962).

282) I. Tabushi, K. Seto and Y. Kobuke, Tetrahedron 37, 863 (1981).

283) R. Orlinski, unpublished results from our laboratory.

284) L. Eberson, Acta Chem. Scand. B 34, 481 (1980).

285) R. Orlinski, J.-P. Wolf, unpublished results from our laboratory.